Stefan Keil

Technology and Practical Use of Strain Gages

Stefan Keil

Technology and Practical Use of Strain Gages

With Particular Consideration of Stress Analysis Using Strain Gages

Ernst & Sohn
A Wiley Brand

Professor Dr.-Ing. Stefan Keil

Library of Congress Card No.: applied for

British Library Cataloguing-in-Publication Data
A catalogue record for this book is available from the British Library.

Bibliographic information published by the Deutsche Nationalbibliothek
The Deutsche Nationalbibliothek lists this publication in the Deutsche Nationalbibliografie; detailed bibliographic data are available on the Internet at <http://dnb.d-nb.de>.

Cover Design: Sonja Frank
Typesetting: Thomson Digital, Noida, India
Printing: Strauss GmbH, Mörlenbach, Germany
Binding: Litges & Dopf Buchbinderei GmbH, Heppenheim, Germany

Printed in the Federal Republic of Germany.
Printed on acid-free paper.

Print ISBN: 978-3-433-03138-4
ePDF ISBN: 978-3-433-60665-0
ePub ISBN: 978-3-433-60664-3
eMobi ISBN: 978-3-433-60782-4
oBook ISBN: 978-3-433-60666-7

Table of Contents

	Preface	XI
1	**Historical review**	1
	References	13
2	**Fundamentals of strain gage technology**	17
2.1	Measurement principle and structure	17
2.1.1	Basic principles	17
2.1.2	Carrier material	30
2.1.3	Measuring grid material	32
2.2	Sensitivity	35
2.3	Transverse sensitivity	42
2.4	Temperature effects	47
2.4.1	Temperature effect on sensitivity	47
2.4.2	Thermal output of strain gages	49
2.5	Mechanics of the strain gage	55
2.5.1	Strain transfer from the measurement object to the measurement grid	55
2.5.2	Creep	57
2.5.3	Hysteresis	64
2.6	Influence of pressure	66
2.6.1	Hydrostatic pressure	66
2.6.2	Operation in a vacuum	72
2.7	Dynamic behaviour	74
2.7.1	First measurements using graphite layer strain gages	74
2.7.2	Fatigue strength	75
2.7.3	Shock load	80
2.7.4	Fatigue-life gage	82
2.8	Heat dissipation	84
2.9	Measuring at elevated temperatures	87
2.9.1	Previews	87
2.9.2	Weldable encapsulated tube strain gages	90
2.9.3	Free-filament strain gages	93
2.10	Stress gages	94
	References	98
3	**Installation of strain gages**	109
3.1	Preparatory work	109
3.1.1	General preparation	109
3.1.2	Preparation of the strain gage	110
3.1.3	Preparation of the measurement point	110
3.2	Methods of fastening	112

3.2.1 Requirements ... 112
3.2.2 Cyanoacrylate ... 114
3.2.3 Methyl acrylate ... 116
3.2.4 Epoxy adhesives ... 117
3.2.5 Phenolic resin adhesives ... 119
3.2.6 Ceramic putty ... 120
3.2.7 Flame spraying ... 121
3.2.8 Spot welding ... 123
3.3 Different measurement object materials ... 124
3.3.1 Glass, enamel, glazed porcelain ... 124
3.3.2 Concrete ... 124
3.3.3 Wood ... 127
3.3.4 Plastics ... 129
3.4 Measurement point protection ... 134
3.5 Quality tests of the strain gage installation ... 140
References ... 142

4 The Wheatstone bridge circuit ... 147
4.1 Circuit principle ... 147
4.2 Basic equation of the bridge circuit ... 147
4.3 Temperature compensation ... 152
4.4 Limit of the bridge signal resolution ... 155
4.5 Examples of some elementary bridge circuits ... 156
References ... 162

5 Adjustment and compensation circuits ... 163
5.1 General ... 163
5.2 Compensation of zero shift with temperature change and zero adjustment ... 166
5.3 Compensation of temperature effects on the sensitivity and linearity measures ... 171
5.4 Adjustment of the characteristic value ... 176
5.5 Creep compensation ... 178
5.6 Full-bridge circuits connected in parallel ... 179
References ... 182

6 Cable between the strain gage bridge circuit and measuring instrument ... 183
6.1 Basic ... 183
6.2 Ohmic resistance of the cable ... 183
6.2.1 Preliminary remarks ... 183
6.2.2 Ohmic resistance of the cable in full-bridge circuits ... 183
6.2.3 Ohmic resistance of the cable in half and quarter-bridge circuits ... 185
6.3 Influence of cable capacitance ... 187
6.3.1 Preliminary remarks ... 187
6.3.2 Capacitive asymmetry ... 188

6.3.3 Phase shift 189
6.3.4 Amplitude drop 191
6.4 Full-bridge connection in four-wire technology 193
6.5 Six-wire circuit 196
6.6 Dual-channel principle 197
6.7 Connection of half and quarter bridges 200
6.8 Protection against disturbing influences 204
6.8.1 Protection against electrical and magnetic fields 204
6.8.2 Electromagnetic compatibility 206
6.8.3 Protection against overvoltage 207
References 210

7 Signal processing 211
7.1 Introductory viewing 211
7.2 Analogue instrumentation amplifier 212
7.2.1 Structure of analogue signal processing 212
7.2.2 Bridge supply with carrier frequency 214
7.2.3 Bridge supply with direct current 219
7.3 Digital amplifier concepts 222
7.3.1 Computer controlled analogue amplifier 222
7.3.2 New methods of analogue-to-digital conversion 223
7.3.3 Structure of digital measuring amplifier 225
7.4 Compensation method 228
7.5 Multipoint measurement 232
7.5.1 Preliminary remarks 232
7.5.2 Serial multipoint measurement 233
7.5.3 Parallel multipoint measurement 236
References 239

8 Calibration of devices for measuring with strain gages 241
8.1 Introduction 241
8.2 Measurement chain 241
8.3 Characteristic and sensitivity 243
8.4 Calibration of the entire measurement chain as a measuring device 245
8.5 Compensators 246
8.6 Transducers 247
8.6.1 Transducers calibrated by the manufacturer 247
8.6.2 Calibrating transducers self-made by the user 248
8.7 Calibration measures 249
8.7.1 Calibration unit 249
8.7.2 Measurement amplifier with built-in calibration device 250
8.7.3 Shunt calibration 251
8.8 Calibration of measuring arrangements with self-installed strain gages 255
8.8.1 Wheatstone bridge circuit 255

8.8.2 Example of the calibration of a measuring device for measuring with self-installed strain gages 256
References 257

9 Determination of mechanical stresses from strains measured with strain gages 259
9.1 Introduction 259
9.2 Terms of stress and strain 260
9.3 Elastic deformation and stress of a tensile rod under uniaxial tensile loading 262
9.4 The biaxial stress state 269
9.5 Mohr's circle 272
9.6 Deformation circle 277
9.7 Types of rosettes and grid notation 280
9.8 Evaluation formulas for 0/45/90° strain gage rosettes 281
9.9 Evaluation formulas for 0/60/120° strain gage rosettes 285
References 288

10 Application examples of elastic deformation 289
10.1 Initial considerations 289
10.2 Principal directions are known 291
10.2.1 Examples of measurement objects loaded longitudinally 291
10.2.2 Measurement objects subject to bending loads 291
10.2.2.1 Initial consideration 297
10.2.2.2 Bending beam clamped at one end 298
10.2.2.3 Bending beam clamped at both ends 302
10.2.2.4 Bending beam freely supported at both ends 306
10.2.3 Measurement objects subject to torsion loads 308
10.2.4 Shear stresses in a beam 312
10.3 Stress analysis for unknown principal directions 319
10.3.1 Example of stress analysis using a 0/60/120° rosette 319
10.3.2 Loading of a shaft with superimposed torsion and bending 322
10.4 Simultaneous measurement of multiple load components 327
10.5 Diaphragm rosettes 334
10.5.1 Basics 334
10.5.2 Example of dimensioning a membrane for pressure measurements 338
References 339

11 Determination of thermal stresses 343
11.1 Emergence of thermal stresses 343
11.2 Sensing the prevented thermal expansion at identical temperature at the compensation gage and the active strain gage 345
11.3 Determination of the restricted thermal expansion by computational correction of the measured values with previously measured thermal outputs on dummies 347

11.4 Determination of the restricted thermal expansion by mathematical correction with the thermal outputs determined at the original measurement object 349
11.5 The 'reversible' strain gage 349
11.6 Separation of the mechanical strain from the thermal strain 350
11.7 Compensated half-bridge strain gage with compensation resistor 352
References 353

12 Strain gages as a means for experimental determination of residual stresses 355
12.1 Preliminary observation 355
12.2 Cutting-down method 357
12.3 Layer removal method 364
12.4 Ring-core method 371
12.4.1 Principle of the ring-core method 371
12.4.2 Application of the method 377
12.4.3 Electrical connection of the ring-core rosette 379
12.4.4 Practical application example 380
12.4.5 Multi-stage application of the ring-core method 388
12.4.6 Modified ring-core rosettes for smaller core diameters 391
12.5 Hole-drilling method 394
12.5.1 Introduction 394
12.5.2 Residual stress as an integrated mean value over the drilling depth 396
12.5.2.1 Introductory remarks 396
12.5.2.2 Stress distribution in the vicinity of a hole 397
12.5.2.3 Strains measured 400
12.5.2.4 Computation of the residual stress state 403
12.5.3 Determination of the residual stress distribution in relationship to the depth 404
12.5.4 Forming the hole 409
References 411

13 Stress analysis using strain gages in the elastoplastic deformation range 419
13.1 Introduction 419
13.2 Equivalent state with elastic deformation 420
13.3 Elastoplastic deformation 422
13.4 Stress analysis 423
13.5 Practical Example of application 425
13.6 Tensorial representation of the elastoplastic deformation 428
References 431

14 Strength theories 433
14.1 Preview 433
14.2 Concept of effective stress 434

14.3 Experimental results 437
14.4 Maximum stress theory 441
14.5 Maximum shear theory 442
14.6 Extended shear theory 443
14.7 Plastic potential theory 444
14.8 Distortion energy (shape changing) theory 446
14.9 Octahedral plane shear stress theory 449
References 450

Index 451

Preface

In 1995, when the first edition of this book was published in German language it soon became very popular, not only in German speaking countries, but also abroad. This first edition has been out of stock for a while. The demand from specialists around the world shows, however, that this book remains to be of interest, because strain gages find application in various fields and in large numbers. They are also often used by beginners in this specialist area. Fast and without much effort these sensor elements can be attached to building components in order to transmit vital information about their strength and operational safety. Comprehensive knowledge of the necessary basics on this topic is a prerequisite to carry-out correct measurements by means of strain gages, and for a proper analysis and evaluation of the measured strain. This is what this book covers. In practice, the measurement procedure and the processing of the readings can be a source of errors. This book points out what can go wrong and how it can be prevented.

A significant number of the first-edition books were purchased by experts from abroad. Repeatedly the question arose as to whether the book would be available in English, since English is world-wide the language most commonly used by engineers. The English version of the book at hand meets this request. The book deals with the technology of strain gages as an independent specialist field. It covers the relevant questions on electrical engineering and elaborates on the required fundamental principles of the strength of materials. Questions on digital signal processing are reviewed comprehensively. Paramount to this book is an application-orientated presentation of the contents.

The measurement by means of strain gages includes a number of individual stages which are elaborately described in this book: preparation of the measuring points, attachment of strain gages, connection and arrangement of cables, selection and settings of devices, and particularly for the strengths analyses: calculation of the mechanical stresses on the basis of the measured strains. The technology of strain gages is also dealt with in detail, since this knowledge will add to the understanding of the limits to what is feasible by the application of strain gages, and what is not. The application of strain gages can lead to amazing results. It is however also limited by the laws of physics.

The book includes a number of genuine examples which can be seen as role models for the solution of similar tasks.

Professor Dr.-Ing. Stefan Keil
Lippstadt, May 2017

1 Historical review

It is an irrefutable fact that the strain gage was invented by two different people at almost the same time. They were situated at widely separated places in the USA and they did not, at that time, have any contact with each another [1.1]. Professor Arthur C. Ruge of the Massachusetts Institute of Technology (MIT) was one of these inventors; the other was Edward E. Simmons. At the California Institute of Technology (Caltech) in 1936, Simmons was investigating the stress–strain behaviour of metals under shock loads. He was at the time a student and worked as a research assistant at the Institute.

To measure the force introduced in the specimens by impact, he used a dynamometer equipped with fine resistance wires made from constantan. The tests carried out by Simmons were part of a research project of Dätwyler and Clark that started in 1936 [1.2]. Although details of the tests and the measurement method used were not published until 1938 [1.3], the start date of the project (1936) shows that it was Simmons who invented the strain gage principle. Publication about the tests and the used measurement method was divulged not before 1938 [1.3]. In 1940, Simmons's invention was patented at the United States Patent Office. Figure 1.1 shows the test equipment that Simmons used for the measurement of shock loads on metal specimens. The dynamometer is equipped with measuring wires made from constantan. The drawing is taken from the Patent [1.4]. This shows that a strain gage based transducer was patented before the strain gage itself.

In 1938 the other inventor, Professor Arthur C. Ruge, with the support of his assistant J. Hanns Maier investigated in the field of engineering seismology the influence of earthquakes on mechanical structures. His test object was a small scale model of an elevated water tank, mounted on a vibration table. However, because the stress was low and the model's skin extremely thin they failed to measure the strain in the tank wall using normal mechanical or optical strain instrumentation available at that time. One day the saving idea came to him, and he attached the thin wire from a potentiometer with Duo household cement to the water tank and was immediately rewarded with excellent and reproducible measurement values.

The resistance change in metallic wires caused by strain due to tensile loading changed the voltage drop in the wire and could be measured with a simple bridge circuit [1.5, 1.6]. The strain gage was now born also on the east coast of the USA.

Figure 1.2 shows a photograph taken by J. Hanns Maier of Professor Ruge carrying out experiments on the model of a water tank using the first strain gages invented by him and his assistant Maier. Strain gage rosettes are cemented to the model's base. Also visible in the photograph is an acceleration transducer made by Ruge using resistance wires.

For easier handling Ruge cemented the measuring wire to a carrier paper that had been stiffened by cementing the paper to two Plexiglas end pieces with a brass spacer bar serving as a handling frame and lead wire holder. The brass spacer bar was

Technology and Practical Use of Strain Gages: With Particular Consideration of Stress Analysis Using Strain Gages. First Edition. Stefan Keil.

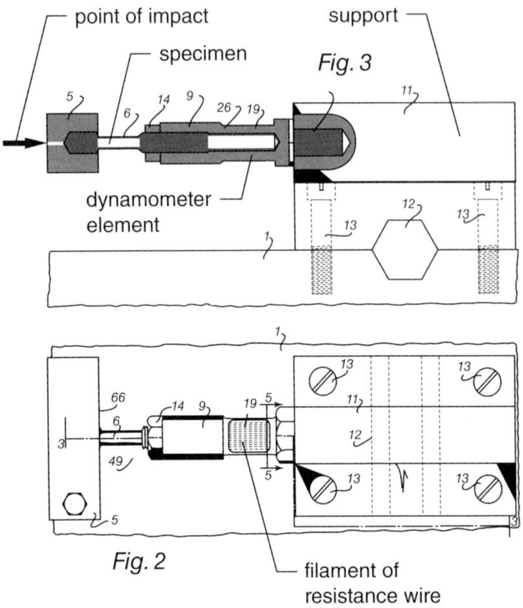

Fig. 1.1 E.E. Simmons's dynamometer equipped with measuring wires for the measuring of shock loads on metal specimens [1.4]

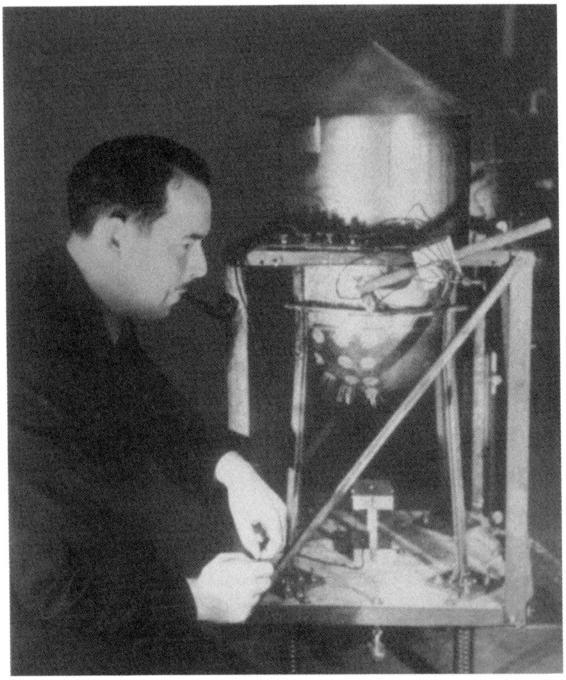

Fig. 1.2 A.C. Ruge carrying out experiments on a small scale model of a water tank fitted with the first strain gages mounted on a vibration table in 1938

when the device is pulled, whereas another section in the other arm of the bridge is relieved.

In 1935, Carlson, upon whose idea the design is probably based, reported that about 1500 of these strain measuring devices in an encapsulated version were cast in concrete constructional elements. However, he also reported on the numerous problems that occurred during measurements, such as through temperature effects and corrosion [1.16].

All the historical examples described above are based on the change in resistance of wires stressed by mechanical strain. In all cases, a Wheatstone bridge circuit was used to measure the change in resistance. This bridge circuit was invented by two scientists in the UK, who independently of one another, were dealing with electrical circuits to measure the electrical resistance of metallic wires. Their names were Samuel Hunter-Christie and Charles Wheatstone. They used the knowledge gained by Simon Ohm [1.17] to develop a circuit, with which it was possible to measure the electrical resistance of wires in spite of the instability of the voltage sources that were then available. Although Hunter-Christie published information about his circuit in 1833 [1.18], and it was ten years before Wheatstone's publication in 1843, the circuit was called after Wheatstone. This circuit became the standard circuit for measurements using strain gages and is still widely used today.

Figure 1.9 shows the Wheatstone bridge circuit in its original form. This circuit enabled Wheatstone to measure the resistance of electrical connecting wires in spite of the instability of his voltage source. He wrote about the response of his galvanometer to changes in the strain of the copper wire that was used [1.19], and the date of these measurements (1843) was very early in the history of strain gages.

In Figure 1.9 we can see the historic measurement device as used by Wheatstone in 1843. It consists of a wooden board with screw clamps Z and C used for connecting the supply voltage, the resistance wires Za and Ca, and screw clamps c, d, e and f for connection the resistances to be measured. Wheatstone tapped the diagonal voltage between a and b, using a sensitive galvanometer. An adjustable short-circuiting arm mn served as a zero balance.

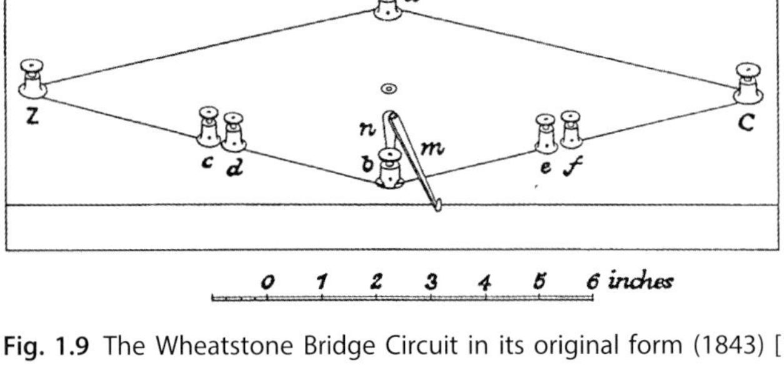

Fig. 1.9 The Wheatstone Bridge Circuit in its original form (1843) [1.19]

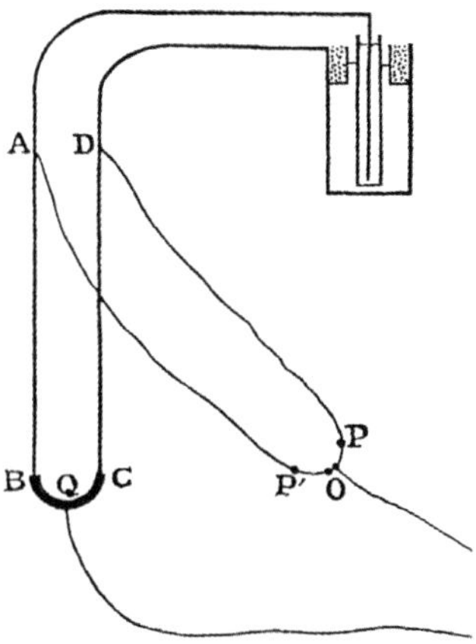

Fig. 1.10 Thomson's bridge circuit for the measurement of changes in the resistance of metal wires under tensile load (1856) [1.20]

Wheatstone explicitly made reference to his circuit's suitability for the measurement of small changes in resistance, and even at this early stage he was able to perceive relative changes in resistance of about two parts per thousand.

The next step was made by Thomson. Also in England, he conducted the first objective investigations into the changes of resistance caused by the strain in wires in 1856. He used a bridge circuit as invented by Wheatstone, in which, as shown in Figure 1.10, thin copper or iron wires were stretched by weights [1.20]. He found proportionality between the strain and the change of resistance and determined several gage factors, as they would be called today, but without specification of exact data.

Thomson used in his test device the bridge circuit as invented by Wheatstone, but without any knowledge of Wheatstone's invention. In this way the Wheatstone bridge circuit was invented again.

Both Wheatstone's and Thomson's paper had long footnotes about their appropriate priorities. For example, Thomson wrote that he found out one hour before his lecture that Wheatstone had already discovered a similar circuit. In those days, it was more a matter of personal honour to be the first with the invention than to protect the invention itself by means of a profitable industrial patent.

The first systematic investigation into resistance change of various wires during tensile loading took place in the 1930s. E. Cerlinsky, working at the German Research Institute for Aviation in Berlin-Adlershof, found that constantan wire was the most suitable material for measurement purposes, a fact which still applies today [1.21].

Figure 1.11 illustrates the research setup used by Cerlinsky. A 30 cm long measuring wire as part of a Wheatstone bridge circuit was strung between a fixed point and the

2 Fundamentals of strain gage technology

2.1 Measurement principle and structure

2.1.1 Basic principles

Strain gages are now employed in many technical fields. They are not used solely for the measurement of strains, being also very suitable for the construction of transducers for the measurement of mechanical quantities. In these transducers strain gages measure the strains arising in the measuring element of the transducer due to the mechanical quantity that the transducer is measuring. In this way it is possible to indirectly measure, for example, forces and torques using strain gages. Apart from transducer construction, experimental stress analysis is a popular field of application for strain gages. This entails the determination of the mechanical stresses from the measured strains obtained during measurement, providing information on the measurement object loading. The same principle is always involved, that is the strain gage detects the mechanical strain and changes its resistance accordingly.

With experimental stress analysis which is carried out to find the loading in the measurement object, the mechanical stress states have to be calculated from the measured strains. The interrelationship between deformation and stresses in a component is given by the material laws relevant to the component material and the state of loading. In the region of elastic deformation Hooke's Law is used, but it cannot be applied in the elastoplastic deformation region [2.1]. With deformation that extends beyond the yield point, material laws must be applied that take into account the degree of plastic deformation [2.2]. These material laws do not have any formal role in the construction of transducers, because they are calibrated directly in the units of the quantity that the transducer is designed to measure.

The clear advantages exhibited by strain gages are based on the conversion of the strain into an electrical signal. Strain, as the mechanical quantity to be measured, produces a change in resistance in the strain gage, resulting in a change of the output voltage from the Wheatstone bridge circuit into which the strain gage is connected. The quantity 'strain' is converted into an electrical quantity which is easy to process. The Wheatstone bridge circuit has excellent methods of compensating for unwanted effects on the strain gage, such as temperature changes. These compensation techniques have been developed to a very high degree, particularly in transducer design. Since transducers are calibrated directly in the unit of the mechanical quantity that is to be measured, the strain gages in these transducers are not used directly for the measurement of strains, but for the indirect measurement of the relevant mechanical quantity.

The basic operating principle of the strain gage is that the changes in the strain on the surface of the measurement object are transferred to an electrical conductor which is bonded to the surface and which changes its resistance accordingly. From these measured changes in resistance the changes in the strain occurring can be found very accurately. The first bondable strain gage, which was produced in the USA in about

Technology and Practical Use of Strain Gages: With Particular Consideration of Stress Analysis Using Strain Gages. First Edition. Stefan Keil.

1940, consisted of a meander-shaped metal wire glued to a paper carrier. Various manufacturing methods were employed, leading to different designs of strain gage.

As shown in Figure 2.1, one of the methods involved gluing a meander-shaped wire flat onto a carrier, so that the active parts of the resistance wire were parallel and aligned in the measurement direction. The *flat-grid gage* so produced has pronounced loop-shaped return points on the ends of the grid.

Another manufacturing method utilizes a flat auxiliary carrier, around which the resistance wire is wrapped, as also illustrated in Figure 2.1. The auxiliary carrier, complete with the wire, is then glued to the carrier. With these *wrap-around strain gages* the active lengths of measuring wire are not parallel and have no pronounced return points at their ends.

Gustafsson in Sweden first introduced the flat-grid gage in the form of a cross-link gage in which the return points of the measuring wire are replaced by transverse links of thicker material as shown in Figure 2.1. These types of *cross-link gages* with their very low transverse sensitivity were marketed under the tradename 'Tepic' by the company Huggenberger in Zurich. Because of their high manufacturing costs they never succeeded, and production ceased in 1994.

A form of wire gage that is still used, particularly for measurements at high temperature, is the *free-grid gage.* These gages are basically flat-grid gages without any carrier and are bonded to the measurement object using insulating bonding agents, such as ceramic putty.

Figure 2.1 shows the structure of a free-grid gage, which has an auxiliary carrier. The auxiliary carrier is only used as a mounting aid and is removed during the bonding procedure.

The introduction of the foil strain gage in the 1950s represented an important development in strain gage technology which, particularly in recent years, has been dominated by the demands of experimental stress analysis and transducer construction. Numerous strain gages of this type, with characteristics matched to the relevant application, are now available. However, foil strain gages are all made in a very similar

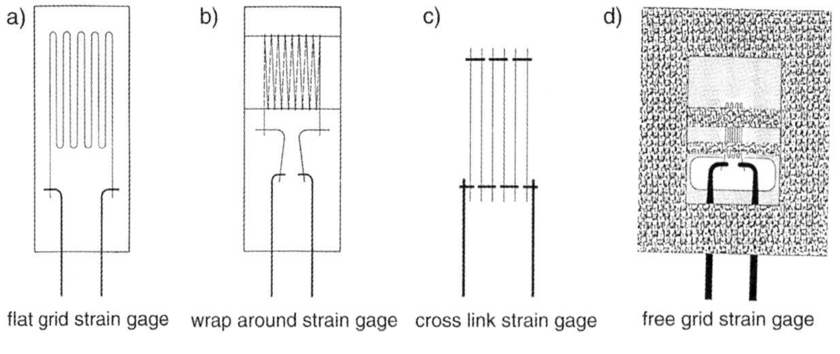

Fig. 2.1 Various designs of wire strain gages

manner and only differ in the shape and size of the grid and in the materials that are used for the carrier and the measuring grid. The basic design of a foil strain gage, as used in huge quantities in many fields of application, is illustrated in Figure 2.2.

An approximately 5 μm thick metal measuring grid foil is supported on a plastic insulating carrier foil about 25 μm thick. The measuring grid shape is produced photo-chemically by an etching process. The measuring grid is protected by a covering layer about 12 μm thick.

A strain gage measures the mechanical strain that occurs in the longitudinal direction of its measuring grid. This strain produces a change of resistance in the strain gage measuring grid that in turn produces a measurable imbalance in the electrical bridge circuit in which the strain gage is connected. Resistance changes in the strain gage can however be caused by effects other than the mechanical strain occurring in the measurement direction of the strain gage. As well as the actual measurement effect, that is the mechanical strain in the measurement direction, thermal effects and creep phenomena generally form the main causes of changes in resistance. If hydrostatic pressure is acting on a mounted strain gage, then a resistance change also occurs. The effect of transverse strain (the change of resistance due to strains acting transverse to the measurement direction) may be significant for precision measurements, but it is of no significance for transducer construction due to the calibration techniques employed. These disturbance variables, which produce unwanted changes of resistance superimposed on those due to the mechanical strain, are discussed in the following sections.

The strain gage must have at least two electrical contact points which enable it to be connected in the bridge circuit used for the measurement, for example by soldering. These connecting pads can be of various shapes; Figure 2.3 shows some typical examples.

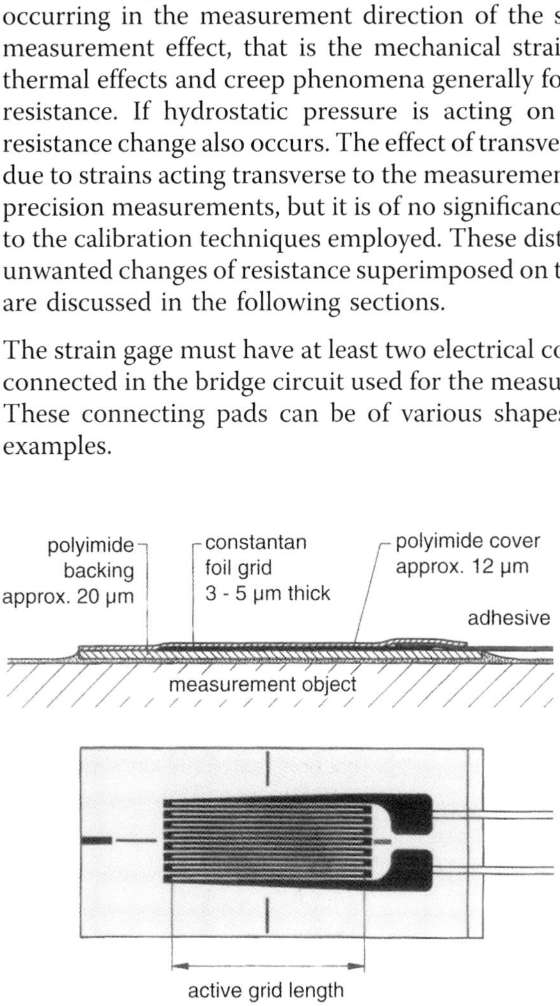

Fig. 2.2 Basic design of a standard foil strain gage with metal measuring grid, as used in large quantities

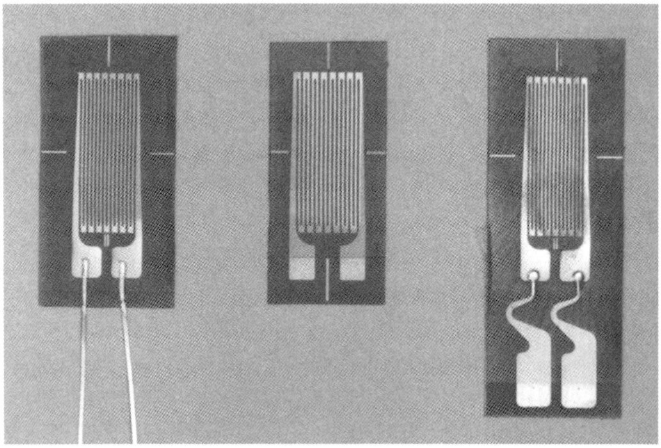

Fig. 2.3 Examples of different designs of connecting pads

In principle, any configuration of measuring grid can be produced by the photochemical method, which means that strain gages can be ideally matched to special geometrical requirements, as required for example in transducer design.

The linear strain gages with metal measuring grids that are available today have an active length of measuring grid from about 0.5 mm to 150 mm. Irrespective of the length of the measuring grid, the nominal resistances are normally between 100 ohm and 1000 ohm. An almost unimaginable number of different grid configurations and grid combinations forming multi-grid strain gages are now available for both experimental stress analysis and for transducer construction. Figure 2.4 shows a few selected examples of different grid combinations.

Strain gages detect the effective strain occurring in the surface of the measurement object at the point of installation, as defined in Eq. (2.1). If strain gradients occur in the measurement object along the axis of the strain gage measuring grid, then the strain gage behaves as an integrating measuring element and measures the mean of the strain arising in the region of the active length of the measuring grid, according to the law of averages in the integral calculation (Figure 2.5). In Eq. (2.1) ε_{SG} is the strain magnitude detected by the strain gage and converted to a change in resistance, x_1 and x_2 are the longitudinal coordinates defining the active lengths of measuring grid in the measurement direction.

$$\varepsilon_{SG} = \int_{x_1}^{x_2} \frac{\varepsilon(x)}{x_2 - x_1} dx \tag{2.1}$$

The strain gage is therefore a genuine strain transducer in contrast to many other devices used for strain measurement which just measure a change in length and refer it to an initial measurement length.

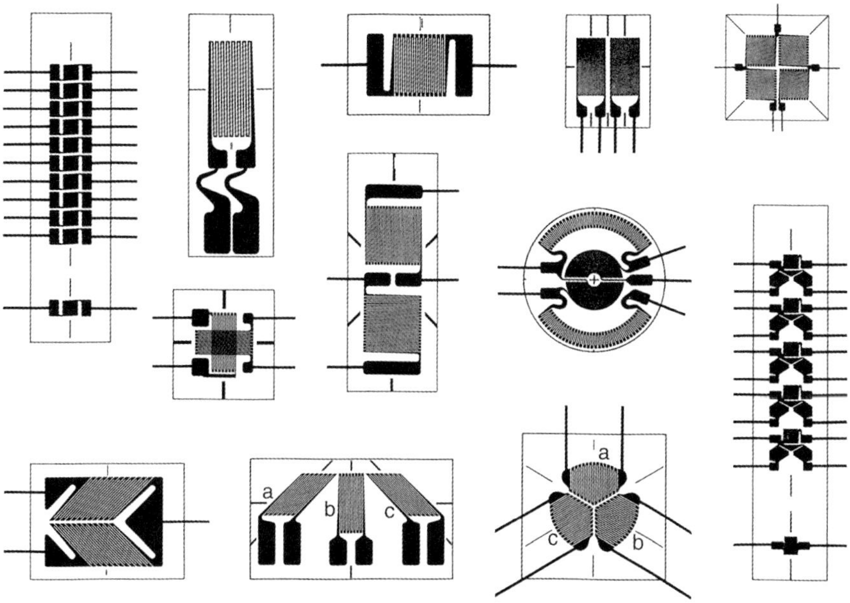

Fig. 2.4 Examples of different grid shapes for foil strain gages

It should be noted however, with regard to Eq. (2.1), that this type of integration takes place along each track of the strain gage and the expression $\varepsilon(x)/(x_2 - x_1)$ applies under the assumption that no strain gradient is present transverse to the measurement direction. The result ε_{SG} is therefore the sum of the integration results along

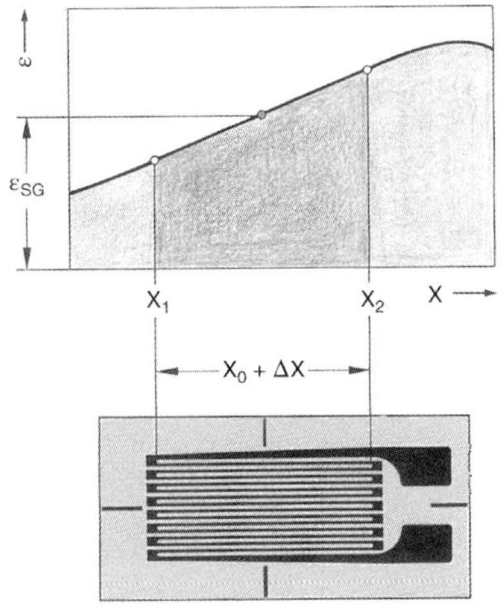

Fig. 2.5 Integrating measurement of the strain gradient over the active length of the strain gage measuring grid

each of the strain gage tracks. This way of looking at the problem becomes more significant if the longitudinal strain to be measured has a gradient transverse to the measurement direction, which is the case, for example, at the edges of holes or grooves. Then different strains act on the separate strain gage tracks and the overall result ε_{SG} is the arithmetic mean from the integration results of all the strain gage tracks.

The integrating characteristic is a significant consideration in the selection of the length of measuring grid of a strain gage to be used for a certain measurement task. If noticeable strain gradients are to be expected in the surface of the measurement object, for example in notches or at changes of cross-section, then short measuring grids should be employed in order to obtain measurements as integrated results over short distances. In this manner, the changing strain that is measured will be as close as possible to the actual strain. Chains of strain gages are available that, for example, combine ten measuring grids, each with 0.6 mm length of measuring grid, over a length of 10 mm. The measurement supplied by each separate grid corresponds to the mean of the strain over a distance of 0.6 mm. Figure 2.6 shows a strain gage chain at the base of a tooth on a straight-toothed gearwheel for the measurement of the strain distribution when the tooth is loaded in a testing device. Figure 2.6 shows the arrangement of the chain in a diagram and also a photograph of the bonded, unwired chain that is still to be covered.

However, there are also applications in which the non-uniformity in the strain distribution caused by inhomogeneities in the measurement object material is

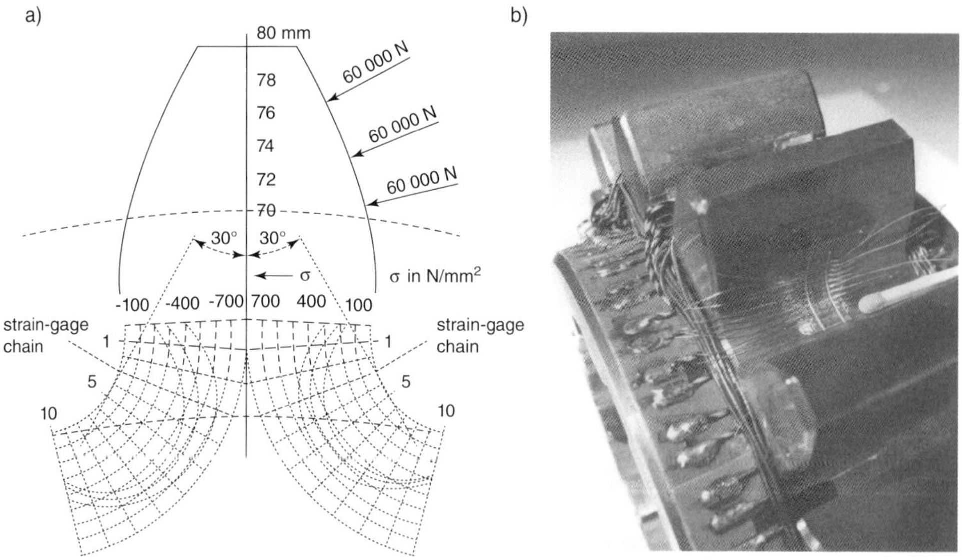

Fig. 2.6 a) The strain gradient occurring in the longitudinal direction of the strain gages; b) A strain gage chain fitted to the rounded base of a tooth on a gearwheel. The chain has extremely short lengths of measuring grid so that, as far as possible, the real strain gradient is measured

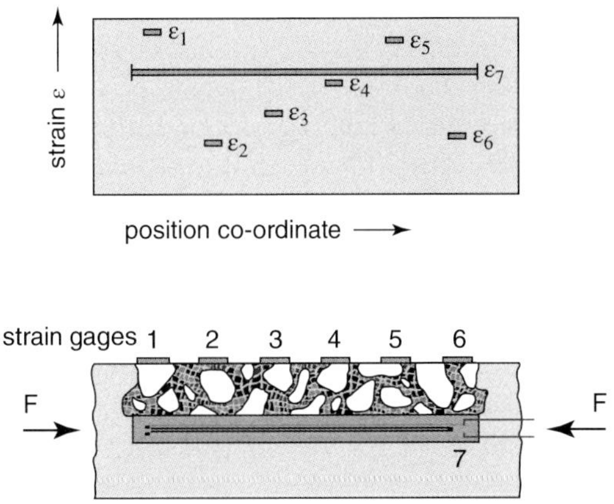

Fig. 2.7 Strain gage results on an inhomogeneous material (concrete) with a number of short strain gages and one long one

regarded as a disturbance, and a representative mean is required as the result. This might, for example, be the case for measurements on concrete, where gravel inclusions produce strain gradients, or for metals with large crystal grains. In these cases strain gages with long measuring grid lengths tend to be preferred and, for example, lengths of up to 150 mm for measurements on concrete are used [2.3].

Measurement results found on concrete with a number of short strain gages and one long one are compared in Figure 2.7. The short gages indicate different strains depending on the position, whereas the long gage supplies a representative mean.

For cases in which the longitudinal strain to be measured has a significant gradient in the transverse direction, narrow strain gages are used, close together, to measure the strain with as little gap as possible. An example here is the use of a strain gage chain to measure the strain distribution in the vicinity of a hole in a tensile beam. It is known that the strain increases towards the edge of the hole due to the stress concentration caused by the hole. To measure this strain gradient a strain gage chain is used as shown in Figure 2.8. The measuring grids on the chain are orientated transverse to the axis of the chain and measure the strain variation due to the stress concentration at the edge of the hole.

As shown in Figure 2.9, each strain gage supplies the arithmetical mean of the strains in the surface region covered by its active measuring grid.

If measurements are carried out on objects in which no strain gradients are expected in their surfaces, then the best length of measuring grid is the one that can be mounted most easily, for example the one that best fits under the thumb of the technician when pressing during bonding! Lengths of measuring grid are generally between 5 mm and 10 mm. For reasons of thermal energy dissipation, strain gages with large grid surfaces are better for taking measurements on poor thermal conductors.

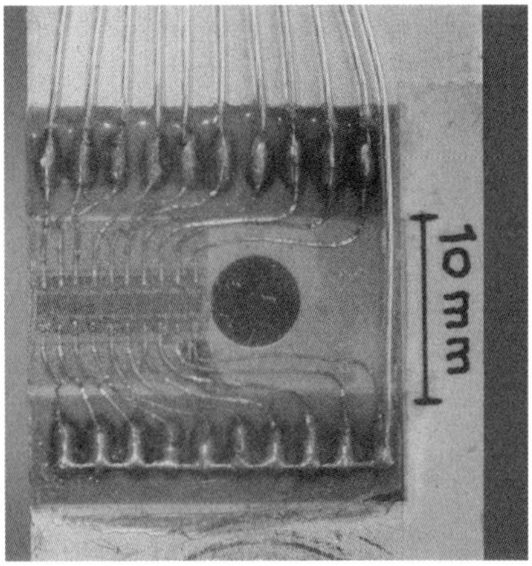

Fig. 2.8 Arrangement of a small strain gage chain for the measurement of strain gradients transverse to the measurement direction in the region of a hole in a tensile beam

It will be apparent from the quoted applications that various types of strain gage chains are supplied by the strain gage manufacturers. Apart from the carrier and grid materials used, the main parameters of these chains are orientation of the measuring directions for the individual grids and the chain dimensions. The smallest chains consist of 10 measuring grids on a length of 10 mm and the measuring directions of the separate grids may be transverse, longitudinal or alternating to the chain axis.

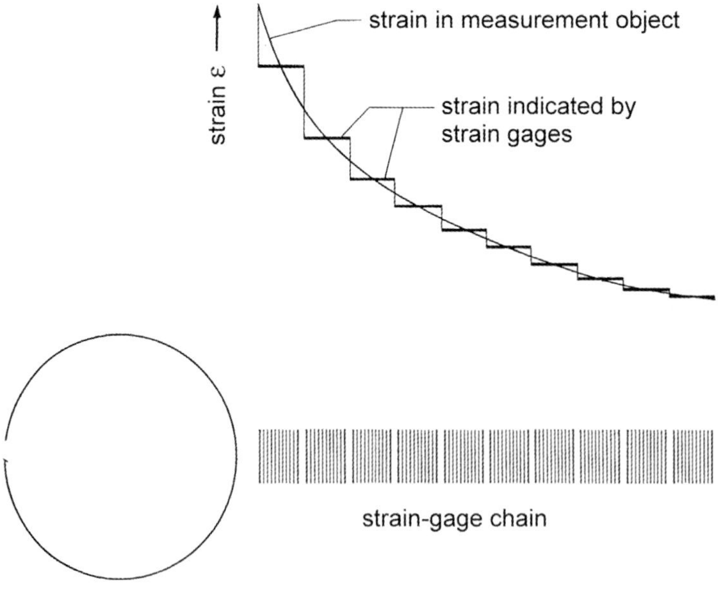

Fig. 2.9 Diagrammatic representation of the actual strain variation at the edge of the hole and of the means supplied by the individual strain gages over each width of measuring grid

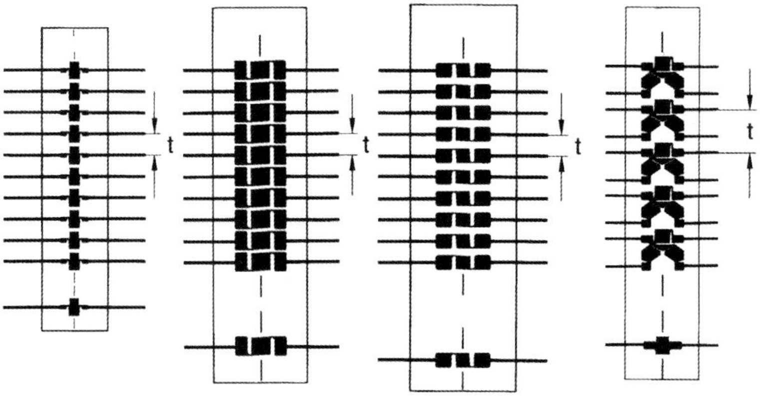

Fig. 2.10 Examples of strain gage chains for the measurement of strain gradients over small areas

The examples given in Figure 2.10 show, starting from the left, a strain gage chain with measuring grids aligned parallel to the axis, one with measuring directions arranged transversely a chain with alternating grids and a rosette chain. The dimension t indicates the pitch of the measuring grids.

For tasks in stress analysis *rosette* chains are supplied which, as shown at the far right in Figure 2.10, combine small adjacent rosettes consisting of three separate gages on one carrier. The big advantage of these chains is that they are relatively easy to mount, due to the common carrier, giving stable mutual alignment of the separate grids.

Strain gages with measuring grids arranged on top of one another are also supplied for carrying out stress analysis. Here, there is a requirement for measuring the strains in different directions at the same point on the surface of the measurement object using two or three stacked grids. During the evaluation the stress state is obtained for the closely bounded position. For strain gage rosettes with grids adjacent to one another, the three grids measure strains at adjacent points which can lead to errors during the stress determination if strain gradients are present. These errors can be markedly reduced by using measuring grids located one on top of the other. The stacked single grids are separated by a foil with good strain transfer and insulation characteristics, such as phenolic resin. A photograph of this type of strain gage is reproduced in Figure 2.11.

There are special cases in which strain gages would normally be bonded to the measurement object using hot-curing cement, but this is prevented by external circumstances, such as where the measurement object cannot be heated and maintained at a constant temperature as required by the cement instructions. Here, weld-on strain gages are used, which are attached by spot welding. They

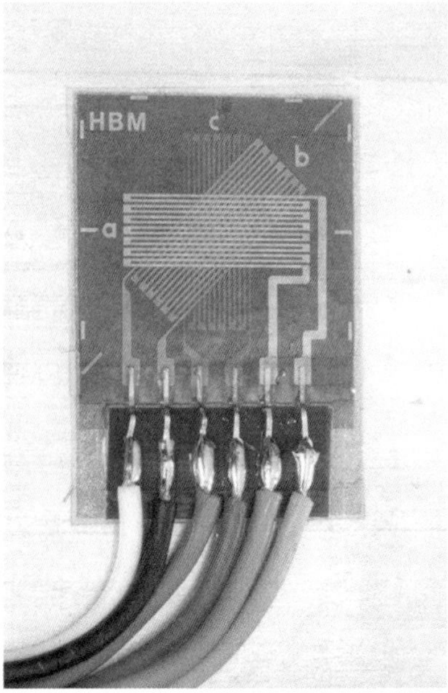

Fig. 2.11 Example of a strain gage rosette with measuring grids positioned over one another

consist of small metal-sheet carriers onto which a strain gage with plastic carrier is bonded according to the curing instructions for the hot-curing cement. This type of weld-on strain gage is supplied by manufacturers in different versions with regard to the covering and the application temperature, with or without connecting leads. However, they can only be used on measurement objects to which it is possible to spot weld. Figure 2.12 shows an example of a strain gage attached to the measurement object by spot welding. It consists of a foil strain gage bonded to a 0.1 mm thick carrier sheet (material X8Cr17) by hot curing. The gage has a 6 mm length of measuring grid and a polyimide carrier.

Encapsulated strain gages are available in various versions, depending on the application requirements, for strain gage measurement tasks under harsh ambient

Fig. 2.12 A weld-on strain gage fixed to the measurement object by spot welding

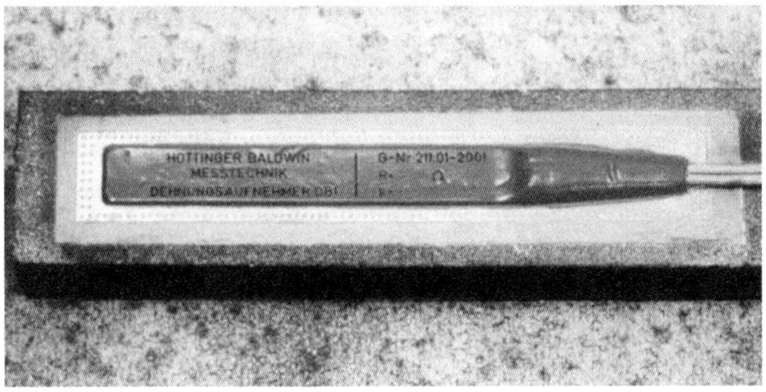

Fig. 2.13 Rubber-encapsulated wire strain gage with 150 mm measuring length bonded to a brass carrier sheet

conditions. These application requirements may be, for example, protection against mechanical or corrosive influences.

Encapsulated strain gages are also available for embedding in media that hardens when applied, such as plastic, asphalt, concrete, and so on. The connecting cable is almost always included in the encapsulation, because the connection point between the cable and the strain gage is a critical point due to the risk of penetration of moisture or other aggressive substance. Today, encapsulated weld-on strain gages are employed particularly for applications at high temperature. Figures 2.13–2.19 show examples of encapsulated strain gages designed for various applications.

A wire strain gage encapsulated in rubber is shown in Figure 2.13. It has a measuring grid of 150 mm length and a 0.2 mm thick brass carrier sheet. The strain gage is particularly suitable for measurements on inhomogeneous materials such as concrete and for measurements on building sites where harsh conditions prevail. Two-component cold-curing adhesives have proved to be good bonding agents for this

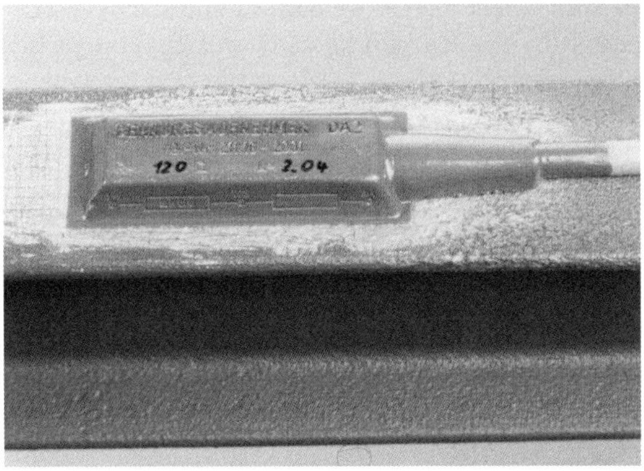

Fig. 2.14 Half-bridge circuit of two foil strain gages with measuring grids of 10 mm length on a brass sheet, encapsulated in rubber

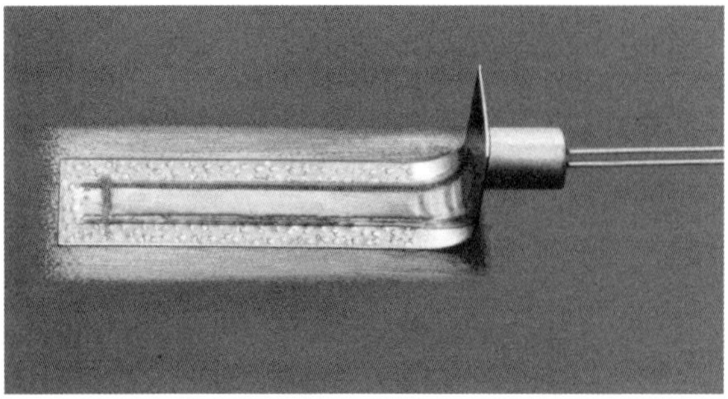

Fig. 2.15 Foil strain gage with 10 mm measuring grid welded between two steel sheets

type of strain gage. The gage shown in Figure 2.13 has a permanently mounted connecting cable several metres long.

The gage shown in Figure 2.14 is also encapsulated in rubber and has a 0.2 mm thick brass carrier sheet. Located under the encapsulation is a half-bridge circuit with integral compensation gage which is bonded to a steel block that remains undeformed during the measurement. The active foil gage has a measuring grid length of 10 mm, and the connecting cable enters the encapsulation.

Apart from normal tasks in strain gage measurement, this type of half bridge gages is especially suitable for the measurement of restricted thermal strain in the determination of thermal stresses. Cold-curing, two-component adhesives are preferred for bonding. Gages of this type need not be discarded after the measurement, because they can be carefully separated from the measurement object so that they can be used again elsewhere.

Figure 2.15 shows a strain gage which is completely encapsulated in stainless steel for mounting on large metal measurement objects such as oil rigs, ships, reactor parts or heavy machines. A foil gage with a 10 mm long measuring grid is located on a 0.1 mm thick carrier sheet. A covering of thin sheet metal welded to the carrier sheet encapsulates the measuring element with the leads. The cable connection passes through a vacuum-tight glass bushing welded with an electron beam. This type of gage is preferably attached by spot welding.

The strain gage illustrated in Figure 2.16 was developed for casting in hardening materials such as concrete or asphalt and has also been used successfully in pack-ice. It consists of a wire grid with a measuring length of 88 mm and embedded on all sides in polycarbonate. The cable, which also enters the encapsulation, is connected using the three-wire technique.

Fig. 2.16 Wire strain gage embedded in polycarbonate, with an 88 mm long measuring grid for casting, for example, in concrete or asphalt

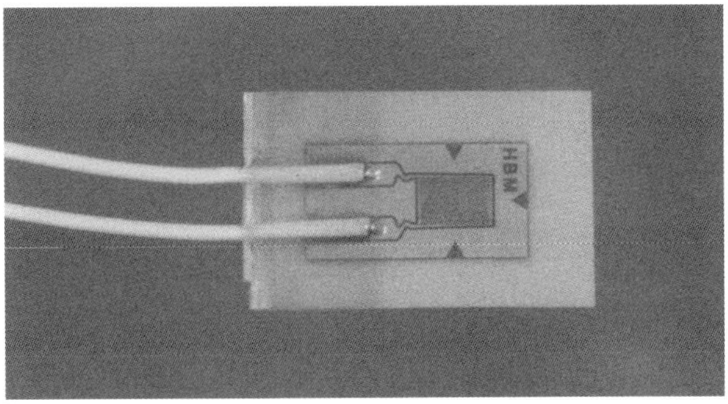

Fig. 2.17 Foil strain gage with a 3 mm long measuring grid enclosed in special plastic for measurement in a moist ambient

Figure 2.17 shows a foil gage which has a measuring grid of length 3 mm, which is completely embedded in a special plastic on a Kapton carrier. The Teflon-insulated stranded connecting lead is included in the encapsulation. Even without any additional protection the gage is insensitive to splashed water and moisture. It can be mounted easily and quickly, also in positions where access is difficult, because mounting does not involve any soldering or covering work. The gage is bonded with cyanoacrylate adhesive or with hot-curing, two-component epoxy resin adhesive.

Figure 2.18 illustrates a weld-on high temperature strain gage. The gage resistance wire is located in a steel tube which is filled with magnesium oxide powder and welded to a thin carrier sheet. This type of gage is supplied in various versions as quarter bridges or as temperature-matched half bridges with resistance wire in NiCr or PtW. A series of types of cable suitable for different application requirements is available with integral connecting cables.

Figure 2.19 shows a modification of the gage illustrated in Figure 2.18. This gage was designed for direct casting in concrete or other castable, hardening materials. The gage, which is encased in a stainless steel tube, has a grid type of flange at each end for anchoring in the measurement object and it is fitted with an integral connecting lead.

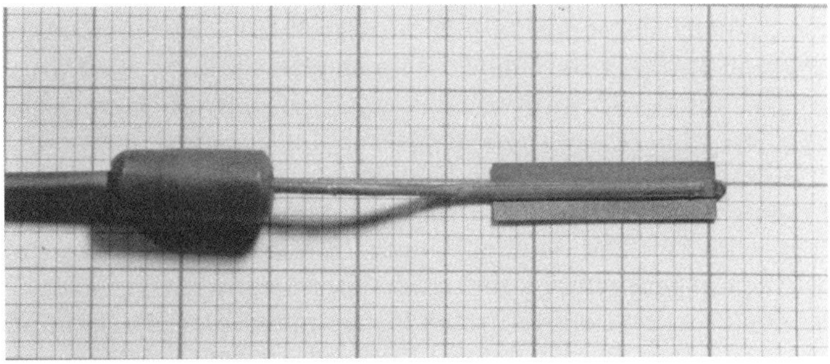

Fig. 2.18 High-temperature weld-on strain gage encapsulated in a stainless steel tube

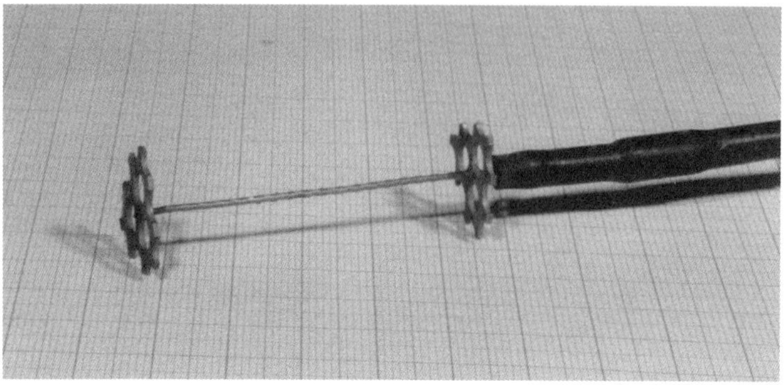

Fig. 2.19 Concrete strain gage encased in a stainless steel tube

2.1.2 Carrier material

The first wire strain gages, which were produced in the USA, had *paper* carriers. Strain gage development in other countries also began based on paper carriers. Examples are Boano in Italy, Philips in the Netherlands, Tinsley in the UK, Nier and TPW in the former GDR, Ebner in Hungary and Aeropal in Czechoslovakia [2.4–2.8]. Gustafsson in Sweden also began with a paper carrier, but he soon replaced it with a sprayed-on acetate film [2.9]. Experiments were also carried out to manufacture strain gages woven like textiles (e.g. Orion in Hungary and Rolls Royce in the UK [2.10, 2.11]) but they were without success. Since paper as a carrier material is very sensitive to moisture, attempts were made to eliminate this disadvantage by soaking the paper in nitrocellulose. Paper carriers can be relatively easily bonded to measurement objects with physically hardening adhesives. They are extendable by about 6% and can be used at temperatures between −100 °C and +80 °C.

It was soon found that paper is not an ideal carrier material for strain gages, and the search began for suitable materials that were more able to fulfil the requirements. The carrier must hold the measuring grid firm and transfer to the measuring grid, without loss, the strain introduced through the adhesive from the measurement object. Because the strain must be transferred as a faithful replica, the carrier must be very thin, but it must possess excellent insulating characteristics to prevent errors due to even the slightest shunt effects. The resistance between the measuring grid and the measurement object should be within an order of magnitude of 10 G Ω. The carrier material, the thickness of which must be as uniform as possible, must be fixed to the surface of the measurement object using a suitable method of fixing, preferably by adhesive bonding. Simple handling during installation, flexibility and adequate strength, particularly against continuous static and dynamic loads, are further requirements. The demands made on the carrier material of strain gages differ between experimental stress analysis and transducer construction. With transducer construction long-term stability is important, the maximum extensibility can be low and even the flexibility of the material is not of primary importance, because with transducers the strain gages are almost always mounted on flat surfaces. With stress analysis the flexibility and, increasingly, the maximum extensibility are of primary importance. To fulfil the demands placed upon it, the carrier material must also have a high shear modulus. The optimum adhesion between the grid and the carrier and also between the carrier and the measuring object must be obtained. But above all, the best possible temperature stability is required, because the temperature limits between which the strain gage can be used depend mainly on the carrier material.

Phenolic resin (Bakelite) was the first carrier material which, apart from paper, was used in the USA. It is substantially more resistant to moisture than paper and it exhibits high resistance to creep. Young's modulus for this carrier material is $E = 7500\,N/mm^2$. The first strain gages to be manufactured industrially in the Federal Republic of Germany in 1956 by Hottinger Messtechnik GmbH were Bakelite gages [2.12] which could not however be attached with physically hardening adhesives. A hot-curing phenolic resin adhesive, Bakelite cement BC6035 from the USA, was used. Phenolic resin can be used at temperatures of −70 °C to +200 °C and with glass-fibre reinforcement at −200 °C to +200 °C. Soon, however, the cold-curing two-part adhesive X57 was developed in Germany and this produced excellent bonding of Bakelite gages. A further development of this adhesive, the X60, is still used successfully today. The dark brown phenolic resin gages produced in Germany in the mid 1950s were only accepted hesitantly by German users. Strain gages with a bright-coloured paper carrier, obtained from other countries, were much preferred. Therefore, a search was conducted in Germany for a bright plastic carrier and in 1957 acrylic resin was found.

Acrylic resin, which as a carrier material resembled phenolic resin in the most important characteristics, achieved a breakthrough on the German market. Acrylic resin gages are very similar to paper gages in appearance, without exhibiting the disadvantages of paper. They can be attached very well with X57 and X60 adhesives and are capable of measurements up to 100 °C. Young's modulus for acrylic resin carriers is about $3500\,N/mm^2$. Acrylic resin gages were successfully manufactured and used in large quantities by Hottinger Messtechnik between 1958 and 1966 under the trademark Impa. Since, apart from their visual appearance, they do not have any advantages over phenolic resin gages, they have now disappeared from the market.

Epoxy resin has proved a particularly good carrier material for strain gages that are used for high quality transducers. It has a relatively high mechanical strength with a Young's modulus of about $3500\,N/mm^2$. Epoxy resin carriers can, however, only be employed between −45 °C and +95 °C. But these temperature limits are adequate for general transducer construction. Young's modulus for epoxy resin drops very rapidly at high temperatures due to softening. The maximum extensibility is about 2%. Strain gages with epoxy resin carriers are only suitable with restrictions for use in experimental stress analysis.

Epoxy phenolic resin is a combination of epoxy resin and phenolic resin, covering a substantially wider temperature range with application limits between −75 °C and +205 °C. With glass-fibre reinforcement, the application temperature limits can be extended from −269 °C to +290 °C. The material is strong, with a maximum extensibility of only about 1.5%.

Polyimide is the carrier material that is most popular today. It is tough and flexible and enables bending radii down to 0.3 mm to be obtained. The application temperature range is −195 °C to +175 °C, and the maximum extensibility is up to 20%. For these reasons, polyimide is an almost ideal carrier material for strain gages for experimental stress analysis.

Metal carriers are particularly suitable for high application temperatures and for encapsulated strain gages which are used under harsh, demanding ambient conditions, such as under seawater. They are usually attached to the measurement object by welding. Their resistance to radioactive radiation is excellent. Today, there are mainly two types of weld-on strain gages in use which differ in their basic construction. With one of them the foil grid or resistance wire is fixed to the metal carrier using insulating putty and with the other, the resistance wire is located in a tube welded to the carrier and filled with insulating powder (e.g. aluminium oxide). These strain gages have no organic constituents and can be employed at temperatures up to about 650 °C.

Free-grid gages are strain gages which are fixed to the measurement object using ceramic putty or the so-called flame-spraying technique without any carrier. They usually have an auxiliary carrier which is just used as a mounting aid and is removed during the mounting process. Free-grid gages were specially developed for high temperature applications.

2.1.3 Measuring grid material

In the early 1930s Czerlinsky carried out experiments in Berlin to find the relationship between the change of resistance and the mechanical strain in metal conductors. This work led to the realization that the Cu-Ni alloy, constantan, was particularly good for the exploitation of this effect for measurement purposes [2.13]. Simmons used a thin constantan wire for his dynamometer in 1936 without being informed of Czerlinsky's work (Simmons, E., Personal communication during meeting of the Western Regional Strain Gage Committee (SEM, USA), April 1988 at Pasadena, Cal.). Following a spontaneous idea, Ruge used resistance wire from a potentiometer that was to hand for his first measurements and it was only after he had contemplated the basic principles involved that he changed to constantan wire.

The first strain gages manufactured industrially by Ruge and de Forest, used metal measuring grids and had constantan as the grid material.

The measuring grid material of a strain gage must satisfy a number of requirements which may in part conflict, depending on the intended application; for example, high extensibility and high fatigue strength are not easy to achieve simultaneously. It is important to obtain the best possible linear, reproducible relationship between the mechanical strain and the change in resistance, which should also have the minimum hysteresis. The sensitivity to strain should be high, that is even slight changes in strain should produce large changes in resistance in order that a suitably large electrical output signal be obtained. A particularly high specific resistance for the grid material enables the manufacture of small strain gages with a high nominal resistance. The resistance should be independent of strain as well as of temperature. The former criterion is approximately achievable with some materials, but not the latter, because the electrical resistance of all metals changes to some extent with temperature. Therefore, a low temperature coefficient of specific resistance is desired, which should be reproducible, so that undesired effects can be regulated by selected compensation measures. The linear thermal expansion coefficient of some grid materials can be influenced within certain limits by heat and deformation treatment, and matched to the measurement circumstances (e.g. to the thermal expansion of measurement object materials). For this matching to be effective, a range as wide as possible between these limits is required. Another significant factor is the application temperature range of a grid alloy, which often sets limits on the range of applications for a strain gage. The importance of the working properties and the price should not be underestimated. The manufacturing method must be economical and produce a good quality product. Difficulties arose formerly, for example, when winding wire strain gages. The processing on automatic machines of the harder types, which were needed for good fatigue strength and low hysteresis, proved to be very difficult. Also, a great deal of experience is still needed today to produce etched foil gages with smooth etched edges. Table 2.1 gives a brief summary of the most important measuring grid alloys and their percentage composition.

Constantan is the alloy most often used as a grid material. It consists of 55% copper and 45% nickel, but the actual percentage composition of the two constituents may vary ±5%. Constantan as a grid material has been and still is available under other trade names: Advance, Cupron, Copel, Alloy 45 and Neutroloy. For applications at temperatures between −50 °C and +180 °C constantan has the most universal characteristics. Its high sensitivity to strain (i.e. the relationship between change of strain and change of resistance) is hardly affected by strain and temperature. The specific resistance ($\approx 0.49\,\Omega mm^2/m$) is high enough to manufacture strain gages with the usual nominal resistances in convenient sizes. The temperature coefficient of specific resistance can be varied within relatively wide limits by selective heat and deformation treatment of the material. This means that the thermal characteristics of

Table 2.1 Common measuring grid alloys and their percentage composition

Material	Alloy constituents
Constantan	50–60% Cu, 40–50% Ni
Nichrome V	80% Ni, 20% Cr
Karma 331	75% Ni, 20% Cr, 3% Fe, 2% Al
Isoelastic	52% Fe, 36% Ni, 8% Cr, 3.5% Mn, 0.5% Mo
Platinum–Tungsten	92% Pt, 8% W

the grid material can be matched to the linear thermal expansion coefficients of almost all measurement object materials, so that self-temperature-compensating strain gages can be manufactured. The relatively high fatigue strength of constantan can be transferred to the strain gage with suitable processing (e.g. smooth etched edges and vibration-resistant connections). Using special heat treatment, a maximum extensibility of up to 20% can be achieved with constantan, so that it is particularly good for the production of high-strain gages.

NiCr alloys are the second most common grid alloys after constantan. The most popular is the alloy known as Karma which, apart from 75% nickel and 20% chrome, also contains some iron and aluminium. An important advantage is the wide application temperature range which is between −269 °C and +290 °C for static measurements. This range can be extended up to 320 °C depending on the duration at such a high temperature [2.14]. The specific resistance of Karma, at ≈1.6 $\Omega mm^2/m$, is more than three times that of constantan, so that very small strain gages for the usual nominal resistance values or very high resistance strain gages (up to 5000 Ω) with a small size can be produced in Karma. The alloy has relatively high fatigue strength. The temperature coefficient of specific resistance can be modified within certain limits by heat and deformation treatment, as with constantan, enabling matching of the thermal expansion to that of various measurement object materials. The strain sensitivity of NiCr alloys reduces slightly with increasing temperature. This effect can be used to advantage in the construction of transducers. Here, the alloy is used to compensate the change in sensitivity that arises due to the reduction in the Young's modulus of the measuring element material which occurs with increasing temperature. In this respect, Karma can be matched to different measuring element materials using special treatment (Modified Karma or K Alloy). The designation Modco (Young's **Mod**ulus **co**nstant) is also used for these matched NiCr alloys. NiCr alloys are also processed without additives to give a further increase in the application temperature. An example there is Nichrome V, an alloy consisting of 80% nickel and 20% chromium. Alloy A, Nikrothal, Tophet A and Protoloy are other trade names for this alloy composition. Measurements above 320 °C can be carried out with this NiCr alloy (assuming that a suitable carrier for the measuring grid is available with adequate temperature stability). However, due to the time-dependent changes of the specific resistance of NiCr alloys at these high temperatures, the application should be restricted to dynamic measurements where a continuous change of resistance is of lesser importance.

Isoelastic is a measuring grid alloy that has a historical development similar to constantan and was employed as early as the 1940s for the industrially manufactured fatigue resistant SR-4 strain gages. The main constituents are iron (52%), nickel (36%) and chrome (8%) with manganese (3.5%) and molybdenum (0.5%) additives. The fatigue strength is excellent. In contrast, the disadvantages are a high temperature coefficient of specific resistance and a nonlinear strain-dependent sensitivity characteristic for strains >0.5%. The sensitivity is just double that of constantan, so that Isoelastic has advantages over other measuring grid materials under extreme conditions of dynamic loading due to its excellent fatigue strength and high sensitivity.

Platinum/tungsten alloy was developed as a material for measuring grids to avoid the disadvantages of NiCr alloys for measurements at high temperatures. The percentage composition is normally 92% Pt and 8% W. Since this alloy resists the etching process used in the manufacture of foil gages, strain gages are now mainly produced with PtW wire. Exceptions here are the die-cut foil gages produced in large quantities in the 1960s by Dentronics in the USA by stamping out from PtW foil [2.15]. Static measurements up to about 500–550 °C are possible with measurement wire in PtW alloy depending on the time spent at high temperature [2.16]. The temperature coefficient of specific resistance is relatively high, but very reproducible. It cannot be matched to the measurement object material as with constantan or Karma. The region of strain is limited to ±0.3% due to the linearity deviations of sensitivity. Strain measurements at high temperatures with PtW wires need special preparation. For example the unwanted drift can be reduced. This is mainly due to undissolved residues of tungsten remaining after the sinter manufacturing process. At high temperatures the tungsten then goes into solution, causing an increase in resistance. The dissolving process of the residual tungsten can be eliminated beforehand by storing commercially available PtW strain gages for 350 hours at temperatures of about 740 °C. Treated in this way, the strain gages do not then exhibit any drift due to the dissolving effect up to temperatures of about 700 °C [2.17].

2.2 Sensitivity

The sensitivity of a measurement instrument is defined as the ratio of an observed change in its output signal to the change in the input signal which causes it [2.18]. The input signal of a strain gage is the strain ε recorded by it and the output signal of a strain gage is the change in resistance $\mathrm{d}R/R$, generated by the strain ε. The relationship will be described by the gage factor k:

$$\frac{\mathrm{d}R}{R} = k\varepsilon \tag{2.2}$$

A metallic wire changes its resistance when it is subject to strain. This phenomenon was discovered in 1856 by Lord Kelvin who first reported on the relationship between strain and the resistance of wire conductors. About 80 years later, Simmons and Ruge made use of this knowledge for measuring strain. They used small diameter wires made of electrical resistance alloy, such as copper-nickel, and bonded them to structures to measure surface strain.

For commercial strain gages the gage factor k is determined experimentally by the manufacturers, according to current guidelines within a strain range between −1000 μm/m and +1000 μm/m [2.19–2.21].

$$\frac{\Delta R}{R} = k\frac{\Delta l}{l_0} \quad \rightarrow \quad k = \frac{\Delta R \cdot l_0}{R \cdot \Delta l} \tag{2.3}$$

The gage factor k of an electrical conductor, which describes the relationship between strain and resistance change, can be derived formally [2.22]. The resistance

R of a wire with a length l_0, cross-sectional area A and resistivity ρ of the wire material is given by

$$R = \frac{\rho \cdot l}{A} \tag{2.4}$$

Partial differentiation and dividing by the resistance R leads to

$$\frac{\Delta R}{R} = \frac{\Delta \rho}{\rho} + \frac{\Delta l}{l} - \frac{\Delta A}{A} \tag{2.5}$$

The term $\Delta A/A$ represents the change in cross-sectional area of the wire due to the transverse strain, and is given by

$$\frac{\Delta A}{A} = -2\nu \frac{\Delta l}{l} \tag{2.6}$$

Introducing Eq. (2.6) into Eq. (2.5) gives

$$\frac{\Delta R}{R} = \frac{\Delta \rho}{\rho} + \frac{\Delta l}{l}(1 + 2\nu) \tag{2.7}$$

From Eq. (2.7) it can be seen that there are two parts of the sum influencing the change in resistance. The part $\Delta l/l(1 + 2\nu)$ represents the change in resistance due to the geometrical alteration of the wire. The part $\Delta \rho/\rho$ represents the change in resistance caused by the alteration of the resistivity of the material due to the change in volume resulting from elastic deformation.

Bridgman found out in extensive research [2.23, 2.24] that in certain materials, for example copper-nickel alloys, there is proportionality between change in volume and resistivity. He introduced the factor of proportionality κ:

$$\frac{\Delta \rho}{\rho} = \kappa \frac{\Delta V}{V} \tag{2.8}$$

Introducing $$\frac{\Delta V}{V} = \frac{\Delta l}{l}(1 - 2\nu) \tag{2.9}$$

leads to $$\frac{\Delta \rho}{\rho} = \kappa \frac{\Delta l}{l}(1 - 2\nu) \tag{2.10}$$

Equation (2.10) describes the change in the resistivity due to the alteration in volume. Introducing Eq. (2.10) into Eq. (2.5) gives

$$\frac{\Delta R}{R} = \kappa \frac{\Delta l}{l}(1 - 2\nu) + \frac{\Delta l}{l}(1 + 2\nu) \tag{2.11}$$

Equation (2.11) describes the relationship between strain and the change of resistance caused by this strain. It leads with $\varepsilon = \Delta l / l$ to the formal definition equation of the gage factor k:

$$k = \frac{\Delta R/R}{\Delta l/l} = \kappa(1 - 2\nu) + (1 + 2\nu) \tag{2.12}$$

The right side of Eq. (2.12) describes the two parts that generate the change in resistance. The part generated by the change in specific resistance is

$$\beta_\rho = \kappa(1 - 2\nu) \tag{2.13}$$

The part resulting from the geometric change is

$$\beta_g = (1 + 2\nu) \tag{2.14}$$

Using the known quantities of κ and ν of the material it is possible to calculate formally a value of the gage factor k, for example $\nu = 0.3$ and $\kappa = 1.13$ (constantan) gives $k = 2.052$.

Note: gage factor calculated this way may not be used as the valid gage factor for carrying out measurements. For measurements always use the gage factor given on the package of the strain gages. The real gage factor will differ from the calculated value due to production technology, heat treatment, construction, attachment points for lead wires, combination with carrier materials and adhesives and so on.

It is interesting to estimate formally the quantities of the parts β_ρ and β_g by introducing the values of k and ν into Eqs. (2.12) to (2.14):

$$k = \beta_\rho + \beta_g \tag{2.15}$$

$$2.052 = \beta_\rho + 1.6 \Rightarrow \beta_\rho \approx 0.45 \tag{2.16}$$

From this result we can see that about 22% of the signal will be created by the change in the specific resistance and about 78% by geometrical alterations.

As mentioned above, the gage factor cannot be calculated with sufficient reliability. Therefore the gage factor of commercial strain gages is determined experimentally by the manufacturer according to guidelines from the relative change in resistance and the relative change in length. For strain gages with a constantan grid the gage factor is usually around 2. The actual gage factor of a strain gage is always given on the package in which it is supplied. Because of the statistical evaluation used to determine the gage factor from the experimental research the results are marked with a specific tolerance, usually ±1% (nowadays also available as ±0.5%). The experimental determination of the gage factor is carried out according to the guidelines between the strain limits of −1000 µm/m and +1000 µm/m on a material which has a Poisson's ratio of $\nu_0 = 0.285$. That means the strain gage is calibrated in a biaxial strain field in which the transverse strain amounts $-\nu_0\varepsilon$.

In [2.22] (1949) the results of extensive experimental investigations of the function $\Delta R/R = f(\varepsilon)$ are published for strains well into the elasto-plastic deformation range. A striking example from [2.22] is given in Figure 2.2 with two diagrams, besides the function $\Delta R/R = f(\varepsilon)$is also the shape of the corresponding stress–strain diagram drawn in order to show clearly the transition from the elastic to the elasto-plastic deformation range.

The experimental results for constantan over the entire investigated range of deformation in a gage factor $k = 2$, with Nichrome the gage factor k decreases with incipient occurrence of plastic deformation from 2.5 to 1.8.

The functions exemplified in Figure 2.20 are valid for thin wires and are specific for the investigated materials. Further strain-dependent changes in resistance are described in [2.25, 2.26]. In the 1950s Bertodo [2.27], during the development of high-temperature strain gages, carried out extensive investigations of gage factors for a variety of measurement grid materials. Figure 2.21 is taken from this paper which reproduces the relationship between strain and resistance change for free tensioned wires of different materials at a temperature of 20 °C. Here, too, constantan shows as the only material which, starting from the origin, has a continuous linear relationship. The gage factors k of all the materials tested, except PtIr and piano wire, tend in the plastic deformation range towards $k = 2$. Nickel shows a negative gage factor, which, when it reaches the yield stress, also becomes $k = 2$ [2.22].

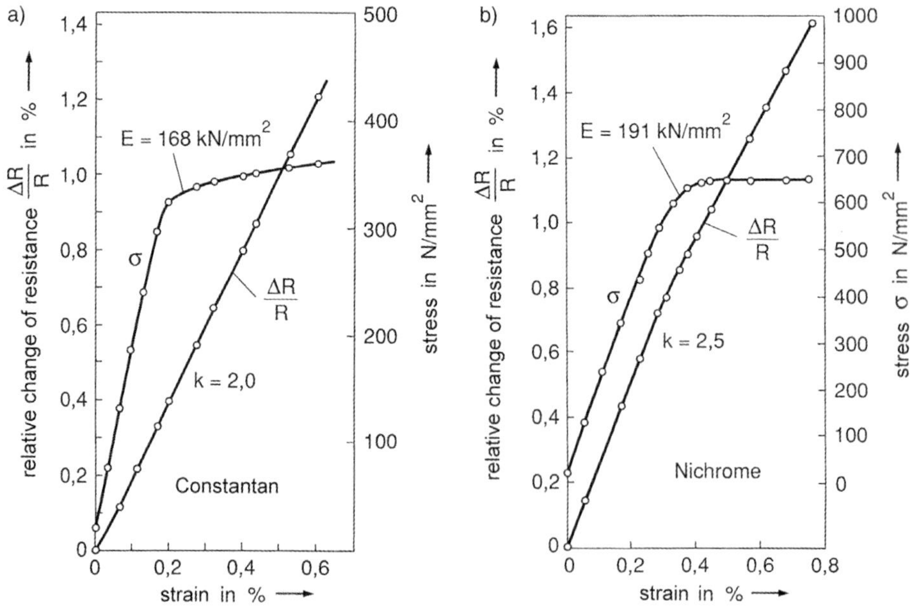

Fig. 2.20 Experimentally determined results of strain-dependent change in resistance of wires (0.025 mmØ) from a) constantan and b) Nichrome to illustrate the transition from elastic to elasto-plastic state, the stress curves are drawn, too [2.22]

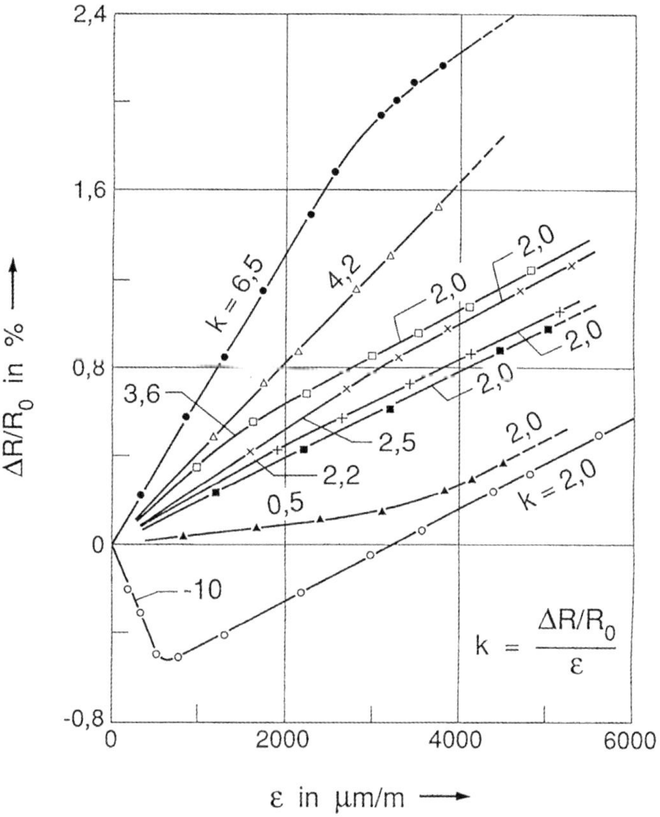

Fig. 2.21 Resistance–strain characteristics of unbonded free tensioned wires [2.27]

○ *Nickel*
▲ *Manganin (83% Cu, 3% Ni, 14% Mn)*
■ *Constantan (56% Cu, 44% Ni)*
+ *NiCr annealed (80% Ni, 20% Cr)*
× *NiCr hard (80% Ni, 20% Cr)*
□ *Soft Iron*
△ *Piano wire*
• *PtIr (95% Pt, 5% Ir)*

Table 2.2 presents a guideline for gage factors (at room temperature) of important grid materials

It is well known from experience that the gage factors of commercial strain gages, as given by the manufacturers on the packages, are reliable for deformations taking place in the elastic deformation range of materials. But if the deformation exceeds the yield point of the material under consideration the gage factor determined in the elastic deformation range needs to be corrected.

In the elastic range of deformation, strain gages have a linear characteristic. That means the relationship between strain and change in resistance is linear. But for

Table 2.2 gage factors of important grid materials [2.27, 2.28]

Constantan 'Advance' (60 Cu, 40 Ni)	≈2.15
Constantan 'Eureka' (56 Cu, 44 Ni)	≈2.0
Isoelastic (36 Ni, 8 Cr, 3,5 Mn, 0.5 Mo, 52 Fe)	≈3.6
Nimonic 90 (46 Ni, 21 Cr, 21 Co, 5 Fe ...)	≈1.8
Karma 331 (74 Ni, 20 Cr, 3 Fe, 3 Al)	≈2.3
Nichrome (80 Ni, 20 Cr)	≈2.1–2.5
Platinum-Tungsten (92 Pt, 8 W)	≈4.0

strain gages with constantan measuring grid a nonlinear characteristic was found in the elasto-plastic deformation range, as can be seen in Figure 2.22 [2.29]. Research was carried out in the strain range up to ±150 000 μm/m (±15%) in order to find out the relationship in this deformation range. The results are represented in Figure 2.22. It can be seen that the gage factor increases with increasing plastic strain, but decreases with increasing plastic compression.

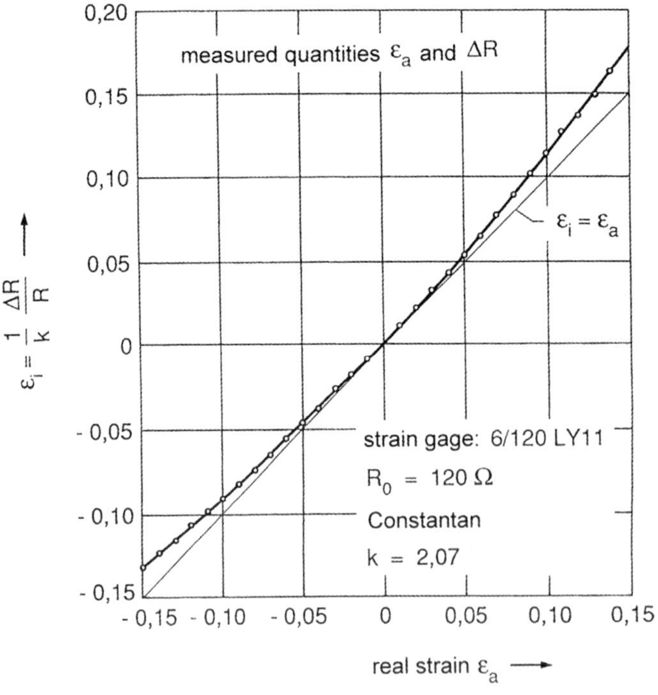

Fig. 2.22 Nonlinearity of the change in resistance in the elasto-plastic deformation range and deviation of the results from linearity

The results given in Figure 2.22 were found on a strain gage that had an elastic gage factor of 2.07. The actual strain ε_a is recorded on the abscissa and the strain value ε_i as calculated using Eq. (2.3) is recorded on the ordinate. The nonlinearity is obvious. The deviation of the results from the straight line is steady and can be formally described by an approximation polynomial of second order:

$$\varepsilon_i = \varepsilon_a + \varepsilon_a^2 \tag{2.17}$$

For practical application, in order to correct strain gage readings obtained in the elasto-plastic deformation range it is useful to introduce the term ε^* as the designation of the corrected strain into Eq. (2.17):

$$\varepsilon_i = \varepsilon^* + \varepsilon^{*2} \tag{2.18}$$

Solving Eq. (2.18) for the corrected strain ε^* one obtains

$$\varepsilon^* = -\frac{1}{2} + \sqrt{\frac{1}{4} + \varepsilon_i} \tag{2.19}$$

Equation (2.19), confirmed by experimental results, can be used to correct the measured strains in the elasto-plastic deformation range. The corrected strain by using Eq. 2.19 can be seen in Fig. 2.23.

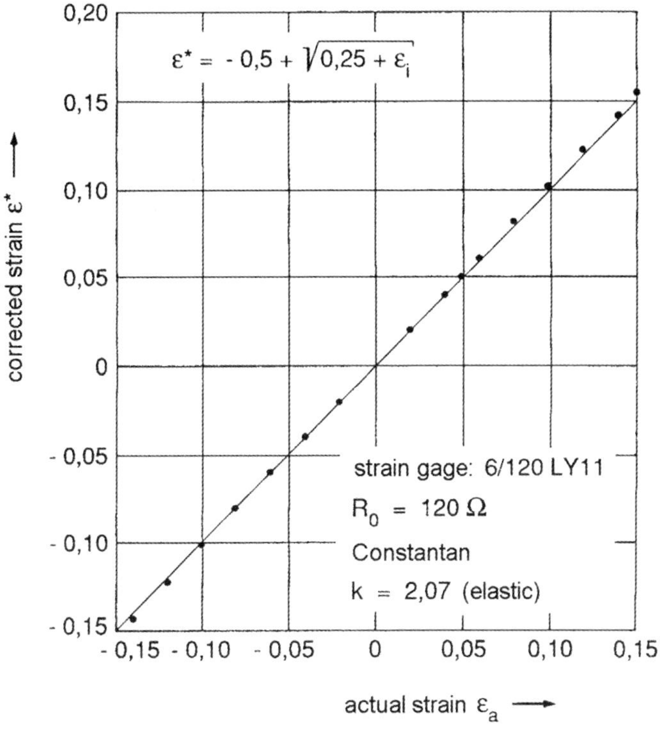

Fig. 2.23 Correction of strain gage readings obtained in the elasto-plastic deformation range by using Eq. (2.19)

The sensitivity of strain gages in the high strain range is the object of investigation of numerous works, for example [2.30–2.34], where also correction suggestions are given. In all these cases, however, the examined strain gages are connected as arms of Wheatstone bridges, and their influence on the linearity of the relationship between change in resistance of the strain gage and electrical output signal of the circuit is included in the consideration.

As will be shown in a later chapter, the output signal of a voltage-fed Wheatstone bridge circuit operating with one active strain gage is not linear. The amount of the linearity deviation occurring between change in resistance on the strain gage and electric signal is just as great as the strain acting on the strain gage. The tendency of this linearity deviation is opposite to that of the k-factor in the high strain range, so that both effects almost compensate.

However, this compensation cannot occur if the output signal of the bridge circuit is linear, such as in half-bridge circuits with two active strain gages, or full bridges. For measurements in the high strain range it is recommended to apply the correction Eq. (2.19) and then carry out, in a second step, any necessary correction of the bridge circuit.

2.3 Transverse sensitivity

Generally the state of strain is biaxial in surfaces of components on which we measure using strain gages. Thus there are strains present in the measuring direction of the strain gage as well as in the transverse direction. Transverse strain also causes, in the strain gage, changes in resistance which affect the result of the measurement by superimposing. Particularly in wire strain gages, which were used in large numbers before the invention of the foil gage, transverse strains can cause remarkable changes in resistance in the reverse loops of the measuring wire. But transverse strain can also generate change in resistance in foil strain gages. Transverse sensitivity q is set to define the influence of transverse strain [2.19].

$$q = \frac{k_q}{k_l} \tag{2.20}$$

In Eq. (2.20) k_q and k_l are the sensitivities in transverse and longitudinal directions respectively.

It is common during the experimental determination of the gage factor k to calibrate the strain gages in the biaxial strain field generated by a uniaxial stress state on the surface of a bending beam. The material of this bending beam has an elastic Poisson's ratio of $\nu_0 = 0.285$. The strain gages to be calibrated will be mounted on the beam with their measuring direction aligned in the principal direction 1 of the beam. That means that during experimental estimation of k, transverse strain $\varepsilon_q = -0.285\varepsilon_l$ is present and k contains the change in resistance generated by this transverse strain. This means that a strain gage with transverse sensitivity $q \neq 0$ can measure correctly only in a uniaxial stress field on a material with $\nu = \nu_0 = 0.285$. In all other cases an error occurs, depending on transverse strain. In order to determine the transverse sensitivity q the strain gage will be installed on a quasi-uniaxial strain field directed transverse to the strain gage's measuring direction. There the change in resistance of the strain gage will be measured. If the acting strain is really uniaxial and acts transverse to the strain gage's measuring direction then its change in resistance becomes

$$\frac{\Delta R}{R} = k_q \varepsilon_q \tag{2.21}$$

$\Delta R/R$ and ε_q will be measured, and from this with Eq. (2.20) the transverse sensitivity q can be determined.

In order to investigate transverse sensitivity of strain gages a test rig was used which could create a uniaxial strain state. The main element was a very wide but short bending beam with thickening at its ends and a thin mid-section, as can be seen in Figure 2.24. Because of the beam's thick ends, transverse strain acting in principal direction 2 of the beam will be restrained. Strain gages to be tested will be installed on the beam with measurement direction aligned with the principal direction 2 of the beam. The transverse strain in principal direction 2 of the beam is severely hampered by the thick bar edges so that, for a bending strain in the principal direction 1of 1000 µm/m a strain in the main direction 2 of 3 µm/m will not be exceeded [2.35]. The bending beam of this test device is made of fine-grained material, so that the strain gages to be tested even for short measuring lengths cover as many grains as possible (homogeneous strain distribution). The strain gages to be tested are installed with their measuring directions in main direction 2 on the bar and then the bar so strongly bent around this direction, so that in principal direction 1 (i.e. transverse to the measuring direction of the strain gages) an elongation of 1000 µm/m occurs.

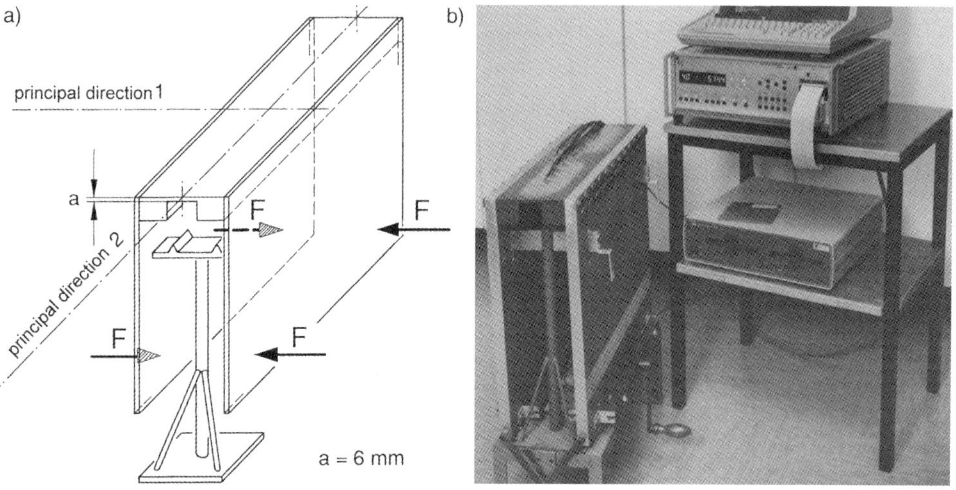

Fig. 2.24 Test equipment for experimental determination of the transverse sensitivity of strain gages; a) schematic, b) photograph

The first formulation of the cross-sensitivity in the form of Eq. (2.20) was carried out by the manufacturer of the first SR-4 wire strain gages, who was the first to derive the necessary relationships for a correction [2.36]. They found that flat-grid gages have positive and wrap-around gages negative cross sensitivities. In these studies the largest cross-sensitivity $q = 4\%$ was found for a short flat grid gage with many end loops, and the lowest cross-sensitivity $q = <0.25\%$ was found for a long flat grid gage with few end loops. A variety of study results were published in 1963 by BLH (Baldwin-Lima-Hamilton Corp.) in [2.37].

In 1946, Gustafsson in Sweden invented the cross-link wire strain gage to avoid disturbing influence on the measurement caused by the end loops. He strung long thin resistance wires parallel to each other and soldered the individual wires together with cross bridge conductors made from thick copper wire. He mounted the entire grid onto a paper or plastic carrier and finally cut off the individual strain gages in pieces. The advantage of this arrangement was very small transverse sensitivity [2.38]. Figure 2.25 illustrates a cross-link strain gage manufactured by Huggenberger.

Later on Gustafsson commenced manufacturing and marketing cooperation with A.U. Huggenberger in Switzerland [2.39]. The cross link strain gages produced in Switzerland by the company Huggenberger in Zurich were marketed under the trade name 'Tepic'. Because of their manufacturing costs they never succeeded. Their production was ceased in 1994.

With the invention of the foil strain gage transverse sensitivity lost significance. The pattern of foil gages could be designed in any desired form and the end loops of the wire gages could be replaced by special etched forms in order to minimize the transverse sensitivity. The free design is limited by consideration of the gage's creep properties which can also be influenced by design of the end loops.

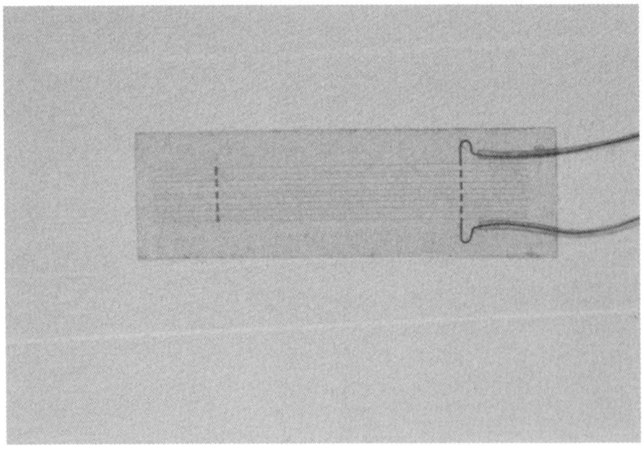

Fig. 2.25 Cross-link strain gage as invented by Gustafsson 1946 in Sweden and manufactured by Huggenberger in 1963 in Switzerland

The error caused by the transverse sensitivity q is of interest in the evaluation of the results. Note that the error is not identical to the transverse sensitivity q; it also depends on the principal strains in the strain field at the measurement point.

To find out the error caused by the transverse sensitivity it is assumed that the change in resistance of the strain gage is caused by the sensitivities k_l and k_q in longitudinal and transverse direction respectively.

$$\frac{\Delta R}{R} = k_l \varepsilon_l + k_q \varepsilon_q \tag{2.22}$$

Introducing Eq. (2.20) leads to

$$\frac{\Delta R}{R} = k_l (\varepsilon_l + q \varepsilon_q) \tag{2.23}$$

Calibration of the strain gage was carried out on material with Poisson's Ratio $\nu_0 = 0.285$ provided

$$\varepsilon_q = -\nu_0 \varepsilon_l \tag{2.24}$$

Then (2.24) introduced into (2.23) leads to

$$\frac{\Delta R}{R} = k_l \varepsilon_l (1 - q \nu_0) \tag{2.25}$$

Equation (2.25) formulates correctly the procedure carried out by the manufacturer to determine the gage factor k. According to the applicable standards the producer usually gives the factor k as the result of the calibration procedure in the form

$$\frac{\Delta R}{R} = k \varepsilon \tag{2.26}$$

In which ε ε_l, because the strain ε is measured mechanically independently of the strain in the strain gage, and the change in resistance $\Delta R/R$ of the gage is measured electrically. Comparison of the results in Eqs. (2.25) and (2.26) leads to

$$k = k_l (1 - q \nu_0) \tag{2.27}$$

The error arising without correction of the influence of the transverse sensitivity is given by

$$f = \frac{\varepsilon - \varepsilon_l}{\varepsilon_l} \tag{2.28}$$

In Eq. (2.28) ε_l is the actually strain in measurement direction and ε is the strain indicated by the strain gage. Introducing Eqs. (2.23) and (2.27) into Eq. (2.26) gives

$$\varepsilon = \frac{\varepsilon_l + q \varepsilon_q}{1 - \nu_0 q} \tag{2.29}$$

Thus Eq. (2.28) gives us the error f

$$f = \frac{q\left(\nu_0 + \frac{\varepsilon_q}{\varepsilon_l}\right)}{1 - q\nu_0} \qquad (2.30)$$

In Eq. (2.30) ε_q and ε_l are the existing transverse and longitudinal strains with respect to the coordinates of the strain gage. In the case of calibration with $\varepsilon_q = -\nu_0\varepsilon_l$ the error f becomes zero.

An example illustrates the inequality of transverse sensitivity q and error: A strain gage with transverse sensitivity $q = 0.01$ (1%), will be installed in transverse direction on a tensile rod in order to find its transverse strain. The material of the tensile rod has Poisson's ratio of 0.3 that means $\varepsilon_q/\varepsilon_l = -0.33$. Use in Eq. (2.30) results to $f = -0.0313$, that means error of −3.1%.

For rapid assessment apart from the correct calculation a rule of thumb is given in [2.40] as

$$f = q\left(\frac{\varepsilon_q}{\varepsilon_l}\right) \qquad (2.31)$$

The diagram given in Figure 2.26 provides a quick way to assess the error caused by transverse sensitivity. With the known coordinates $\varepsilon_q/\varepsilon_l$ and q we can find the error from the diagram.

Modern foil strain gages can be produced without transverse sensitivity due to the quality of the design. But here the technological requirements in respect to creep of the gage limit the scope for free design.

Nowadays almost all foil strain gages have some transverse sensitivity, even if very small. Using a new invention, an existing transverse sensitivity can be eliminated by constructional means, without making fundamental changes in the geometry of the strain gage. The transverse sensitivity can be minimized by introducing small additional gridlines between the gage gridlines. These gridlines don't have electrical contact to the strain gage grid. But they increase the mechanical stiffness of the gaps [2.41]. Prototypes of strain gages with this modification as can be seen in Figure 2.27 were designed by M. Stockmann and manufactured by HBM [2.42, 2.43]. Figure 2.28 shows the new layout compared with the standard gage.

The additional introduced gridlines stiffen the gaps between the active gridlines of the strain gage and change the transmission of transverse strain by their higher modulus of elasticity.

The strain in the gaps between the active grid and the additional gridlines will be increased and the transverse sensitivity will be compensated. When the additional gridlines are made from the same material as the grid material the extra effort is very small.

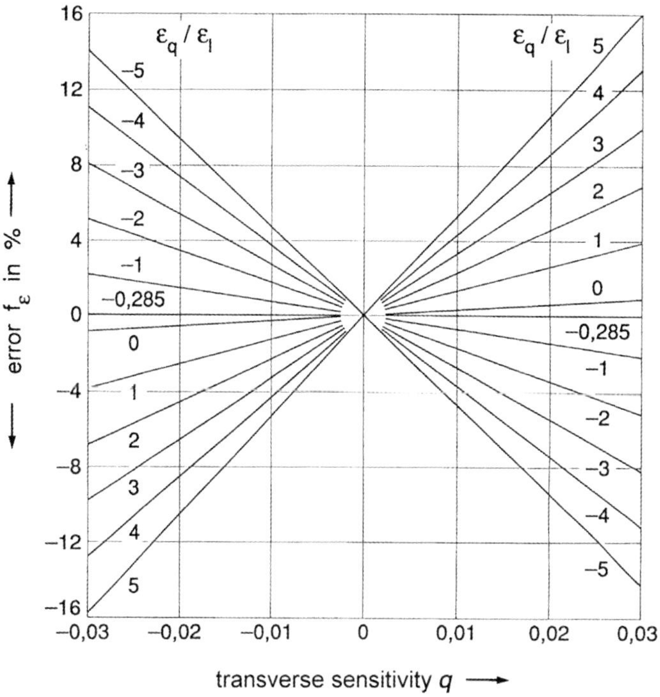

Fig. 2.26 Diagram for finding the measurement error due to a strain gage's transverse sensitivity

2.4 Temperature effects

2.4.1 Temperature effect on sensitivity

The values for the gage factor k, as given by the manufacturer, are valid for room temperature. Strain gages show an alteration in the k factor with temperature, especially if the carrier material and the adhesive used are not suitable for the measurement temperature. Alterations of the mechanical properties influence the strain transfer from the measurement object on the strain gage.

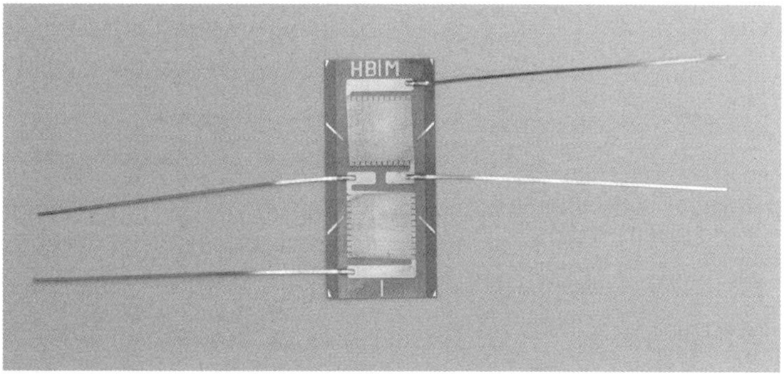

Fig. 2.27 Strain gage with self-compensation of transverse sensitivity

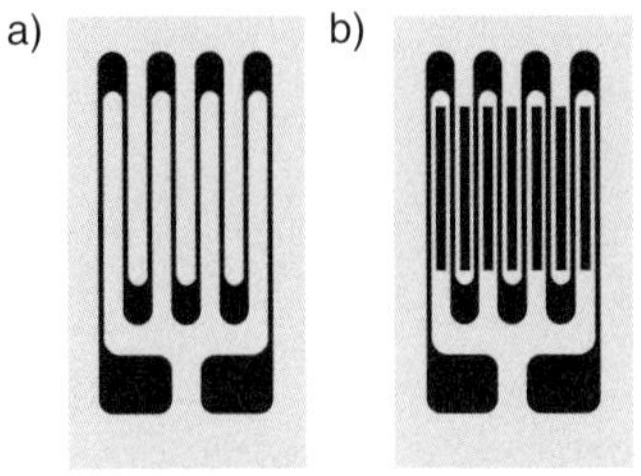

Fig. 2.28 Schematic layout of a) standard strain gage and b) special strain gage with transverse sensitivity compensation [2.41]

The relationship between sensitivity and temperature can be investigated if the components used are adapted to the temperature. Figure 2.29 shows the relative change of the gage factor $\Delta k/k$ in dependence on temperature for two of the same strain gages bonded using different adhesives on the same measuring object material. Figure 2.29 shows linear behaviour within the temperature range under consideration. That is

$$\frac{\Delta k}{k} = \alpha_k \Delta T \text{ or } k_T = k(1 + \alpha_k \Delta T) \tag{2.32}$$

Here ΔT is the temperature difference to the reference temperature (here 20 °C), k_T is the gage factor at any temperature T within the linear range and α_k is a coefficient to be determined experimentally. Guidelines for experimental determination of the gage factor are given in [2.19]. Especially observed in the determination are the close limited times in order to avoid disturbing influences by creep.

The temperature-induced change of the gage factor is relatively low from the perspective of stress analysis using strain gages, as can be seen from Figure 2.29. However, for the construction of transducers this change is relatively high. When cyanoacrylate adhesive is used, a temperature increase from room temperature to

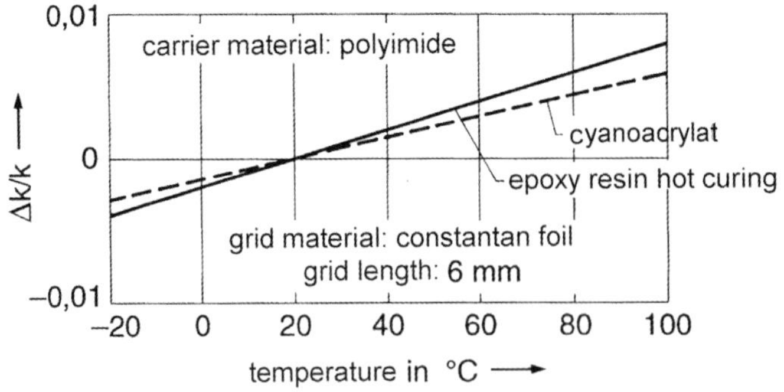

Fig. 2.29 Experimentally determined curves of the temperature dependence of the gage factor *k* for two strain gages with 6 mm grid length installed with different adhesives

100 °C causes a relative change of the gage factor of 0.6%. The Wheatstone bridge circuit in the transducer offers opportunities to compensate the impact of temperature changes on the sensitivity, as will be explained in a later section.

2.4.2 Thermal output of strain gages

If temperature changes occur while taking strain gage measurements, they cause changes in the resistance of the strain gage which are superimposed on the measurement signal produced by the mechanical strain, leading to a shift of the zero point signal. These resistance changes due to temperature are independent of the strain being measured. If the resistance changes occurring in a mounted strain gage are plotted in relation to the temperature, then the strain gage's thermal output characteristic is obtained. Often instead of the change of resistance for the ordinate in the thermal output diagram, the mechanical strain for the corresponding change of resistance is used.

This is known as the apparent strain, since in reality it does not exist and is only created by the temperature effect. Figure 2.30 shows the thermal output diagram of a strain gage enclosed in the strain gage packing.

During measurement the change in resistance given in the thermal output diagram does not, however, normally cause errors in the measurement signal, because its effect can be simply compensated by the Wheatstone bridge circuit using a dummy gage, that is a compensating strain gage. Manufacturers of strain gages endeavour to supply temperature compensated strain gages which are able to provide reliable measurement results within a given temperature range without needing to use dummy gages for compensation.

The resistance changes observed as the result of temperature variation only take on significance if they occur during the measurement procedure, that is zero balancing of the measurement system cannot be carried out at the current measurement temperature. On the other hand, for measurements where zero balancing of the

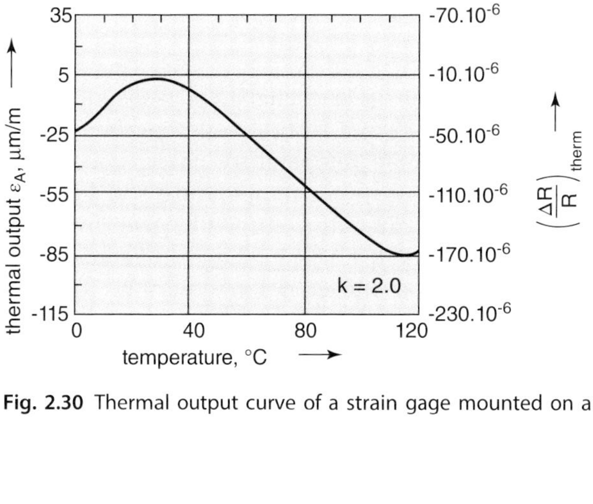

Fig. 2.30 Thermal output curve of a strain gage mounted on a given test object

system at the measurement temperature can be made, the resistance changes caused by temperature variations are insignificant.Three factors affecting thermal output

The resistance changes in the mounted strain gage occurring due to a temperature change and the ensuing zero point shift are mainly due to three causes:

- the resistance change α_R of the measuring grid material as a result of temperature variations
- the linear thermal expansion coefficient α_M of the measuring grid
- the linear thermal expansion coefficient α_O of the measurement object material.

The temperature coefficient of electrical resistance α_R for the measuring grid material includes both the change in the grid's material resistivity and the change in resistance due to its thermal expansion. The temperature coefficient α_R is itself temperature dependent and is usually a nonlinear function of the temperature T.

The coefficient α_R does not, however, only depend on the temperature, but also on the grid's alloy composition and its pre-treatment, for example cold working and heat treatment. At first sight this dependence may appear detrimental, but it can be used to advantage in the manufacture of temperature compensated strain gages. By specific treatment of the grid material the value of its temperature coefficient can be modified within the given material limits, and the strain gage can therefore be matched to a given measurement object material. Hence, the term 'temperature compensated' means the matching of the strain gage during manufacture to a certain measurement object material, that is to its linear thermal expansion coefficient α_O.

A strain gage's thermal output is not only dependent on the temperature coefficient of resistance for the measuring grid material, but also on the thermal expansion coefficients of the constituent materials. The complete temperature response of a mounted strain gage is obtained if the individual changes of resistance occurring in the strain gage due to temperature change are superimposed as

$$\left(\frac{\Delta R}{R}\right)_{therm} = \alpha_R \Delta T + k_T \Delta T(\alpha_O - \alpha_M) \tag{2.33}$$

The left-hand side of Eq. (2.33) represents the relative change in resistance; that means that the change in resistance ΔR is referred to the strain gage's nominal resistance R_0. In the equation ΔT signifies the temperature change occurring during the measurement with respect to the reference temperature and k_T is the gage factor applicable at the measurement temperature. It can be seen from Eq. (2.33) that the change of resistance resulting from the change in temperature consists of two parts. The first summand in the equation describes the part influenced by α_R.

The second summand represents the difference that occurs between the linear thermal expansion coefficients: α_M of the measuring grid material and α_O of the

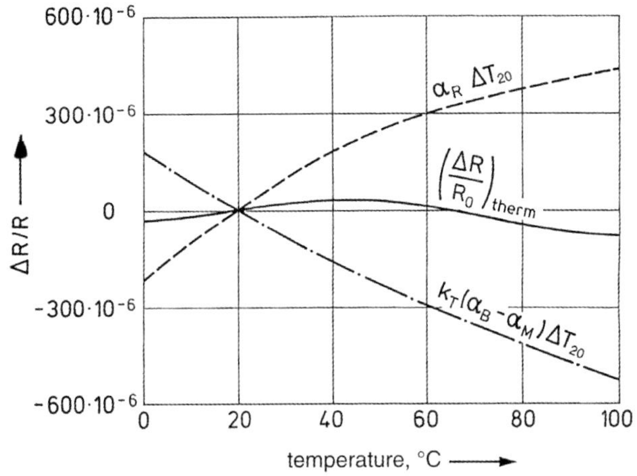

Fig. 2.31 Thermal output of a mounted strain gage showing the components due to α_R and due to the linear thermal expansions; from [2.44]

measurement object material. Figure 2.31 shows graphically the shape of each of the terms in Eq. (2.33).

Temperature compensated strain gages

The condition for a strain gage which is ideally matched to a measurement object material with regard to thermal output can be taken from Eq. (2.33) if the relative change of resistance due to temperature is set to zero:

$$\left(\frac{\Delta R}{R}\right)_{therm} = 0 \tag{2.34}$$

By combining Eqs. (2.33) and (2.34) the condition is obtained for a strain gage to have complete temperature compensation:

$$\alpha_R = -k_T(\alpha_0 - \alpha_M) \tag{2.35}$$

Equation (2.35) expresses the requirements that the manufacturer must fulfil with respect to the temperature coefficient α_R in order to produce a temperature compensated strain gage. It can be seen that, for strain gages compensated in this way the adjustment of the temperature coefficient α_R must be carried out appropriate to the thermal expansion coefficient α_O of the measurement object material. But the thermal coefficient α_O is normally temperature dependent, as is the gage factor k_T, so that the requirements expressed by Eq. (2.35) can only be satisfied to a close approximation.

The broken line in Figure 2.31 represents the change in resistance of a constantan measuring grid caused by the temperature dependence of α_R. The curve $\alpha_R \Delta T$ was obtained in the shape reproduced in the diagram by specific pre-treatment of the measuring grid material in order to match the strain gage to the thermal expansion coefficient of mild steel.

The chain line in Figure 2.31 shows the relative change in resistance which is produced by the temperature dependence of the different thermal expansion coefficients α_M and α_O.

The continuous line in Figure 2.31 indicates the thermal output of a strain gage mounted on a material having a thermal expansion coefficient α_O. This is obtained, corresponding to Eq. (2.33), from the summation of the broken and chain lines in the diagram.

The effect on α_R produced by a specific treatment during the strain gage's manufacture gives the disadvantage of an unavoidable spread in the values for α_R within a given production lot of pre-treated foil material. This spread is then apparent in the thermal output of the strain gages, but can be kept within low limits if the manufacturer has sufficient experience. This spread is responsible for the tolerance given for the thermal output of strain gage batches.

Figure 2.32 shows the unavoidable spread as shown by experimental research on constantan wires carried out by the author. This spread is the reason why the values given for the thermal output in the specification of strain gages are provided for a tolerance region with a certain indication of tolerance.

A typical example is shown in Figure 2.33, which illustrates the spread in the thermal output of 30 strain gages that were taken at random from a production lot and then mounted on the same measurement object material in identical manner. This spread may be greater for strain gages from different production lots. It is therefore

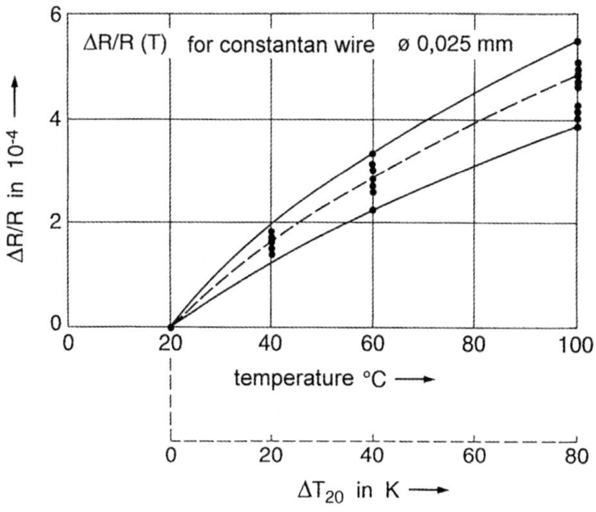

Fig. 2.32 Result of experimentally determined values: Spread range of the temperature coefficient of resistivity α_R of constantan wire

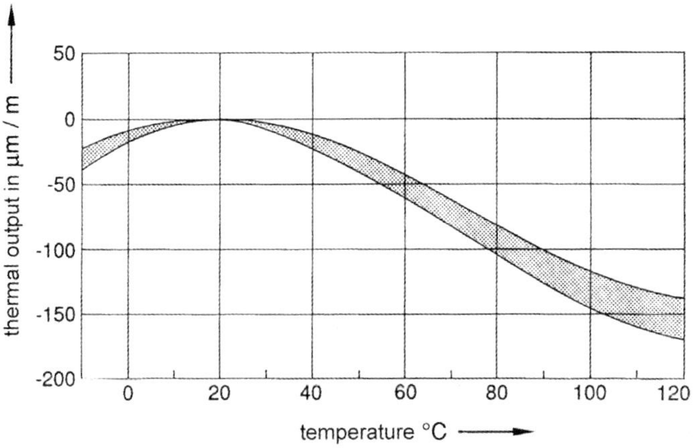

Fig. 2.33 Spread in the thermal outputs of 30 strain gages selected at random from a production lot; from [2.46]

recommended that the dummy gages used for the compensation of temperature effects are always taken from the same production batch as the active strain gages. The best results are obtained if all strain gages used in a bridge circuit originate from the same pack.

In Figure 2.32 the apparent strain ε_A is used as the ordinate. The apparent strain can be determined from the relative change in resistance with the aid of the relationship

$$\varepsilon_A = \frac{1}{k_T}\left(\frac{\Delta R}{R_0}\right)_{therm} \tag{2.36}$$

From the temperature responses shown in the diagram it can be seen that the manufacturer attempts to 'adjust' the strain gages such that the curve's maximum point is located at room temperature. This is because measurements with strain gages are mainly made at room temperature and, since the tangent at the curve's maximum point is horizontal, the temperature dependence of the change in resistance at this point is relatively low.

Another way to determine the effect of temperature changes on the resistance of the measurement point (not only the measurement grid) is obtained by summarizing the right side of Eq. (2.33):

$$\alpha^{*}\Delta T = \alpha_R \Delta T_{20} + k_T(\alpha_B - \alpha_M)\Delta T_{20} \tag{2.37}$$

The coefficient α^{*}can be determined experimentally for the measuring point. In order to exclude the influence of the measuring object material the examined combination of strain and adhesive will be installed on Zerodur; this installation will be subjected to temperature changes and thereby the relative change in resistance $\Delta R/R$ will be measured. Zerodur (Jena Glass Schott & Gen. Mainz) has a vanishingly small thermal expansion coefficient $<0.05 \times 10^{-6}$/K [2.47], so that this small value can be neglected. The scattering occurring can then be a measure of the scatter of α_R. The thermal output of the considered installation on the measurement object material is achieved by correcting the measured thermal output on Zerodur with $k_T\alpha_B$ of the corresponding object material.

The effect of connecting strips

In order to evaluate the spread of thermal output values given in Figure 2.32 for a typical production lot, it is important to realize that the diagram only includes mounted strain gages. When wiring active and dummy strain gages in half and full bridge circuits, it should be noted that all solder points and leads are constituent parts of the bridge circuit and also contribute to the thermal output. Therefore, the wiring of the individual arms of the Wheatstone bridge circuit should be made as close to identical as possible, that is the same type of lead with the same length. This also applies to the connection strips which are present on some types of strain gages to simplify mounting.

Figure 2.3 shows examples of various types of connections used on strain gages. The connection strips also influence the thermal output due to their temperature dependent resistances and this effect increases with decreasing nominal resistance of the strain gage. During manufacture of strain gages with connection strips the influence of the strips, as can be seen on the left-hand sample in Fig. 2.34, has to be taken in account when the thermal output 'adjustment' is made and so cutting a strain gage's strips alters the thermal output. This change does not have any effect on measurements if the strips of all strain gages in the relevant bridge circuit are cut in the same manner.

Figure 2.35 compares the temperature responses of strain gages with and without connection strips. The degree of displacement of the curve with connection strips relative to that without them is directly proportional to the length of the strips.

Recommendation

Experience from taking measurements shows that dummy gages should be used in the bridge circuit if measurements are made at varying temperatures and if circumstances permit it. There are two reasons for this recommendation:

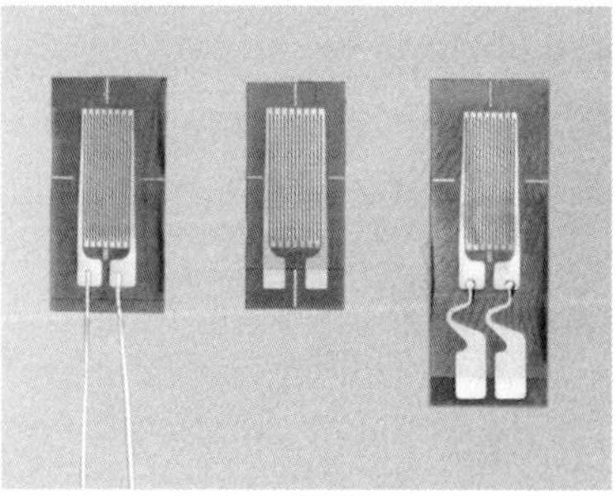

Fig. 2.34 Various types of connections used on strain gages: connection strips; open ends of the grid terminals (pads); integral solder terminals

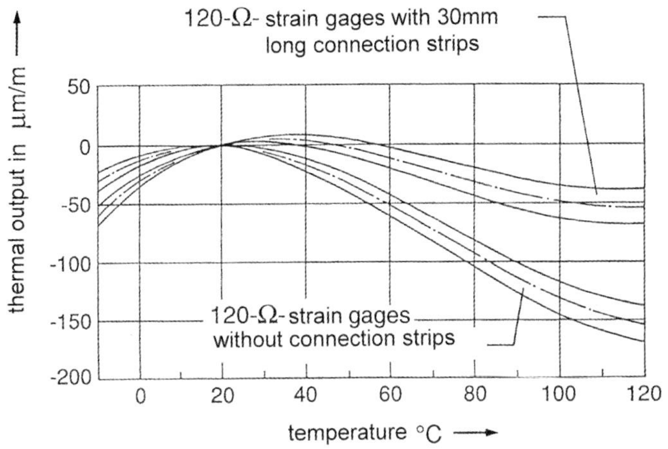

Fig. 2.35 Influence of a 30 mm length of connection strip on the thermal output of 120 Ω strain gages

1. The thermal expansion coefficient of the measurement object material can vary by a relatively large amount and may deviate from values upon which the manufacture of the temperature compensated strain gages are based.
2. With larger temperature variations during the measurement the spread in the temperature responses of the strain gages within a production lot is substantially smaller than the temperature dependent changes in resistance that still occur in spite of the matching that has been undertaken.

2.5 Mechanics of the strain gage

2.5.1 Strain transfer from the measurement object to the measurement grid

Prerequisite for an accurate detection of the strain in the test object by the strain gage is a genuine transfer of this strain on the measurement grid. Media of this transfer are the adhesive of the strain gage on the measurement object and the carrier foil of the measurement grid. The strain transfer is not uniform along the measuring grid length, since at the measuring grid's end loops transitional zones of lower strain are formed. Detailed studies of these transmission mechanisms were carried out in 1989 by Rohrbach and Czaika [2.48], and then by Hönisch [2.49] taking over the transmission model by Rohrbach and Czaika.

Figure 2.36 shows the model developed in [2.48] of the strain transmission from the measurement object via the bonded carrier film on the measurement grid. The transition region T_{OC} can be seen at the carrier edges where the strain is transferred from the object to the carrier, within which the carrier strain assumes the object's strain.

Shortly ahead of the measurement grid, the strain in the carrier increases because, relative to the carrier foil, the measuring grid material is very stiff, with a modulus of

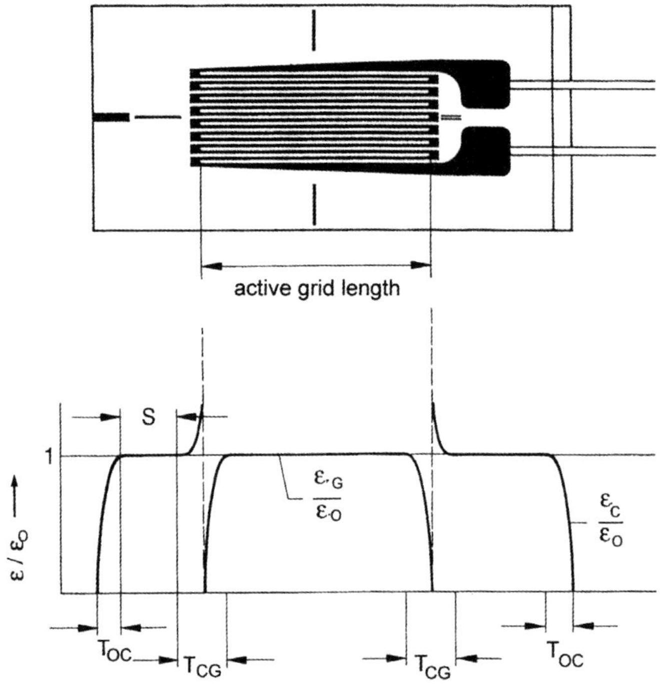

Fig. 2.36 Strain transfer from the measurement object to the measuring grid [2.48]

elasticity more than a hundred times that of the carrier. The rigid measuring grid must take the strain of the easily deformable carrier. So at the measuring grid end loops there is the transition region T_{CG} from the carrier to the measuring grid.

There is an overlapping region of identity – not covering the entire grid length – between the strain of the measuring grid ε_G and the strain of the measurement object ε_O in the middle of the measuring grid. At the end loops there are increasing differences between strain ε_C of the carrier and strain of the grid. An equilibrium state is established between the resetting force of the elastically strained measuring grid and the shear stress in the boundary surface between the carrier and the measuring grid. In cases where the carrier material and the measuring grid material have the same modulus of elasticity, as may be approximately true for ceramic carriers, the transition zones T_{CG} will disappear.

Looking at the two transition regions formed at the ends of the carrier foil and the grid T_{OC} and T_{CG} there is an area in between where $\varepsilon_C = \varepsilon_O$. The length of this area, designated S in Figure 2.36, depends on the deformation behaviour of the carrier material as well as on the length of the carrier protruding over the end loops of the grid. To make the installation of the strain gage easier, the carrier length may be reduced by reducing S, and it will not have an adverse effect on the measurement result.

However, cutting back the carrier foil by more than S, the sensitivity (gage factor k) of the strain gage will be impaired. This effect is worse for short strain gages than for long ones. Strain gage manufacturers usually design the carrier foil to be slightly longer than is absolutely necessary, because the transition zones increase (S decreases) due to relaxation effects with long-term measurements and with increasing temperature. A minimum of 1.5 mm should be maintained for the end-side projection of the carrier foil on the measuring grid. Strain gages can safely be cut laterally until close to the lateral edges of the measuring grid [2.50].

2.5.2 Creep

If a strain gage will be strained for a sufficiently long time, relaxation phenomena can occur that cause an increase in the transition zones of the measuring grid carrier material at the end loops. The resetting force of the measuring grid will overcome the opposing holding force of the carrier material and the measurement grid can relax partially. This long-term process is called creep of the strain gage; this creep is of particular interest for long-term measurements or at higher temperatures or for high precision measurements, such as with weighing cells in custody transfer or with force transducers in the calibration of material testing machines.

Creep is promoted by increasing temperature, so creep is time and temperature dependent. The creep phenomena are influenced by a range of other parameters.

So, for example, long strain gages tend to suffer less creep than short ones owing to the lower share of the transition zones of the entire measuring length. Favourable design of the measuring grid end loops and the selection of the carrier foil to be as thin as possible provide more opportunities for reducing the creep.

By the end of the 1950s detailed investigations had been carried out by Rohrbach and Czaika [2.51, 2.52] to capture the phenomenological creep in strain gage installations, developed on the basis of a model for computational detection of creep.

Onnen and Fritz who dealt in the 1960s with similar studies extended model concepts [2.53–2.55]. A purely mathematical determination of the creep behaviour of strain gages is described in [2.56]. In [2.57] the qualitative models of Rohrbach and Czaika and of Onnen and Fritz are justified quantitatively on the base of an analytical consideration of the stress conditions in the layer between strain gage and measurement object.

The formulation of a general quantitative relationship for the creep is made more difficult by various complicating factors such as the elastic moduli of carrier foil, grids and adhesive, geometric dimensions and shape design of the strain gage, thickness of carrier foil and adhesive layer, temperature dependencies of the elastic moduli of all materials involved and the differing time dependencies of creep of the various components.

In Figure 2.37, the phenomenon of strain gage creep is shown schematically. In the diagram, the strain of the measuring object and the strain recorded by the strain gage installed thereon are shown versus time.

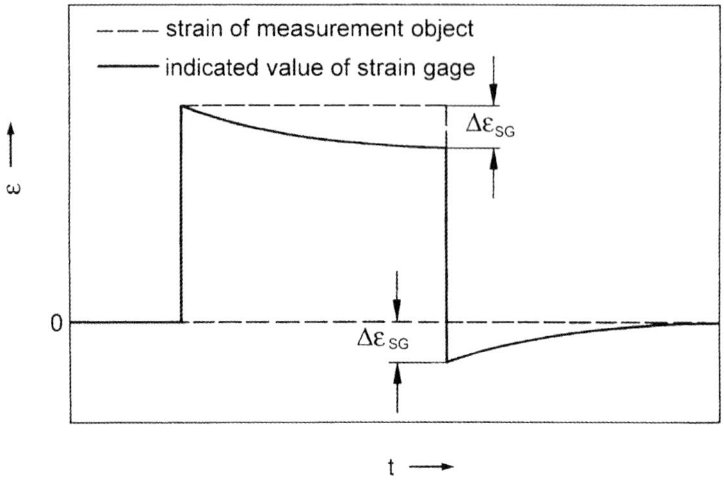

Fig. 2.37 Temporal change of the indicated strain caused by creep, in a greatly exaggerated representation

At time t = 0 the measuring object is loaded. The strain in the test object and the measured strain rise synchronously to the same maximum value. While the strain in the object remains constant the measured strain decreases slowly because of the creep of the strain gage. The result is a time-dependent strain difference $\Delta\varepsilon_{SG}$. If, after a certain period of time, the measurement object is suddenly discharged to zero, the strain of the object goes back to zero and the strain of the strain gage decreases by the same amount. But since the strain of the strain gage by the amount $\Delta\varepsilon_{SG}$ was smaller than the object strain, the strain of the strain gage is situated after the discharge by the amount $\Delta\varepsilon_{SG}$ below zero. After a certain period of time it will creep back to zero.

As shown in Figure 2.36, the grid ends are of particular importance due to the presence of the transition zones for strain introduction into the measuring grid. Since the creep is associated with an enlargement of the transition regions, short strain gages are more affected by creep than long ones. In short strain gages the transition zones have a larger share of the total measuring grid length than in long. The geometric design of the end loops of foil strain gages is of particular importance with regard to creep.

Figure 2.38 shows the creep behaviour of two foil strain gages of the same series with different lengths that were installed under the same conditions [2.58]. It can be seen very clearly the higher creep sensitivity of the shorter strain gage.

Figure 2.38 shows results of investigations of creep in the form of the relative creep $\Delta\varepsilon/\varepsilon$, where $\Delta\varepsilon$ is the change of creep strain and ε is the original total strain. For experimental studies, this relative form of representation has proved to be advantageous, since the ratio assuming stress proportional creep is constant. Thus, the creep is independent of the relative creep and can be experimentally determined for different temperatures and depicted as a function of time.

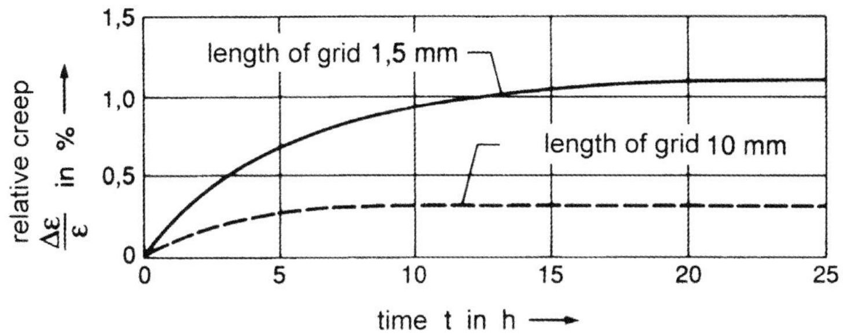

Fig. 2.38 Creep behaviour of two foil strain gages of different lengths

To obtain reliable information on the creep behaviour of strain gage installations, one has to rely on experimental studies. A device for this purpose is described in [2.52]. The device consists of a U-shaped bracket. A bending moment acting on the two legs of the clamping apparatus created a defined strain state in the straight connecting part between the two legs. The strain gages to be tested are installed on the connecting part in the tensioned state. After installation, the device is relaxed and thus the installed strain gages are strained by the relaxation strain of the connecting part, which now is strain free. So the measurement results are not influenced by the non-existent creep in the test object.

Various tests with this device were carried out all at the same temperature. The results are representative of the relative creep for various temperatures over time, as shown in Figure 2.39.

Note, however, that creep under load also occurs in the component on which the strain gage is installed. This quantitatively very low component creep has hitherto received little attention in mechanical engineering. However, the highly sensitive strain gage takes on this component creep and the strain to be measured. The effect of the component's creep gage tends to counteract the creep of the strain gage. Despite constant mechanical stress in the component, the strain in the component increases marginally. It results in the superposition of the two oppositely running creep events in the component and in the strain gage. The opposite effect of creep in the strain gage and in the component opens up possibilities – especially for the manufacture of transducers – to adapt the strain gage's creep properties to the component's creep so that the effects of the two creeps cancel each other out. The free design possibilities of measuring grid geometry of foil strain gages affect this favourable effect.

Figure 2.40 schematically depicts the superposition of the two creep processes, and for a better understanding the creep strain is selected somewhat smaller than that of the component. After the spontaneous load the component creeps in the loading direction and the strain gage experienced the total strain of the component as drawn by the dash-dotted lines. The strain gage creeps opposite to the direction of its

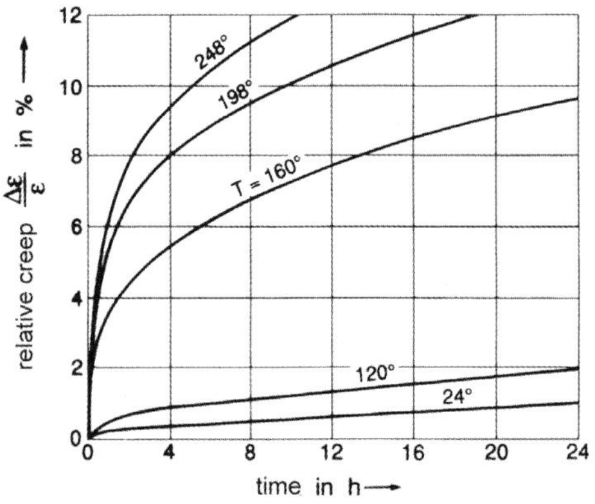

Fig. 2.39 Time dependent representation of relative creep of a strain gage for different temperatures [2.51]

loading, thereby making sure that the component's creep not or as shown in the picture only partially enters into the measurement result.

These opposing time-dependent creep processes in the strain gage and in the component material open up the possibility of complete creep compensation for the construction of precision transducers. This requires the availability of strain gages with creep behaviour adapted to the creep of the measuring body material. Since the creep phenomena mainly occur at the measuring grid ends, the creep behaviour of a strain gage can be enhanced or reduced by their design.

It is interesting in this context that the first strain gages produced in 1940 in the UK in the testing department of the LMS Railway Company for its own use were designed with a view to the avoidance of creep. The first strain gages they manufactured consisted of a 25 µm thick single wire (constantan or NiCr) laid between two layers of cigarette paper fixed with Durofix (cellulose acetate).The connections were copper wires fixed at right angles to the measuring wire. They were soldered not to the respective measuring wire ends, but as shown in Figure 2.41 further inwards, so that a piece of measuring wire protruded at each side of the junction.

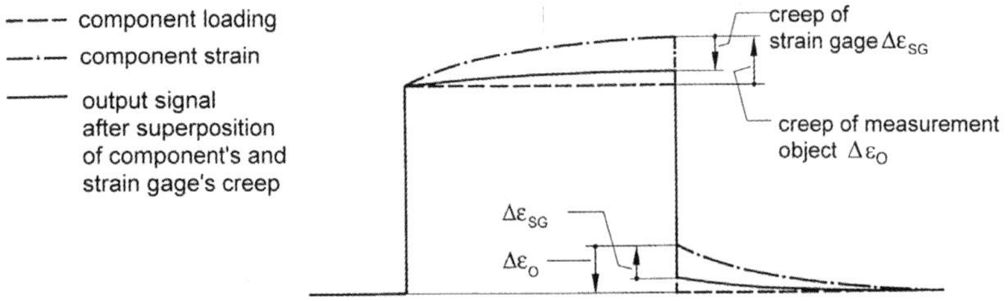

Fig. 2.40 Superposition of the counteracting creep effects of the measurement object and the strain gage

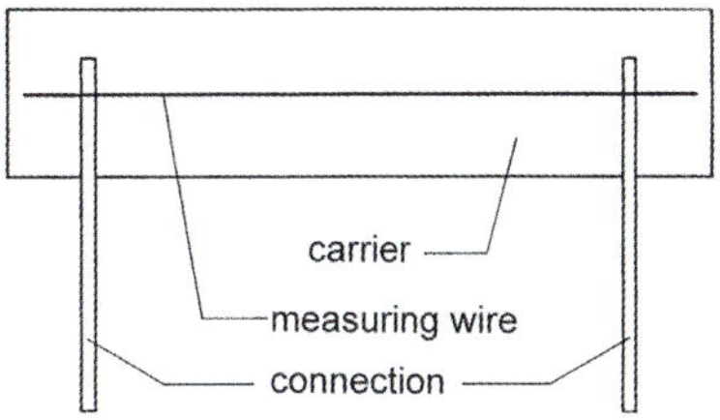

Fig. 2.41 Single-wire strain gage from the UK (1940); the creep is shifted to the parts of the active measuring wire outside the junction points

The purpose of this arrangement was to shift the creep to the parts located outside the active length of the measuring wire. The idea stemmed from experiences with long welded rails, in which the creep caused by temperature changes occur only at the ends (Swindells, B. Personal communication (Letter of April 1989)).

The same idea can be realized in modern foil strain gages. For example, the creep of a strain gage can be considerably reduced, as shown in Figure 2.42, by the grid ends being extended beyond their end loops to so-called dead ends.

However, for the construction of transducers the aim in the development of strain gages is not the minimization of the strain gage creep but an optimal compensation of creep of the measuring body material by the creep of the strain gage, so that the output signal of the transducer does not show any time-dependent drift.

Thus, for example, a displacement transducer that detects displacement by means of the deflection of a bending beam fitted with strain gages, needs absolutely creep free strain gages, since relaxation in the material of the bending beam causes reduction in the reaction force and has no influence on the strain, which here only depends on the beam's deflection and not on the reaction force.

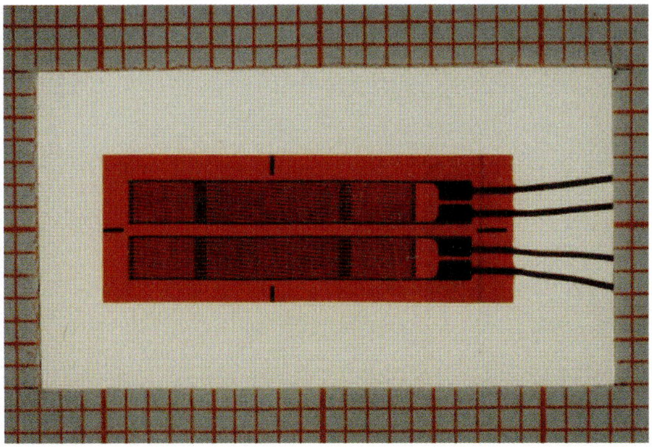

Fig. 2.42 Foil strain gage with 'dead ends', an idea conceived for use in transducers

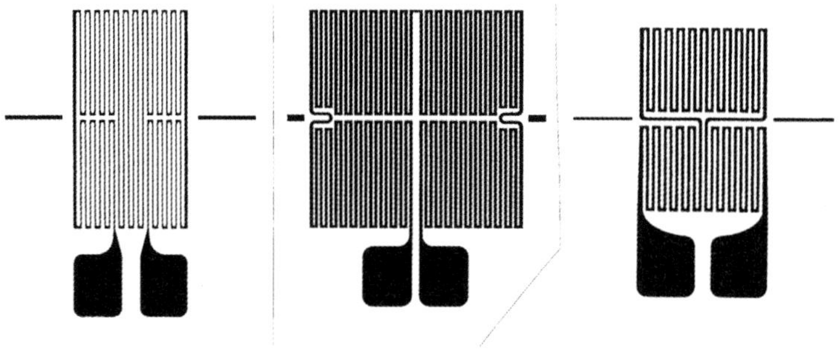

Fig. 2.43 Special strain gages for the construction of transducers with special end loops in order to increase creep

With the aim of a long-term constant transducer output signal of a transducer made from strongly creeping material can be required, for example, strain gages with strong creep behaviour. This can be achieved by arranging additional end loops that are dimensioned such that the creep of the strain gage is compensated for by the creep of the remaining parts of the measuring device [2.59].

In addition, it is possible to vary the proportion of creep compensating end loops by means of corresponding taps. Specially designed grids for the construction of force transducers provided with additional end loops are shown in Figure 2.43.

Another, now much used option is the more or less compact design of the end loops of foil strain gages, which can be adapted in its creep behaviour to different measuring object materials. In Figure 2.44, two such foil strain gages are shown with different designed measurement grid end loops.

Nowadays these strain gages are offered with stepped lengths of the end loops and corresponding creep behaviour in standard catalogues of self-made

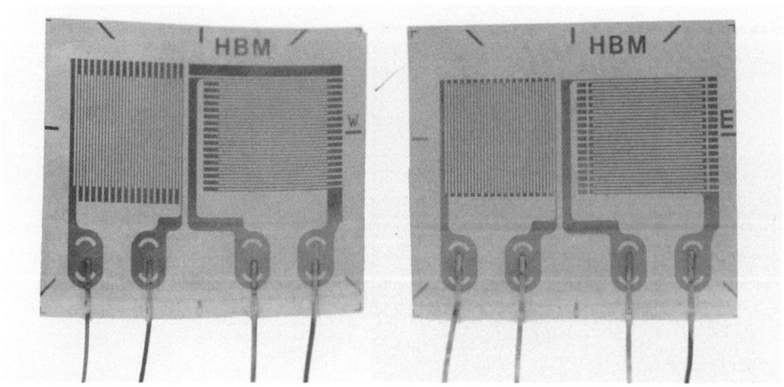

Fig. 2.44 Two foil strain gages with differently designed end loops for the construction of transducers with different creeping measuring body materials

transducers [2.60]. The length of the end loops varies in steps from the width of a grid line to 12 times. With increasing length of the end loops the creep of the strain gage decreases. The optimal selection of the strain gage suitable for the creep behaviour of the transducer material can only be made empirically.

The process consists of choosing a strain gage with end loops of medium length and performing a creep test, as outlined in [2.61] with a test time of 30 minutes. If the measuring point displays positive creep (increasing output) the transducer material is creeping more than the strain gage and so a strain gage must be chosen that has shorter end loops, and the creep test must be carried out again. The selected installation conditions in a creep test affect the test result, so they must be the same as will later be performed on the transducer: the same measurement site preparation, same adhesive, same pressure, same curing time and same temperature In this way a creep adaptation in steps of approximately 0.04% to the rated output of the considered transducer can be achieved. Experience shows that the creep behaviour of the transducer material varies slightly from batch to batch, so a new creep adaption may be required for each batch.

The lamination technique provides another method of creep adaption. In this technique the photochemical etching process produces in one operation a complete full-bridge circuit including traces, as well as balancing and compensating elements. Additional grid regions are arranged with defined creep values which can optionally be included in the circuit by disconnecting traces [2.62]. A foil grid with adjustable creep is shown in Figure 2.45. These grid parts are dimensioned during manufacture such that the total creep behaviour is positive, that is the output signal rises slightly with time.

Creeping grid parts can alternatively be switched on by separating the conductor loops as indicated in Figure 2.45 with A_1 to A_4. The creep behaviour of these switchable grid parts is binary weighted so that in this way a strain gage bridge circuit can be reached which is very well adapted to the creep behaviour of the measuring body. This method allows the production of transducers with a creep of less than 0.005% of the nominal output signal over a period of 30 minutes [2.63].

On the other hand, nowadays total creep-free strain gages can be produced using specific technologies. One example is the thin-film technology, in which all

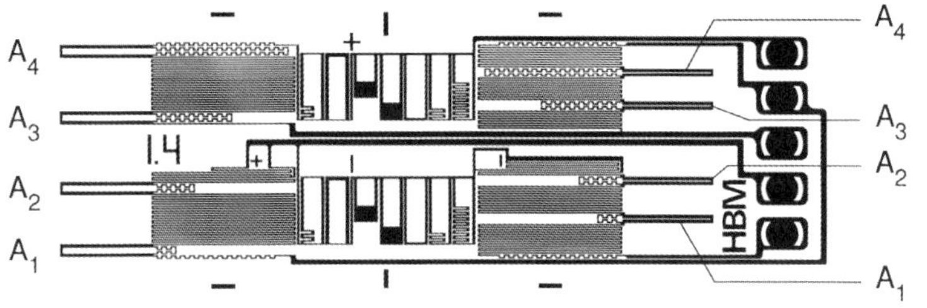

Fig. 2.45 Full bridge circuit made using a laminating technique with adjustable creep behaviour by separating the conductor loops A_1 to A_4

components of the strain gage installation are vapour deposited on the measuring body of the transducer. These components consist exclusively of thin layers of inorganic materials. With creep-free strain gages a long cherished desire of the strain gage developers has been met, although for transducers they have the disadvantage that compensation for creep of the measuring body material is excluded.

Under certain conditions it is possible to circumvent the creep of strain gages by using a trick. This is possible if, prior to installing, it is known what amount of strain in practical use will be acting for a long time on the strain gages. Then one installs the strain gages on the measurement object which is artificially pre-stressed at the operating load, if that is technically possible (this depends on component size, the possibility of restraint in loading the device, etc.). In use under operating load, the strain gages are then unloaded and therefore not subject to creep.

This method has successfully solved the measurement task of monitoring the tensile force in anchor bars that are used in civil engineering. Firstly, a strain gage full bridge circuit is installed on the non-loaded bar. Then the bar is stressed in a tensile test device to service load and a second full bridge is installed under load. The prepared anchor bar is then mounted into the building and stressed to the intended tension while adjustment can be made using the bridge circuit installed on the stress-free bar. The long-term monitoring of the tension force in the bar is then performed with the bridge circuit that was installed on the loaded rod, and whose strain gages are therefore unloaded during operation and are thus creep-free. Strain gages that serve to monitor the tensile force in the rod can be already attached to the tensioned anchor in the installed condition when local conditions allow this.

Another trick used in the construction of precision transducers is to install the strain gages in two bridge circuits on the transducer body. In this case, the strain gage bridges differ in geometry in such a way as to generate, in one case positive creep and in the other negative creep. Interconnecting the bridges results in a very low mean creep, which can be reduced by attenuating a bridge with further resistors if necessary [2.64, 2.65].

2.5.3 Hysteresis

Another issue closely related to the creep characteristic of the strain gage is the mechanical hysteresis occurring between the strain of the test object and the strain of the measuring grid. Since the strain of the measuring grid is directly proportional to the change in resistance of the strain gage, this hysteresis also occurs between the resistance change and the strain of the measurement object and can be determined by measurement.

Figure 2.46 shows schematically the mechanical hysteresis in exaggerated form. It arises in a strain cycle as the difference of changes in resistance between increasing and decreasing strain of the measurement object. Reference [2.19] defines as a characteristic for the mechanical hysteresis (as shown in Figure 2.46) the difference between the resistance changes at the ascending and the descending branch of the

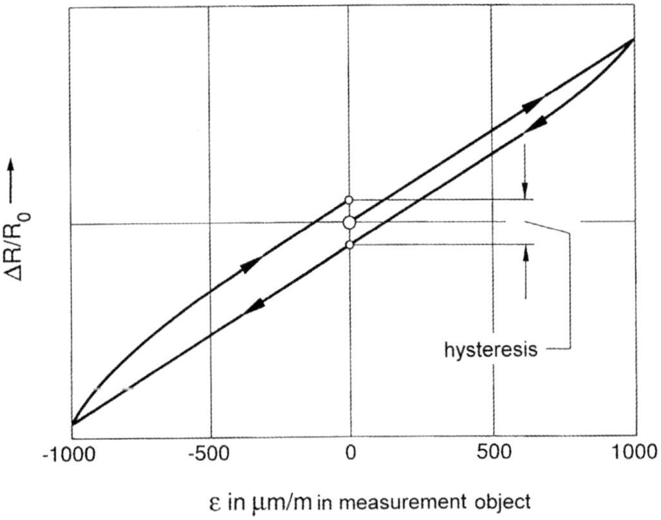

Fig. 2.46 Schematic representation of the hysteresis occurring between strain in the test object and change in resistance of the strain gage

curve for a strain cycle of ±1000 µm/m measured at the zero point of strain. From experience the hysteresis decreases with the number of applied load cycles, so it is specified for both the first and third load cycles.

It is almost impossible to define the hysteresis of the strain gage independent of associated characteristics. The hysteresis will always be influenced by all the components involved in the measuring point (including the cover).

Figure 2.47, which reproduces the model developed in [2.48] is used to explain the origin of the hysteresis. As the illustration shows, in the model is the measurement grid on top of a thrust prism in the transition zone between the carrier and the measuring grid connected to the measurement object. In this thrust prism there is elasto-plastic deformation when strain occurs in the measurement object, transmitting the strain to the measuring grid. If the strain of the object to be measured is relieved to zero, the plastic deformation remains in the prism, and causes a slight compression of the measuring grid, and therefore the hysteresis.

Hysteresis is also influenced by any existing residual stresses in the strain gage, which usually decrease after the initial strain cycles. Therefore, in cases where a disturbance of the measurement is expected by the hysteresis, this can be greatly reduced by repeated preloading of the measuring point.

However, it should be noted that such preloading cycles may also reflect back to the stress and deformation state of the measurement object. Systematic studies of the hysteresis of strain gages has revealed that in foil gages with increasing measuring grid length the hysteresis decreases [2.66].

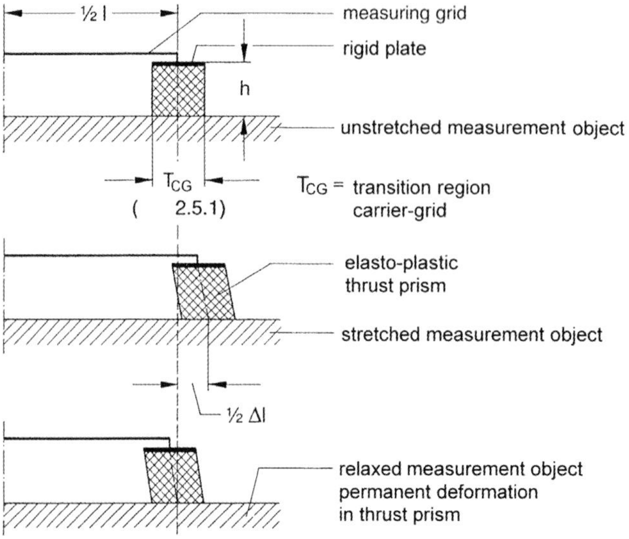

Fig. 2.47 Model of the explanation for the hysteresis of a strain gage [2.48]

Figure 2.48 shows, for four different measuring grid lengths, the hysteresis of the first and third strain cycles with ±1000 μm/m, ±2000 μm/m and ±3000 μm/m. This gives an indication of the magnitude of the hysteresis to be expected nowadays in properly installed high quality strain gages.

Figure 2.49 shows the results of an investigation into the influence of different adhesives on hysteresis. The figure shows particularly clearly the fall-off of the hysteresis after the third strain cycle. It also demonstrates how small the hysteresis is in a high quality foil strain gage with 6 mm active gage length.

2.6 Influence of pressure

2.6.1 Hydrostatic pressure

With increasing frequency, particularly in the context of stress analysis on the inner walls of pipes and pressure vessels, strain measurements must be performed in which the strain gages are exposed to hydrostatic pressure. Knowledge of the effect of hydrostatic pressure on the signal-generating resistance change of a strain gage is required for measurement results to be reliable as possible. This leads to the question of how sensitively an installed strain gage reacts to hydrostatic pressure, and how much of the signal is induced by the pressure superimposed compared to the signal generated by the strain in the component.

One of the earliest works dealing with the influence of hydraulic pressure on strain gages is a 1946 paper, focused on the determination of stresses in the wall of a steam boiler. In this work they tried to eliminate the influence of pressure by placing (a number of?) compensation gages (at different places?) in the pressure chamber [2.67].

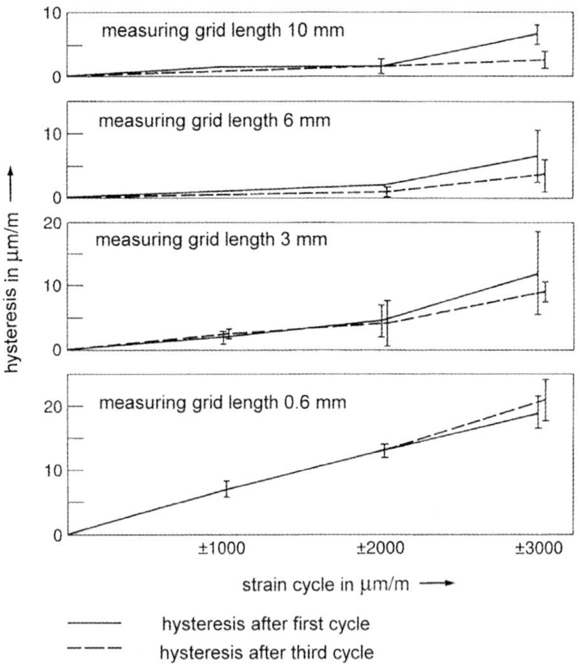

Fig. 2.48 Mechanical hysteresis of foil strain gages with different measuring grid lengths, bonded with cyanoacrylate adhesive Z70 [2.66]

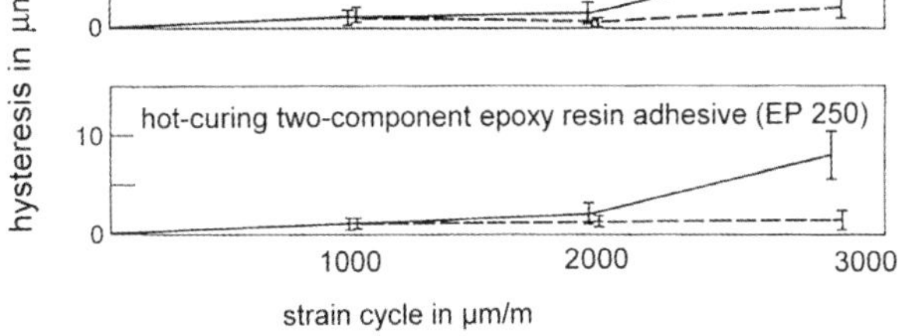

Fig. 2.49 Mechanical hysteresis of a foil strain gage with 6 mm measuring grid length, bonded with three different adhesives [2.66]

This was followed by numerous papers from other authors about systematic studies of pressure effects, with details of the correction factors for wire and foil strain gages [2.68–2.83]. Amongst wire strain gages, the wrap around strain gages prove to be unsuitable for measurements under pressure [2.71]. A summary of the results of all investigations is that, assuming perfect installation, the functionality of flat grid strain gages and foil strain gages is not affected by hydrostatic pressures of up to 10 000 bar. In all studies, a linear change in resistance (apparent strain) yields a linear function of the amount of pressure which slope, however, is dependent on the selected parameters, such as strain gage type and adhesive.

The influence of adhesives was investigated in [2.72] using four different adhesives (cyanoacrylate, cold-curing and heat-curing epoxy adhesive and epoxy resin). It was shown that the adhesive layer should be thin and of uniform thickness. In terms of the reproducibility of results at multiple loading and unloading with hydrostatic pressure and of the ease of handling in order to obtain a uniform thin layer, the cyanoacrylate adhesive was found to be most suitable. In [2.73] it was found that even with curved surfaces for installation of the strain gages, the relationship between pressure and resistance variation is linear. On a convex surface, the pressure influence affects slightly more than on a concave surface.

A first analysis of the influence of pressure on the basis of a theoretical model with the mathematical formulation of its effects was carried out in [2.74].

The previously cited papers are concerned with the reaction of the unstretched strain gage on the hydrostatic pressure. In [2.78] the influence of pressure on the sensitivity of the strain gage is tested. For this purpose, in an experimental device, strain gages were exposed to hydrostatic pressures up to 10 000 bar and loaded with positive and negative strains, and it was found that the gage factor was not affected by pressure. Another study [2.78] shows that pressures up to 10 000 bar have a negligible effect on sensitivity. In [2.79] in the pressure range up to 10 000 bar a sensitivity increase of 0.4% per 1000 bar was found. A study conducted at the BAM Berlin [2.79] yielded a result of equal magnitude, indicating an uncertainty in the gage factor of ±0.5% per 1000 bar. With regard to current industrial applications in which measurements are primarily carried out at pressures of <1000 bar, these results show that the influence of pressure on the sensitivity is negligible.

The results of the cited papers show that the slope of the linear relationship between pressure and strain gage resistance is mainly dependent on the particular combination of strain gage type and adhesive. A series of experiments was therefore performed with strain gages made and quite often used in Germany, in order to provide their users with information for their practical needs [2.82]. According to the practical requirements, this study covered a pressure range up to 500 bar.

Two different foil strain gage types were tested, each with 3 mm and 6 mm measuring grid length and two wire strain gage types (flat grid) with 6 mm and 10 mm measuring grid length. The research involved types that were in greatest demand at the time of investigation. Table 2.3 gives an overview of the investigated combinations of strain gages and adhesives.

Table 2.3 Research objects in the pressure sensitive tests

		Acrylic resin adhesive	Cyano-acrylate adhesive	hot curing epoxy-resin adhesive
foil strain gages (constantan)	Polyimid resin carrier	3/120 6/120	3/120 6/120	3/120 6/120
	Phenolic resin carrier with glass fibre		3/120 6/120	3/120 6/120
Wire strain gages (constantan)	Acrylic resin carrier with cellulose fibre	6/120 10/120	6/120 10/120	
	Phenolic resin carrier with cellulose fibre	6/120	6/120	6/120

The following adhesives were used: two-component acrylic resin adhesive X60 in a layer thickness of 65 µm ±15 µm which chemically cures within one hour at room temperature; the single-component cyanoacrylate adhesive Z70, which is applied in a layer thickness of 8 µm ± 2 µm and chemically cured within half a minute under thumb pressure (capable of measurement after about 30 minutes); the heat-curing two-component epoxy adhesive EP 250 with a layer thickness of 20 µm ± 15 µm.

The strain gage installations to be examined were carried out on small steel blocks and placed without cover in the pressure chamber. The fluid used as pressure medium was a water-free and acid-free oil, known from previous studies (BII of Aral)to be harmlessness to strain gages and adhesives. The compensation gages were located outside the pressure chamber in an open container, which was filled with the same oil as used for the pressure medium. For each parameter combination, eight samples were analysed and used to determine the mean and standard deviation. The experimental setup and details of the experimental procedure are described in [2.82].

The results obtained on two parameter combinations as representative of all test results are given in graphical form in Figure 2.50. The upper graph shows the combination with the smallest scatter of measurement results (foil strain gages with polyimide carrier and hot curing epoxy resin adhesive) and below, the combination with the largest scatter of the measurement results (wire strain gage with acrylic resin carrier and cyanoacrylate adhesive).

Both diagrams show with dash-dot lines the pressure-dependent strain of the steel block to which the strain gage under consideration is installed. The hydraulic induced strain in the block results from Hooke's law for the hydraulic stress state to

$$\varepsilon_{Pr} = \frac{1}{E}(2\nu - 1)p \tag{2.38}$$

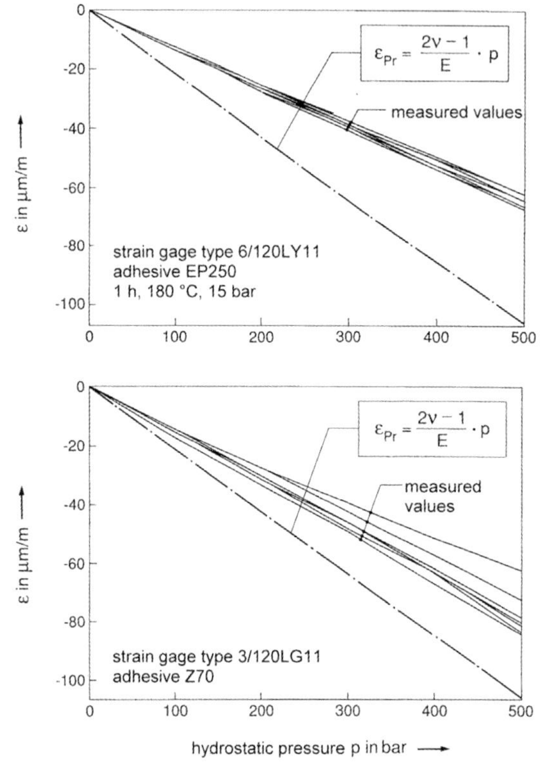

Fig. 2.50 Examples of the effect of hydrostatic pressure on the resistance change (apparent strain) of strain gages [2.82]

where E is the modulus of elasticity, ν is Poisson's ratio and p is the hydrostatic pressure. For a correct evaluation an exact knowledge of the modulus E and Poisson's ratio ν of the material used for the small blocks is required. It can be seen that the hydrostatic pressure causes an increase in resistance in the strain gage corresponding to the strain

$$\varepsilon_P = \varepsilon_{SG} - \varepsilon_{Pr} \tag{2.39}$$

From the apparent strain measurement result ε_{SG} must be subtracted the sample strain ε_{Pr}, to obtain the apparent part of the strain ε_P, generated by the pressure.

All results of the investigation are summarized in this form as means and standard deviations s in Table 2.4. The table shows that in all cases the pressure influence is less than $10 \times 10^{-6}/100\,\text{bar}$. It is evident that the correction values for wire strain gages are smaller than for foil strain gages. This means that, for wire, the pressure effect has less impact. Significant advantages of an adhesive (for extremely careful handling) are not recognizable. Adhesives that contain solvents and emit gaseous substances during curing are unusable.

Table 2.4 Influence of hydrostatic pressure on various strain gage installations in µm/m per 100 bar for pressures up to 500 bar

Strain gages				$\varepsilon_P = \frac{10^{-6}}{100}\text{bar} \pm s$	
		type	bonded with X60	bonded with Z70	bonded with EP250
foil strain gages	LY11	3/120	6.6 ± 1.2	7.9 ± 1.0	8.1 ± 1.0
		6/120	7.3 ± 1.0	8.1 ± 1.0	8.2 ± 0.4
	LG11	3/120		7.0 ± 1.5	5.7 ± 1.0
		6/120		7.9 ± 1.0	6.8 ± 0.6
wire strain gages	LA11	6/120	5.7 ± 0.8	6.5 ± 1.1	
		10/120	4.0 ± 0.8	5.1 ± 1.6	
	LP11	6/120	5.6 ± 0.7	5.3 ± 1.0	5.9 ± 0.6

In the course of the investigations it was found that the installation technique under just the loading of the hydrostatic pressure is of far greater importance than in any other application case. Too thick or uneven adhesive layers under the strain gage have a negative effect, since the adhesive layer is deformed by hydrostatic pressure too. The adhesive layer must be as thin as possible and free from air pockets. Therefore it is important to make sure that during mixing of multi-component adhesives that no air bubbles are entrained. Hydrostatic pressure debunked installation errors inexorably, as an example given in [2.82] clearly shows.

Strain gage installations that are intended for measurements under pressure should be checked before actual use by an installation inspection in order to avoid unpleasant and costly surprises during use. Here is a further example: For the hot curing adhesive EP250 is sufficient under normal conditions during curing already a pressure of 1 bar. It turns out that this pressure is too low for applications in the high pressure range, because the adhesive layer becomes too thick and too uneven. Therefore the pressure was increased to 15 bar. Table 2.4 and the upper part of Figure 2.50 show that the results obtained have the smallest standard deviation, which indicates their reliability.

In practical applications, strain gages are often exposed to pressurized water or other pressurized media against which the installations must be carefully protected by suitable covers to maintain the high insulation resistance between the measuring grid and the measurement object. A paper published in 1985 [2.83] shows that covers act on the installed strain gages attenuating on the influence of the pressure. The effects of four different covers against pressurized water at pressures up to 1700 bar were examined and it was found that the pressure influence was reduced by more than 50% by the covers used. For coverage with a rosin mixture resulted in a decrease to 20% compared to the uncovered measuring points (from 7 to $1.4 \times 10^{-6}/100\,\text{bar}$).

This attenuating effect can negatively influence the results, if they were corrected using correction values found from experiments using uncovered strain gages.

The lowest measurement uncertainties can be achieved if reliable correction values are available, which have been determined using all components including the protection in compression tests before performing the actual measurements. It is recommended in these experiments as well as in the actual measurements (as described in [2.82]) to work with compensating strain gages arranged outside the pressure chamber. This allows the pressure influence on the leads in the bridge circuit to be taken into account and to be eliminated by self-compensation. This can be done using the three-wire connection or by connect the compensating strain gages with a cable loop consisting of the same material and length as the cable loop of the active strain gage fed through the pressure chamber. Such an arrangement was used in [2.82] to facilitate the exchange of the samples.

If measurements are to be carried out without covering the strain gages, for example when an oil is used as a pressure medium that has good insulation properties (e.g. transformer oil), then a measurement signal correction is recommended with the values given in [2.82], which are also a guide for the expected pressure effects when the strain gages are properly installed.

The problem of pressure-tight cable entry through the wall of the pressure chamber will occur in the context of strain gage measurements under hydraulic pressure.

Furthermore, note that in the calculation of stress from under hydraulic pressure measured strain the triaxial stress state has to be taken into consideration. As well as the two principal stresses occurring in the measuring plane, a third principal stress in the form of a compressive stress related to the hydrostatic pressure appears perpendicularly to the measuring plane.

2.6.2 Operation in a vacuum

A demonstrative example of the successful applicability of strain gages in a vacuum is given with the soft landing of the NASA Surveyor in 1966 on the moon. There were 24 foil strain gages, cemented using special epoxy cement to Bakelite carriers, fitted to the engine mounts and shock absorbers of the three landing legs, and connected to Wheatstone bridge circuits. The signals were amplified and transmitted to earth via telemetry equipment, and recorded as force–time data.

The composite graph of Figure 2.51 shows strain gage readings for each of the spacecraft's three legs before, during and after impact. Vertical divisions, representing time intervals reveal that the Surveyor made an almost simultaneous three-point landing, with only 0.01 second intervals between touchdowns of each leg. Following initial impact, about six secondary impacts were recorded before the spacecraft assumed a static relationship with the lunar surface [2.84].

In addition, the measurement results together with photographs of the surface indentation caused by the landing gear permitted accurate measurement of lunar surface bearing strength, a hitherto unknown factor and one which is tremendously

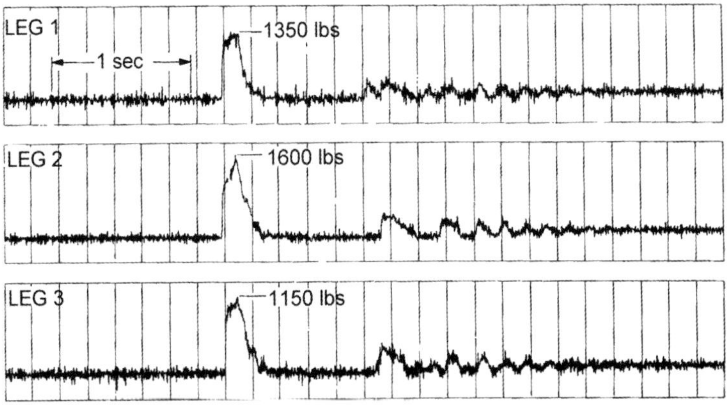

Fig. 2.51 Force–time graphs recorded by strain gages cemented to the shock absorbers in the three legs of the spacecraft Surveyor I during touchdown on the Moon [2.84]

important for future manned explorations. This provided important information about the sustainability of the lunar surface which was important for the planned manned flights of the Apollo programme (Fig. 2.52).

Fig. 2.52 Landing pad and support legs on Survivor 3 were photographed on 19 November 1969, by Apollo 12 astronauts Conrad, Jr and Bean during their moon walk. The strain gage can be seen (in circle), attached to the landing gear shock absorber [2.85]

The example shows that strain gages are functional under the conditions prevailing in space of maximum vacuum, and are thus generally suitable for measurements in any vacuum range. For measurements in closed vacuum chambers in which infinitely large pump down capacity of the space is not available, as with all operations in the vacuum range, the question of disturbance of the vacuum by outgassing of the materials involved is of great importance. In [2.86] the levels of outgassing of various materials are compiled in different treatment conditions.

With regard to strain gage measuring points in closed vacuum chambers [2.87, 2.88], recommended the use of open grids on carriers made of pure epoxy resin as possible, or pure polyimide. The carrier should be edge trimmed to the allowable minimum dimensions and free of plasticizers, fillers or colour additives. The adhesive should be made of pure resins and be hot curing adhesives, a complete cure with several hours post cure is recommended. Solid cables must be used to interconnect the strain gages, as strands are unsuitable for use in vacuum chambers. Working successfully in a high vacuum requires extreme care. The general problems of vacuum technology are more prominent than the strain gage technique.

2.7 Dynamic behaviour

2.7.1 First measurements using graphite layer strain gages

In the 1930s, particularly in the developing aviation industry, needs arose for measurement methods that enabled the determination of dynamic stresses in aircraft components such as propellers during operation. In 1937 in the USA, C. M. Kearns solved the problem with thin graphite strips, which he obtained by grinding carbon resistors [2.89, 2.90]. He stuck his measuring elements, which he referred to as strain gages, to his measurement objects and recorded strain-proportional changes in resistance. The invention was patented in 1941 in the USA [2.91]. The graphite strip provided very good results for measuring dynamic strain, but the zero signal stability, which played a minor role in these measurement tasks, was unsatisfactory. Also DeForest used in 1937, without knowing of the work of Kearns, graphite strips for the measurement of dynamic stresses of aircraft propellers [2.92]. Shock waves studies with graphite strips carried out in 1940 in the USA are described in [2.93].

During World War II, in Germany, thin carrier sheets coated with graphite layers were used for dynamic strain measurements. In 1941 Theis, who worked in the laboratories of AEG in Berlin at that time, described the first commercially produced strain gages of this type and their successful application in dynamic strain measurements on propellers for aircraft [2.94]. An insulated graphite layer was applied to a steel plate several μm thickness and provided at both ends with lead terminals.

Figure 2.53 shows, on the left, graphite strain gages as described by Theis, which were sold by AEG in relative large quantities in 1940 (at a price of 30 RM) [2.95]. A further developed graphite strip of the AEG is shown on the right side of Figure 2.53. The dimensions of the graphite strips can be identified from the graph paper selected as a base (bolted lines are 10 mm squared). The gage factor of these strips is quite high at 15 to 20.

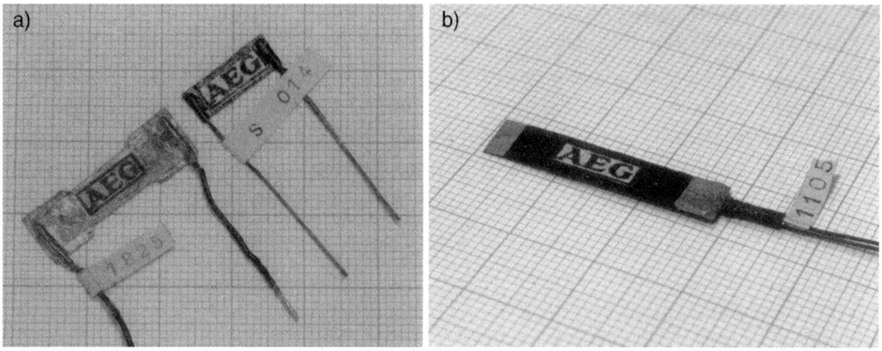

Fig. 2.53 a) and b) Graphite layer strain gages, marketed as 'Streifengeber' by AEG in 1940

Theis examined the properties of graphite strain gages and compensated for the undesirable drift caused by supply current, ageing and temperature effects in measuring rings, each with four single gages interconnected to form a full bridge circuit. But because of the disadvantages caused by drift and large hysteresis, the graphite strips didn't find wide application as single strips. In 1942 graphite gages were used at Junkers in Dessau for vibration measurements on aircraft components, such as during flight operation of the world famous Junkers Ju 52 [2.96, 2.97]. These used cylindrical carbon resistors of 6.5 mm diameter and 45 mm length ($R = 5000\ \Omega$) manufactured by the American company Hamilton Standard propeller. They were cut in half in the longitudinal direction and ground on the plane side up to a resistance of $20\,000\ \Omega$. These sensors were then cemented with a special adhesive to the measurement object. Such strips were manufactured by Siemens Plania and used for numerous dynamic measurements on aircraft during flight operations. Larger distribution has not been found because of the invention of wire strain gage with their much fewer disadvantages. The wire strain gage, however, was known in Germany only after the end of World War II.

2.7.2 Fatigue strength

Metallic resistance strain gages are destined to measure dynamic strain. Owing to their small mass, even in the presence of large accelerations they don't have retroactive effects on the characteristics of the measurement object in most cases. Strain gages are able to convert instantaneously the recorded strain into a change in resistance of their measuring grid, added to which, it is easy to attach them to the measurement object using a suitable adhesive. Even the invention of the strain gage by Simmons can be attributed to the solution of a dynamic measurement task.

The invention of wire strain gages opens up new possibilities for dynamic measurements. In strain gages the transformation takes place almost instantaneously and the main metrological problem is in the faithful reproduction of this transformation by electrical instrumentation [2.98, 2.99]. Of interest is the question of what is the upper frequency limit at which proper signal conversion in the strain gages takes place. Fink investigated this question and used strain gages to study shock waves at the Max Planck Institute for Iron Research, in Düsseldorf [2.100]. Fink calculated the course of elastic shock waves in cylindrical metal rods and compared the calculated results with experimental values. From the experimental results Fink found an upper frequency limit for the measurement capability of strain gages of at least 50 kHz, a value that is still valid today. It is interesting that Fink used graphite strain gages from Theis as trigger elements.

The main applications for strain gages were and are strength analysis of components under static or varying service loads and the construction of transducers for mechanical quantities. In transducers or components under vibrating load the occurring oscillating strain will be recorded by the strain gage installations. This leads to the question about the fatigue strength of the whole strain gage measuring system, adhesive, solder and cable connections, as well as the protection layer. Also, strain gage installations show that from the material technology known fatigue behaviour leads from increasing number of load cycles to increasing damage and at sufficiently large strain amplitude to failure.

Dynamic loads of strain gages can occur in the form of often repetitive quasi-static strains, such as with strain gages in load cells by repeated weighings or they take the form of oscillating strains resulting from fatigue stresses of measurement objects. In both cases a drift of zero signal occurs as a characteristic feature, resulting from fatigue damage caused by dynamic stress in the strain gage after a certain number of load cycles. The amount of zero signal drift depends on the number of load cycles and the strain amplitude.

Figure 2.54 shows a typical result of fatigue tests on foil strain gages with four different strain amplitudes and the mean strain $\varepsilon_m = 0$ [2.101]. In the region of incipient damage scattering occurs as typical of experimental results of fatigue tests.

Rohrbach and Czaika [2.102] carried out a series of systematic investigations about the fatigue strength of strain gages and its clear presentation. They displayed the results specifically for different strain gage types in diagrams.

They introduced the zero signal drift in these fatigue diagrams as characteristic parameters for the fatigue strength, as can be seen in the fatigue diagram of a wire strain gage (Constantan) in Figure 2.55. In this diagram are, as parameters, lines drawn for the zero signal drift of 10 μm/m, 30 μm/m, 100 μm/m, 300 μm/m and 1000 μm/m.

From the lines of equal zero signal drift it can be seen that with decreasing strain amplitude larger numbers of cycles will be endurable. A diagram like Figure 2.56 is only valid for one particular strain gage type, provided optimum installation conditions are used. As pointed out in [2.102], no effect of mean strain on the respective fatigue life can be determined in wire strain gages for mean strains between −3000 μm/m and +3000 μm/m. However, the mean strain will cause by their static creep an effect in the relevant strain gage type own magnitude.

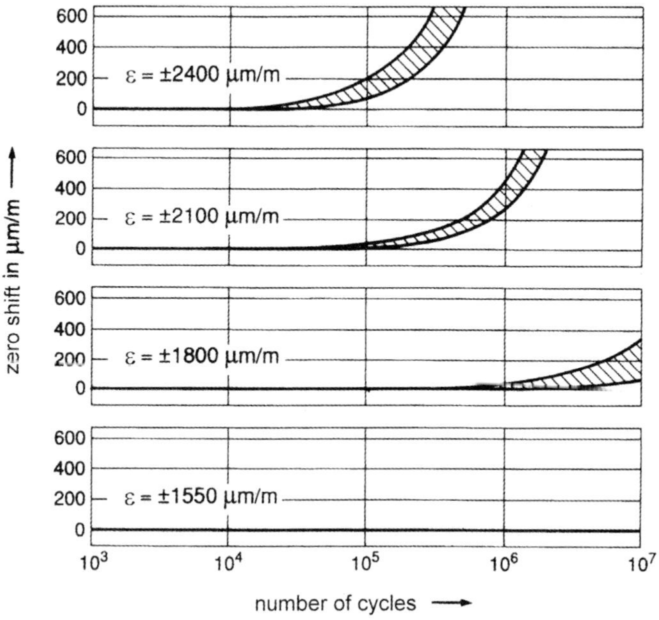

Fig. 2.54 Zero shift of four different foil strain gages as a result of fatigue strain loading [2.101]

A fatigue diagram for a foil strain gage (Constantan) is shown in Figure 2.56 [2.103]. Basically in view of the fatigue behaviour no significant difference between the wire and foil strain gage can be detected.

The growing zero signal drift with increasing vibration stress is not caused by inadmissible changes of the insulation resistance, it is not caused by creep and no

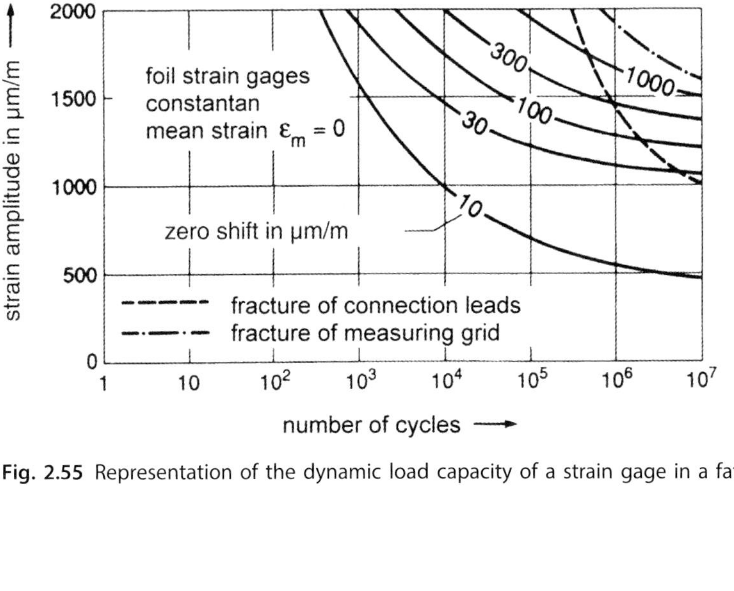

Fig. 2.55 Representation of the dynamic load capacity of a strain gage in a fatigue diagram [2.102]

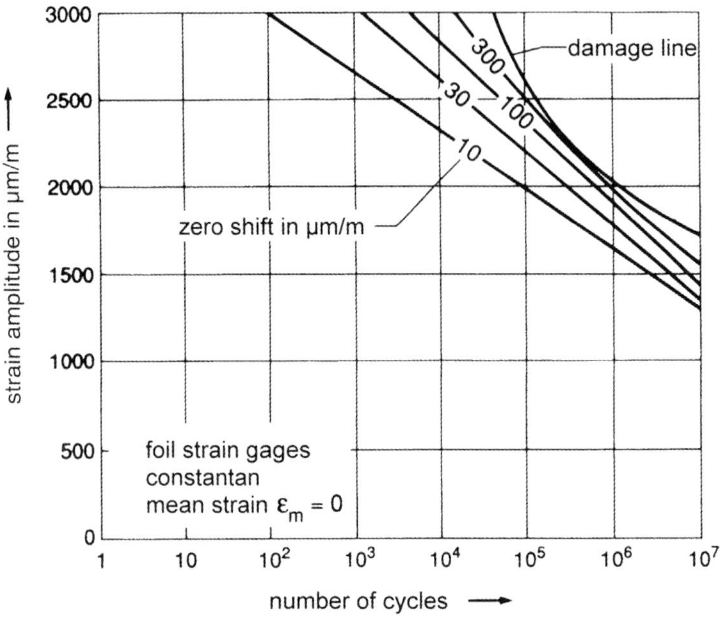

Fig. 2.56 Fatigue diagram of a foil strain gage (constantan) [2.103]

result of macroscopic cracks in the grid material. Rather, its causes lie at the submicroscopic range of the grid material, where increased dislocation density (strain hardening) and formation of lattice vacancies (submicroscopic cracks) influence the resistance value [2.102].

In purely dynamic measurements, in which the goal of the measurement is the detection of the amplitude of the time-dependent change in strain, the zero signal drift is of minor importance gage. Even after the onset of zero signal shift, the strain gage still has a significant lifespan, as can be seen from Figure 2.54. It was only after the occurrence of signal distortion that measurements debilitating degree of damage is reached. This distortion in the output signal, which can be seen with an oscilloscope, occurs at the end of lifetime as unmistakable sign of the failure of the strain gage. Distortion is caused by macro-cracks which widen in the pull cycle of the alternating stress and thus produce an increase in resistance, while they are compressed in the pressure cycle and cause little change.

This event may occur in the pull cycle up to the complete transection of the conducting track, resulting in the oscilloscope display to an open at the top pull cycle. Thus, an oscilloscope display can provide information about an already occurred damage in the strain gage. Closely related to this phenomenon is supersensitivity [2.104], which manifests itself in an increase in the gage factor *k* shortly before complete failure of the strain gage occurs. The scope of this supersensitivity is more pronounced in foil strain gages due to the slower crack growth in the foil than in wire strain gages [2.105].

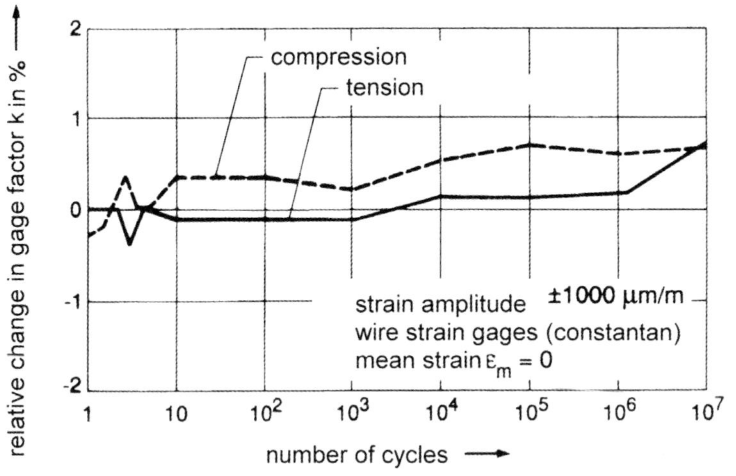

Fig. 2.57 Influence of fatigue loading with a strain amplitude of ±1000 µm/m on the gage factor k of a wire strain gage [2.102]

In the load cycle range before the inception of supersensitivity so far influence of the fatigue stress on the sensitivity of the strain gage has not been ascertainable. Any effects always remained within the tolerance limit of the gage factor of ±1%. Figure 2.57 presents a test result from [2.102]. The carrier materials and adhesives used today possess sufficiently high immunity against fatigue that this change of the *k* factor is largely excluded.

There have also been studies performed about the influence of fatigue loading on the thermal output [2.104]. In the investigated temperature range between RT and 80 °C constantan (Advance) in the annealed (soft) state, showed a noticeable change of thermal output. In the cold-rolled (hard) state, no influence was observed, not even in the state after the heat treatment for the adjustment of the thermal output of self-compensating strain gages. Due to their better fatigue strength hard foil is used for strain gages for vibration measurements. The soft foil with its higher ductility is particularly suitable for the production of high-extension strain gages. As a result, strain gages with annealed foil should not be used for fatigue tests.

Recently strain gages have been increasingly used as sensing elements in low cycle fatigue tests. These are fatigue tests with relatively large strain amplitudes in the time range of fatigue strength. For such application cases comparison tests showed that of various combinations of different strain gages and adhesives the optimal measuring components were constantan foil gages (120 Ω) with phenolic resin carrier installed with epoxy adhesive to a measurement object made of steel [2.106]. This installation easily survived the required number of cycles (10^5) with an amplitude of about ±2000 µm/m with a mean strain of about 4000 µm/m. The occurring zero signal drift of about 430 µm/m could be eliminated from the measurement results by performing zero calibrations at certain time intervals. The sensitivity (gage factor *k*) of the measuring point was not affected by the load cycles. In experiments with such high mean strains it is recommended to install the strain gages at the test object under a static load equal to the mean strain. It then passes during the fatigue stress as a burden on the strain gage only the amplitude and creep through the (static) mean strain is avoided.

It became apparent during the investigation of the fatigue behaviour and also in numerous practical applications that the critical points of the strain gage installation in fatigue loading are the leads and the lead terminals (e.g. soldering). These almost always experienced more fractures than the measuring grid. This can be seen from Figure 2.55 in which the lines of weakness for the leads and the measuring grating are drawn. This means that in measurement practice special attention must be paid to the connection components. Here the use of highly flexible connecting cables is recommended and also placing the soldering points in the low stress areas.

2.7.3 Shock load

With strain gages you can even monitor momentary shocks. Examples are given in detecting the pressure waves from explosions [2.107], in determining the action of the force of hammer blows [2.108] or in experimental studies in fracture mechanics [2.109]. Here, the main questions whether the gage factor k specified in the static calibration for a strain gage is valid even for very short-term strains, and whether the strain gage also faithfully reproduces the temporal strain curve in its output signal. The first question can be answered in the affirmative on the basis of experimental investigations. In their experimental efforts on the shortest times, Fink realized in 1950 about 17 µs [2.100], Cunningham and Goldsmith in 1959 about 7 µs [2.110] and Oi in 1966 approximately 1 µs [2.111] as the time for an abrupt change in strain in the test object, and they found that the gage factors k determined under static load were valid for these extremely short-term strain loads.

Answering the question about the temporally faithful reproduction of the strain jump in the test object by the output signal of the strain gage, a thorough examination of the parameters involved is required. A strain jump, for instance caused by a blow, a detonation or the rupture of a component under tensile stress is moving at the speed of sound through the test object, and the strain gage is installed on this. The strain jump itself has a certain rise time. If this rise time is zero, this means that the change in strain takes place in an ideal rectangular shape, and the change needs to pass through the strain gage in a time t, which depends of its length l

$$t = l/c \tag{2.40}$$

Where c is the speed of sound in the test object. The output of the strain gage builds up within this time t. This shows that a strain gage needs a time that is dependent on its length l to monitor a rectangular jump. For steel with a speed of sound of about 5100 m/s with an installed strain gage with a gage length of 6 mm we calculate the time t according to Eq. (2.40) giving a value of 1.18 µs. That means that the 6 mm long strain gage can reproduce the rectangular pulse at best with this rise time. An ideal rectangular wave shock will hardly occur in practice, as the previously cited efforts show to achieve the shortest possible rise time in the experiment. Thus, there has to

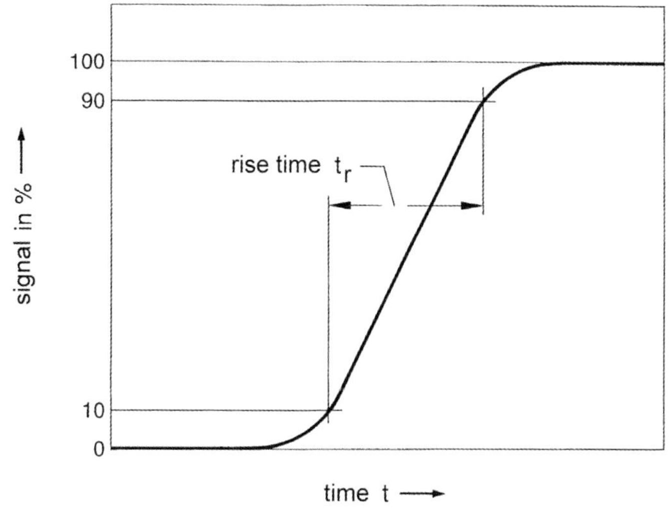

Fig. 2.58 Step response of a strain gage on a rectangular shock wave with finite velocity; technical definition of the rise time

be added to the rise time caused by grid length and sound speed the rise time of the shock actually occurring in the measurement object. These considerations show that the image correctness becomes worse with increasing grid length. On the other hand, the time periods occurring have to be assessed for practical relevance. In this respect, the order of a few microseconds is sufficient in many applications.

However, the theoretical considerations made above require practical supplement. In practice, one used as a parameter for a step response corresponding to Figure 2.58 the defined period of time, which characterizes the increase of 10% to 90% of the total jump height and refers to this time as the rise time t_R. Incorporating the recent approach this results in

$$t_R = 0.8\, l/c \tag{2.41}$$

The step response delivered experimentally by the strain gage is not identical to the analytical result. Through capacities between grid and carrier, through the thickness of the carrier material and the mechanism of resistance change in the grid enter additional delays that are formally very difficult or even no detectable. Oi [2.111] applies this delay as an additional term in the definition of the rise time and finds in his experiments for the rise time τ of the step response provided by the strain gage

$$\tau < 0.2\,\mu\text{s} + 0.8\, l/c. \tag{2.42}$$

This equation describes a limit below which the duration of the step response of the strain gage is to be expected. It provides no clear statement about the strain course to be recorded, because with increasing $l\tau$ also becomes larger, without change in the strain course to be measured. Bickle [2.112] researched this problem and developed a correction algorithm for the expression 0.8 l/c. Using this correction algorithm the distorting effect of a larger grid length can be compensated. This was proved on a grid length of 48 mm. The reference value used by Bickle was the step response of a short grid length of 0.4 mm.

In summary it can be stated that strain gages with usual grid lengths, that is between 3 mm and 10 mm, can monitor technical shocks very well. At 10 mm gage length and with steel as the test object material, gages from Eq. (2.42) strain gages ll cause a delay of 1.77 µs. This time is relatively much lower than the rise time of most technical impacts to be measured. So, for example, a generated shock wave from striking with a hammer the front end of a 3 m long steel rod (20 mm Ø) a signal period of 170 µs (rise plus drop) would be produced that can be captured easily in an experiment with a 6 mm long strain gage [2.113].

When the strain oscillates periodically, for example sinusoidally, at a high frequency, a small wavelength can cause a strain gage signal not to reflect the full amount of the signal to be measured. The ratio of the displayed strain value to the true strain value is referred to as the transmission ratio. The transmission ratio depends on the length of the strain gage and on the wavelength of the measured strain. If the two lengths are identical, then the strain gage indicates the average of the positive and negative oscillation portions. For the truest representation of the oscillation, the wavelength must be much larger than the grid length of the strain gage. This is, in technical applications, almost always the case. Short wavelengths of the order of the grid length are relevant only in exceptional cases. So, for example, the shock wave in the steel bar (as described in the preceding paragraph), has a wavelength of 0.867 m (the product of the speed of sound $c = 5100$ m/s and the signal duration of 170 µs).

During the construction of transducers with strain gages for detecting transient or varying mechanical quantities it is important to be sure that the natural frequency of the transducer body, as far as possible (factor > 4) exceeds the frequency of the signal to be measured. A typical example is the pressure transducer described in [2.107] used to measure the explosion waves resulting from the explosion of gas-air mixtures. The dimensions of the measuring element serving as a receiver in this transducer are chosen so that the natural frequency of the diaphragm was adjusted with respect to the frequency of the pressure waves to be measured. The rise time of the explosion waves to be detected were the order of magnitude of 10 ms.

2.7.4 Fatigue-life gage

In 1962, Harting at the Boeing Company in Seattle had the idea of using the zero signal drift of a strain gage as an indicator of the progressive fatigue of the strain gage and thus indirectly to use it as a measure of the damage accumulation in the component on which the strain gage is installed [2.114]. Harting's idea led Boeing to the development of a fatigue-life gage, indicating that the resistance change caused by hardening during repeated stress was a measure of the increasing material damage and that after a certain damage accumulation it is going to break [2.115].

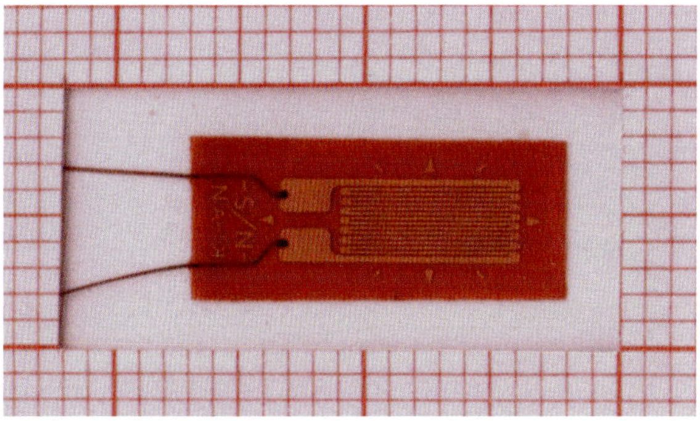

Fig. 2.59 S/N-gage for monitoring the damage accumulation occurring by fatigue stress in components

To achieve quantitatively reliable measurement results was difficult because of the different fatigue behaviour of different measurement object materials. However, the first fatigue-life gages were well suited as indicators for the comparison of damage incident on aircraft under service loads and thus provided useful inspection hints. The idea was patented [2.116]. In 1964, Micro Measurement Inc. took over Boeing's rights for the manufacture and sale of fatigue-life gages.

Figure 2.59 shows one of the S/N-gages, as used in the 1960s on aircraft parts and power plant components [2.117, 2.118]. A detailed description of the application of these gages and the evaluation of the signals thus obtained is given in [2.119].

The fatigue-life gage looks like a foil strain gage. It consists of an etched constantan grid on a hard carrier. This resistor (about 100 Ω) changes its value beginning at an amplitude-dependent number of load cycles with increasing number of load cycles. The dependence of the resistance change on the number of load cycles is not linear.

Figure 2.60 shows the characteristic curves of resistance change of a fatigue-life gage as a function of the number of load cycles for different strain amplitudes [2.120]. It can be seen from the diagram that relatively high strain amplitudes are required to generate the resistance variations in the fatigue-life gage. If these do not occur at sufficient size at the measurement point it is then necessary to use the fatigue-life gages in combination with mechanical multipliers, which ensure that the effective strain amplitude at the fatigue-life gage is larger than that at the actual measurement point. In principle, this multiplier consists of a clip bonded at both ends to the measurement object. In its central portion the cross-section is tapered in order to create an increasing strain. In [2.121] the successful application of fatigue-life gages at strain amplitudes occurring in collective form and the practical application of a mechanical strain multiplier is described. The surprisingly good reproducibility of the calibration curves of the fatigue-life gages is highlighted. A restriction should be noted, however, that the successful application of the fatigue-life gages requires intensive training which is not readily feasible as a normal strain measurement with strain gages. Fatigue-life gages can be used to monitor the accumulation of stress

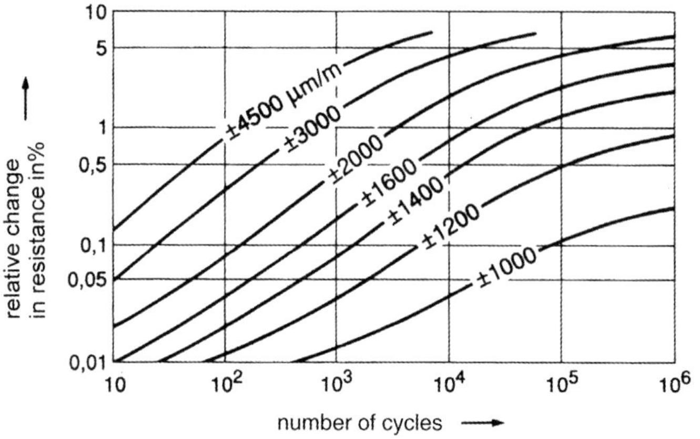

Fig. 2.60 Calibration curves of a fatigue-life gage type; representation of the change in resistance as a function of number of load cycles with the strain amplitude as a parameter

under service load in components, but they are not an adequate tool for the prediction of the life of a component [2.122].

A completely different method which is successfully applied in fatigue tests on aircraft structures as a measure of damage accumulation is the changing of the surface structure of tin and aluminium foils due to changing load. These films were stuck in a bright polished state on the measurement object and their optical reflectivity was measured. The mechanical cyclic loading forms slip bands and extrusions in the film that lead to macroscopic visible roughening of the surface and affect their optical reflectivity. With an increasing number of load cycles, the reflectivity of an originally highly polished metal surface decreases steadily [2.123]. With decreasing strain amplitude, the beginning of the decay of the reflection intensity shifts to higher numbers of cycles.

In practice, life monitoring can be performed both with one level cyclic load as well as with cyclic stress with several load levels ('programme cycles' or 'blocks'). Information about the stress distribution in structures can be obtained by employing several reflection gages. The gages have small dimensions and are cemented like normal strain gages. No disturbing influence of temperature changes was found between −50 °C and +100 °C. The measurements are made by a portable measuring head connected to a measurement device [2.123].

2.8 Heat dissipation

Flow of current creates thermal energy in any electrical conductor. This thermal energy is called Joule's heat loss after the discoverer of the electric equivalent of heat. Strain gages are current-carrying electrical resistors, in which electrical energy is converted into thermal energy.

The output of a strain gage circuit proportionally not only depends on the change in resistance, but also on the applied supply voltage. To achieve the largest possible signal effect it seems desirable to apply the highest possible supply voltage. On the other hand a high supply voltage can result in undesirable heating effects in the strain gage. This can cause changes in the properties of the materials involved in the measuring point, and thus lead to variations in the result. When measurement objects are made of materials with good thermal conductivity disturbance by self-heating of the strain gage are not normally to be expected since the heat generated is dissipated immediately. With poor heat conductors, such as glass, ceramics, plastics or wood attention must be paid to the heating of the measuring point by the strain gage.

The resulting measurement error in most cases can be kept small by choosing a low supply voltage. A time-dependent variation of the zero signal is observed as the main consequence of measuring point heating. Due to the local heating, the strain gage and the measurement object both expand until a stationary temperature state is reached. The time-dependent zero signal shift caused by heating is not noticed when using a compensation gage that is installed on the same material in the strain gage bridge circuit. This compensation will, for reasons of economic efficiency, not be realized in measuring objects with many measuring points, such as found in model measurement technology [2.124]. But if a multipoint measuring system is being used, a common compensating strain gage can be applied for a multi-active strain gage group.

With some measurement instruments the strain gages at the measuring points can be pre-heated to avoid any shift of the zero signal when a strain gage is in action. The pre-heating supply voltage is applied to the strain gage even when it is not selected by the measurement instrument. This method has the disadvantages of constant heat production at the strain gage which can, with very poor thermal conductivity, lead to relatively strong heating of measuring objects as well as the difficulty of reproducing a zero signal as a reference for an automatic zero calibration. When measuring on metals, pre-heating is only useful when working with bridge excitation voltages >2.5 V, but even then only when the pre-heating voltage is exactly the same as the bridge supply voltage and when the changeover between pre-heating and bridge supply is carried out without any break. A switching interval of even 5 ms can generate, with 5 V, a variation which is as great as when a supply voltage is switched from 2.5 V without preheating [2.125]. A better solution is to use instruments that retrieve the measurement signal according to a precisely reproducible time interval after switching the measuring point (applying the supply voltage).

Another undesirable effect of self-heating of strain gages is increasing the creep and hysteresis of the measurement point. In addition, it can disturb the self-temperature-compensating effect in the strain gage. In stress analysis the temperature-dependent change of the modulus of elasticity of the test object material is an additional source of error.

In a strain gage with applied supply voltage V and current I there is an electrical power P given by the product of voltage and current:

$$P = VI \tag{2.43}$$

If the voltage is in volts and the current in amperes, the power is in watts. Substituting Ohm's law in Eq. (2.43) leads to

$$P = I^2 R = V^2/R \tag{2.44}$$

Supplying strain gages in a Wheatstone bridge circuit with bridge excitation voltage V_B, then the single strain gage gets a voltage of $V_B/2$. Substituting this voltage in Eq. (2.44), then the electric power P_{SG} converted into heat at the individual strain gage is obtained.

$$P_{SG} = V_B^2/4R \tag{2.45}$$

If one feeds a bridge circuit of 120 Ω strain gages with 5 V excitation voltage, then the power of 0.052 W per strain gage is converted into heat output, as given by Eq. (2.45). Since 1 W = 1 J/s (joule per second), then in the strain gage under consideration each second a quantity of 0.052 J of heat is released. If you carry out the same calculation for a 350 Ω strain gage, then it is found that the amount of heat released is only 0.018 J/s. It is apparent that the use of strain gages with a higher resistance reduces the flowing current, and thus the released amount of heat can be reduced: in the first case A flow of 21 mA, and in the second case only 7 mA.

Good dissipation of the generated heat is required, since heating of the strain gages during the measurement must be avoided. The thermal conductivity of the test object material and the area available for heat dissipation play the main roles. In a strain gage with large geometrical dimensions the heat generated dissipates faster than with a small strain gage of the same resistance. In the larger gage the amount of heat generated is distributed over a larger area. Therefore the area covered by the measuring grid must be included in the analysis and the released amount of heat from the strain gage must be related to this area. The amount of heat released by a measurement site per unit time depends on the thermal conductivity, which in turn is a function of specific heat capacity and the density of the observed material.

In [2.126] recommended values are tabulated for optimal excitation of strain gages in the unit milliwatts per square millimetre. Values are given as parameters for the thermal conductivity of the test object in a scale graduated from very good to very bad and the demands on measurement accuracy in static or dynamic measurements. The maximum value of 78 mW/mm^2 is specified for dynamic measurements with low accuracy requirements on very good heat conductors (such as copper or aluminium), the minimum value of 0.016 mW/mm^2 for static high precision measurements on very poor heat conductors such as filled acrylic resin. This minimum value is however technically realizable. Assuming normal available modern equipment with smallest supply voltage of 0.5 V and a strain gage resistance of 350 Ω, then Eq. (2.45) gives the power of 0.178 mW converted into heat, so to comply with the specified minimum value of 0.016 mW/mm^2 would require an area of at least 11.2 mm^2. If one uses a strain gage with the nominal resistance of 1000 Ω, then this would require an area of about 4 mm^2. Thus also the limit of miniaturization is reached with respect to tests on plastic models. It is recommended to carry out preliminary tests before measurements with strain gages on poor heat conductors, in

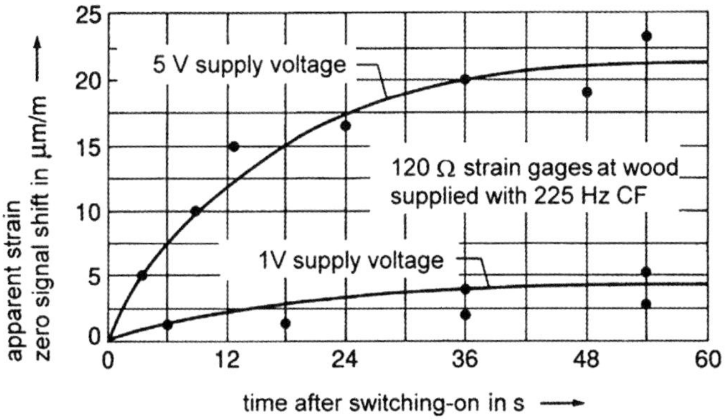

Fig. 2.61 Zero signal of a single strain gage installed on wood at two different supply voltages as a function of time [2.129]

which the parameters strain gage resistance, strain gage dimensions and supply voltage are tested separately. A suitable criterion here is the behaviour of the zero signal.

At switching-on, the strain gage begins a time-dependent zero signal shift. This process is referred to as the running-in process. The course of the time-dependent change in zero signal with a correct installation corresponds to an exponential function and asymptotically approaches a limiting value which represents the equilibrium between generated and dissipated heat. Such processes are described for example in [2.127, 2.128] for plastics and in [2.129] for wood. Figure 2.61 shows an example of the running-in process for two different supply voltages for a 120 Ω strain gage cemented on wood [2.129].

An analysis of the running-in behaviour of strain gages on steel and aluminium found that the zero signal shift during the running-in process was proportional to the power dissipation, as expected. It increases quadraticly with the applied supply voltage and decreases linearly with the increase of the strain gage's resistance. A difference in the running-in behaviour when supplied with DC voltage and with a 1 kHz carrier frequency was not observed. When adhesives have been a cold-curing cyanoacrylate adhesive (Z70), a cold-curing two-component adhesive based on polymethacrylate (X60) and a hot-curing epoxy resin (EP250), there was no difference in running-in behaviour of the two measuring sites with the cold-curing adhesives. With hot-curing adhesives, however, the running-in error decreased significantly, it decreased to about 30% compared to the cold-curing adhesives with both steel and aluminium. In general, the running-in errors for the same measuring point parameters were about 40% to 50% less for aluminium than for steel.

2.9 Measuring at elevated temperatures

2.9.1 Previews

There is a continuing need to use strain gages at higher temperatures. The common standard strain gages with constantan grid and organic carriers with cold-curing adhesives are generally applicable up to about 70 °C without problems. From this limit, however, the individual components of a measuring point must be considered separately in terms of their suitability for the intended application temperature. These components are the measuring grid material, the carrier, the adhesive, the connection cables including their insulation, the soldering and the solder used and the cover.

For the upper limit of the operating temperature often two specifications are given, one for static and one for dynamic measurements, since the temperature effect has its main impact on the zero signal, that is the reference point of the measurement values. It is better to refer to static measurements as 'zero-related measurements' since each zero shift is fully affects the measurement result. Therefore, the upper temperature limit is lower there than it is for dynamic (better called non-zero-point related measurements), in which only the amplitudes of a time-varying measuring quantity should be recorded.

The data sheets published by manufacturers don't always refer to the time period over which the specified temperature can act without loss of signal stability. The upper temperature limit, however, depends on this time period. If reliable data is not available it may be useful before the actual measurement to examine the drift behaviour (time-dependent zero signal change at constant temperature) of the intended strain gage type. In general, this test results in diagrams as shown as an example in Figure 2.62, which show the drift behaviour of foil gages with measuring grids of a common nickel–chromium alloy and a carrier of glass fibre reinforced epoxy-phenolic resin. In such diagrams is to be noted that as an essential parameter of the characteristic of the curves the measured length of the strain gage may occur. However, the drift can be largely compensated for when using Wheatstone bridge circuits.

For strain gages with measuring grids of constantan there can be seen to be an upper application temperature limit at about 200 °C. Ref. [2.131] describes how a constantan grid on glass-fibre reinforced phenolic resin carrier and heat curing adhesive as a force measuring element was used successfully up to 230 °C. Operating

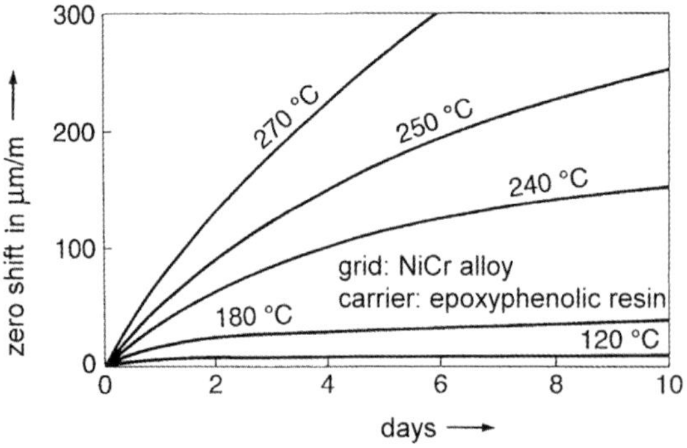

Fig. 2.62 Schematic representation of the zero signal drift of strain gages suitable for temperatures up to about 280 °C over several days at a constant temperature

temperatures up to about 320 °C are reachable with measurement grid alloys of nickel and chromium (e.g. Nichrome V, 80% Ni and 20% Cr).

For even higher temperatures alloys of about 92% platinum and 8% tungsten are used, with which zero-point-related measurements up to about 550 °C (here the time of use plays a role) and dynamic measurements up to 800 °C are possible.

The question of the application temperatures always includes the issues about the accompanying conditions such as time and the surrounding medium. Long-term measurements in aggressive environments require a lower upper temperature limit than for short-term measurements with hermetically sealed gages. Also the timing of the temperature changes plays a significant role.

Especially for measuring at higher temperatures strain gages have been developed without organic components. Typical forms were free-filament strain gages that are attached with ceramic cement or by flame spraying on the measurement object as described in Section 3.2. Strain gages have also been developed that are encapsulated in small stainless steel tubes on thin metal carrier sheets which are bonded by spot welding onto the measurement object. Figure 2.63 shows photographs of a free-filament wire strain gage and a tube gage attached by spot welding.

Both types use resistance wires of either NiCr alloy or PtW alloy, about which some explanatory notes have already been given in Section 2.1.3. Nickel-chromium alloys, like constantan, have a gage factor of $k \approx 2$ and they are thermally stable up to about 340 °C. PtW wires have a gage factor of $k \approx 4.5$ and allow application temperatures for static measurements up to about 600 °C and for non-zero-point-related measurements up to 1000 °C. However, at these temperatures, owing to drifting, the measuring time plays a significant role.

The results of extensive studies of the influence of temperature on the gage factor k for different measurement grid materials were published as early as 1959 [2.132]. In general, it can be stated that in all measuring grid materials under consideration with increasing temperature the gage factor k decreases and the resistance increases. Here, the effect of mechanical and thermal load cycles on the

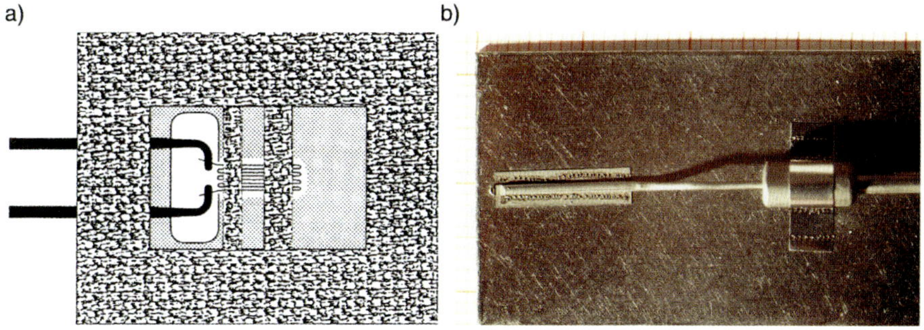

Fig. 2.63 Typical designs of high-temperature strain gages a) free-filament strain gage gage; b) welded tube

reproducibility of the measurement results becomes important. In [2.133] constantan, Nichrome V and PtW are considered in their behaviour under repeated temperature cycles.

This shows very clearly, especially when comparing cyclical heated constantan in an argon atmosphere and in air, that corrosive influences occur besides temperature-induced metallurgical changes. In [2.133] is also described a test facility developed at BAM in Berlin, with which gage factor and thermal output of high-temperature strain gages can be tested and calibrated. Results are given in [2.134] of detailed investigations into the metallurgical behaviour of the mainly used grid materials, NiCr and PtW, for high temperature strain gages.

Due to the ongoing demand for suitable high-temperature strain gages the development of high-temperature strain gages can be referred to as not complete. To improve measurement capability of strain gages in the high temperature range, tests have been carried out on specially developed new alloys, such as Fe-Cr-Al [2.135] and Pd-Cr-base [2.136] and even the idea of combining wire of two different materials to compensate for the temperature effects has been revisited [2.136].

Today in many cases strain gages are the only possible means of measuring operating stresses, without too much effort, for thermally highly stressed components such as internal combustion engines, gas turbines, turbochargers and reactor components. Often a variety of unusual factors are connected with these applications, which can greatly complicate the application of strain gages: in addition to high temperatures high accelerations can occur; the surrounding atmosphere can contain aggressive components; installations must often take place within a closed housing and require complicated measurement cable runs or telemetric signal transmission; the measurement points are often accessible only in the disassembled state of the measured object and can't be replaced without great expense.

Measurements in these environments can be very complex and therefore require reliable and secure knowledge of the metrological characteristics of the strain gages to be used and especially information about their long-term behaviour under the acting influences [2.137]. Experience shows that, owing to the wide diversification of demands made by different operating conditions, the characteristics given by the strain gage manufacturers' are not always sufficient for the intended application. In these cases, the user must identify for themselves the additional information needed for the solution of the measurement task. Today this is common practice, as shown below for working with high-temperature tube-strain gages.

2.9.2 Weldable encapsulated tube strain gages

In these specifically developed strain gage elements for measurement under difficult environmental conditions (high and low temperatures, vapour atmosphere, liquid sodium, salt water, totally desalinated water, chemicals) the active resistance wire is encapsulated in a thin casing tube filled with high-purity magnesium oxide. The highly compressed magnesium oxide transmits the positive and negative strains of the tube to the active wire without slippage, and it also serves as insulation. The

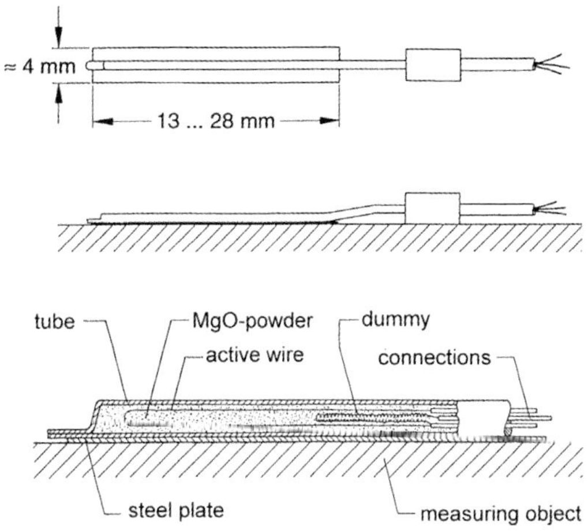

Fig. 2.64 Basic structure of a tube strain gage for high temperature applications

housing tube is welded to a thin carrier sheet that is then spot welded to the measurement object as can be seen in Figure 2.63b) [2.138].

Tube strain gages are manufactured both as a quarter-bridge and half-bridge gages. The half-bridge design contains, as well as the looped measuring wire, a spirally wound compensation wire which remains mechanically unloaded and forms the compensation branch of a half-bridge. The quarter-bridge version does not contain this compensation wire. Figure 2.64 shows the construction principle of the tube gage. Due to the instability of the NiCr alloy above 300 °C half-bridge types should not be used above this limit. The connection cables are integrated into the total encapsulation of the gages. Tube gages are also available with different cables, and with welded tubular steel cables and flexible cables from different materials.

Measuring with these tube gages requires skill and some preparatory work. Under the influence of relatively high temperatures a thermal output (variation of the zero signal) is to be expected directly superimposed on the measurement result if there is little or compensation. Furthermore, the characteristic properties of the strain gage change with temperature, and so, without knowledge and correction of these changes, interpretation of the obtained results is not possible [2.139]. The thermal output of a measuring point can only be determined if it is possible to heat the measurement object without load and with the installed strain gage. That's rarely possible with the large components in conventional power plants.

However, here the tubular strips offer a way out, since before the actual installation you can perform preliminary investigations on the component under test with the strain gage on the measurement object material. These preliminary investigations reduce the measurement error caused by uncertainties in the specifications of the thermal expansion coefficients of component material and strain gage. They also allow a reliable determination of the actual temperature response and sensitivity, and they enable the early detection of possible defects on or in the strain gage. These preliminary investigations can be done either with a clamp [2.139–2.144] or by twice welding of the strain gages [2.139].

The clamping device has to be made of the measurement object material in order to verify its thermal expansion. In the device, the flange of the strain gage is gripped from the top, along both sides of the tube, by means of a clamping head on a base plate. Even with different thermal expansion coefficients of the clamp and the clamped strain gage the clamping must be strong enough so that the strain gage can't move relative to the clamp. Thermal outputs of tube strips determined using such clamping devices on gages with NiCr wire up to 320 °C and with PtW wire up to 600 °C are published in [2.143].

Calibration of the tube bars using the method of twice welding is described in detail in [2.139], where the strain gages to be examined are welded onto a measuring beam made of the measurement object material. With this measuring beam the thermal output and also the gage factor k can be determined in a specially constructed device [2.133]. The strain required for the determination of the strain gage factor is produced by bending (four-point bearing) of the measuring beam. As the material creeps under bending stress at elevated temperatures and also the modulus of elasticity changes depending on the temperature, the beam deflection is not generated in a force-controlled way but by moving the bearing of the beam in order to set the desired deformation. Via a bearing shifting step motor, a computer controls the desired rod deflection; a displacement transducer provides the information about the deformation [2.145].

After this preliminary examination, which must be performed on each strain gage, the gage is removed with a scalpel from the measuring beam. The individual welding points will be cut between the flange and the beam surface. After separation, the underside of the flange of the strain gage is straightened and then smoothed with fine sandpaper. This is performed on the strain gage with the tube pushed down into a device with a cutout for the tube. After this intermediate treatment the strain gage is welded to the measuring object, with the spot welds placed in between the points used for the first weld. With proper execution of the treatment a satisfactory agreement of the thermal output after the first and the second welds can be reached [2.139].

In [2.146] an interesting application example is described carried out with tube strips on the compressor and turbine disks of a gas turbine during operation. In areas with expected temperatures of about 320 °C quarter bridge gages with NiCr measuring wire (Eaton SG 125) were used and in areas with temperatures of about 450 °C half-bridge gages with PtW wire (Eaton SG 425). Prior to their installation in the turbine each strain gage used was welded to a measuring beam and calibrated in the apparatus as described above. In addition to the high temperatures there are other difficulties with the measurement task: high centrifugal forces resulting from the rotation of the rotor; the required telemetric transmission of strain and temperature readings from rotor to stator; temperature readings are needed for the corrections based on the preliminary determined calibration values; and the difficulty of installation of the measuring cables within the turbine from the measuring points to the transmission ring on the shaft end.

2.9.3 Free-filament strain gages

Free-filament strain gages are attached to the measurement object without a carrier, either by ceramic putty or by the flame-spray method. They are available in different measuring grid lengths (~1.5–6.5 mm) with different wire materials (NiCr alloy, PtW alloy, Karma etc.). The installation requires a lot of skill and practice. As supplied, free-filament strain gages are fixed on a self-adhesive glass fibre reinforced Teflon carrier (Figure 2.64), by means of which they are fixed prior to attachment to the measurement object which has first to be provided with an insulating layer. After the provisional fixing the final installation is performed. In Section 3.2.6 the individual installation steps for bonding a free-filament strain gage with ceramic putty are described with reference to Figure 3.5.

In contrast to the tube strain gages free-filament strain gages can't be separated from their base and then installed again on a measurement object after calibration. Thus, calibration can only be performed on the measurement object itself. This can be carried out in order to determine the thermal output when an oven of sufficient size is available to heat the mechanically unloaded measurement object with the installed free-filament strain gages to measuring temperature and to determine the thermal output as a function of also being measured temperature. The completion resistor that forms the half-bridge is held at a constant temperature outside the oven. The application of the three-wire circuit is absolutely necessary to minimize the influence of cables. Furthermore, the heating rate of the measuring object in the calibration should be slow enough to avoid thermal stress caused by temperature gradients.

With the thermal outputs thus determined the actual measured values can be corrected if their temperature readings in temporal association with the strain readings for each measurement point are also available. This is achieved by using thermocouples located at each measuring point. In this type of correction the measurement uncertainty is influenced strongly by the uncertainty of the temperature measurement. In [2.147] a corresponding application example from the field of power plants is described, in which with free filament strain gages (PtW wire, installation with ceramic putty) zero-point-related strain measurements at temperatures up to 350 °C were performed successfully in notches (notch radius 10 mm) at cyclic thermal loaded junctions of pipes.

In principle, with free-filament strain gages it is possible to compensate the thermal output of a half-bridge circuit using a mechanically unloaded compensation gage from the same manufacturing batch in addition to the active gage. A problem with this arrangement, however, is that in many cases it is difficult to ensure identical temperature experience for both strain gages, especially in transient and multidimensional temperature fields, since the compensation gage can't be located at the same site as the active gage, and because the connection of the compensating strip to the measurement object is different from that of the active gage. Especially rapid temperature changes can also cause large errors.

Generally non-zero-point-related measurements are less problematic than zero-point-related, since they superimpose the thermal output signal component on the mean strain, so the amplitude won't be affected. Also the signal frequency of the time varying strain is unaffected by the temperature. A very good example of this are vibration measurements on turbine blades of turbochargers described in [2.148] with free-filament gages (PtW wire, flame spray installation) at temperatures up to 600 °C and radial accelerations at the measurement points up to 160 000 *g*. The temperature responses of the original measuring points were on the turbine wheel (88 mm diameter) previously determined in an oven. Consideration of the influence of temperature on the gage factor was based on the manufacturer's specifications. The recorded frequency–rotational speed diagrams allow comparisons of the effects of different turbine housing designs on the resonant stresses of the blades.

2.10 Stress gages

In principle, it is possible to determine directly, with an appropriately designed strain gage, the stress in any direction on a measurement plane. The stress in an arbitrary direction x depends not only on the strain in the x direction, but is also affected by the transverse strain occurring in the y direction perpendicular to the x direction, as explained in Chapter 2.9. From Poisson's ratio ν of the measurement object material the effective stress in the x direction can be given by

$$\sigma_x = \frac{E}{1-\nu^2}\left(\varepsilon_x + \nu\varepsilon_y\right) \tag{2.46}$$

If the strain gage is attached at an angle φ to the measurement axis, then at a given angle φ its resistance change is proportional to the stress in the measuring direction x [2.150], since its measurement grid receives a strain proportional to the sum $(\varepsilon_x + \nu\varepsilon_y)$.

The question of the angle φ at which the measuring grid of the strain gage to the measuring direction must be inclined to achieve the desired proportionality, via a relatively complex derivation [2.150–2.153] results in

$$\cos 2\varphi = \frac{1-\nu}{1+\nu} \tag{2.47}$$

The associated relationship between stress σ_x and recorded by the measuring grid of slanted stress gage strain ε^* is given by

$$\sigma_x = \frac{E}{1-\nu}\varepsilon^* \tag{2.48}$$

With the definition of the angle φ, the most important design feature for the preparation of a stress-measuring gage is then known.

It determines that the conductor tracks must be at an angle φ to the measuring direction, where the angle φ must be adapted to the Poisson's ratio of the material of

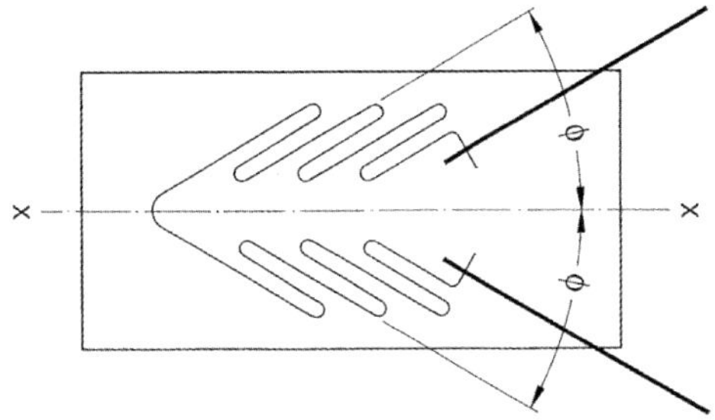

Fig. 2.65 Basic measuring arrangement of a stress gage in V shape

the measurement object. This results, for example, in an angle of 28.71° for the Poisson's ratio $\nu = 0.3$. The arrangement of the conductor tracks is then in principle as shown in Figure 2.65.

Such stress gages were prepared as wire gages in the early 1940s in the USA. The photograph in Figure 2.66 shows two historic wire stress gages. The two gages differ in the design of the connection leads, which are located either on opposite sides or on the same side. Stress gages were later manufactured as foil gages in a V-shape. Figure 2.67 shows an example.

If you have properly installed such a V-shaped stress gage then you obtain as a measurement result the change in resistance that unbalances the Wheatstone bridge and causes a signal like a normal strain gage. In a normal strain measurement, the strain of the measuring grid generates a change in resistance proportional to its gage factor:

$$k\,\varepsilon = \frac{\Delta R}{R} \tag{2.49}$$

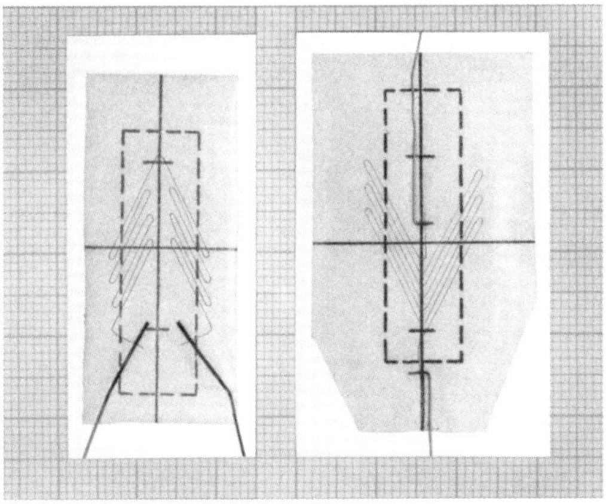

Fig. 2.66 First stress gages with wire grids from the early 1940s

Fig. 2.67 Stress gage with foil grid in V-shape $2\varphi = 57°$ with $\nu = 0.295$

With stress gages, the stress σ_x in the measuring direction x is proportional to the strain ε^* of the slanted grid, as described by Eq. (2.48). Combining Eqs. (2.48) and (2.49) when setting $\varepsilon = \varepsilon^*$, the relative change in the resistance of the stress gage becomes a function of this stress:

$$\frac{\Delta R}{R} = \frac{1 - \nu}{E} \sigma_x k \tag{2.50}$$

Thus, the desired correlation has been found. One can easily determine the stress from the measured change in resistance. In order to facilitate further evaluation, the manufacturer of a stress gage can use the expression $k(1 - \nu)$ and summarize it as a gage parameter where for $\nu = 0.3$ and $k = 2$ it takes a value of 1.4.

The possibility of free configuration of the foil gage led to a further form of stress gage containing two orthogonal measuring grids. Figure 2.68 shows such an arrangement (BLH, 1961) [2.154]. The ratio of the resistances of the two grids is determined by Poisson's ratio of the measurement object material; for example, with a Poisson's ratio of 0.28 and a resistance of the longitudinal grid of 350 Ω the resistance of the cross-grid would be 98 Ω.

The two grids are connected in series for stress measurement. The gage shown in Figure 2.68 also offers the possibility of measuring the single strains in both grid

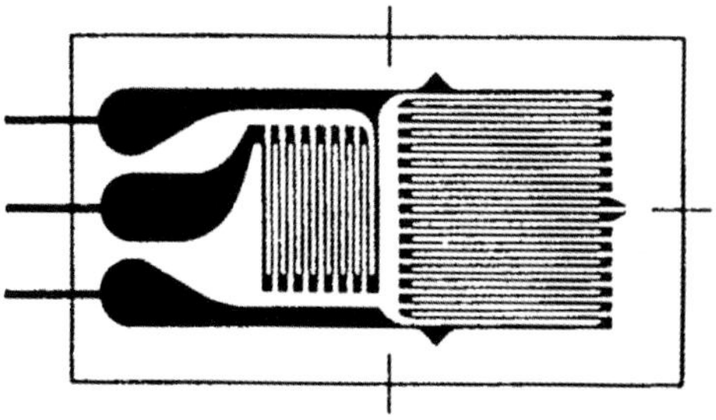

Fig. 2.68 Stress gage with T-shaped arranged foil grids

directions by appropriately arranged solder tabs. As parameters of such strips a 'stress gage factor' F_C and two gage factors for the individual grids are specified. The stress to be determined in longitudinal direction x of the strip is given by:

$$\sigma_x = \frac{\Delta R}{R} \cdot \frac{E}{F_C} \tag{2.51}$$

with $F_C = k(1 - \nu)$. Such gages also were offered as self-temperature-compensating gages in adaptation to thermal expansion coefficients of different measurement object materials.

Stress can be measured directly using stress gages with only one measuring channel. It is not necessary to combine a plurality of strain results measured by multigrid rosettes to obtain the stress. This brings advantages, especially for dynamic measurements, since one can also detect fast time-dependent changes in stress with one analogue measurement channel. The disadvantage of the unknown principal directions and thus also the ignorance of the principal stresses must therefore be taken into account. Currently available instruments with faster simultaneous high-resolution analogue-to-digital conversion and digital signal processing are able to process synchronously recorded data and to display the results as time functions. Therefore stress gages are rarely used today.

Stress can also be measured directly with normal strain gages by installing two strain gages as shown in Figure 2.67 according to the arrangement of the two individual grids of the stress gage. One must first calculate the required angle matched to the Poisson's ratio of the measurement object material according to Eq. (2.47) according to which the two strain gages must be inclined to the direction of the stress to be measured. The two strain gages are then placed in series in an arm of the Wheatstone bridge circuit. In the compensation arm of the bridge circuit is then to provide an appropriate arrangement. The nominal resistance of the bridge arms will come together from the sum of the nominal resistances of the two strain gages connected in series.

But it is also possible to arrange two identical strain gages with the same nominal resistance in a T-shape on the measurement object, so that one is in alignment with the direction of the stress to be determined and the other is transverse to it. This arrangement is most easily accomplished with two-grid strain gages whose grids are offset by 90° to one another.

By means of a shunt resistor R_S on the transverse strain gage, as indicated in Figure 2.69, the resistance will be changed from that detected to the transverse strain reduced by the factor ν. The required value of the shunt resistor R_S is dependent on ν and is given by

$$R_S = \frac{\nu}{1 - \nu} R_N \tag{2.52}$$

where R_N is the nominal resistance of the strain gages used. For example, for strain gages with nominal resistance $R_N = 350\,\Omega$ and a Poisson's ratio $\nu = 0.3$, the shunt resistance $R_S = 150\,\Omega$.

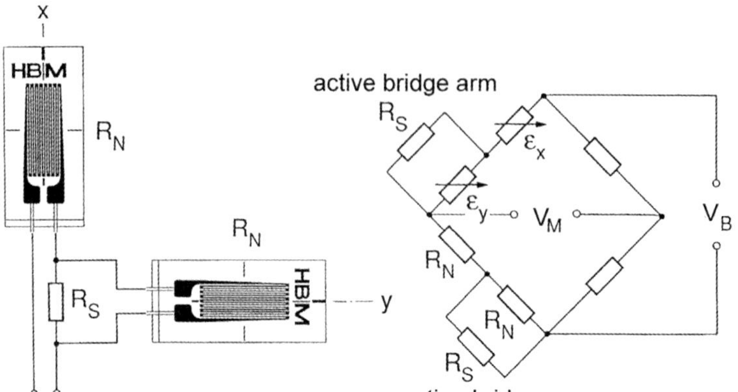

Fig. 2.69 Direct measurement of stress with two strain gages in a T-shape arrangement gages and with shunt resistor R_S across the transverse strain gage

By switching over the shunt resistor R_S from one gage to the other, with the described arrangement the stress can be measured in both directions [2.155]. Here, too, it should be noted that in the compensation arm of the Wheatstone bridge circuit the same combination of resistors must be used as in the active branch. Temperature stable resistors have to be used as shunt resistors.

References

[2.1] Keil, S. (1987) Stress calculation from measured strains in the elastic deformation ramge. *RAM*, **3**, 56–63.

[2.2] Keil, S. and Benning, O. (1979) On the evaluation of elasto-plastic strains measured with strain gages. *Experimental Mechanics*, **19**, 265–270.

[2.3] Müller, R.K. (1966) Der Einfluß der Meßlänge auf die Ergebnisse der Dehnungsmessungen an Beton. *MTB*, **2**, 37–41.

[2.4] Lukas, J. (1988) Development of resistance strain gauges in Czechoslovakia, *Proc. IMEKO XI World Congress Houston*, 50-year Jubilee, pp. 35–39.

[2.5] Thamm, F. (1988) Early strain gage instrumentation and application in Hungary, *Proc. IMEKO XI World Congress Houston*, 50-year Jubilee, pp. 113–121.

[2.6] Wieringa, H. (1988) The history of strain gauges in the Netherlands, *Proc. IMEKO XI World Congress Houston*, 50-year Jubilee, pp. 179–184.

[2.7] Bray, A. (1988) Accounts witnessing the use of strain gages in Italy, *Proc. IMEKO XI World Congress Houston*, 50-year Jubilee, pp. 179–184.

[2.8] Maslen, K. (1988) Early days with strain gauges, *Proc. IMEKO XI World Congress Houston*, 50-year Jubilee, pp. 179–184.

[2.9] Fristedt, K. and Scherling, G. (1988) The resistance strain gauge in Sweden, the early history, *Proc. IMEKO XI World Congress Houston*, 50-year Jubilee, pp. 229–238.

[2.10] Ludvig, G. (1988) The woven strain gage, *Proc. IMEKO XI World Congress Houston*, 50-year Jubilee, pp. 119–130.

[2.11] Jones, H. and Wilde, J.R. (1955) Improvements in strain gauges; UK-Patent, 723,370.

[2.12] Keil, S. (1988) On the strain gage's 50th Jubilee – a review of its evolution and of 33 years strain gage production at Darmstadt. *RAM*, **4** (2), 39–43.

[2.13] Czerlinsky, E. (1938) Untersuchungen über die Widerstandsänderungen von Drähten durch Zug; Jb. der Dtsch. Luftfahrt-Forschung, Abt. II, pp. 377–380.

[2.14] Amberg, C. and Czaika, N. (1981) Über das langzeitige Drift- und Kriechverhalten gekapselter aufschweißbarer Dehnungsmeßstreifen mit NiCr-Meßdraht bei Temperaturen bis 320°C. *VDI-Berichte*, **399**, 105–111.

[2.15] Denyssen, I.P. (1963) Strain Gage and Method of Manufacture; US-Patent 3,078,431, applied for 8.7.59, granted 1963.

[2.16] Amberg, C. and Czaika, N. (1982) On long-term behavior of encapsulated, weldable half-bridge strain gages with PtW-wires at temperatures up to 550°C, *Proc. 7th Int. Conference on Stress Analysis Haifa*, pp. 237–255.

[2.17] Andreae, G. and Niessen, G. (1985) Über Möglichkeiten und Grenzen der Hochtemperatur-Dehnungsmeßstreifen mit Platin-Wolfram-Leiter. *Materialprüfung*, **27**, 344–346 and 381–383.

[2.18] DIN 1319 (1980) Grundbegriffe der Meßtechnik, Teil 2, Januar.

[2.19] VDI/VDE-Richtlinie 2635: Dehnungsmeßstreifen mit metallischem Meßgitter, Kenngrößen und Prüfbedingungen; Beuth-Vertrieb GmbH Berlin und Köln (1974)

[2.20] ASTM Specification E-251 (1980) *Tests for Performance Characteristics of Bonded Resistance Strain Gages*, ASTM.

[2.21] OIML Intern. Recommendation No. 62: Performance Characteristic of Metalic Resistance Strain Gauges (1985)

[2.22] Kammer, E.W. and Pardue, T.E. (1949) Electric resistance changes of fine wires during elastic and plastic strains. *Proceedings of the Society for Experimental Stress Analysis*, **7** (1), 7–20.

[2.23] Bridgman, P.W. (1932) The effect of homogeneous mechanical stresses on the electrical resistance of crystals. *Physical Review*, **42**, 858–863.

[2.24] Bridgman, P.W. (1957) Effects of pressure on binary alloys. *Proceedings of the American Academy of Arts and Sciences*, **84**, 131–177.

[2.25] Druyvestein, M.J. (1951) The variation of the resistance of some metals with elastic deformation. *Physica* **XVII**, (8), pp. 748–760.

[2.26] Kuczinski, G.C. (1954) Effect of elastic strain on the electrical resistance of metals. *Physical Review*, **94**, 61–64.

[2.27] Bertodo, R. (1959) Development of high-temperature strain gauges. *The Proceedings of the Institution of Mechanical Engineers*, **173**, 605–622.

[2.28] Rohrbach, Chr. (1989) *Handbuch für Experimentelle Spannungsanalyse*, VDI-Verlag, Düsseldorf, p. 493.

[2.29] Hoffmann, K. (1976) Zum Linearitätsfehler bei Dehnungsmessungen mit Dehnungsmeßstreifen im Hochdehnungsbereich. *MTB*, **12, 3**, 53–57.

[2.30] Rees, D.W.A. (1966) The sensitivity of strain gauges when used in the plastic range. *International Journal of Plasticity*, **2** (3), 295–309.

[2.31] Andreae, G. (1971) Über das Verhalten von Dehnungsmeßstreifen bei grossen Dehnungen. *Materialprüfung*, **13**, 117–123.

[2.32] Perry, C.C. (1984) The resistance strain gage revisited. *Experimental Mechanics*, 286–299.

[2.33] Kockelmann, H. and Diem, H. (1988) Untersuchungen zum Linearitätsfehler von Dehnungsmeßstreifen im Hochdehnungsbereich und Vergleich mit anderen Meßmethoden. *VDI-Berichte*, (679), 145–160.

[2.34] Khan, A.S. and Huang, S. (1991) Variation of electrical resistance strain gage factor during large dynamic plastic deformation, *Proc. SEM Spring Conf. Milwaukee*, pp. 791–795.

[2.35] Transverse sensitivity of strain gages; unpublished results of internal research by HBM, Darmstadt (1972)

[2.36] Baumberger, R. and Hines, F. (1944) Practical reduction formulas for use on bonded wire strain gages in two dimensional stress fields. *Experimental Stress Analysis*, **I** (1), 113–127.

[2.37] Wu, C.T. (1963) *Transverse Sensitivity of Bonded Strain Gages*, Electronics Division of Baldwin-Lima-Hamilton Corp., Waltham, Massachustts.

[2.38] Gustafsson, G.V.A. (1950) Strain gauges. *Ingeniøren*, 546–549.

[2.39] Huggenberger, A.U. (1952) Über die Querdehnungsempfindlichkeit von Dehnungsmeßstreifen und ihre Beseitigung, in *Grundlagen und Anwendung des Dehnungsmeßstreifens* (ed. K. Fink) Verlag Stahleisen, Düsseldorf, pp. 60–78.

[2.40] Perry, C.C. (1969) Errors due to transverse sensitivity in strain gages, *Proceeding of SESA*, 9–19.

[2.41] (1998) German Patent DE 198 25 411 C2 Dehnungsmesssteifen und Verfahren zur Querempfindlichkeitskompensation; applied.

[2.42] Stockmann, M. (2005) *Mikromechanische Analyse der Wirkungsmechanismen elektrischer Dehnungsmessstreifen*, Habilitationsschrift TU Chemnitz.

[2.43] Stockmann, M. (2008) Cause, compensation and determination of the transverse sensitivity of strain gages; WRSG paper.

[2.44] Keil, S. and Hoffmann, K. (1973) Verstimmung der Wheatstoneschen Brückenschaltung beim messen mit Dehnungsmeßstreifen. *VDI-Z,* **115**, 904–911.

[2.45] Keil, S. (1988) The thermal output of a strain gage. *RAM,* **4** (1), 15–17.

[2.46] Ort, W. (1988) Bemerkungen zur neuen OIML-Richtlinie IR 62: Toleranzen von Folien-DMS-Kenngrößen. *VDI Berichte,* **679**, 161–171, VDI-Verlag, Düsseldorf.

[2.47] Prospectus of Jenaer Glaswerke Schott & Gen., Mainz.

[2.48] Rohrbach, Chr. and Czaika, N. (1959) Deutung des Mechanismus des DMS und seiner wichtigsten Eigenschaften anhand eines Modells. *Materialprüfung,* **1**, 121–131.

[2.49] Hönisch, G. (1966) Der Einfluß der Dehnungsübertragung zwischen Bauteil und Meßdraht auf den k-Faktor von DMS. *Maschinenbautechnik,* **15**, 61–68, 133–136.

[2.50] Hoffmann, K. (1987) *Eine Einführung in die Technik des Messens mit Dehnungsmeßstreifen,* HBM, Darmstadt, p. 45.

[2.51] Rohrbach, Chr. and Czaika, N. (1959) Über das Kriechen von Dehnungsmeßstreifen unter statischer Zugbelastung. *ATM,* 255–258. Feb. 1960 pp. 35–38, March 1960 pp. 55–56.

[2.52] Rohrbach, Chr. and Czaika, N. (1960) Das Kriechen von DMS als rheologisches Problem. *Materialprüfung,* **2**, 98–105.

[2.53] Onnen, O. and Fritz, H. (1966) Über die Bestimmung der mechanischen Spannungen bei DMS und ihr Einfluß auf die Meßgenauigkeit. *Feinwerktechnik,* **70**, 466–474.

[2.54] Onnen, O. and Fritz, H. (1968) Kriecherscheinungn bei DMS als Folge der mechanischen Spannungen in der Klebstoffschicht. *Feinwerktechnik,* **72**, 409–413.

[2.55] Onnen, O. (1969) Verbesserung der Meßgenauigkeit in der Dehnungsmeßstreifen-Technik. *Feinwerktechnik,* **73, 2**, 62–69.

[2.56] Kronmüller, H. and Onnen, O. (1974) Dehnung und Kriechen bei Dehnungsmeßstreifen. *Feinwerktechnik+Meßtechnik,* **82**, 285–288.

[2.57] Lenk, A. (1975) Dehnungsübertragung von Meßobjekten auf Dehnmeßstreifen. *Feingerätetechnik,* **24**, 535–536.

[2.58] Hoffmann, K. (1986) Die Steigerung der Meßgenauigkeit durch gezielte Beeinflussung des Kriechens von Dehnungsmeßstreifen, insbesondere beim Bau von Meßgrößen-Aufnehmern, *Proc. Simpozion National de Tensometrie, Brasov, Romania,* pp. 131–138.

[2.59] Paetow, J. Dehnungsmeßstreifen mit mäanderförmigem elektrischen Leiter; German Patent DBP, 2 049 820, applied for 10.10.1970 granted 1973.

[2.60] Technical Note G 24.01.6: Dehnungsmeßstreifen mit Zubehör; Hottinger Baldwin Messtechnik GmbH, Darmstadt (1990)

[2.61] VDI/VDE Richtlinie 2638: Kenngrößen für Kraftaufnehmer, Begriffe und Definitionen, April (1989)

[2.62] Ort, W. Meßumformer mit einer Feder und einer darauf applizierten Dehnungsmeßstreifenanordnung; Patentschrift DE, 9 16 427 C2, applied for 23.4.1979 granted 2.12.1982.

[2.63] Ort, W. (1983) New developments in foil strain-gage transducers; Experimental Techniques, August, pp. 19–23.

[2.64] Hellwig, R. (1987) Precision force transducer for international comparison measurements on force standard machines. *RAM*, **3** (1), 17–22.

[2.65] Hoffmann, K. (1986) Zur Herstellung moderner Foliendehnungsmeßstreifen und den dabei gegebenen Korrekturmöglichkeiten für Kiechen und Querempfindlichkeit. *MTB*, **22** (2), 41–46.

[2.66] Hoffmann, K. (1989) *An Introduction to Measurements using Strain Gages*, Hottingr Baldwin Mrsstechnik GmbH, Darmstadt, p. 100.

[2.67] Brunot, A.W. and Schmittner, W.G. (1946) Stress study of a fabricated steam chest. *Proceeding of SESA*, **I** (1), 49–55.

[2.68] Clough, W.R., Shank, M.E. and Zaid, M. (1953) The behavior of SR-4 wire resistance strain gages on certain materials in the presence of hydrostatic pressure. *Proceeding of SESA*, (2), 167–176.

[2.69] Majors, H. (1955) Influence of fluid pressure on SR-4 strain gages. *Proceeding of SESA*, **XII** (1), 13–24.

[2.70] Steele, M.C. and Eichberger, L.C. (1955) The use of foil gages to measure large strains under high fluid pressure. *Proceeding of SESA*, **XII** (1), 151–160.

[2.71] Guerard, J.P. and Weissmann, G.F. (1958) Effect of hydrostatic pressure on SR-4 strain gages. *Proceeding of SESA*, **XV** (1), 151–156.

[2.72] Gerdeen, J.C. (1963) Effects of pressure on small foil strain gages. *Experimental Mechanics*, 73–80.

[2.73] Milligan, R.V. (1964) The effects of high pressure on foil strain gages. *Experimental Mechanics*, 25–36.

[2.74] Brace, W.F. (1964) Effect of pressure on electric-resistance strain gages. *Experimental Mechanics*, 212–216.

[2.75] Milligan, R.V. (1965) The effects of high pressure on foil strain gages on convex and concave surfaces. *Experimental Mechanics*, 59–64.

[2.76] Montague, P. (1966 /2) The effect of hydrostatic pressures on the behaviour of foil strain gauges. *Strain*, 17–22.

[2.77] Milligan, R.V. (1967) Hydrostatic pressure effect as related to strain gages, *Proc. WRSGC Fall Meeting*, pp. 33–42.

[2.78] Kular, G.S. (1972) Use of foil strain gage at high hydrostatic pressure. *Experimental Mechanics*, 311–316.

[2.79] Schock, R.N. and Duba, A.G. (1973) Pressure effects on the response of foil strain gages. *Experimental Mechanics*, 43–44.

[2.80] Frank, W. and Krumbiegel, E. (1973) Messen mit Folien-dehnungsmeßstreifen an innendruckbeanspruchten Bauteilen. *Schweißtechnik*, **23** (10), 466–469.

[2.81] Andreae, G. (1974) Über den Einfluß hydrostatischen Drucks auf Dehnungsmeßstreifen. *Materialprüfung*, **16** (4), 98–102.

[2.82] Hoffmann, K., Jost, D. and Keil, S. (1978) Experimentelle Untersuchung des Einflusses hydrostatischen Drucks auf DMS-Applikationen und die Ermittlung von Korrekturwerten. *VDI-Berichte*, **313**, 553–558.

[2.83] Khan, A.S. and Chen, J.C. (1985) Further study on the use of foil strain gages under extremely high-pressure water environment. *Experimental Mechanics*, 123–128.

[2.84] NN (1966) BLH strain gages check forces of surveyor's landing on moon. *BLH Measurement Topics*, **4** (3)

[2.85] NN (1979) Astronauts photograph BLH strain gages on Moon. *BLH Measurement Topics*, **8** (1)

[2.86] Telinde, J. (1968) Strain gages in cryogenics and hard vacuum, *Proc WRSGC Spring Meeting*, pp. 45–54.

[2.87] Caris, R.F. (1968) Strain gages in hard vacuum, *Proc. WRSGC Spring Meeting*, pp. 29–35.

[2.88] Hoffmann, K. (1987) *Eine Einführung in die Technik des Messens mit Dehnungsmeßstreifen*, HBM GmbH, Darmstadt, p. 111.

[2.89] Kearns, C.M. and Guerke, R.M. (1937) Vibration stress measurement in strong centrifugal fields. *Journal of Applied Mechanics*, A-156–A-159.

[2.90] Kearns, C.M. (1939) Vibration characteristics of aircraft engine-propeller systems. *SAE Transactions*, **45**, 5540–5548.

[2.91] Kearns, C.M. and Guerke, R.M. Stress measuring means; US Patent Nr. 2,252,464, applied for 6.12.1937, granted 12.8.1941.

[2.92] de Forest, A.V. (1937) Measurement of propeller stresses in flight. *Journal of the Aeronautical Sciences*, **4**, 227–232.

[2.93] Fanning, R. and Bassett, W.V. (1940) Measurement of impact strains by a carbon-srip extensometer. *Journal of Applied Mechanics*, A–24–A-28.

[2.94] Theis, A. (1941) Bestimmung von Materialbeanspruchung und Untersuchung mechanischer Schwingungsvorgänge mit Streifen- und Ringgebern. *Zeitschrift für Technische Physik*, (11), 273–280.

[2.95] Prospectus with pricing: "AEG-Streifengeber"; AEG, Technisch-Physikalische Werkstätten Berlin, April (1940)

[2.96] Schmidt, R. and Klein, H. Über einige Meßgeräte und Meßmethoden für Beanspruchungs- und Schwingungsmessungen an Triebwerksteilen mit Anwendungsbeispielen; Jahrb. 1941 der deutschen Luftfahrtforschung, vol II, pp. 473–490.

[2.97] Schmidt, R. and Klein, H. (1942) Das Hamilton Meßverfahren und seine Anwendung auf Luftschrauben-Schwingungsmessungen. *Luftwissen*, **9** (3), 75–81.

[2.98] Grover, H.J. (1943) The use of electric strain gages to measure repeated stresses. *Proceeding of SESA*, pp. 110–115.

[2.99] Dohrenwend, C.O. and Mehaffey, W.R. (1943) Measureement of dynamic strain. *Journal of Applied Mechanics*, A85–A92.

[2.100] Fink, K. (1950) Eine dynamische Eichung von Dehnungsmeßstreifen. *Archiv für das Eisenhüttenwesen*, **21**, 137–142.

[2.101] Hannah, R.L. and Reed, S.E. (eds) (1992) *Strain Gage Users Handbook*, Elsevier Applied Science, p. 57.

[2.102] Rohrbach, Chr. and Czaika, N. (1961) Über das Dauerschwingverhalten von Dehnungsmeßstreifen. *Materialprüfung*, **3**, 125–136.

[2.103] Hoffmann, K. (1987) *Eine Einführung in die Technik des Messens mit Dehnungsmeßstreifen*, HBM, Darmstadt, p. 84.

[2.104] Wnuk, S. (1964) Performance characteristics of bonded strain gages under repeated cyclic strain, *Proc. SESA Meeting Cleveland, Ohio, Oct. 1964.*

[2.105] Drew, D.A. (1961) Super-sensitivity of foil gages. *Strain Gage Readings*, **I** (3), 14.

[2.106] Byre Gowda, C.V. and Topper, T.H. (1979) Performance of miniature resistance strain gages in low cycle fatigue. *Experimental Mechanics*, 27N–38N.

[2.107] Bartknecht, W. (1954) Die Aufnahme des zeitlichen Druckverlaufs von Explosionen mit Hilfe von Dehnungsmeßstreifen. *Industrie-Elektronik*, **2** (6)

[2.108] Bieber, E. (1957) Schlagkraftmessungen an Dauerschlagwerken mittels Dehnungsmeßstreifen. *Industrie-Elektronik*, **5** (1)

[2.109] Seidl, W. (1980) Möglichkeiten und Grenzen zur Messung dynamischer Spannungsintensitätsfaktoren. *Materialprüfung*, **22**, 254–259.

[2.110] Cunningham, D.M. and Goldsmith, W. (1959) Short time impulses produced by longitudinal impact. *Proceeding of SESA*, **XVI/2**, 153.

[2.111] Oi, K. (1966) Transient response of bonded strain gages. *Experimental Mechanics*, 463–469.

[2.112] Bickle, L.W. (1970) The response of strain gages to longitudinally sweeping pulses. *Experimental Mechanics*, 333–337.

[2.113] Hoffmann, K. (1987) *Eine Einführung in die Technik des Messens mit Dehnungsmeßstreifen*, HBM GmbH, Darmstadt, pp. 89.

[2.114] Harting, D.R. (1963) Strain gage as fatigue failure warning device. *Strain Gage Readings*, (6), 3–4.

[2.115] Harting, D.R. (1966) The S/N fatigue life gage: a direct means of measuring cumulative fatigue damage. *Experimental Mechanics*, 19A–24A.

[2.116] Harting, D.R. (1966) Fatigue Life Gaging Methods; US Patent 3,272,003. patented 13.th Sept.

[2.117] Beyer, G.A. (1968) Fatigue life gage evaluation on operational aircraft, *Proc. ISA Conf. New York City*, Paper 68–527.

[2.118] Krevitt, R. (1968) Fatigue life gages on a 46,000 kVA generator under pulsed loading, *Proc. ISA Conf. New York City*, Paper 68–526.

[2.119] NN (April 1969) *S/N Fatigue Life Gage Applications Manual*, 2nd edn, Micro-Measurements, Romulus.

[2.120] Dorsay, J. (1971) Engineering concepts in fatigue life gage use, *Proc. SESA Session Salt Lake City*, pp. 1–19.

[2.121] Gerharz, J.J. (1977) Unterlagen für die Verwendung von Ermüdungsmeßstreifen in der Anlagentechnik; LBF-Bericht 3748, Laboratorium für Betriebsfestigkeit Darmstadt.

[2.122] Sheth, N.J., Bussa, S.L. and Mercer, N.M. (1973) Determination of accumulated structural loads from S/N gage resistance measurements, *Proc. Int. Automotive Engineering Congress, Detroit.*

[2.123] Rieger, H., van Rensen, E. and Sahm, K.F. (1979) Ein neuartiges Verfahren zur Lebensdauerüberwachung wechselbelasteter Bauteile. *Zeitschrift für Werkstofftechnik*, **10**, 353–360.

[2.124] Müller, R.K. (1964) Ein Beitrag zur Dehnungsmessung an Kunststoffmodellen; habilitation thesis TH stuttgart.

[2.125] Kreuzer, M. (1977) Untersuchung über das Einlaufverhalten von DMS infolge der Aufheizung nach dem Anschalten; unpublished research report HBM GmbH.

[2.126] Chalmers, G.F. (1982) Materials, construction, performance and characteristics, in *Strain Gage Technology* (eds A.L. Window and C.G. Holister), Applied Science Publishers LTD, p. 30.

[2.127] Procopec, M. (1974) Strain gauge measurement on plastics. *Strain*, 17–21.

[2.128] Zhang, T., Evans, J.R.G. and Rees, D.W.A. (1992) The heating effect of electrical strain gauges applied to low thermal conductivity materials. *Strain*, 107–112.

[2.129] Bröker, F.-W. (1985) Dehnungsmessungen an Holz mit direkt applizierten DMS. *MTB*, **21** (1), 18–23.

[2.130] Little, E.G. (1982) Effects of self heating when using a continuous bridge voltage for strain gauging epoxy models. *Strain*, 131–135.

[2.131] Grünbaum, P. (1992) Measurement of forces in a rock-wool hardening oven at 230°C. *RAM*, **8** (2), 25–30.

[2.132] Bertodo, R. (1959) Development of high-temperature strain gauges. *The Proceedings of the Institution of Mechanical Engineers*, **173** (23), 605–622.

[2.133] Rohrbach, Chr. and Knublauch, E. (1968) Dehnungsmessung mit Meßstreifen bei hohen Temperaturen. *Materialprüfung*, **10** (4), 105–115.

[2.134] Andreae, G. and Niessen, G. (1982) Über die Ursache des Driftens von Hochtemperatur-Dehnungsmeßstreifen, insbesondere solcher mit PtW-Leiter. *Materialprüfung*, **24**, 431–436.

[2.135] Hobart, H.F. (1985) Evaluation results of the 700°C Chinese Strain Gauges. *NASA Technical Memorandum*, **86973**.

[2.136] Lei, J.-F. (1991) A resistance strain gage with repeatable apparent strain to 800°C. *Experimental Techniques*, 23–27.

[2.137] Ziegler, K. (1982) Einsatzproblematik und Meßsicherheit von Hochtemperatur-Dehnungsmeßsreifen unter verschiedenen Temperaturgrenzbedingungen bis 1000 °C. *VDI-Berichte*, **439**, 31–38.

[2.138] NN: (1980) *Schweißbare und einbettbare Ailtech-Dehnungsmeßstreifen*, Catalogue of Micro-Epsilon Measurement GmbH.

[2.139] Böhm, W. (1983) Hinweise zur Anwendung des 2fachen Anschweißens von EATON-Hochtemperatur Dehnungsmeßstreifen zur Voruntersuchung von DMS-Eigenschaften. *VDI-Berichte*, **480**, 167–169.

[2.140] Amberg, C. and Czaika, N. (1973) Über die Anwendbarkeit von Hochtemperatur-Dehnungsmeßstreifen und Transformator-Wegaufnehmern für die Instrumentierung von Spannbeton-Reaktordruckbehältern, *Proc. 2. Int. Konf. "Structural Mechanics in Reactor Technology" Berlin*, Paper H 4/8.

[2.141] Hofstötter, P. and Weichsel, J. (1980) Einsatz einer Klemmvorrichtung zur Kalibrierung von Hochtemperatur-Dehnungsmeßstreifen. *TÜ*, **21**, 147–150.

[2.142] Hofstötter, P. (1982) Calibration of high-temperature strain gages with the aid of a clamping device. *Experimental Mechanics*, **22**, 223–225.

[2.143] Amberg, C. and Czaika, N. (1983) Zur Vorherbestimmung und Reproduzierbarkeit des Nullpunkttemperaturganges aufschweißbarer DMS bis 320 bzw. 600°C. *VDI-Berichte*, **480**, 145–150.

[2.144] Nightingale, C. and Twa, G. (1982) Evaluation of precalibration techniques for Ailtech SG 425 Strain Gages, *Proc. WRSGC, Febr.*

[2.145] Böhm, W. and Rasche, N. (1980) Bestimmung von DMS-Eigenschaften bei höheren Temperaturen mit einer rechnergesteuerten Prüfeinrichtung. *VDI-Berichte*, **366**, 51–53.

[2.146] Böhm, W., Böres, H.J. and Zemann, R. (1985) Statische Hochtemperaturmessungen mit Telemetrie-Meßwertübertragung an einem Gasturbinenläufer. *VDI-Berichte*, **522**, 195–211.

[2.147] Thiem, H. and Lemcke, A. (1992) *Dehnungsmessung bei thermischer Wechselbelastung*, Messen & Prüfen, pp. 20–24.

[2.148] Ludewig, H. (1969) Schwingungsmessungen an Turbinenschaufeln von Abgasturboladern mit Hochtemperatur-Dehnungsmeßstreifen. *MTB*, **5** (1), 8–11.

[2.149] Tschierschke, P. (1968) Messung der behinderten Wärmedehnung bei Temperaturen bis zu 650 °C. *MTB*, **4** (3), 46–49.

[2.150] Kern, R.E. (1946) The stress gage. *Proceedings of SESA*, **4**, 124–129.

[2.151] Perry, C.C. (1953) Principal stresses measured directly with SR-4 strain gages. *BLH Testing Topics*, **8** (4).

[2.152] Perry, C.C. and Lissner, H.R. (1955) *The Strain Gage Primer*, McGraw-Hill Book Company, p. 146.

[2.153] Sevenhuijsen, P.J. (1984) Stress gages. *Experimental Techniques*, 26–27.

[2.154] NN: New SR-4 stress-strain gage; Product Data 4323, BLH, May (1961)

[2.155] Hehn, K.-H. and Schulz, W. (1966) Brückenschaltungen mit DMS zur unmittelbaren Messung der Material-Spannung. *MTB*, **2** (2), 18–21.

3 Installation of strain gages

3.1 Preparatory work

3.1.1 General preparation

A strain gage can only perform its task properly if the strain to be measured is transferred faultlessly and free of loss. For this purpose, an intimate connection between the strain gage and the measurement object is required. This connection is best performed by using special adhesives. Some bonding agents and methods are limited to special applications such as ceramic bonding agents for high temperature applications or spot welding. Consequently the quality of an installation greatly influences the accuracy and reliability of the measurement.

The important characteristics of a strain gage, such as gage factor and thermal output and creep are basically characteristics of the installation as a whole, including all the components that are involved at the measurement point: the strain gage itself, the cleaning agent used, the adhesive, the solder and the kind of lead connection, and finally the material and the method of protection. A reliable result requires an error-free installation of the strain gage, in which case the proper use of tools and careful preparation of the measuring point play a crucial role. Mistakes are often momentous, and they can lead to erroneous measurement results or failure of the measuring point. This can require, particularly for complex large-scale tests, considerable additional effort.

This raises the question of the quality of the strain gages used. The use of strain gages from a high quality production method pays off in the majority of applications, even if these strips are somewhat higher in cost due to the higher production expenses. Overall, the cost of a strain gage is small compared to the total cost of the installation of a measuring point. What is more, the failure of a measurement can be expensive if it interrupts an experimental test procedure and requires new technical installation skills, both of which can be acquired in training courses and seminars offered by various institutions and companies.

Before starting the actual installation work – colloquially 'the bonding' – a number of preparatory works should be done. This includes setting a measurement point, laying down a wiring plan and creating a favourable workplace for the installation with easy access to all the necessary tools. In this case, the pursuit of cleanliness should be prioritized, because even the slightest contamination of the surfaces at the test site or of the strain gage reduces the adhesive effect and can lead to incorrect measurements or even failure of the measurement point. Covering layers adhere imperfectly on unclean surfaces and allow the infiltration of moisture or other contaminants at the boundary layer. All required tools should be ready to use. Experienced measurement technicians carry an installation case, whose contents are often compiled specifically and individually on the basis of their own experience.

Technology and Practical Use of Strain Gages: With Particular Consideration of Stress Analysis Using Strain Gages. First Edition. Stefan Keil.

3.1.2 Preparation of the strain gage

Preparatory work needs to include the preparation of the strain gage, which can be run in the clean environment of a laboratory. This preparation should allow the strain gage, at the measuring site, to be removed from a storage container in a state 'ready for installation' and fixed on the test object by means of a previously attached installation aid. As an installation aid a tape is normally laid across the strain gage, protruding gage on one side by about 15 mm. On the other side of this tape it must release the edge of the strain gage so that the edge can be accessed in the subsequent bonding process by the adhesive. During the preparation the strain gage is put on a clean, not too thin plastic sheet and stuck with the transverse tape in such a way that it can be folded like a hinge. At the measurement site the strain gage is removed from the plastic sheet with a pair of tweezers, aligned on the prepared measurement surface of the test object and fixed there using the tape across the strain gage.

Strain gages without integrated solder tabs have connecting ribbons, which can then be connected to solder terminals. The solder terminals are used for stress relief of the connections. The solder terminals should be attached to the strain gages as part of the preparatory work and then be installed with the strain gage. The carrier film of the appropriately tailored solder terminals should be pushed between the base film and the connecting ribbons of the strain gage. Neither the strain gage nor the solder terminals should be touched with the bare hand. Preferably this work should be carried out with the aid of tweezers. Then, the connection ribbons are trimmed to length. The soldering terminal has to be fixed with a narrow tape to the strain gage. This tape should completely cover the solder points. The edges of the carrier must remain free. The carrier foils of the strain gage and of the terminal should overlap to avoid contact of the connector ribbons to the measurement object. The strain gage, thus prepared, together with the terminal will be stuck provisionally on an auxiliary foil. Thus the strain gage is ready for easy installation. Figure 3.1 shows the prepared strain gage ready for installation. This brings particular advantages when strain gages need to be installed in industrial environments or outdoors under aggressive environmental conditions in the shortest possible time.

Good quality strain gages, when delivered, usually have clean adhesive surfaces and can be installed without cleaning measures. They should therefore only be touched with tweezers and not with the fingers. If a strain gage on its adhesive surface is touched by fingers or otherwise contaminated, it should be cleaned with a solvent-soaked cotton swab.

3.1.3 Preparation of the measurement point

The basic condition for the preparation of a measurement point is a smooth clean bonding surface. This surface should be roughened to some extent, depending on the bonding agent and the material. The bonding surface does not need to be flat, but it must be free of any irregularities such as small bumps or pores. Depending on the strain gage type the installation is also possible on a more or less curved surface. Always the bonding surface must be clean and absolutely free of grease.

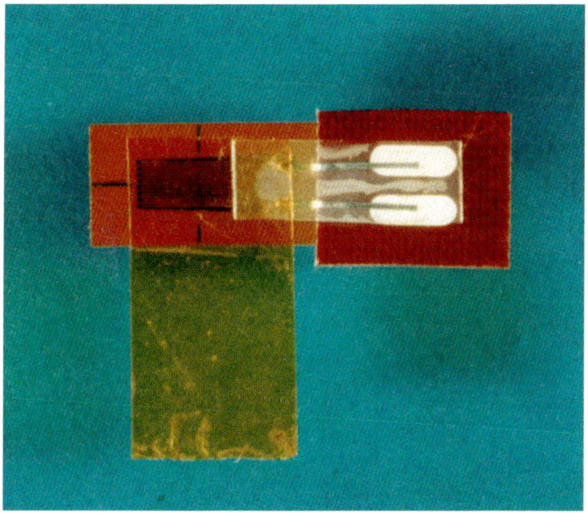

Fig. 3.1 Strain gage with connection ribbons prepared for installation in conjunction with soldering terminal

Strain gages should always be attached to the base material of the test object. Layers of paint, rust, roll scale, grease, and so on must be carefully removed from the surface at the point of measurement with plenty of outstanding region on all sides. Before starting with the coarse mechanical smoothing, traces of oil, grease and lubricants must be completely removed. Bumps have to be smoothed away by sanding. The smooth cleaned measuring surface should be slightly roughened to improve adhesion. An almost perfect device is sandblasting with corundum, and on particularly hard surfaces by using boron carbide. Required for sandblasting are oil-and water-free compressed air and unused abrasive. You should never sandblast a not completely degreased surface, since particles hammered by the blasting agent into the surface are very difficult to remove.

The mean roughness height of the roughened surface should, for the use of cyanoacrylate or epoxy resin adhesive, be 2–4 µm, for methyl acrylate 4–10 µm and for ceramic putty 10–20 µm. However, it is not always possible to roughen the bonding surface by means of sandblasting. For sandblasting, the appropriate device must be accessible, but in the vicinity of the measuring point there may be certain parts such as bearings or sliding surfaces which prohibit the use of sandblasting. In that case sandpaper has to be used, making sure that there is no preferred direction in the grinding pattern. The question of grain sizes is dependent on the measurement object material and the adhesive applied; these lie in the 80–360 µm range. For such preparations of the bonding surface experience and personal judgement of the technician play a major role, typical values are therefore always specified with some reservation.

At this mechanically prepared measurement surface a very careful cleaning process now has to be performed. This should be carried out with solvents such as trichloroethane, acetone or isopropyl alcohol in a chemically pure form. When working with these solvents it is important to know that they can be unhealthy or flammable and that the solvent must not be contaminated. Thus, one should pour the necessary amount of cleaning fluid from the stock bottle into a Petri dish, in which the cleaning swab can be immersed. Each swab is only immersed in the solvent once, since this otherwise would bring contamination back into the fluid; the swab must be thrown away after it has been used on the measurement surface. As swabs are preferably suitable cellulose tabs which are wound around the tip of tweezers. Particularly critical are placing on the bonding surface any type of silicone grease, which are often components of skin creams. Under no circumstances should you put your fingers in the solvent. Due to the latent heat of evaporation of the solvent, moisture can condense on the measurement surface. This can best be eliminated with a hair dryer. The strain gages should be applied to the cleaned surface as soon as possible gages. It is definitely not a good idea to take a break from work after the successful cleaning, and then make the bond later.

Caution is advised when cleaning plastic parts, as the cleaning agent can damage the plastic surface by partially dissolving.

For porous surfaces, such as cast iron and titanium, it is recommended that impurities that are possibly seated in the pores (e.g. hydrocarbons, cutting oil, lubricants) be driven out by heating. This should be done before the final cleaning. It is recommended to heat the surface for at least two hours at 175 °C.

3.2 Methods of fastening

3.2.1 Requirements

Bonding of the strain gages on the test object with an adhesive is the most obvious method of attachment. When the strain gage was invented by Prof. Ruge, he used household cement to bond resistance wires to the test object [3.1]. In addition to its fixing function, the insulating property of the adhesive was also used. The first commercially produced strain gages had a paper carrier and were fixed with nitrocellulose which cures physically by evaporation of the solvent. With the introduction of new carrier materials based on plastic, chemically cured adhesives were then required, because the impermeable backing material did not allow evaporation of the solvent. The first strain gages in Germany produced gages used phenolic resin carriers. For their installation a cold-curing chemically cured adhesive was then urgently needed, since the available cellulose adhesives were unsuitable for phenolic resin strain gages. In 1960, in search of suitable adhesives, the polymath acrylate adhesive X57 was found, which is regarded as a forerunner of the later adhesive X60, which was very successful in Germany. The path to these adhesives came through dentistry, where similar cold-curing, two-component, cold-hardening putty is used as filler for holes in human teeth [3.2].

Today proven adhesives are available in sufficient numbers for strain gage installation. These adhesives are exclusively chemically cured adhesives, which are characterized by good resistance to moisture. Strain gages with cellulose carriers are now only used in exceptional cases and fixed with physically cured adhesive. An adhesive must be able to faithfully transfer the strain changes in the surface of the measurement object, without elastic or plastic after-effects, to the carrier of the strain gage. The adhesion of the carrier on the measurement object is based not on the primary chemical bonds, but mainly on secondary bonds, which are attributable to the valences of the molecules involved. The Van der Waals forces acting thereby arise from interactions between involved molecules that form electric dipoles which cause adhesion in the boundary layers. The mechanical adhesion by capillary action or as a result of the teeth of rough surfaces is only of very minor importance. Too high roughness of the surfaces is rather detrimental, since thereby there is a risk that too low in pores or grooves air remains trapped, and prevents the adhesive to the full filling of the deepenings. A fine grain, however, increases the usable surface area, and thus the Van der Waals forces [3.3].

The adhesion caused by the gluing is limited to thin molecular layers in the border zones of the components involved. The transfer of deformation within the adhesive layer must not be influenced by creep or slip. Ideal for this is a 'hard' set adhesive in as thin a layer as possible. The finite value of the modulus of elasticity of the adhesive and any plastic deformation can cause reductions in the transferred strain, particularly in the areas of grid ends and the ends of the carrier. These reductions increase with the thickness of the adhesive layer. For transferring very large strains in the percentage range 'soft' set adhesive is required because hard glue becomes brittle at large strains and tends to crack.

The main criteria for the selection of an adhesive are the carrier material of the strain gage, the material of the measurement object, the measuring temperature and possible special requirements such as the recording of large strains, or installation on highly curved surfaces. The required measurement certainty and the long-term stability are also criteria for the installation of strain gages. The cost of strain gage installations for long term use of precision transducers will be relatively high compared to a strain gage installation only used for short-term stress analysis.

The requirements of an ideal adhesive for strain gage installation are, in no particular order:

- optimum transmission of strain in the specimen to the strain-sensing grid
- long-term stability
- good workable consistency to form a uniform thin adhesive layer without air inclusions
- compatibility to both the gage carrier material and the material of the specimen
- acceptable pot time and fast curing
- applicable to all kinds of carrier and test object material
- simple and straightforward processing at room temperature without complex combination and mixing of different components
- possibility of positioning correction of the strain gage during fixing
- capable of high elongation, free of creep and hysteresis
- operation over a wide temperature range from cryogenic to high temperature
- high resistivity for good isolation resistance between grid and specimen.

The ideal adhesive fulfilling all these requirements has not yet been invented. Depending on the measurement task the user has to make compromises which do not affect the quality of the measurement in most cases, but can lead to increased costs in individual cases, such as when using a hot-curing two-component adhesive that requires contact pressure for curing. In any case the user should follow carefully the operating instructions of the manufacturers. Especially care is needed to avoid technologically incompatible combinations of adhesive and carrier or measurement object material.

To best meet the different application requirements various types of adhesives are available, which differ not only in their material composition but also their processing temperature and number of components. There are cold- and hot-cured adhesives, which may consist of one or more components.

In the following the main types of adhesives and mounting methods are briefly described.

3.2.2 Cyanoacrylate

Cyanoacrylates are fast polymerizing cold-curing one-component adhesives for strain gages. They are relatively runny and easy to use on horizontal surfaces. Their application on steeply sloping areas and overhead is problematic due to the thin liquid. They harden under thumb pressure in less than a minute in a very thin layer – less than 10 μm. Cyanoacrylates are compatible with almost all metals and many plastics. They can be used with strain gages with polyimide and phenolic resin carriers. Cyanoacrylates need a relatively smooth surface and are not suitable for porous measurement object materials such as ceramics or concrete as they can't be used as a filler for smoothing irregularities. They only completely cure in thin layers, so too roughened surfaces should be avoided.

Humidity acts as a catalyst for the polymerization of cyanoacrylates. For the processing of this adhesive the relative humidity should be between 40% and 70%. At values below 30% the reaction is much slower, and with dry air it is almost impossible. With too high humidity (80% or more), the curing is accelerated to such an extent that the application of the adhesive is almost impossible. The normally short curing time of less than one minute requires a high processing speed, which in turn requires good accessibility to the measuring point and a good installation routine.

The strain gage is immovably stuck immediately after the first contact with the adhesive layer, so an alignment during the pressing process is not possible. One must therefore provide appropriate aids (e.g. hinge-like fixation with tape before applying the glue layer). Furthermore, it should be noted that basic components of the surface accelerate the polymerization, but acids retard and may even prevent it. For acidic surfaces, therefore, a neutralizer should be applied.

The majority of all currently (2013) performed strain gage installations carried out on metal uses cyanoacrylate adhesives. Therefore, in Figure 3.2 a standard example is

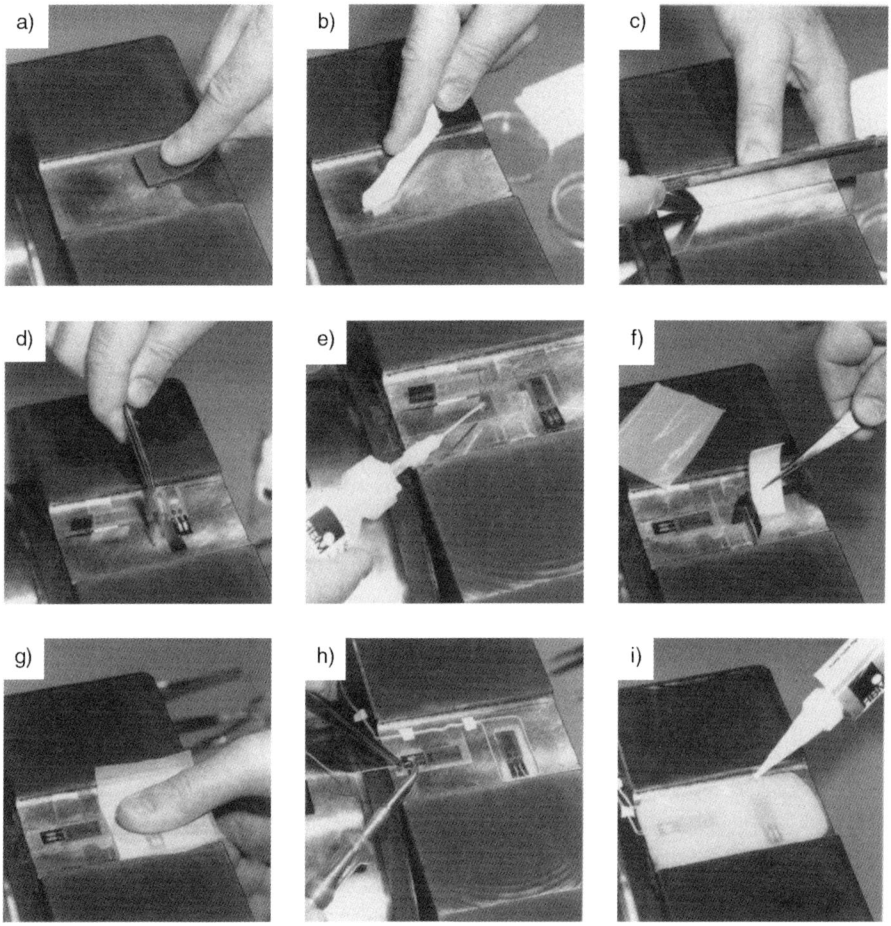

Fig. 3.2 Installation steps when sticking a foil strain gage to a test object made of steel using cyanoacrylate adhesive: a) roughen the cleaned and degreased measuring point b) careful clean the area c) mark the intended strain gage position d) position the strain gage prepared accord to Figure 3.1 gage e) apply the cyanoacrylate adhesive f) spread the adhesive to a thin layer with a Teflon strip g) press the strain gage with the thumb (interposed with a Teflon film) h) solder the contact ribbons and the lines on the solder terminal i) cover after careful removal of the residues of the flux

shown, in the form of a series of images, of the work steps to be performed when installing a foil strain gage with cyanoacrylate adhesive.

Cyanoacrylate adhesive should not be used for long-term measurements. Their application temperature range is between −30 °C and 100 °C. The maximum elongation allows measurements of strain up to about 15%. Measurement points with cyanoacrylate should be very well protected against moisture. When working with this adhesive it should be noted that it also has excellent adhesive properties with the human skin! Carelessness can result in an adhesive joint between two fingers that is only releasable by sub-skin damage. When pressing the strain gage with the thumb, always a thumb protective Teflon film to interpose. The instructions attached to the adhesive pack by the manufacturer should be carefully followed.

In a recent biomechanical study [3.4] the cyanoacrylate adhesive has been successfully used for the installation of strain gages on fresh bovine bones. Due to an interruption of studies, a bone was frozen with its measuring points for four weeks. After thawing the readings determined before freezing were reproducible again.

3.2.3 Methyl acrylate

In 1957, in search of a suitable adhesive for strain gages with phenolic resin carrier, methyl acrylate adhesive was discovered being used in the dental field gages. Since then, under the symbol X60 displaced cold-curing two-component adhesive is applied to the strain gage technique in large dissemination. It consists of a liquid and a powder component which are stirred with one another just before use. Getting the correct mixture ratio is quite simple, according to the instructions supplied with the adhesive. The adhesive must have, when ready to use, a 'stringy' consistency.

The advantages of methyl acrylate adhesive are fast measuring readiness, ease of use and wide applicability to almost all structural materials such as metals, and many plastics such as acrylic, PVC, polyester resin (also GF reinforced), polystyrene, epoxy resin, phenolic resin with a sanded surface and Teflon that has been made tacky by sodium treatment. Polyethylene and untreated Teflon won't stick. With methacrylate adhesive, strain gages can be installed on surfaces of glass, porcelain and enamel, and you do not have to roughen the surface, but it does need to be absolutely free of grease. Also concrete, wood, bone and ceramics strain gages can easily be stuck with methacrylate adhesive. Wood should be sanded dry with glass or flint paper before gluing. In concrete the possibly existing to plug up pores with adhesive avoiding air bubbles. In other porous surfaces such as ceramics, the adhesive can be used by brushing and grinding it smooth so as to form a closed planar surface, onto which the strain gage may be stuck. However, it should be noted how the reinforcing effect of the adhesive layer reacts to the mechanical properties of the measurement object. The elastic modulus of the cured methacrylate X60, depending on the mixing ratio, is between 4500 and 6000 N/mm^2 [3.5].

Measuring points installed with methacrylate adhesive can be operated at temperatures in the range between −200 °C and 50 °C. This adhesive is preferably used for measurements in stress analysis. Advantages are the small amount of effort on mixing the two components and the pot life in room temperature for about three minutes, that is long enough to perform even complicated installations at locations with restricted access and not too long in terms of cost-generating waiting times [3.6]. The adhesive does not cure immediately like cyanoacrylate, but allows slight corrections of the orientation of the strain gage during pressing. Depending on the ambient temperature, curing takes place in a matter of minutes, but at 20 °C the thumb pressing the strain gage can be lifted off after about one minute, by which stage the adhesive has sufficient grip that it can be to itself. As a release agent between the thumb and strain gage a cellulose film should be used, which can be easily pulled away from the joint after lifting the thumb. The incipient hardening can be readily seen by the consistency of the adhesive residue remaining in the mixing pot.

Full curing is only completed after about 30 minutes at room temperature, after which the measuring point is ready for measurement. At lower temperatures, the full cure takes a little longer; at higher temperatures it is faster. (Reference values: at 0 °C about 90 mins; at 30 °C for about 10 mins). When used properly, the thickness of the adhesive layer after curing is about 60 µm. After removal of the cellulose film, the soldering for connecting the soldering terminals and cables can be started. Connecting ribbons covered by the adhesive can be melted free with the soldering iron (the adhesive softens when heat) and then separated with tweezers from the adhesive.

The maximum extensibility of methyl acrylate adhesive X60 when used properly is well over 10%. The adhesive surface of the test object should be lightly brushed, for example by sandblasting with sandblasting grit 80 . . . 100 or with 180 grit sandpaper.

In addition to its use as an adhesive methyl acrylate adhesive X60 is also excellent for fastening measuring cables and underlay plates with compensating strain gages and so on.

3.2.4 Epoxy adhesives

Epoxy adhesives are always two-component adhesives. They consist of the reactive epoxy resin component and a hardener, which causes the liquid or paste-like resin to polymerize. A number of different resin modifications with different curing agents are offered as adhesives for strain gages, each of which is processed by different methods. The results obtained satisfy different physical requirements depending on the components used. Their main advantages over cyanoacrylate and methacrylate adhesives are the much wider setting limits for application temperatures and their long-term stability. However, installation is a bit time-consuming and almost always requires curing at elevated temperature in an oven as well as the application of contact pressure.

Epoxy resin adhesives are offered solvent-free and also in dilute form with solution. They cure without noticeable shrinking. With the diluted components, which offer better workability the solvent has to be evaporated by pre-drying before curing. With epoxy resin, glued joints are resistant to moisture and many chemicals as well as to outgassing in high vacuum. The application temperature range is between −260 °C and 315 °C. Strain gage installations on measuring bodies of precision transducers are preferably carried out with epoxy adhesives. They have particularly good long-term stability.

With most types of epoxy resin, the hardener only acts at elevated temperatures and can even be mixed at room temperature with the resin, without polymerizing. Only when heated does polymerization. A certain amount of contact pressure is always required during curing. The most important parameters for the curing are temperature and time. In order to save time, you have to work with higher curing temperatures. The upper applicable temperature is limited by the measurement object, then at lower curing temperatures the curing time is extended. The time–

temperature combinations for each adhesive will be given in the manufacturer's instructions probably in diagrammatic form. In practice, the temperature should be applied during curing to at least 30 K above the temperature at which the measuring point will later be used. The heating to hardening temperature should be done slowly to avoid any formation of bubbles in the adhesive layer. The clips generating contact pressure are not removed before cooling down to at least 50 °C. The durability of the bond can be further improved by post-curing of the measuring point for two hours at about 30 K above the application temperature.

Measurement objects are not always small enough to put in an oven. Often the measurement engineer faces the problem of having to install strain gages with hot-curing adhesives on larger measuring objects, where the measuring points are, during the measurement, exposed to higher operating temperatures. There are then two options to solve the problem: avoid the problem by using welded strain gages attached by spot welding to the test object or the measuring point can be heated by some special procedure outside but not in an oven. This can be done by radiant heat such as infrared radiators or by inductive heating of limited areas of the test object. Here the ingenuity of the technician is crucial.

As an example an application case is cited, in which a 1.5 m long test rod needed measuring points to be installed for the highest demands of precision and long-term stability. The test rod was made of special steel with a diameter of 240 mm and was used as a measurement element for a force standard machine.

As shown in the photograph in Figure 3.3 the heating of the installation area was carried out by induction. On both sides of the installation area suitably made half-shells of copper were placed around the bar, and a current of 3600 A with an alternating voltage of 1.25 V and 50 Hz was applied to the measurement site.

For curing, the measurement site was heated to 180 °C. A thermostat was mounted at the measuring point, which interrupted the current when this temperature was

Fig. 3.3 Generation of heat required for hot curing on a large measurement object by induction

exceeded and this provided an almost constant temperature around the site. The high current required correspondingly large cross-sections for the connecting cables of the copper half-shells which are clearly visible in the photograph. The low voltage to the high current intensity is generated in a power transformer, whose primary coil is supplied with mains voltage (220 V, 50 Hz).

3.2.5 Phenolic resin adhesives

Phenolic resin adhesives are one-component adhesives that cure at temperatures of 140–200 °C. They were widely used in the 1950s and 1960s as a means of fastening strain gages with the phenolic resin carriers then much used. One example is the phenol resin adhesive BC6035 mentioned in the previous chapter. The disadvantage of the unmodified phenolic resin adhesives is the evaporation of volatiles and water vapour occurring during curing. Therefore, a very high contact pressure is required during curing, but the possibility of leakage must be given at the same time to allow the evaporating portions to escape, thereby avoiding the formation of gas bubbles or water droplets.

Modified phenolic resins are still readily used today as adhesives. Their use in solvent-thinned form is quite simple as one-component adhesive. The measuring site at the test object and the adhesive surface of the strain gage will be thinly covered with the adhesive. Then you have to wait to allow the solvent to evaporate. Subsequently, the heat sealing is done under contact pressure. To use the advantage of this possibility, the 'self- adhesive' strain gage with a glass fibre reinforced phenolic resin carrier has been developed. It is coated on its sticking surface with a phenol resin adhesive. Figure 3.4 shows a photograph of this gage.

According to the manufacturer's instructions [3.7] this strain gage has to be fixed with heat-resistant tape to the measurement site, the curing of the adhesive is then

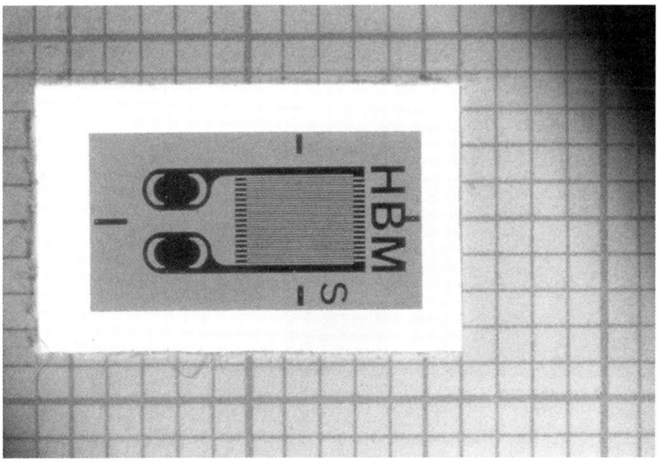

Fig. 3.4 Self-adhesive strain gage with dry phenolic resin coat

carried out for 3.5 h at 165 °C under a pressure of 2.0–3.5 bar. It is recommended after cooling under pressure, a pressure-free post-curing for one hour at 165 °C. The great advantage of these coated strain gages is not just the ease of application but also the uniform thickness of the adhesive layer and the reproducibility of the bond.

3.2.6 Ceramic putty

Ceramic putty as a means of fastening is used exclusively for free-filament strain gages. They are intended for use at the higher temperatures that organic adhesives cannot withstand. Organic carriers and adhesives are also unstable to nuclear radiation, which changes or destroys the molecular structure of plastics. The measuring points become unusable as a function of radiation intensity and time. The upper temperature limit for the application of ceramic putty for static measurements is about 650 °C. The usable temperature range is limited at the top by the falling insulation resistance of the putty.

Ceramic putty offered by the strain gage manufacturers is available in different versions, both as a two-component or even as a single-component. Again, the user should follow the specific application instructions carefully. Some of the ceramic cements may contain hazardous components such as chromium trioxide CrO_3. Appropriate protective measures (polyethylene gloves, good ventilation, protective goggles, etc.) must be used.

As an example, the installation procedure for the single-component ceramic putty CR760 [3.8] is described below. This cement consists mainly of finely powdered silicon and aluminium phosphate in a water soluble liquid. Since the solid particles settle at the bottom of the vessel during storage, the putty must be mixed well before use and during the processing steps. Ceramic putty adhesive surfaces at the test object should have a roughness of 10–20 µm, which can be achieved with emery paper of 180 grit or by sand blasting with sandblasting grit of 80–120, the coarser grain is used with harder materials and the finer grain with softer ones.

The individual installation steps are briefly shown in Figure 3.5. A thin insulation layer of about 25 µm is carefully laid at the measuring point (Figure 3.5, left). A

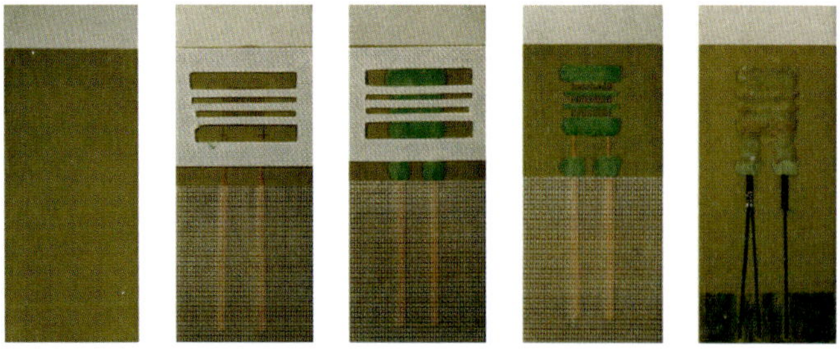

Fig. 3.5 Installation phases of a free-filament strain gage with ceramic putty, from left to right: deposited insulation layer, attaching the free filament strain gage with its auxiliary carrier, fixing the accessible grid parts with putty, detachment of the auxiliary carrier, covering the measuring point and welding of cables

practical tip to achieve a uniform distribution is to immediately wipe off the first deposited layer with tissue paper and then apply again. The primary layer is allowed to dry at room temperature (about 15 minutes), and then cured for one hour at 180 °C. On this foundation the free-filament strain gage is aligned and attached with its self-adhesive auxiliary carrier as shown in Figure 3.5 (second from left). Then, the exposed parts of the measurement grid are covered, using a brush, with ceramic putty (Figure 3.5 middle). After a drying time of 15 minutes at room temperature, curing follows in an oven (approximately 1 hour at about 100 °C). Then, the still warm auxiliary carrier is removed. This auxiliary carrier is still flexible and thus easier to take off (Figure 3.5 second from right). After cooling, the remaining positions, that were previously covered by the auxiliary carrier, are coated with putty, dried and cured. This process can be repeated a number of times to achieve as thick a cover layer as required. Generally two layers are sufficient. Subsequently, a one-hour final cure of the entire measuring point should be carried out at 180 °C. Figure 3.5 shows, at the right, the measuring point ready for measurement with connections prepared for the three-wire circuit.

Ceramic putty can also be applied to strain gages with a normal organic carrier as a cover layer to protect it, that is against short-term hot impacts from outside. Note that the maximum elongation of the ceramic putty is about only 0.5%, and that, due to its hygroscopic properties, ceramic putty can absorb moisture from the surrounding atmosphere. Therefore, is recommended to apply, immediately after curing, a waterproof protective lacquer to prevent moisture-related zero signal interference. Also, ceramic putty can crack when the temperature rises suddenly; to avoid this, temperature changes during the measurement should not be faster than about 3 K per minute. The installation of a free-filament strain gage with ceramic putty requires a lot of skill and practice.

3.2.7 Flame spraying

Flame spraying has been specially developed for the installation of free-filament strain gages and thermocouples for use at high temperatures or where there is nuclear radiation. As can be seen in Fig. 3.6 in a special spray gun, a ceramic rod is melted in a gas flame and the melt is sprayed by an air jet as fine particles onto the measuring site [3.9]. The ceramic particles strike in the plastic state on the test object and solidify immediately. The fuel gas consists of an oxygen-acetylene mixture and this generates, in the spray gun, a temperature of about 2750 °C. The spraying material consists of sintered rods of aluminium oxide, which are offered at different levels of purity. The insulation resistance between the measuring grid and the test object depends on the purity of the ceramic rod, and it decreases with increasing temperature [3.10].

The solidifying ceramic particles form a hard adherent layer which has almost ideal properties for the installation of sensors. The solidification of the sprayed particles takes place without significant heat transfer to the measurement object. Onto the cleaned and sanded surface of the measurement object a first thin insulating layer of about 80 µm is sprayed. The free-filament strain gage is attached with its self-adhesive auxiliary carrier on this layer, orientated in the desired measurement direction. The removal of the protective film from the adhesive layer of the auxiliary carrier must be made in the direction of measurement, so as not to damage the measuring grid. On the strain gage fixed by the auxiliary carrier another ceramic layer is sprayed in order to fix the measuring grid. Then the auxiliary carrier is removed. The withdrawal of this carrier is best done at 45° to the direction of measurement [3.11]. Then, a further ceramic layer is sprayed over the entire measuring point which serves as a cover layer. This cover, however, is porous and absorbs almost all materials that come into contact with it, such as moisture, oil, solder flux and fumes, and this contaminates it. Therefore, an additional protective layer is necessary, which has to meet the intended measurement temperature. The strain gage measuring point installed in this way is immediately ready for measurement. No additional heat treatment is required. Depending on the purity of the ceramic rods, strain gages installed by flame spraying can be used at temperatures from absolute zero up to almost 1000 °C. In this case, however, the working temperature substantially depends on the temperature stability of the strain gage used.

The flame spray method offers these advantageous properties:

- cryogenic, intermediate and high-temperature capabilities
- nuclear radiation resistance
- thermal and mechanical shock resistance
- no need for cement preparation, heat curing or clamping
- reliable bonding with less critical installation techniques
- rapid installation on a variety of materials.

The main advantage of this method is its ease of use. A measuring point can be installed within 10 minutes, which is then immediately ready for use. The sprayed ceramic layer has excellent resistance to thermal shock and mechanical stress. The method can also be easily applied on curved surfaces such as compressor blades. The maximum elongation is about 1%. The benefits of the method are offset by disadvantages by the consuming flame spraying device. Experiences of the application of flame spraying process are reported in [3.12, 3.13].

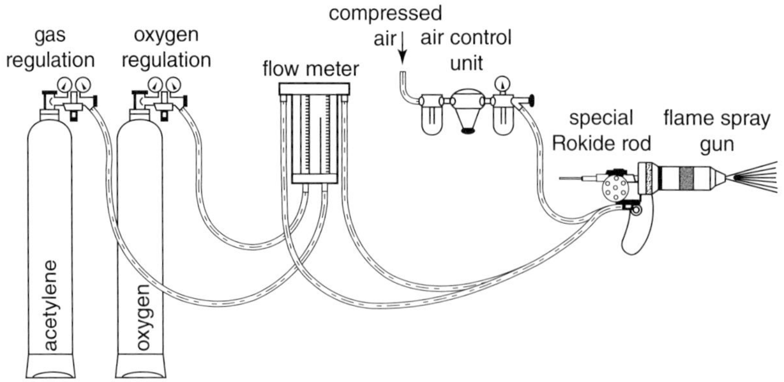

Fig. 3.6 Schematic drawing of flame spray equipment

3.2.8 Spot welding

There are two main groups strain gages that can be fixed by spot welding. One group consists of metallic encapsulated strain gages for use at high temperatures or in extreme environmental conditions against which they must be encapsulated (e.g. seawater). The other group consists of strain gages with organic carriers, which would have to be bonded for their use on the measurement object at medium elevated temperatures up to 300 °C by hot curing adhesive, but due to the circumstances can't be fixed in this way. Often, the measurement objects are too large to heat them for the strict observance of the prescribed curing in an oven. Help comes here from weldable strain gages which are made of thin strips of sheet metal onto which the appropriate foil strain gages are already glued in compliance with all the curing conditions. These sheets can then be fixed by spot welding even to very large measurement objects. For attaching the welding points, these gages have a wide margin around the measuring grid, as can be seen in Figure 3.7. However, spot welding of course limits the applicability of these strips to weldable materials. In addition, the reinforcing effect must be considered – the carrier sheet may alter the mechanical properties of the test object. Despite these limitations, these weldable strain gages have found wide application. The carrier sheets used are about 0.01 mm thick.

A typical example is shown in Figure 3.8. Here for strain measurements on warm pipes with hot-curing adhesive attached strain gages are required. Hot-curing

Fig. 3.7 Strain gage attached by spot welding on a hot measurement object

a) b)

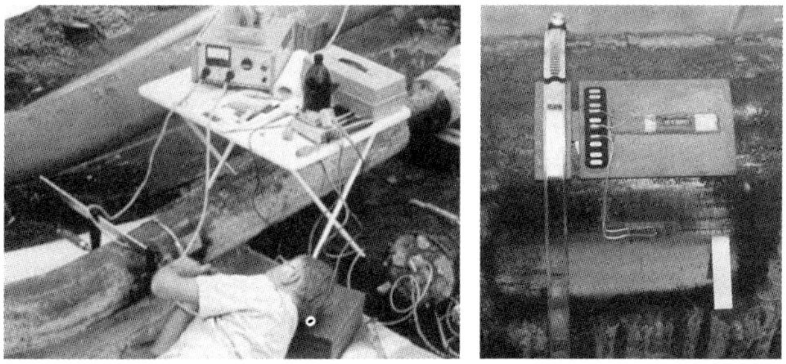

Fig. 3.8 Strain gage installation using spot welding on a hot pipe, a) The measurement technician doing the installation, b) Installed half bridge with compensation strain gage

adhesion, however, can't be performed on the test object without considerable effort. Therefore, strain gages are used that have previously been fixed by hot-curing onto carrier sheets which are now spot welded to the measurement object [3.14]. A full bridge comprising two diametrically opposed half bridges on the pipe circumference was mounted at each measuring point on the pipe. The active gages in these bridges were welded in the longitudinal direction directly onto the pipe and the compensation gages were welded to mechanical unloaded panel sections. The panels with the compensation gages were previously prepared in the workshop so that they just needed fastening to the pipe with a band clamp. Figure 3.8b) shows one half of the full bridge circuit installed on the pipe. The measurement technician can be seen in Figure 3.8a) welding an active strain gage. The picture also shows the spot welding equipment on the table and the grounding point attached to the pipe with a screw clamp. Users are often concerned about the behaviour of these gages during compressive strains of the test object, fearing buckling of the carrier sheet. This risk is, however, generally overestimated. In an extensive investigation [3.14] a comparison was made between directly glued and spot welded strain gages on a bending beam. The gage factor of the welded gages showed a slight decrease with increasing temperature, the maximum was 3% at 290 °C (550 °F).

3.3 Different measurement object materials

3.3.1 Glass, enamel, glazed porcelain

Bonding of strain gages onto the smooth surfaces of glass, enamel and glazed porcelain has been proved to be done easily (e.g. [3.15].). Suitable adhesives are cyanoacrylate, methyl acrylate and hot-curing epoxy resin. As with other object materials, a thorough cleaning and degreasing of the measurement surface is essential. Roughening of the bonding surface is not required. The bonded strain gages can usually be removed from the smooth hard surfaces, without residue, after implementation of the measurement task.

3.3.2 Concrete

Bonding on gages to concrete, subject to a few basic rules, is easy and generally less complicated than on metals. Almost exclusively the cold-curing two-component methyl acrylate adhesive is used, which easily fills existing pores in the surface of the concrete. In general, you first need to level out existing unevenness and produce a smooth concrete surface. If oiled formwork had been used while pouring the concrete, the oily layer has to be sanded at the measuring point. After grinding the measurement surface and the thorough removal of the sanding dust the pores present in the surface have to be filled with methyl acrylate adhesive to produce a smooth surface on which the strain gages can be bonded. The adhesive binds well in spite of rest moisture. When the still wet concrete has to be measured, the strain gages should not be directly attached to the concrete surface. It is advisable in this case to stick a thin aluminium foil as an insulating layer on the concrete and then mount the strain gages onto the foil. However, such measurement tasks are best solved by using encapsulated strain gages [3.16]. When measuring on concrete it is important to use strain gages with sufficient measuring grid length because concrete is an inhomogeneous material because of its aggregate. The glued strain gage should cover as representative a length as possible. Basic investigations indicate the need to adapt the strain gage length to the diameter of the largest aggregates in the investigated concrete [3.17].

Strain measurements are more problematic in the interior of concrete. Because of the way concrete components are made, it is possible to place the strain gages on prepared measurement elements within the fluid concrete as it is poured into the component. It can be used to investigate the solidification process of the concrete and also for the detection of external loads. As mentioned in Chapter 1 encapsulated strain transducers have been included in the poured concrete which had interior freely tensioned resistance wires. This was done in large numbers in parts of buildings made of concrete in 1931 by Carlson in the USA.

Different procedures are possible for strain measurements in the interior of concrete structures:

- embedding of cylindrical hollow bodies, inside which are measurement elements with strain gage installations
- bonding of strain gages to parts of the reinforcement or specially cast steel measurement bars
- embedding of strain gages glued onto a stiffening girder and fixed with wires or otherwise in the space
- embedding of specially encapsulated strain gages.

A robust version of an embeddable strain gage transducer is described for example in [3.18, 3.19]. It was designed for embedding during pouring the concrete. There are also references given to the adjustment of the geometry of this transducer with respect to the modulus of elasticity of the surrounding concrete in order to prevent detrimental effects on the internal stress distribution in the concrete structure. It is also possible to use industrially manufactured force transducers for strain measurements in concrete, because their signal is proportional not only to the force, but also to the deformation of the transducer in measuring direction.

If this transducer deformation relates to a defined measuring length, the load cell becomes a strain transducer. By appropriate selection of the transducer with respect to its rigidity, it can be matched to the modulus of elasticity of the concrete and used as an embedded extensometer. However, one condition for the application of this measurement method is that the concrete dimensions are large relative to the transducer so that no significant changes in the load-bearing cross-section occur.

One such successfully used measurement arrangement is described in [3.20]. A force transducer was inserted of the type U2 (HBM). Transducers of the U2 series are available with different rigidities depending on rated load; the maximum deformation is about 0.1 mm. These transducers made of stainless steel have excellent zero signal stability and are usable on rough construction sites. Figure 3.9 shows the drawing of the assembly with a force transducer enclosed in a PVC capsule and linked by its force introduction parts to the reinforcement and a photo of its incorporation into the reinforcement just before pouring the concrete.

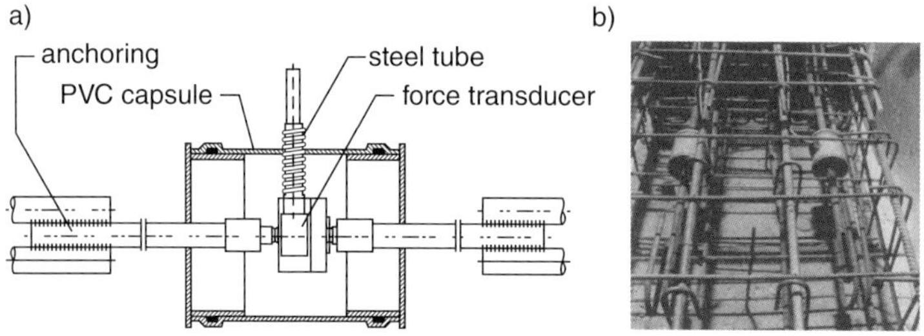

Fig. 3.9 Measurement element for measuring strains inside concrete produced on the basis of a force transducer [3.20], a) Installation drawing, b) Photograph of sensors before pouring the concrete

Another preferred method used particularly in recent years for the measurement of strain in concrete is the measurement of strains in reinforcing bars which specifically for these measurements were inlaid into the concrete. This method is discussed in detail in [3.21]. An 800 mm long steel rod is provided at its two ends with transverse welded round steel dowels to enable slip-free elongation, and at its centre is a strain gage full bridge circuit. Such transducers can be calibrated prior to installation. Due to their effective measuring length of 500 mm, they were superior to all previously known embeddable strain gages for use in fracture-critical areas. In a development of this idea, crosses from such rods were formed, and in addition to the strain gages thermocouples were also placed. With these 'transducer crosses' strains could be recorded in two mutually perpendicular directions and temperatures, too. In [3.21] an application example is described, in which are a group of these crosses, one above the other, were placed in the middle of a slab bridge and used immediately after the concreting to measure values of elongation and temperature. Comparisons between the measured values in the concrete and at the surface with rosettes showed good agreement. Figure 3.10 shows photos of the prepared rods and crosses for the measurements in concrete.

Another method more focused on laboratory conditions fixes strain gages on rigid carriers by using aid systems inside the empty space, which is then filled with

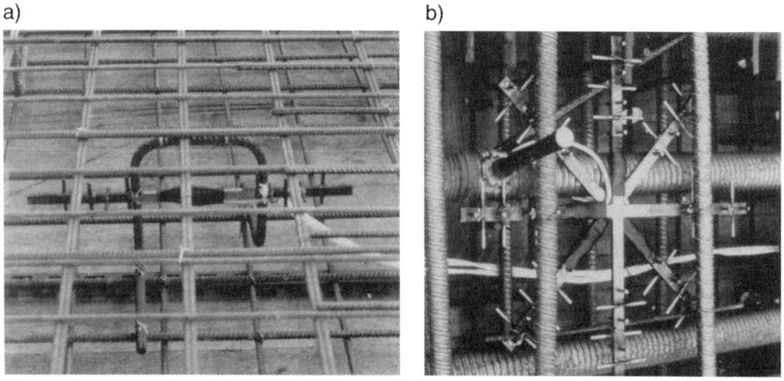

Fig. 3.10 Rod elements prepared for strain measurements in concrete with strain gage full bridge circuits: a) rod element, b) two superimposed transducer crosses

concrete. In [3.22] strain gages were installed on orthogonally orientated carrier steel plates and given a waterproof covering of epoxy resin. These prepared measuring elements are aligned with thin greased steel wires to their intended positions before pouring the concrete. In another work [3.23], the stresses which occur in the pile heads during driving of foundation piles of concrete are determined by means of instrumented pile heads which were made of epoxy resin mortar. The epoxy resin mortar of the pile head was adjusted so as to be equal to the modulus of elasticity of the pile concrete. Ten strain gages were carefully positioned in the head with their orientation parallel to and transverse to the longitudinal axis of the pile. The positioning of the strain gages prior to pouring the concrete was performed on plates that were connected by rods to an open skeleton. Plates and rods were made of the same epoxy resin mortar with which the skeleton was then cast into a measuring body. In this way, the fixing elements have no disturbing effects on the stress distribution.

3.3.3 Wood

Wood is an organic material with inhomogeneous internal structure of very different mechanical behaviour depending on the type of wood. Also, even for one type of wood there are, within the annual rings, significant differences in physical properties. The strong anisotropy of wood also causes a strong directional dependence of the technological characteristics. These peculiarities of wood require special consideration for experimental stress analysis using strain gages. The technological properties of some 30 000 wood species on Earth can be very different. In addition the anatomical structure of the wood causes differences in the properties in the various test directions [3.24]. So the strength values and the modulus of elasticity in a given direction decrease with increasing deviation of the loading direction from the fibre direction. The strength values transverse to the fibre direction can decrease to as little as 10–15% of the values in the longitudinal direction. Note that there is also swelling and shrinkage due to moisture changes that can overlap the strains generated by mechanical loads.

Since there are significant differences in density and hence in properties within the annual rings, the grid length of the strain gage used should cover several annual rings in order to obtain a representative mean of strains occurring along the grid length. For this, strain gages with grid lengths of 10 mm and more are used.

Wood is a poor heat conductor and therefore has only a limited ability to dissipate the heat generated by the strain gages. To avoid unwanted heating of the measurement points the smallest possible supply voltage (e.g. 1 V) should be used when measuring on wood. Running-in behaviour due to the heating effect will be observed when switching on a mechanically unloaded measuring point. The result is a time-dependent apparent strain. At low supply voltage, after a short time, a steady state is reached that is characterized by the apparent strain no longer changing with time. These values of the apparent strain should be recorded at the individual measuring points before the loading tests take place, in order to make the appropriate corrections to the measured values later in the load tests (subtracting the apparent

Table 3.1 Reference values for the elastic moduli of different types of wood in the longitudinal direction [3.24]

Species	Spruce	Oak	Beech	Hickory	Bongossi
Modulus of elasticity in N/mm^2	10 000	13 000	14 000	15 000	17 000

strains of the measured values under load). In the cases where a waiting time for running-in is too costly, for example, with multi-point measurement units, it is possible for each individual measurement point to be switched on after a specified period of time after switching on the supply voltage. Modern multipoint measuring units enable reproducible measurements to be made for all points by identical switching-on times.

Wood is a poor conductor of heat and therefore possesses only a limited ability to dissipate the heat generated by the strain gages. To avoid unwanted heating of the measurement points, the smallest possible supply voltage (1 V) should be used when measuring on wood. Running-in behaviour due to the heating will be observed when switching on a mechanically unloaded measuring point. The result is a time-dependent apparent strain.

Since wood has a relatively small modulus of elasticity, the strain gage and the adhesive selected for its attachment can exert a stiffening effect on the test object and cause a reduction of the strain recorded by the strain gage. Table 3.1 gives some typical values for elastic moduli of different wood species [3.24]. The error associated with the stiffening effect is greater, the greater the proportion of the deformation energy is consumed by the measuring point to the total strain energy at the test object [3.25, 3.26].

The error means that the measured value of the strain will be too small. Comparative studies between measurements with strain gages and other strain measuring methods confirm that the values obtained with strain gages are smaller than they should be. It was found that parallel to the fibre measured strains show smaller deviations from the nominal values than transversely measured strains, and that the variations in compressive load are greater than tensile load [3.27]. For solid spruce wood samples with foil strain gages installed directly on the face areas, it was observed that in the best case the deviation of the strain gage readings was −13% (tensile test) compared to inductively measured strains, and the maximum deviations of −23% have been found when compressive load is applied [3.24].

The strain gages are glued onto a wood surface that is as clean as possible smoothed by grinding. The sanding dust should be carefully removed. Wood types with a higher content of resin or fatty acids the measurement site should be cleaned with ethanol. Cyanoacrylate adhesive is well proven as an adhesive, when it is thinly applied in a first step, then smoothed and hardened. The adhesive penetrates slightly into the cell-wall cavities and provides a level basis for the second layer of adhesive. When fixing the strain gages to face planes it is best to apply several layers, since low viscosity adhesives such as cyanoacrylate are very quickly absorbed into the pores.

Then, after sanding with very fine sandpaper, the final bonding of the strain gages can be carried out [3.24].

Another method that avoids the direct adhesion of the strain gages on the timber is to use clamp-on strain measuring elements. This method can't always be used due to the geometric requirements. In [3.28] a strain measurement bracket is described, which has been specifically designed for carrying out elongation measurements of wood. Today, for this purpose industrially produced displacement transducers (can be used as strain sensors) are available that are very well suited for measurements on wood, because of their long measurement ranges extending over several millimetres and the negligible restoring force.

3.3.4 Plastics

Plastics we find today in an extraordinary diversity, in terms of both their areas of application and their properties. A very clear summary of current industrially used plastics, including the characteristics and the main areas of application, can be found in [3.29]. There are also many common abbreviations of the names of the plastics shown and explained.

In general for a description of their mechanical behaviour the information given by a stress–strain diagram is not sufficient as it is with most metals. Plastics are viscoelastic materials, which show flow phenomena even at low loads. Parameters such as load time, loading rate and temperature have a much larger influence on the mechanical behaviour of plastics than with metals. In plastics significant warming can occur under dynamic load by irreversible energy absorption (damping). Due to the low thermal conductivity, the generated heat is not sufficiently dissipated to the environment, and the resulting heat build-up can significantly affect the strain measurement.

The elastic moduli of plastics extend over a wide range, depending on the type of material considered. At the lower end are the thermoplastics with moduli of the order of 1 kN/mm^2. Towards the top, elastic moduli can be achieved >200 kN/mm^2 with carbon or graphite fibre-reinforced plastics. In between are the duroplastics and glass fibre reinforced plastics.

Important for measurements with strain gages is the superposition of elastic deformation and time-dependent creep caused by the viscoelastic behaviour. Even at low constant load there can be increasing creep with time. With load relief, the strain spontaneously returns to a predetermined value, from which, in the unloaded state, a more slow decrease occurs. Time thus plays a major role in the description of the stress–strain behaviour. For this reason an unambiguous description of the deformation behaviour has been established with isochronous stress–strain curves such as shown in Figure 3.11. This also introduces the creep modulus E_C which describes, similar to the modulus of elasticity, the slope of the stress–strain curve in the initial region. The creep modulus E_C decreases with increasing load time. Figure 3.12 shows the effect of time and temperature on the creep modulus of a glass fibre reinforced polyamide.

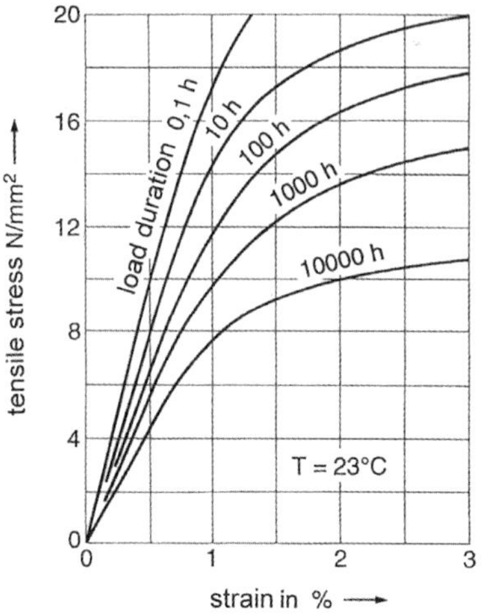

Fig. 3.11 Isochronous stress strain curves of ABS plastic [3.30]

The Poisson's ratio, which is influenced by the share of plastic deformation of the total deformation, may already have values above 0.3 at room temperature. With increasing temperature, the Poisson's ratio increases due to the increasing plastic shares on the total deformation, and reaches for pure plastic deformation its limit of 0.5. Figure 3.13 shows values of Poisson's ratio for some plastics, depending on the temperature.

In fibre-reinforced plastics the mechanical properties strongly depend on the direction considered. This so-called anisotropy affects the strain measurement itself less, but it does affect the calculation of the effective stresses from the measured strains. The constitutive law required for this calculation is based on direction-dependent material parameters, which must be known for an evaluation of measured strains. In [3.32] a method is described for the experimental determination of the complete constitutive law for anisotropic fibre-reinforced plastics.

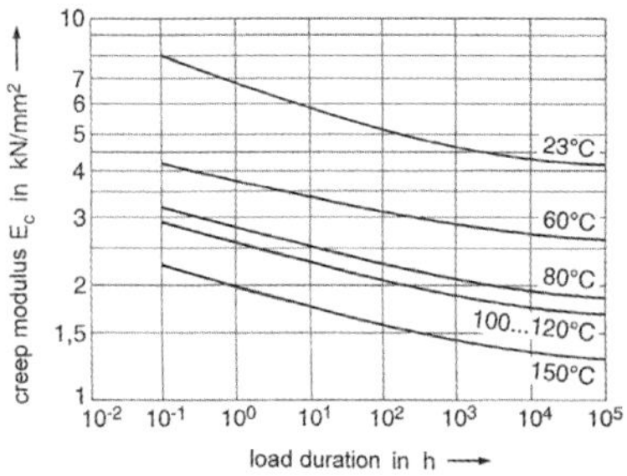

Fig. 3.12 Creep modulus for glass fibre reinforced polyamide PA6 GF30 determined in tensile creep test [3.31]

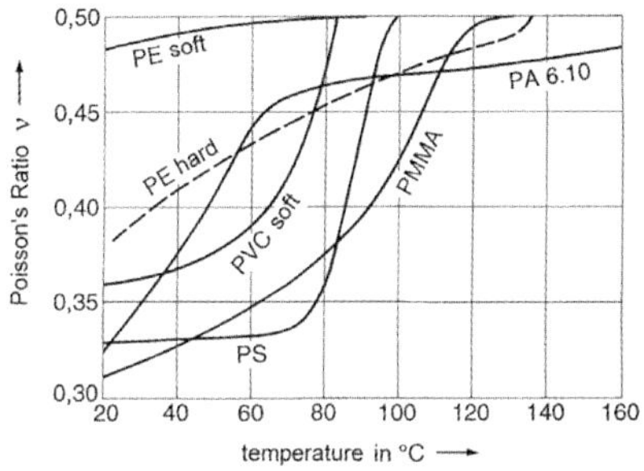

Fig. 3.13 Poisson's ratio as function of temperature for various plastics [3.31]

The thermal expansion coefficients of plastics vary widely for the different materials, in general they are 5–10 times as large as that of steel. Because of these large differences it is not possible to manufacture self-temperature compensated strain gages adapted to all plastics. The linear coefficient of thermal expansion is highly dependent always at room temperature on the temperature. Thus, the careful temperature compensation with compensating gages is particularly important. In addition, in fibre-reinforced plastics, thermal expansion is anisotropic.

Another influencing factor can be the humidity. Plastics absorb moisture from the environment by diffusion and swell. This process is reversible. The different volumes associated with the diffusion profile varies with the material cross-section and leads to residual stresses [3.33]. But they also lead to strain on the material surface, which is detected by the strain gage and – similar to the thermal output – overlaying the measured mechanical strain. The effects of these strains caused by varying moisture content on the strain results can be compensated by applying a compensation strain gage attached to a load-free remaining piece of the same plastic which is exposed to the same environmental conditions and switched into the neighbouring branch of the Wheatstone bridge circuit. This strip may be simultaneously used for the temperature compensation [3.34].

For successful measurement with strain gages on plastics and especially for the determination of stresses from measured strains it is recommended to become more familiar with the material to be tested. Prerequisite for successful stress analysis of plastics is a sound knowledge of their specific material properties.

The bonding of strain gages to plastics is almost always with cold curing adhesives, since the curing temperatures of the hot-curing adhesives typically used for plastics are too high.

Highly soluble polymers, which are mainly amorphous constructed plastics such as polystyrene (PS) and its modifications (e.g. ABS), polyvinyl chloride (PVC), polymethyl methacrylate (PMMA), cellulose acetate butyrate (CAB), and so on can be bonded pretty well with the two mainly used cold-curing adhesives, methyl acrylate and cyanoacrylate. In [3.31] it is found that cyanoacrylate adhesives show a slightly greater aggressiveness than methyl acrylate, which can however be mitigated by pre-treatment with a neutralizer. Before gluing, the surface of the measurement object must be properly cleaned, which also includes the thorough removal of the separation or lubricant residues left over from the manufacturing process.

Difficult to bond are the slightly soluble or insoluble plastics, such as the semi-crystalline plastics: polyethylene of high and low density (HDPE and LDPE), polypropylene (PP), polyoxymethylene (POM) and polyamide (PA). Here pre-treatment methods are recommended to activate the molecular structure of the surface. It is sometimes advantageous to scrape the measurement object surface with broken glass or roughen it with fine emery paper (grit 220–360). In most cases, however, more in-depth treatments are required, which alter the structure of the plastic surface and thereby increase adhesiveness. For example, short term melting of the surface of polyethylene with a Bunsen burner flame has proven effective [3.31]. In this case, however, the mechanical properties may be impaired.

With all plastics caution is advised in the use of detergents, which can lead to stress corrosion and swelling. Acetone dissolves ABS and even pure alcohol, which is less aggressive for plastics, swells CAB. Pure gasoline (but not with polystyrene!), isopropyl alcohol and Freon TF have proven to be largely harmless [3.35]. However, a test should always be performed for new applications, possibly over a long period of time, in order to avoid unpleasant surprises. The rates in the following examples of successfully implemented bonds to plastics are taken from [3.35]:

- *Polyethylene structural foam (HDPE) with cyanoacrylate adhesive Z70.* Preparation: the adhesive surface is roughened curling with clean, dry sandpaper of 320 grit and dust removed carefully. No solvent is used. The bond strength up to $\varepsilon = 2\%$ is good.
- *Polypropylene (PP), both Z70 and adhesive with X60.* Preparation: roughen the bonding surface with clean, dry sandpaper grit 320. No solvent is used. It works up to 6% strain.
- *Polyoxymethylene (POM) with adhesive Z70.* Preparation: dry roughen with emery paper grid 220. The obtained binding strength is very good.
- *Polymethylmethacrylate (PMMA acrylic glass) with adhesive Z70 or X60.* Preparation: only degreasing. Very good binding strength.
- *Plasticizer-free polyvinyl chloride (PVC) with adhesive Z70 or X60.* Preparation: only degreasing. Very good binding strength.
- *Polycarbonate (PC) with adhesive Z70 or X60.* Preparation: dry roughening with sandpaper grit 220 to 320 Good bond strength.
- *Polyester resin (cured, even glass fibre or carbon fibrereinforced) with adhesive Z70 or X60.* Preparation: dry roughening with sandpaper grit 220–320. Good binding strength.
- *Epoxy resin (EP) with adhesive Z70 or X60.* Preparation: dry roughening with sandpaper grit 220–320. Very good binding strength.
- *Polystyrene (PS) with adhesive Z70.* Preparation: dry roughening with sandpaper grid 220 to 320. Very good binding strength.

- *Phenolic resins, cresylic resins, melamine resins with adhesive X60.* Preparation: dry abrade with emery paper grade 220.

Two phenomena should be noted when measuring with strain gages on plastics which are much less important for metals. This is due to the poor thermal conductivity of plastics and the possible heating of the measuring point, but they can be avoided without major difficulty in almost all cases by appropriate selection of influencing parameters. As already explained in Section 3.2.8, by reducing the supply voltage, the selection of a strain gage with high resistance and a large grid area it possible to avoid undesirably heating of the measurement point. The other undesirable effect of strain gages on plastics can be a local reinforcement of the measurement object by the mechanical properties of the adhesive and the strain gage. This stiffening begins when the modulus of elasticity of the installed measuring point (combination of adhesive and strain gage) is greater than the modulus of elasticity of the measurement object material. This leads to incorrect measurements. The strain that would be present at the measuring point, if there were be no strain gage installed, is larger than the measured strain. In [3.31] a method for the estimation of the resulting measurement error is given. This method requires knowledge of the moduli of all the components involved and the thickness of the adhesive layer and strain gages. In Table 3.2 reference values are given for elastic moduli of foil strain gages with different carrier materials.

The stiffening effect of the strain gage installation should be considered when strain gages are selected. Preferably foil gages with a low modulus of elasticity (e.g. foil strain gages with polyimide carrier) should be installed with thin layers of adhesive (e.g. cyanoacrylate). However, the above-mentioned application limits must be observed for cynoacrylate. Due to stiffening, errors up to 20% can occur. Particularly in studies on plastic models this influence will be significant because often miniaturized components with small cross-sections are used as test objects in models. Since the reinforcing effect of an installed measurement point causes a systematic measurement error, this error can be determined by targeted studies. The quantitative influence of the effect on the measurement results can be found under the assumption of some unchanged parameters such as strain gage type, adhesive and material of the measurement object. In [3.36] is an explanation of how to determine correction values for strains measured with rosettes on plastics with lower stiffness.

Table 3.2 Reference values for elastic moduli of strain gages with different carrier materials [3.31]

Carrier material	Modulus of elasticity
Polyimide	10 kN/mm^2
Epoxy resin	14 kN/mm^2
Phenolic resin film	18 kN/mm^2

3.4 Measurement point protection

A properly glued strain gage provided with electrical terminals is immediately ready for measurement, but it must be protected by a cover adapted to the application. This cover serves to maintain the relevant properties of the measurement technique of the strain gage for the period of time over which it has to do its job. A strain gage must be protected against physical and chemical effects, and this includes not only the gage, but also its cable connections, including the solder joints. The type of cover necessary for the protection of a measurement site depends essentially on the existing or anticipated environmental influences as well as on the time period of the measurement. Important environmental factors include water, steam, oils and fats, and other chemicals according to the application case. The temperature has to be considered as an essential element which can limit the effective protection of the application.

Protection against damage to the measuring point and the cable connections by mechanical effects must also be noted. These possibilities of mechanical damage are often underestimated on construction sites or installations of large objects. In special cases test points must be covered by welded enclosures as described in [3.37] concerning the example of the research vessel 'Polarstern' with measuring points installed on the propeller nozzles. Here steel plates were welded frusto-pyramidal to a protective box as seen in Figure 3.14. The wiring box was filled up with a sealing compound. To prevent the robust protective cover from being a local reinforcement of the measurement object, the sheets were cut by separating flex sections in the longitudinal and transverse direction of the load transfer. The joints were finally smoothed with sealing compound.

A lack of information about installed measuring points for people carrying out other work at the test object can lead to damage, too. A well-covered measuring point can be inconspicuous and therefore easily overlooked. Mechanical damage is especially likely when welding in the vicinity of the measuring point or when moving heavy

Fig. 3.14 Measuring point on the propeller nozzle ring of the MS 'Polarstern' during installation, showing the protective cover with welded steel box and cable connection pipe before filling with potting compound [3.37]

parts over measuring installations. Therefore, a clear, highly visible marking of the measuring points is also part of an effective protection.

The protection should be adapted to the measuring task associated with the external conditions. Thus, for example, for measurements under laboratory conditions at room temperature just a simple cover with a special lacquer is sufficient, while for long-term measurements of machine parts under operating conditions multiple cover layers of different cover materials may be required. The undisputed best protection of a measurement site is achieved by hermetic encapsulation, as is practised in the construction of precision transducers. Such measures, however, due to their technical requirements and for economic reasons are usually not practical for measuring points for stress determination.

While building a protective cover, it should be noted that critical points occur where the covering material comes into contact with the components of the measuring point, especial with the measuring lines. Therefore, the prepared area of the adhesive surface on the measurement object should protrude on each side by 10–20 mm in order to also provide good adhesion to the cover. The cover of the measuring point should be put in place quickly, immediately after installation to prevent contamination of the contact surfaces for the cover. Scratches or grooves in the contact area are hidden paths for the ingress of moisture or corrosive substances. The contact surfaces should not be touched with fingers, as grease from your skin can cause problems, and possible corrosion can lead to rusting under the cover. It is also important that the installation is dry before covering, because the humidity enclosed in a covered measuring site is part of the damaging moisture that you want to protect from it. If it can't be avoided that a measuring point is set up in a humid environment, such as in rain or fog outdoors (preferably under a protective tarpaulin), then the measuring point before covering needs to be dried with a hair dryer or similar.

Another problem is the cable routing through the protective layer. In this case, the inclusion of a possibly long cable run is advantageous. The cover must enclose the cable on all sides, even upside down! The mistake of hitting the cover of the measuring point with the cable lying on the object to be measured must be strictly avoided. The cable must also be embedded from below into the protective layer. The material of the outer cable sheath must have a good binding capacity with covering and should be thoroughly cleaned and degreased. In the construction of long-term measurement sites note that materials can change over time, for example, cable insulation of soft PVC can become brittle due to leakage of the plasticizer. Moisture introduction by the cable itself must be avoided, that is within the outer cable jacket. With regard to the suitability of certain cable types for their inclusion in a cover, experience plays a big role, the user should be as familiar as possible with already successfully used material combinations. For example, it has been reported that due to the influence of pigments for colouring, admixed cable glands with white PVC cable fail more often than those with black PVC [3.38].

When experience (your own or from the literature) is not available for the intended application, then the suitability of the covering materials to be used should be tested in preliminary experiments. Basically, the substances considered as a covering must be water-resistant and free of moisture. They should react neutrally and not have a corrosive effect on the metallic components of the measuring point, as for example, can be the case with silicone rubber. Various silicone rubbers, although they have very good temperature resistance, can even damage the measuring point at room temperature following the elimination of acetic acid. Covering materials must be compatible with the selected adhesive. They need to be constantly within the relevant temperature range for the measurement and must not reinforce the measurement object. Also important for the covering are good binding properties to the materials of the test object and the cable sheath. Application and handling of the cover materials should not be too complicated and storage should be easy. They must not contain any hazardous substances. Limits of use and application for the covering are specified in the data sheets offered specifically for strain gage measurements.

Another important aspect in this context is the execution of installation works by the measuring technician. The care of the execution of each step in the composition of a measurement site has a significant contribution to successful measurement. Even a little carelessness, often caused by undue haste, can lead to premature failure of the measuring point. The application instructions recommended by the strain gage's manufacturers are strictly to be followed.

Measuring points operated under laboratory conditions almost always get away with little effort in terms of the cover. Here often a coat of polyurethane varnish is sufficient, as it has good abrasion resistance and provides adequate protection against normal humidity, sweat and oil. The temperature resistance is between −30 °C and 100 °C. Polyurethane varnish is often used as the bottom layer in multilayer covers.

For measurements at very low temperatures such as occur in liquid gases, nitrile rubber varnish is recommended as a cover, which is applicable down to −269 °C, although it must not come into contact with liquid oxygen. Another very easy to handle covering is silicone rubber, which is usually used from a tube, and quickly cures to a transparent elastic rubber coating. The mechanical protection is very good, as is the protection against moisture at room temperature.

The two-component epoxy resin adhesive, UHU-plus, has been effective as a simple covering. It offers good resistance to oils, gasoline, dilute acids, dilute alkalis and many solvents. It should not be used in hard setting. The temperature limits for the application can be expanded by heat curing of the applied coating layer. UHU-plus offers in the cured state a stable support for upward directed connecting wires with good insulation. In [3.39] it describes, in the context of residual stress investigations with the cutting method used with welds, how the strain gage connections were stabilized at the top with UHU-plus which simultaneously acted as a covering; thus the saw cuts could be made very close to the strain gage installation.

Strain gage measuring points can be covered with suitable non-conductive liquids. This method is particularly appropriate when the measurement has to be made on the inner walls of pressurized containers, and this special liquid can be used as a pressure medium for the time of measurement. To determine the sensitivity of strain gages against hydraulic pressure load, experiments have been carried out with uncovered test points. A water-and acid-free oil was used as a pressure medium, because it is known to be harmlessness to strain gages and adhesives as shown in previous studies (white oil BII of Aral), as described in [3.40]. The compensation strain gages were located outside the pressure chamber in an open vessel, which was also filled with the oil used as a pressure medium. Other suitable liquids are pure petroleum, paraffin oil or polyisobutylene (trade name Oppanol). Liquid covered test points are also known from the long-term monitoring of buried pipelines [3.41]. Here, a lead cap was applied with a soldered lead pipe on each of the three measuring points attached to the circumference of the tube. The attachment of the lead caps on the steel pipe was made with corrosion-resistant adhesive. The test cables were led through the lead pipes up to a terminal. To protect against moisture the lead caps and the tubes were filled with oil. The entire assembly was secured to protect it against earth pressure by special caps, and the lead pipes by upward supporting steel tubes. The terminal was accessible from above so that at any time the measurement readings could be taken.

For deformation measurements in the Antarctic ice shelf, strain gages were glued in provided depressions at deformation bodies made of Plexiglas. The depressions were filled with silicone rubber after installing the strain gages. These deformation bodies were frozen without further cover at about 40 m deep in the ice shelf, the ice extent during the freezing process caused for the force-and form-fitting embedding of the sensor in the ice. The ice itself formed an excellent additional coverage [3.42].

Manufacturers of strain gages offer different covering methods with special properties for different applications. The user is well advised to carefully review these pieces of advice about the technical characteristics. A good deal of experience has been gained with the permanently plastic putty AK22 (HBM), which is kneaded into the measuring point and has excellent adhesion through its strong tackiness. The putty has excellent weather resistance, provides excellent moisture protection and can also be used for measurements underwater. The temperature resistance is between −50 °C and +170 °C. The duration of the efficacy of the putty can be increased by combining it with a kneaded aluminium foil as a diffusion barrier.

A good long-term protection is often reached by the combination of several different covers, which are arranged in layers one above the other at the point of measurement. An example is given in [3.43], where the main landing gear of an Airbus A300 was fitted with strain gages for measurement of chassis forces over a long period during normal flight operations. In the installation area of the measurement points the protective coatings and layers were removed. The strain gages (with polyimide carrier) were glued with cyanoacrylate adhesive. The first protection layer consisted of polyurethane coating. The second layer was plastic putty AK22 (HBM). The third layer consisted of the special protection agent PRC1422, which had proved in preliminary experiments to be the only one of a number of concealers that had resistance to jet fuel kerosene and Skydrol (hydraulic oil). After curing the final coat, the measurement points were painted over by the chassis manufacturer. There were a total of 42 strain gage full-bridge circuits on the chassis, with changes in temperature of 70 K (about 20 °C at sea level and about −50 °Cat 10 km altitude), and they were successfully used in operation for three years.

Another example of a multi-layer cover is given in [3.44], where torque measurements were carried out on a diesel engine by using strain gages. Temperatures ranged between 40 °C and 80 °C at the measurement points lying in an oil mist. The cover corresponding to Figure 3.15a (polyimide carrier stuck with cyanoacrylate adhesive) consisted of a layer of polyurethane varnish applied with a soft brush on the measuring point (using paint from spray cans is not recommended because of possible propellant residues). On this layer an approximately 1 mm thick layer of permanently plastic putty was laid. The next also approximately 1 mm thick layer consisted of a solvent-free silicone rubber. Chlorine rubberized paint was used as a final layer. Characteristic of this multi-layer cover, in addition to their good protective effect, are their overall permanently good elastic properties. It is worth noting that due to the very small area available for installation, the strain gages (3 mm grid length, 120 Ω) were cut laterally by 1 mm parallel to the direction of measurement to have around the carrier enough free metallic surface for the adhesion of the coverage.

A similar cover to that in Figure 3.15a is shown in 3.15b. In this case, the entire measurement point was finally covered with protection tape consisting of 50 μm

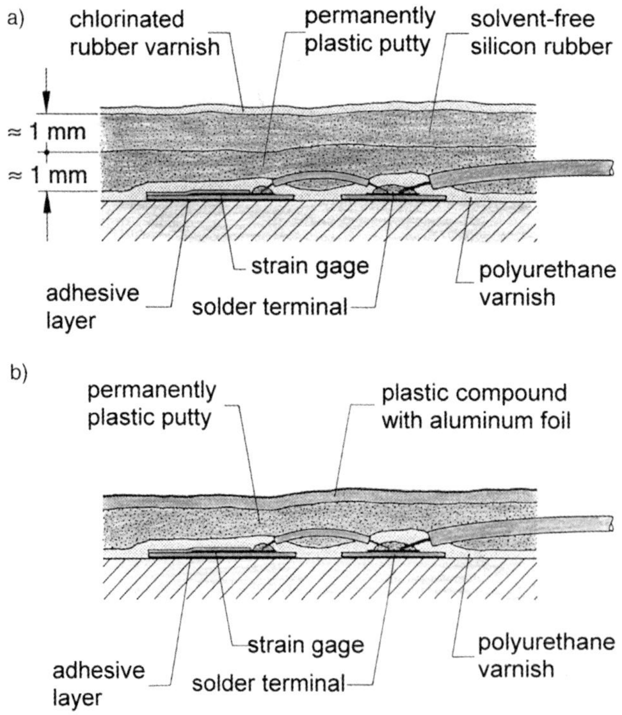

Fig. 3.15 Schematic representation of measuring point protections consisting of several layers a) permanently high elastic cover [3.44] b) cover with mechanical and electrical protective effect [3.45]

thick aluminium foil combined with permanently plastic putty. This tape has been used to increase the mechanical and electrical protective effect [3.45] instead of silicone rubber SG250. These measurement points were used for the measurement of forces on the roller bearings on a belt-filter press in ore processing under harsh environmental conditions. Figure 10.8 shows a photo of the measuring points in use, covered in the manner described.

For measuring points on test objects buried in the earth, bitumen is particularly well suited as a covering layer. In [3.46] are described pole force measurements on concrete injection piles. Strain gages are installed and coated with permanently plastic putty on the steel pipes used as reinforcement for the injection piles. A heated tar blanket was wrapped around this and after cooling a silicone hose was placed over this, as final protection.

In [3.47] is described a large coverage of several measuring points on a propeller blade (about 4.5 m Ø) made of bronze. In the field of measurement points (six rosettes), including the wiring, a large plate (60 cm × 90 cm) made of a special rubber compound was glued and then vulcanized in an autoclave for six hours at 127 °C and about 7 bar with the absence of air. The application of this type of covering requires the availability of appropriate aids and the possibility of the preparation of the test object in the laboratory. After 3000 hours of the propeller operating in salt water, all the measuring points on the propeller blade were still functional.

Measurements using strain gages at elevated temperatures provide increased demands on the protection design. Here, the inventiveness of the measuring technician is required, especially if beside the temperature other extreme environmental conditions occur. In [3.48] is a description of the coverage of heat-cured measuring points with measurements at 230 °C in the atmosphere of a hardening oven for rock wool products. Epple sealant 37 is used as a masking compound. This sealant is resistant to coolant, boiling water, fuels and hot fats and oils. It remains elastic up to 120 °C and is strongly hardened at higher temperatures. Its upper temperature limit is 400 °C. Since this sealant is acetone-soluble and can swell strain gages, the measuring point was coated, prior to the application of this cover, with a thin layer of silicone rubber SG250 (this silicone rubber is suitable for measuring points as it does not release any harmful acetic acid). After hardening of the SG250, a 3 mm thick layer Epple sealant 37 was placed over the measuring point and then solidified without cracking in stepwise thermal cycles up to 250 °C in an oven.

A particularly critical environmental condition for strain gage installations is produced by pressurized hot water. In [3.49] is a description of an effective multi-layer cover at a measurement point under pressures up to 100 bar and hot water of 150 °C consisting essentially of a thick layer of plastic putty AK22 with an integrated kneaded neoprene plate. Over all this is set a preformed cap of stainless steel sheet with prepared holes for the measuring lines, which is connected at the edges by spot welding to the measurement object. Over the steel cap a silicone rubber layer forms the outer jacket, and Teflon insulated cables were used.

3.5 Quality tests of the strain gage installation

It is annoying when a newly installed measurement point fails after careful coverage and access to its location may be difficult. The failure of a strain gage installation can have very costly consequences. Often a new installation is only possible after a large amount of time. In some cases, achieving the measurement objective can be called into question by a failed measurement point. Therefore, at each installation step the quality of this work has to be carefully checked. So one should also use good quality strain gages, especially in measurements where important decisions depend on the results. If we consider the total expense required for the installation of a measuring point compared to the price of a strain gage, and include in the analysis the extra expenditure that could be caused by a poor quality strain gage, then the value of using a high quality strain gage is obvious.

Quality assurance begins with the components used for the installation. All components, including the strain gages, should be stored properly. In particular, adhesives should not be used beyond their expiration dates, and instructions supplied by the manufacturers should be followed carefully.

If you have bonded a strain gage according to the application requirements onto the object and wired it correctly, then a first visual testing is recommended, which is best carried out with a magnifying glass. Included air bubbles in the adhesive layer and unreliable solder connections can be detected. Flux residues between the soldering points on the strain gage connections are not acceptable, as these may cause short circuits.

Before including the glued and wired strain gages in the cable network of the entire test setup, one should measure its resistance with an ohmmeter. Strain gages are matched in their preparation in close tolerance to their nominal resistance. This nominal resistance as well as the tolerance is specified in the packaging of the strain gage. During the installation the resistance shouldn't change markedly, but resistance changes of $\pm 0.25\%$ can be tolerated.

Especially important for reliable measurements is good electrical isolation between the measuring grid and the test object. The ideal resistance between the two would be infinite, since every finite resistance represents a shunt, which causes a voltage drop in the respective bridge arm. A properly installed strain gage should have at room temperature, depending on the application conditions, an insulation resistance to measuring object between 2000 MΩ (difficult installation conditions, e.g. outdoors) and 20 000 MΩ (installation in the laboratory). It is important that the insulation resistance for the duration of the measurement remains constant, because any change of this resistance causes a zero signal change. This is important for long-term measurements, where there is no opportunity to adjust the zero signal between measurements.

If adjusting the zero signal is possible at certain time intervals, then low zero signal changes effected by insulation resistance changes can be adjusted. In particular, moisture adversely affects the insulation resistance, since most plastics used as support material for strain gages absorb moisture (swell) and thus lose some of their insulating properties. A good protective covering of the measuring point should also include a conservation of the insulation resistance. If forced to install strain gages under unfavourable climatic conditions outdoors, then the measuring point should be heated, before covering, with a hair dryer or any other suitable means of expelling the moisture, in order to achieve a stable insulation resistance of sufficient amount.

Measuring the insulation resistance can be done with a standard ohmmeter. One should note that the voltage applied to the measuring point is of the order of the supply voltage used during the measurement. When measuring points are installed for long-term measurements one should log the measured insulation resistance in order to have a comparison value for later checking. For a measurement point that is to be used for long periods of time in operation the insulation resistance is an essential criterion for its functioning. Measuring the insulation resistance is an essential part of checking the functionality of a measuring point that is a long time in operation. This can occur, for example, on large outdoor situated measurement objects with weather-protected surfaces where you can't find a point of contact to the conductive surface of the material of the measurement object in the vicinity of the covered measuring point. Scraping the protective layer to form a contact point for the material is almost always not available for the sake of the consequent long-term protection of the measurement object. In these cases, if the measurement object is grounded, you can alternatively measure the resistance between the strain gage connecting cable and safety earth.

A relative simple method for measuring the insulation resistance is proposed in [3.50]. Used to carry out the procedure is an instrument for strain measurements (e.g. a measuring amplifier). The observed insulation resistance will be switched in parallel to a bridge arm of a balanced bridge circuit (half or full bridge). The insulation resistance acts as a shunt resistor and does not imbalance the bridge when it is infinite. If a small imbalance is caused by the shunt, the changed value can be read in this test of the device. It is identical to the zero-signal, which is caused by the insulation resistance of the investigated measurement point in the bridge circuit. The amount of imbalance gives information about the quality of the installation.

The impact of a shunt resistor on the total change in resistance of the bridge circuit depends on the amount of the base resistance (nominal resistance of the strain gage used in the circuit). This means that a constant shunt resistance has less effect on a bridge with 120 Ω strain gages than on a bridge with 350 Ω strain gages or with even larger resistors. The equations for calculating the effect of a shunt resistor on the bridge imbalance are given in Section 8.7.3 in the treatment of shunt calibration.

The insulation resistance of the connecting cable needs attention, too. The resistance between the wires or between the screen and the wires should be of the same order of magnitude as the insulation resistance of the strain gage. In checking the functioning of a measuring point the insulation resistance of the cable should also be measured. From the measurement results of insulation resistance it can be ascertained whether a malfunction is due to a defective cable or to a failure of the measuring point itself.

a) b)

Fig. 3.16 Implementation of checking the measuring point; a) eraser pressure test, b) measuring the insulation resistance after installation

Another recommended check of the installation is known by practitioners as the eraser test. This is performed on the bonded and wired strain gage which is connected to the measuring instrument before the cover is installed. By pushing down hard with a pencil eraser on the top of the strain gage the display of the instrument will be observed. One limitation of this pressure force is simply given by the test object, which must not be damaged. After pressing the eraser, the display must go back to its initial value. Deviations of more than 2 µm/m point to deficiencies in the bond, the measurement grid having been pressed into unevenness of the surface or in the adhesive layer. Faulty installations must be replaced. Figure 3.16a shows a photo of the execution of such a test *in situ*, and in the right photo can be seen measuring the insulation resistance between installed strain gages and the test object with a standard ohmmeter.

References

[3.1] Tatnall, F.G. Tatnall on testing. *American Society for Metals*, second printing 1967, p. 92.

[3.2] Keil, S. (1988) On the strain gage's 50th Jubilee – a review of its evolution and of 33 years strain gage production at Darmstadt. *RAM*, **4** (2), 39–48.

[3.3] Mittrop, F. (1965) *Das Kleben als Befestigungsverfahren für Dehnungsmeßstreifen*, Seminar Publication No. 31, pp. 15–27 HdT Essen.

[3.4] Walther, M., Wölfel, R., Stangl, R. and Beck, H. (1993) Meßmodell zur Überprüfung der Biomechanik nach Implantation einer Endoprothese. *MTB*, **29** (2), 27–32.

[3.5] data sheet: (1992) *Instruction for use X 60*, HBM GmbH, Darmstadt.

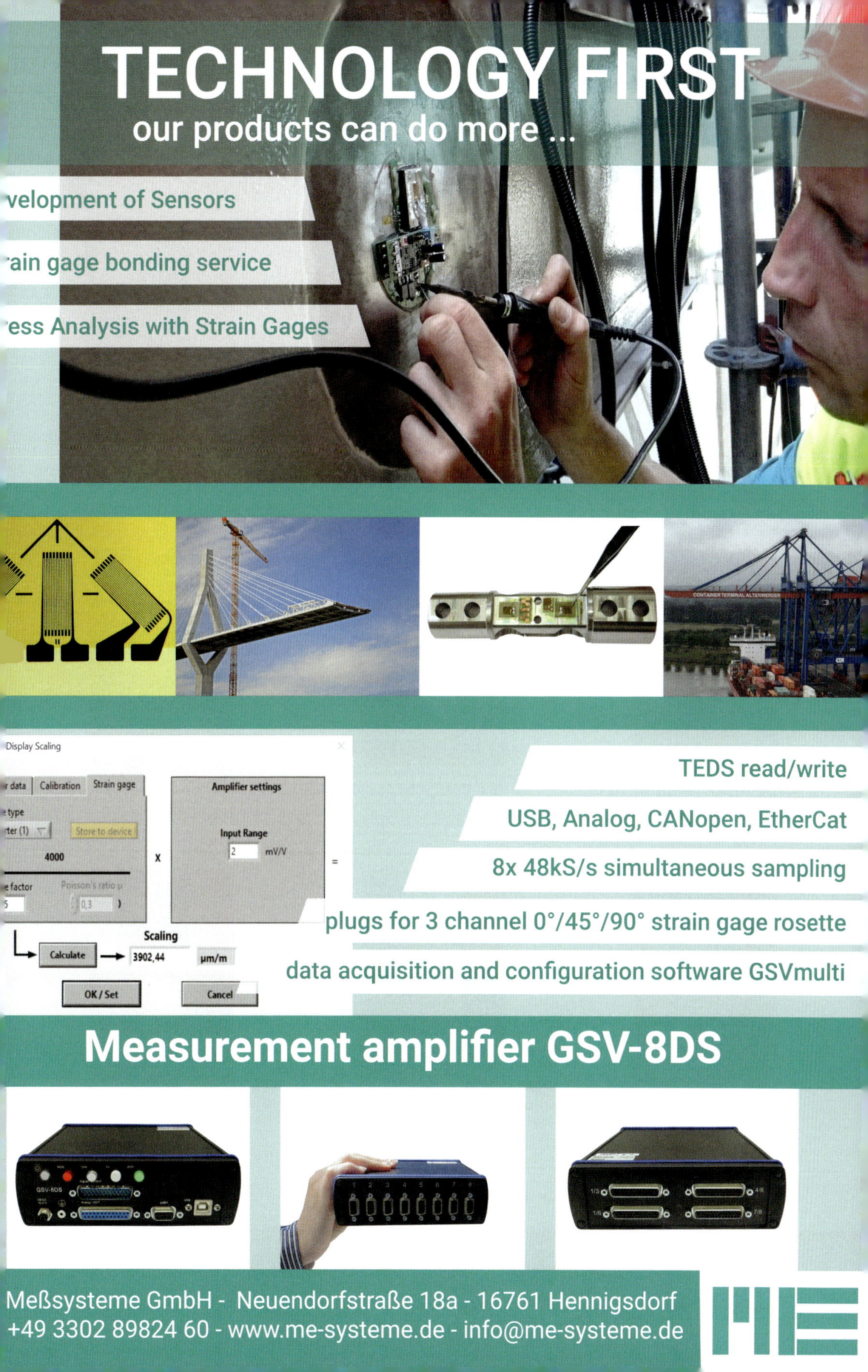
TECHNOLOGY FIRST
our products can do more ...
velopment of Sensors
ain gage bonding service
ess Analysis with Strain Gages
CONTAINER TERMINAL ALTENWERDER
Display Scaling
Strain gage
Calibration
Amplifier settings
Input Range
2 mV/V
4000
Store to device
Poisson's ratio
0,3
Scaling
Calculate
3902,44
µm/m
OK / Set
Cancel
TEDS read/write
USB, Analog, CANopen, EtherCat
8x 48kS/s simultaneous sampling
plugs for 3 channel 0°/45°/90° strain gage rosette
data acquisition and configuration software GSVmulti
Measurement amplifier GSV-8DS
GSV-8DS
Meßsysteme GmbH - Neuendorfstraße 18a - 16761 Hennigsdorf
+49 3302 89824 60 - www.me-systeme.de - info@me-systeme.de

[3.6] Hoffmann, K. (1984) *Hinweise zum Applizieren von Dehnungsmeßstreifen,* Techn. Note vd 84005, HBM GmbH, Darmstadt, p. 7.

[3.7] N.N. (1993) *Stick-on DMS,* Techn. Note I 24.36.0, HBM GmbH.

[3.8] N.N. (1990) *Keramischer Kitt CR 760 für Freigitter-Dehnungsmeßstreifen,* Instruction for use, HBM GmbH.

[3.9] N.N. Flammspritz-Verfahren für die Applikation von Dehnungsmeßstreifen. *MTB,* **1** (2) (1965) 15.

[3.10] N.N. (Oct. 1969) *The Rokide process and equipment,* BLH Product Data 200–6.

[3.11] Wnuk, S.P. and MacLean, J.G. (1967) Five ways to avoid troubles in rokide installations. *BLH Measurement Topics,* **5** (3), 4.

[3.12] Wolf, H. and Schlichting, H.-D. (1967) Zur Messung statischer Beanspruchungen grosser Bauteile bei hohen Temperaturen mit Dehnungsmeßstreifen. *ATM Sheet V,* **123-23**, 261–266.

[3.13] Ludewig, H. (1969) Schwingungsmessungen an Turbinenschaufeln von Abgasturboladern mit Hochtemperatur-Dehnungsmessstreifen. *MTB,* **5** (1), 8–11.

[3.14] Hines, F.F. (1960) *Weldable bonded resistance strain gages,* unpublished research results, BLH-report 1960.

[3.15] Schönbrunn, G. and Hellwig, R. (1971) Schwingungsuntersuchungen an Hohlgläsern bei Schlagbeanspruchung. *MTB,* **7** (2), 29–33.

[3.16] Rohrbach, Chr. (1958) Dehnungsmeßstreifen mit metallischem Träger als schnell meßbereites, feuchtigkeitsunempfindliches Meßelement für Dehnungsmessungen auf Beton. *Der Bauingenieur,* **33**, 265–268.

[3.17] Müller, R.K. (1966) Der Einfluss der Meßlänge auf die Ergebnisse bei Dehnungsmessungen an Beton. *MTB,* **2** (3), 37–41.

[3.18] Rohrbach, Chr. (1969) *Dynamische Dehnungsmessungen an Beton mit Dehnungsmeßstreifen,* Schriftenreihe über zivilen Luftschutz, No. 13, Verlag Ziviler Luftschutz Dr. Ebeling K.G. Koblenz.

[3.19] Rohrbach, Chr. (1962) *Spannungs- und Dehnungsmessungen an Beton, part I and part II,* Archiv für Technisches Messen, July 1962, pp. 159–160 and August 1962, pp. 183–186.

[3.20] Schmidt, K.W. and Brockhaus, U. (1980) Baubegleitende Messungen bei der Erstellung des Bauloses 3 der Stadtbahn Rhein-Ruhr in Duisburg. *MTB,* **16** (1), 1–8.

[3.21] Lierse, J. (1980) *Dehnungs- und Durchbiegungsmessungen an Massivbauwerken,* Werner-Verlag GmbH, Düseldorf.

[3.22] Klink, S.A. (1974) Embeddable strain gage units for measuring strains in concrete. *Cement and Concrete Research,* **4**, 223–230.

[3.23] Dijst, G. (1979) Messen der beim Rammen von Gründungspfählen aus Beton auftretenden Pfahlkopf-Beanspruchungen. *MTB,* **15** (3), 60–63.

[3.24] Bröker, F.-W. (1985) Dehnungsmessungen an Holz mit direkt applizierten DMS. *MTB,* **21** (1), 18–23.

[3.25] Kufner, M. (1982) Messungen mit Dehnungsmeßstreifen auf Holz und Holzwerkstoffen, *Proc. 8th Congr. on Material Testing, Budapest,* pp. 889–894.

[3.26] Kufner, M. (1987) Messungen mit Dehnungsmeßstreifen auf Holz. *VDI/GMA-Bericht,* **16**, 1–11.

[3.27] Sliker, A. (1971) Resistance strain gages and adhesives for wood. *Forest Products Journal,* **21** (12), 40–43.

[3.28] Keylwerth, R. and Höfer, W. (1962) Untersuchungen an Dehnungsmeßbügeln. *Holz als Roh- und Werkstoff,* **20**, 57–61.

[3.29] NN (1993) *Kunststoff Kautschuk Produkte '93, Jahreshandbuch für Kunststoff- und Kautschukanwender,* Verlag Hoppenstedt & Co., Darmstadt.

[3.30] Oberbach, K. and Heese, G. (1974) Dehnungsmeßstreifen – Hilfsmittel zum Prüfen von Kunststoff-Formteilen. *Kunststoffe,* **64** (9), 488–493.

[3.31] Heese, G. (1983) *DMS-Technik an Kunststoffen,* manuscript of the VDI-Bildungswerk BW 2564.

[3.32] Scharr, G. (1985) Experimentelle Bestimmung des kompletten Stoffgesetzes von anisotropen faserverstärkten Kunststoffen. *MTB,* **21** (1), 7–11.

[3.33] Menges, G. and Rogalla, D. (1982) Messen von Austrocknungseffekten bei Kunststoffen. *MTB,* **18** (3), 53–58.

[3.34] Perry, C.C. (1987) Strain gauge measurements on plastics and composites. *Strain,* Nov. 1987, pp. 155–156.

[3.35] Hoffmann, K. (1984) *Hinweise zum Applizieren von Dehnungsmeßstreifen,* HBM Techn. Note vd 84005.

[3.36] Glücklich, D. (1976) Die Versteifung von Kunststoffmodellen durch elektrische Dehnungsmeßsreifen. *VDI-Z,* **118** (17/18), 829–834.

[3.37] Westram, A. (1987) The measurement instrumentation of MV *Polarstern* for an icebreaking expedition. *RAM,* **3** (1), 23–29.

[3.38] Pople, J. (1982) Gauge installation and protection in hostile environments, in *Strain Gauge Technology* (eds A.L. Window and G.S. Holister), Applied Science Publishers Ltd., p. 117.

[3.39] Rappe, H.-A. (1973) Messung von Schweißeigenspannungen mit Dehnungsmeßstreifen. *MTB* **9**, 31–37.

[3.40] Hoffmann, K., Jost, D. and Keil, S. (1978) Experimentelle Untersuchung des Einflusses hydrostatischen Drucks auf DMS-Applikationen und die Ermittlung von Korrekturwerten. *VDI-Berichte,* **313**, 553–558.

[3.41] Gisen, F. (1967) Messung der Beanspruchung von starren erdverlegten Rohrleitungen mit Hilfe elektrischer Dehnungsmeßstreifen. *Erdöl und Kohle, Erdgas, Petrochemie,* **20** (7), S505–S509.

[3.42] Weigelt, E., Sobesiak, M., Aulbach, E. and Berckhemer, H. (1992) Elastic deformation measurements in the ice shelf at the georg von neumayer station in Antarctica. *RAM,* **8** (1), 1–6.

[3.43] Ladda, V. and Zaschel, M. (1985) Long-term measurement of service loads on the landing gear of a commercial airliner. *RAM,* **1** (2), 21–25.

[3.44] Grünbaum, P. (1985) Drehmomentmessungen an Nockenwellen von Dieselmotoren. *MTB,* **21** (2), 39–46.

[3.45] Grünbaum, P. (1987) Applying force transducers and strain gages for rollbearing measurements on continuous dewatering pressure filters. *RAM,* **3** (2), 40–44.

[3.46] Bernuth, W., Pfirschke, H. and Vogler, W.-D. (1992) Sanierungsbegleitende meßtechnische Untersuchungen am Turm der mittelalterlichen Runneburg in Thüringen. *MTB,* **28** (2), 36–40.

[3.47] Dean, M. (1977) Vulcanized-rubber protection for strain gages in a seawater environment. *Experimental Mechanics,* August 1977, pp. 303–307.

[3.48] Grünbaum, P. (1992) Measurement of forces in a rock-wool hardening oven at 230°C. *RAM,* **8** (2), 25–30.

[3.49] Hosaluk, L.J. (1985) A technique for protecting bondable foil strain gages in pressurized water to 150 °C. *Experimental Techniques March,* 1985, pp. 29–32.

[3.50] Hoffmann, K. (1987) *Eine Einführung in die Technik des Messens mit Dehnungsmeßstreifen,* HBM GmbH, Darmstadt, p. 137.

4 The Wheatstone bridge circuit

4.1 Circuit principle

Even about 100 years before the invention of the strain gage the English scientists Hunter-Christie and Wheatstone were working on the measurement of the resistance of metal connecting wires, resulting in the invention of the bridge circuit with which it was possible to measure resistances even with unstable voltage sources [4.1, 4.2]. A brief account of the historical background was given in Chapter 1. The circuit which was later named after Wheatstone became the most important basic circuit in widespread use for strain gage measurements.

Figure 4.1 shows schematically the basic structure of the Wheatstone bridge circuit in the two graphical forms usually found today. The two forms are electrically identical. The bridge circuit consists of four bridge arms with the resistances R_1 to R_4 which are here arranged on the four sides of a parallelogram. The effect of the circuit is that if it is excited with a voltage applied to opposite points, the voltage on the other two corner points, that is the output voltage, is dependent on the ratios of the resistances in the four arms of the bridge. For example, if all four resistances in the arms of the bridge are identical, then the output voltage is equal to zero. If a change occurs in one resistance, the effect on the output voltage is a measure of this change in resistance.

This effect is used for measurements with strain gages by measuring the change of resistance of a strain gage connected in one arm of the bridge in the manner described above. The output voltage of the circuit is then a measure of the strain.

The supply voltage V_B is normally between 1 V and 10 V for strain gage measurements. If this voltage is applied to points 2 and 3 of the bridge circuit illustrated in Figure 4.1, then the bridge output voltage V_M between points 1 and 4 is zero when identical resistances are present in all four arms of the bridge, that is if $R_1 = R_2 = R_3 = R_4$. The bridge is therefore said to be balanced when the output voltage $V_M = 0$ for an applied supply voltage of V_B.

During the installation and wiring of bridge circuits, cables of different colours are used to provide a defined assignment between the cable cores and the bridge connection points. The preferred colour arrangement assigns white and red to the signal leads, connected to points 1 and 4 respectively, and black and blue to the supply leads, which are connected to points 2 and 3 respectively.

4.2 Basic equation of the bridge circuit

The output voltage V_M in an unbalanced bridge circuit changes according to the bridge supply voltage V_B, and this change is proportional. Therefore in considering the Wheatstone bridge circuit, it is of advantage to start with the ratio V_M/V_B of these two voltages. The ratio V_M/V_B is independent of the magnitude of the supply voltage.

Technology and Practical Use of Strain Gages: With Particular Consideration of Stress Analysis Using Strain Gages. First Edition. Stefan Keil.

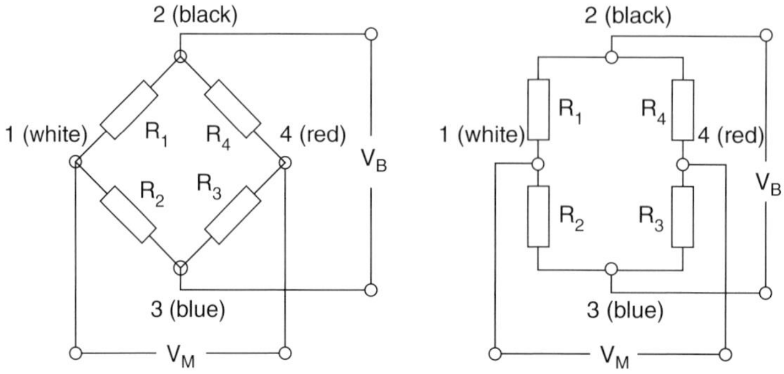

Fig. 4.1 The usual forms of representing the Wheatstone bridge for measurements using strain gages

The application of elementary principles of electrical engineering (Kirchhoff's Laws) to the Wheatstone bridge circuit gives the following relationship:

$$\frac{V_M}{V_B} = \frac{R_1}{R_1 + R_2} - \frac{R_3}{R_3 + R_4} \tag{4.1}$$

It can be seen from Eq. (4.1) that the bridge is balanced under the following conditions:

- $R_1 = R_2 = R_3 = R_4$
- $R_1 = R_2$ and $R_3 = R_4$
- $R_1/R_2 = R_3/R_4$

If resistance changes of ΔR_i are allowed to occur in the separate bridge arms, which is desired for strain gage measurements, then Eq. (4.1) becomes

$$\frac{V_M}{V_B} = \frac{R_1 + \Delta R_1}{R_1 + \Delta R_1 + R_2 + \Delta R_2} - \frac{R_4 + \Delta R_4}{R_3 + \Delta R_3 + R_4 + \Delta R_4} \tag{4.2}$$

In order not to complicate the further treatment unnecessarily, the nominal resistance R_0 of the strain gages used is introduced here, that is it is assumed that all strain gages in the four arms of the bridge have the same resistance before resistance changes occur. With

$$R_1 = R_2 = R_3 = R_4 = R_0 \tag{4.3}$$

Equation (4.2) becomes

$$\frac{V_M}{V_B} = \frac{R_0 + \Delta R_1}{2R_0 + \Delta R_1 + \Delta R_2} - \frac{R_0 + \Delta R_4}{2R_0 + \Delta R_3 + \Delta R_4} \tag{4.4}$$

Rearrangement of Eq. (4.4) gives the general equation for the Wheatstone bridge circuit

$$\frac{V_M}{V_B} = \frac{R_0(\Delta R_1 - \Delta R_2 + \Delta R_3 - \Delta R_4) + \Delta R_1 \Delta R_2 - \Delta R_3 \Delta R_4}{2R_0(2R_0 + \Delta R_1 + \Delta R_2 + \Delta R_3 + \Delta R_4) + (\Delta R_1 + \Delta R_2)(\Delta R_3 + \Delta R_4)} \tag{4.5}$$

The resistance changes ΔR_i occurring on the individual strain gages during the measurement process are very small compared with R_0, so that Eq. (4.5) can be used in the simplified form given in Eq. (4.6) without any detrimental effects on the measurement evaluation. Equation (4.6) represents the basic equation for the Wheatstone bridge circuit for measurements with strain gages:

$$\frac{V_M}{V_B} = \frac{(\Delta R_1 - \Delta R_2 + \Delta R_3 - \Delta R_4)}{2(2R_0 + \Delta R_1 + \Delta R_2 + \Delta R_3 + \Delta R_4)} \tag{4.6}$$

Equation (4.6) represents the combined effect of the resistance changes in the four arms of the bridge. It will be appreciated that resistance changes of the same magnitude in all arms of the bridge have no effect on the output voltage V_M. Therefore, the Wheatstone bridge provides excellent possibilities for compensation of the effects of disturbances which affect the complete circuit. For example, if the four strain gages in the bridge circuit are subjected simultaneously to the same temperature change, then this will cause the same resistance change in all the strain gages and the output voltage of the bridge circuit remains unaffected. From Eq. (4.6) it follows that an extreme signal is obtained when resistance changes of equal magnitude, but with opposite signs, occur in neighbouring arms of the bridge circuit. For example, $\Delta R_1 = -\Delta R_2$ and $\Delta R_3 = -\Delta R_4$. This phenomenon is exploited in transducer construction by designing measuring elements in which positive and negative strains occur simultaneously during loading with the measurement quantity. These strains are then measured by strain gages wired in a full bridge configuration. Here, typical examples are bending beams with strain gages mounted on the tension side and on the compression side as well as shear beams in which the principal stresses occur with the corresponding positive and negative principal strains at an angle of $\pm 45°$ to the plane of shear stress.

It also follows from Eq. (4.6) that the relationship between the bridge output signal and the resistance changes in the bridge circuit need not necessarily always be linear. Linearity deviations occur when the sum of all the resistance changes is not equal to zero, so that in the denominator in Eq. (4.6) a sum remains composed of the nominal resistance R_0 which is independent of strain and a strain dependent change in resistance ΔR. This is the case, for example, when only one of the four arms of the bridge contains an active strain gage, that is a change of resistance occurs in only one bridge arm. The Eq. (4.6) then simplifies, with $\Delta R_2 = \Delta R_3 = \Delta R_4 = 0$, giving

$$\frac{V_M}{V_B} = \frac{\Delta R_1}{2(2R_0 + \Delta R_1)} \tag{4.7}$$

The result given by Eq. (4.7) describes the effect on the voltage ratio V_M/V_B due to the change in resistance in only one arm of the bridge. It can be seen that the relationship between ΔR_1 and the voltage ratio V_M/V_B is not linear due to the sum in the denominator. However, since with strain gage measurements on metallic materials in the elastic deformation range $\Delta R_1 \ll R_0$, the following approximate equation is used in practical measurement instead of Eq. (4.7)

$$\frac{V_M}{V_B} = \frac{\Delta R_1}{4R_0} \tag{4.8}$$

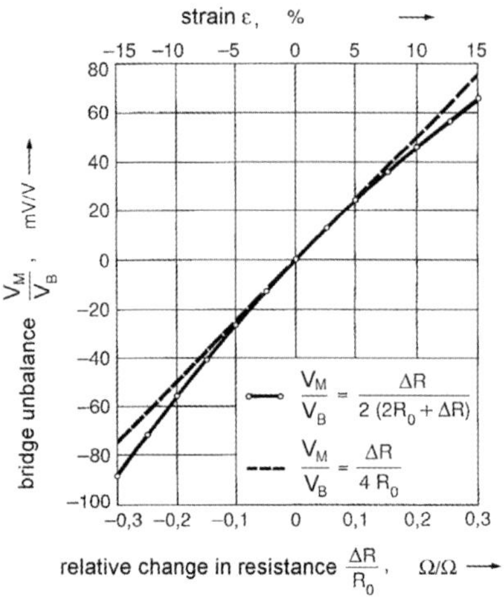

Fig. 4.2 Nonlinear characteristic of a Wheatstone bridge circuit with only one active strain gage

With this equation there is no true linearity between the measurement signal and the change in resistance.

Figure 4.2 shows graphically the deviation of the approximate solution given by Eq. (4.8) from the correct one of Eq. (4.7). The error that arises with the application of Eq. (4.8) is as large as the strain on the strain gage, which can be simply shown by considering the error.

With the measurement of large strains, such as on metals in the elastoplastic deformation range or on plastics, the approximate character of Eq. (4.8) should be appreciated and either evaluation should be carried out with Eq. (4.7) or – usually easier in practice – or the measurement result based on the approximate solution should be corrected.

Measurement with strain gages in the Wheatstone bridge configuration requires the presence of all four bridge arms, but all four arms need not be fitted with strain gages. Therefore, it is possible, for example, to substitute fixed resistances or 'dummies' in two or three of the bridge arms, leaving the remaining arms to be occupied by strain gages.

This gives rise to the various possibilities presented by the application of the full, half and quarter bridge configurations, as they are commonly known. These designations, however, only refer to the part of the bridge in which the strain gage is used.

The completion of the half and quarter bridge configurations to form complete Wheatstone bridges follows as shown in Figure 4.3 using fixed resistances which are switched into circuit in the measurement instrument.

When using the Wheatstone bridge circuit for strain gage measurements, the orders of magnitude are of interest in which the voltages and resistances occur. Assuming that commercially available strain gages are used with a nominal resistance of $R_0 = 120\,\Omega$ and a gage factor of $k = 2$, the equation $\Delta R/R = k\varepsilon$ gives a relative change

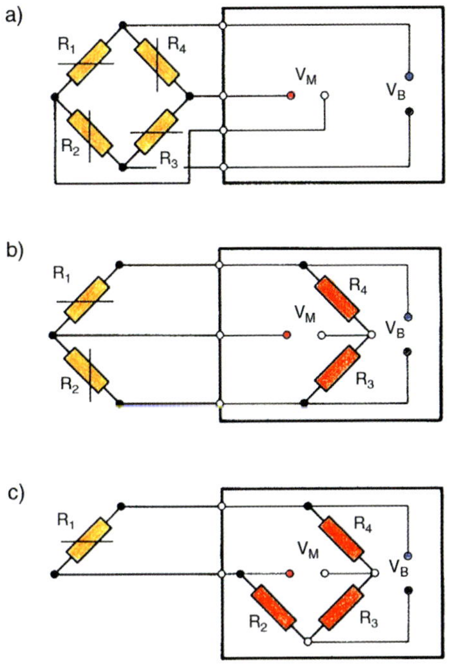

Fig. 4.3 Full, half and quarter bridge configurations for strain gage measurements

in resistance of $\Delta R/R = 0.002$ for a strain of $\varepsilon = 1‰$ (i.e. $\varepsilon = 1000 \times 10^{-6}$). This signifies an absolute change of resistance in the strain gage of $\Delta R = 0.24\,\Omega$. If the relative change of resistance $\Delta R/R = 0.002$ is substituted in Eq. (4.8), then a voltage ratio of $V_M/V_B = 0.0005$ (i.e. $V_M/V_B = 0.5$ mV/V) is obtained. For a supply voltage of $V_B = 5$ V a relatively low signal voltage of $V_M = 2.5$ mV is produced. The low order of magnitude for the signal voltage demands that the precautions regarding the measurement leads outlined in Chapter 6 are followed in order to keep the effects due to possible interference as low as possible.

The basic or nominal resistance of the strain gages used is also the nominal resistance of the complete bridge circuit to which the supply voltage is applied. Normally, bridge circuits are supplied with a constant DC voltage or with an AC voltage of constant frequency and amplitude. The term 'carrier frequency excitation' has been adopted for the supply with AC voltage. In this case the carrier frequency is the frequency of the AC voltage used for the supply. The imbalance in the bridge circuit, that is the output signal, is modulated on the carrier frequency. Common frequencies for the supply of strain gage bridge circuits are between 225 Hz and 5 kHz. With an AC supply the bridge circuit output voltage is also an alternating voltage, the amplitude of which forms the measurement signal. The Wheatstone bridge therefore forms a modulator which impresses the measurement signal onto the carrier frequency as an amplitude variation. This means that with carrier frequency supply a carrier-frequency modulated measurement signal is available at the transducer output. The demodulation of the measurement signal takes place in the following measurement amplifier.

The type of supply normally used in the supply of strain gage bridge circuits in transducers is the low frequency AC voltage supply. DC voltage supply must be used if the frequency of the quantity to be measured exceeds the upper limiting frequency of the carrier-frequency measuring amplifier. The disadvantages of the DC measuring amplifier with regard to noise, instability and thermal voltages must be taken into account, but these can usually be tolerated for dynamic measurements. In all other cases the advantages of the carrier frequency method prevail, the lower carrier frequency being preferred provided the dynamic characteristics of the process to be measured permits this [4.3].

4.3 Temperature compensation

From the basic equation for the Wheatstone bridge circuit given in Eq. (4.6) it follows that resistance changes of the same sign in adjacent arms of the bridge cancel one another out, but they add if the changes are in opposite arms of the bridge. This property of the Wheatstone bridge configuration is especially favourable for compensation of effects caused by temperature.

As already explained, each strain gage has a temperature response describing the undesired temperature-dependent change of resistance which is additively superimposed on the change of resistance due to strain, leading to temperature-dependent shifts in the zero-point signal (apparent strain). These undesired resistance changes in the strain gage caused by temperature changes cannot be avoided, but their effects can be almost completely compensated within the Wheatstone bridge circuit.

If, for example, a full bridge circuit formed from four strain gages having the same temperature response is subjected to a temperature change, all four strain gages will exhibit the same change of resistance (the same apparent strain). These equal resistance changes in all four arms of the bridge cancel one another out in the bridge output signal as indicated by Eq. (4.6). The same also applies to the half bridge circuit. This consists of two adjacent bridge arms in which equal resistance changes occur during temperature changes, again having no effect on the bridge output signal, since they cancel one another out.

This mutual compensation of equal resistance changes in neighbouring bridge arms is exploited for temperature compensation by using one (or two for a full bridge) temperature compensating gages. This is achieved by applying an active strain gage to the measurement object and a compensation strain gage on a mechanically unloaded small block or sheet of the same material as the measurement object. Both strain gages are wired to form a half bridge.

To ensure that both strain gages have the same temperature response, they must be taken from the same production batch (the production batch number is stated on the strain gage package). For proper measurement without any effects due to temperature changes the compensation strain gage with its carrier must always be at the same temperature as the active strain gage and its measuring point. This is best achieved by thermally coupling the carrier for the compensation gage to the measurement object as close as possible to the measuring point. The procedure for temperature

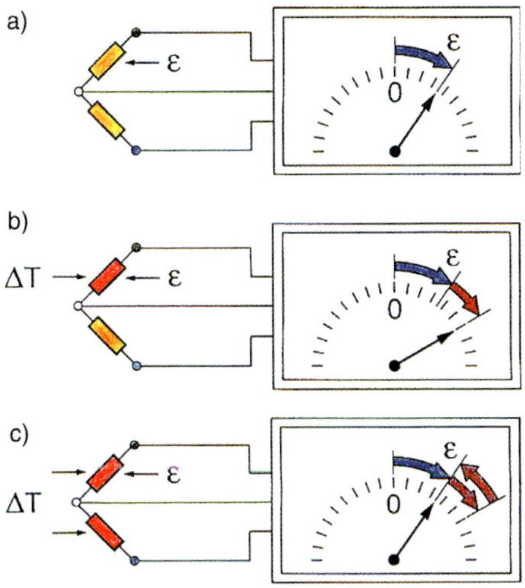

Fig. 4.4 Diagram of the compensation for the effect of temperature changes in a half bridge circuit using an active strain gage and a compensation strain gage; a) no temperature changes, b) temperature change only affecting the active strain gage, c) temperature change equally affecting both active and compensation strain gages

compensation in a half bridge circuit is clearly shown in the diagram below in Figure 4.4.

The mutual compensation of temperature effects in neighbouring arms of the bridge also opens up the possibility of providing temperature compensation just with active strain gages.

A classic example here is the bending beam with two strain gages applied to each of the upper and lower sides. A strain acts on the strain gages on the upper side with the same magnitude as the strain on the strain gages on the lower side, but the signs of the strains are opposite. If these strain gages are wired in a full bridge configuration so that the gages located on one side of the beam are located in opposite arms of the bridge circuit, then the full bridge configuration provides a signal corresponding to the sum of the absolute magnitudes of the four strains. If temperature changes occur during the measurement, affecting all strain gages equally, then effects of these temperature changes are automatically compensated by the alternating signs in the numerator of Eq. (4.6).

A clear example of a full bridge circuit with two active strain gages and two strain gages used for temperature compensation is provided by a measuring point which was applied to a 90 m high steel chimney to monitor the loading on the external wall. Both active strain gages were mounted at the measuring point next to one another on the external wall in the axial direction of the chimney. The two compensation strain gages were mounted on a sufficiently large section of sheet in the same material as the chimney's external shell, giving the same linear thermal expansion coefficient. This sheet with the compensation strain gages was then bonded to the external shell of the

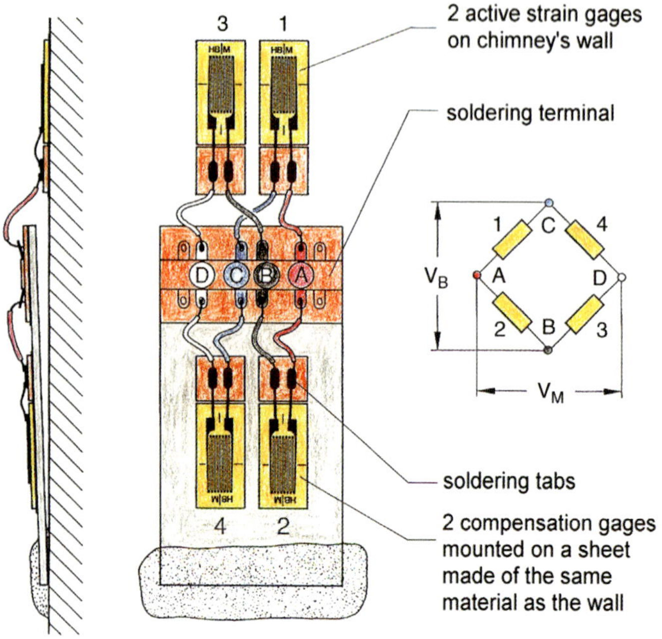

Fig. 4.5 Basic structure of a full bridge circuit with two unloaded but thermally coupled compensation gages and its circuit diagram

Fig. 4.6 Photograph of a full bridge circuit with two active and two mechanically unloaded compensation strain gages mounted on a steel chimney

chimney directly adjacent to the measuring point, so that it remained mechanically unloaded, but assumed the temperature of the chimney wall due to the direct contact. Figure 4.5 shows a sketch of the basic arrangement of this measuring point.

All four strain gages were then wired crosswise so that both active and both compensation strain gages were each arranged in opposite arms of the bridge. This means that an active and a compensation strain gage are always neighbouring one another, providing excellent temperature compensation.

Figure 4.6 is a photograph illustrating the strain gages which were mounted on the chimney wall and wired to form a full bridge circuit.

4.4 Limit of the bridge signal resolution

For a Wheatstone bridge in transducer production, as well as the measurement accuracy gage – a property that is closely connected with the technology of the complete transducer – there is also the question of the possible measurement signal resolution of the bridge circuit. This resolution is limited by interference signals which are superimposed on the measurement signal.

Interference signals arise, for example, from thermal voltages, inductive and capacitive pickup and noise voltages due to the resistor and semiconductor components used. These interference signals can be transformed in frequency and filtered out by using the carrier frequency technique. Actual interference is only caused by those interference voltages with frequencies near the carrier frequency and which therefore cannot be separated from the measurement frequency. Thermal voltages and pickup can be prevented by taking appropriate measures, but the thermal noise due to the irregular thermal movement of electrons in resistors and semiconductors cannot be easily prevented. Each Wheatstone bridge circuit with strain gages creates noise voltages which are present as interference voltages at the input of the following measuring instrument, setting a physical limit to the resolution of the measurement signal. The individual steps in the resolution of the measurement signal must be larger than the noise voltage amplitude so that they are not swamped by the noise.

The rms (root-mean-square) value e_n of the thermal noise can be found using the equation

$$e_n = \sqrt{4kTRB} \tag{4.9}$$

Where k is Boltzmann's constant ($k = 1.38 \times 10^{-23}$ J/K), T is the absolute temperature in K, R is the resistance of the bridge circuit in ohms and B is the frequency bandwidth of the following measurement arrangement [4.4].

The frequency distribution of the momentary values of the noise voltage is a Gaussian normal distribution as shown in Figure 4.7 in which the standard deviation is identical to the rms value e_n. We see from this that 95.4% of the momentary values of noise voltage occur in the region bounded by two standard deviations, so that this limit is used in equipment practice for the definition of the peak-peak value e_{pp} of the noise voltage [4.5]. This gives a peak-peak value e_{pp} that is four times the rms value. This

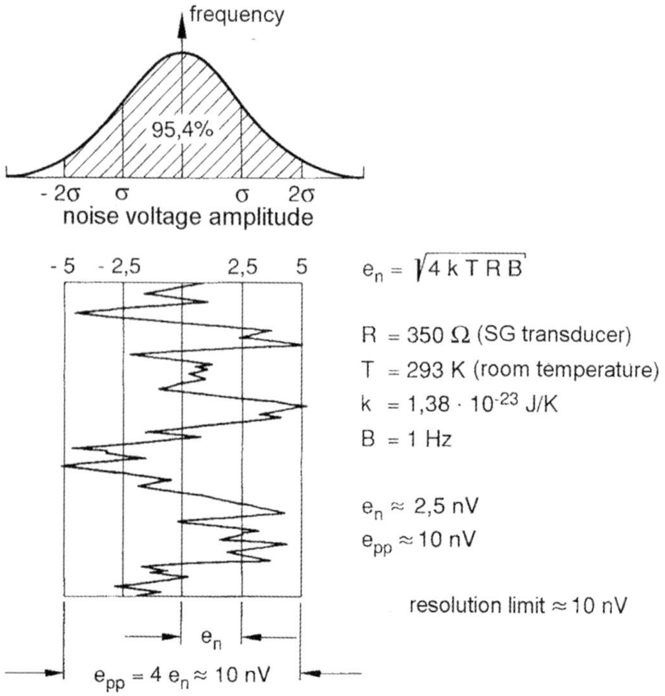

Fig. 4.7 Gaussian normal distribution for the frequency of thermal noise voltage amplitudes in a strain gage circuit and the limits that it defines for the resolution of the circuit measurement signal

value of e_{pp} defines the figure for the physical limit of the possible resolution, because signal voltages lower than e_{pp} can no longer be separated from the noise voltage.

If the Wheatstone bridge circuit of a transducer with a resistance of 350 Ω is taken as an example, then it produces an rms noise voltage of $e_n = 2.5$ nV and a peak-peak value of $e_{pp} = 10$ nV at room temperature and for a measurement bandwidth of 1 Hz in the following measurement equipment. If a transducer is taken with a nominal output signal of 2 mV/V at a bridge supply voltage of 10 V, then a nominal measurement voltage of 20 mV is produced which can be resolved in steps down to 10 nV. Therefore in this example, the absolute physical limit on the resolution is 2×10^6 indicated steps. This limit is due only to the thermal noise in the Wheatstone bridge circuit, assuming that the following measurement equipment is completely free of error and noise voltages. The best equipment of today enables, for example, the measurement of force with a resolution of up to 10^6 indicated steps, approaching the absolute physical limit by a factor of 2.

4.5 Examples of some elementary bridge circuits

The signal summing and compensating effects of the Wheatstone bridge circuit open up a multitude of possibilities of connecting strain gages to fulfil certain measurement tasks, depending on whether the resistances producing the signal are located in opposite or adjacent arms of the bridge. A few basic strain gage arrangements on beams with rectangular and round cross-sections are described as standard examples in the following, together with the reactions of the bridge circuits to the longitudinal forces F, bending moments M_b, torsional moments M_t and the temperature changes which occur.

I. One strain gage mounted on a beam in the longitudinal direction

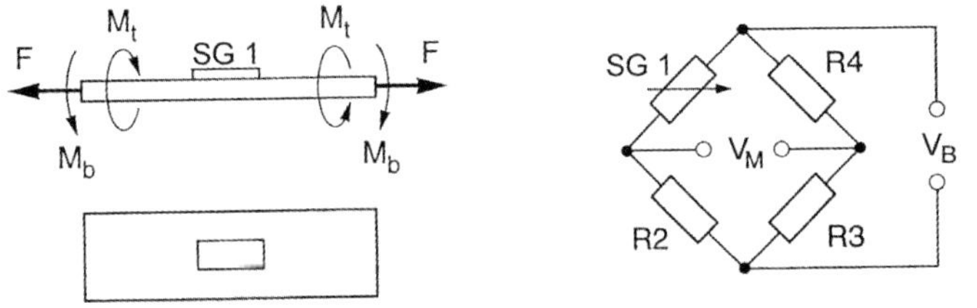

This circuit does not include compensation gages and relies on the self-compensating effect of self-compensating strain gages, if they are used. The circuit can only be recommended for measurements at constant temperature, because the temperature response of the strain gages that are used is included additively in the measurement result, even with the application of the Kreuzer circuit (connection of single strain gages with four leads). The output signal V_M/V_B of the circuit reacts to strains due to longitudinal forces (ε_F), bending moments (ε_{Mb}) and apparent strains (ε_S) due to temperature changes, but it is insensitive to twisting (ε_{Mt}):

$$\frac{V_M}{V_B} = \frac{k}{4}(\varepsilon_F + \varepsilon_{Mb} + \varepsilon_S) \tag{4.10}$$

Where it is used on its own, this circuit can only be employed for stress analysis if the strain gage is bonded in the principal direction 1, for example with the beam loaded on one axis in the longitudinal direction, and the temperature not changing during the measurement. With general applications this assumes that the principal direction is known before the strain gage is mounted and it is also known that the stress state is uniaxial. For plane stress states, at least two of these circuits must be used with two strain gages arranged perpendicular to one another, because the computation of the stresses from the measured strains requires the inclusion of Poisson's ratio for the material. Temperature compensation is therefore not obtained.

II. Strain gage mounted on a beam in the longitudinal direction with a compensating gage on a separate, unloaded piece of the same material used for the beam

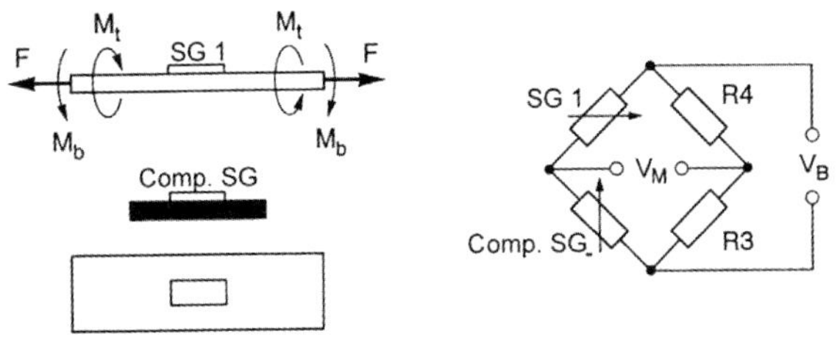

This arrangement is most often used in stress analysis. It measures the mechanical strain occurring in the measurement direction and, because of the compensation gage, it does not react to changes in resistance due to temperature changes. The active and compensating gages should be taken from the same production batch, since the degree of temperature compensation depends on the extent to which their temperature responses match. The compensation gage is located on a mechanically unloaded piece of sheet or a small block in the same material as the measurement object. It is important that the compensation gage is always subjected to the same temperature as the active gage. The circuit output signal V_M/V_B reacts to strains due to longitudinal forces (ε_F) and bending moments (ε_{Mb}). It is insensitive to apparent strains (ε_S) due to temperature changes and to strains due to twisting loads (ε_{Mt}):

$$\frac{V_M}{V_B} = \frac{k}{4}(\varepsilon_F + \varepsilon_{Mb}) \tag{4.11}$$

When used with modern multipoint instruments, this circuit can take the form of a group of closely arranged strain gages (subject to the same temperature) associated with a compensation gage, also located in the vicinity. When the measuring point is scanned, the compensation gage is switched in, together with the selected measuring point. For plane stress states *one* of the described bridge circuits is normally not sufficient for stress analysis. At least two of these circuits with active strain gages must be mounted perpendicular to one another. When using strain gage rosettes with three measuring grids, each rosette needs three of these bridge circuits.

III. Measuring grids wired in adjacent arms of a half bridge in the longitudinal and transverse directions on a beam

This circuit is used if the arrangement of a separate compensation gage is difficult or impossible. Here, the two single strains are not measured, but rather the sum of the magnitudes of the two strains. With a uniaxial stress state, this sum is determined by Poisson's ratio ν, because the strain under the transversely bonded gage is ν times the longitudinal strain with reversed arithmetic sign. If a longitudinal force and a bending moment are acting on the beam as in the figure, then the circuit output signal V_M/V_B reacts to the strains (ε_F) produced by the longitudinal force and the strains (ε_{Mb}) caused by the bending moment, but it is insensitive to apparent strains (ε_S) due to temperature changes and to strains due to twisting loads (ε_{Mt}):

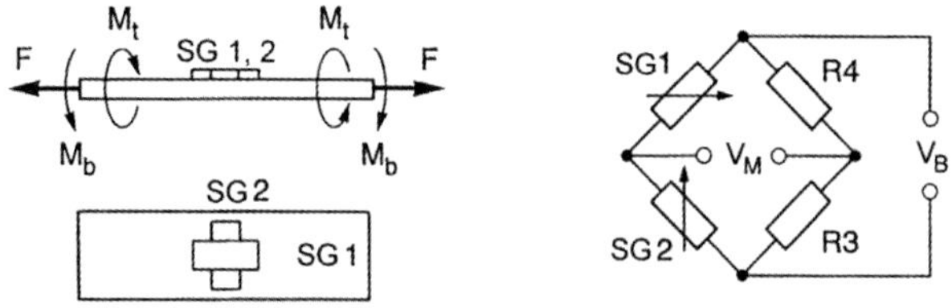

$$\frac{V_M}{V_B} = \frac{k(1+\nu)}{4}(\varepsilon_F + \varepsilon_{Mb}) \tag{4.12}$$

This circuit is only partially suitable for stress analysis, because the unambiguous determination of the longitudinal and transverse strains from the circuit output signal is only possible with a uniaxial stress state with measuring grids aligned in the principal directions.

IV. Strain gages wired in adjacent arms of the bridge and arranged on the upper and lower sides of the beam in the longitudinal direction

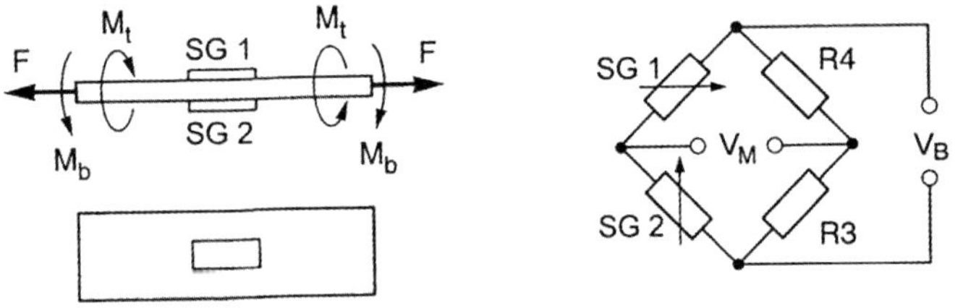

This circuit represents the ideal half bridge circuit for measuring bending loads on beams. With this circuit the magnitudes of the two strains measured on the upper and lower sides ($|\varepsilon_{Mb}|$) are added. The circuit compensates the strains (ε_F) caused by the longitudinal forces in the beam in the same manner as the apparent strains (ε_S) produced by the temperature changes. The circuit is insensitive to twisting loads. The circuit output signal V_M/V_B then becomes

$$\frac{V_M}{V_B} = \frac{k}{4}(2\varepsilon_{Mb}) \tag{4.13}$$

V. Strain gages wired in opposite arms of the bridge and arranged on the upper and lower sides of the beam aligned in the longitudinal direction

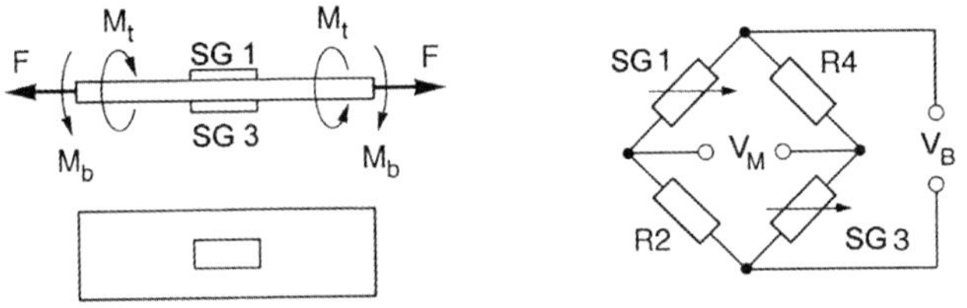

Owing to the strain gages located in the opposite arms of the bridge, this arrangement requires two completion resistors for the full bridge circuit, but it does not exploit the possibilities of temperature compensation in the adjacent arms of the bridge. Therefore with this circuit, the apparent strains (ε_S) due to temperature changes are not compensated. The strains (ε_{Mb}) resulting from the bending of the beam cancel each other out due to their opposing signs in opposite arms of the bridge, whereas the strains (ε_F) produced in the beam due to the longitudinal forces appear with double their values in the output signal. This circuit is insensitive to twisting loads. The circuit output signal V_M/V_B is represented by:

$$\frac{V_M}{V_B} = \frac{k}{4}(2\varepsilon_F + 2\varepsilon_S) \tag{4.14}$$

VI. Active strain gages wired in opposite arms of the bridge and arranged on the upper and lower sides of the beam aligned in the longitudinal direction. Separate compensation gages are wired in the neighbouring arms of the bridge

This circuit uses the possibilities of compensation offered by the full bridge and prevents apparent strains due to temperature changes from affecting the signal. To achieve this, the two compensation gages must be bonded to mechanically unloaded blocks or pieces of sheet from the same material as the measurement object, and they must be subject to the same temperature changes.

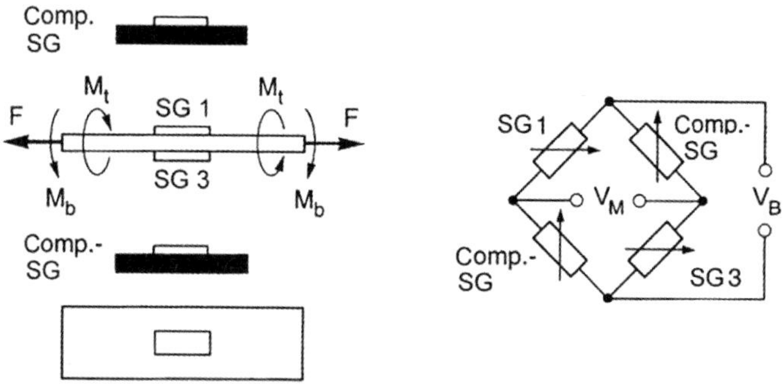

Due to the active gages being located in opposite arms of the bridge, the circuit is insensitive to strains resulting from bending and is doubly sensitive to strains (ε_F) in the beam due to longitudinal forces. The circuit is insensitive to twisting loads. The circuit output signal is represented by:

$$\frac{V_M}{V_B} = \frac{k}{4}(2\varepsilon_F) \tag{4.15}$$

VII. Two strain gages applied to both the upper and lower sides of a beam and wired diagonally opposed to one another in a full bridge

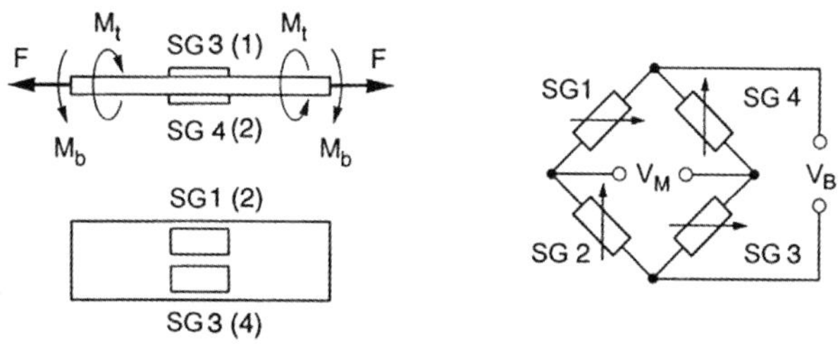

This full bridge circuit is ideal for measuring elements or transducers subjected to bending. Due to the diagonally opposed wiring in the bridge, strain gages loaded in the same direction are located in opposite arms and gages loaded in opposite directions are located in adjacent arms of the bridge. The apparent strains due to temperature changes are compensated, as are the strains due to longitudinal forces in the beam which give the same resistance changes in all four arms. The circuit is insensitive to twisting loads. The circuit output signal V_M/V_B corresponds to four times the magnitude of the bending strain ($|\varepsilon_{Mb}|$) occurring in the beam surface:

$$\frac{V_M}{V_B} = \frac{k}{4}(4\varepsilon_{Mb}) \tag{4.16}$$

VIII. Strain gages applied in each of the longitudinal and transverse directions on both the upper and lower sides of a beam with those gages aligned in the same direction wired diagonally opposed to one another in a full bridge

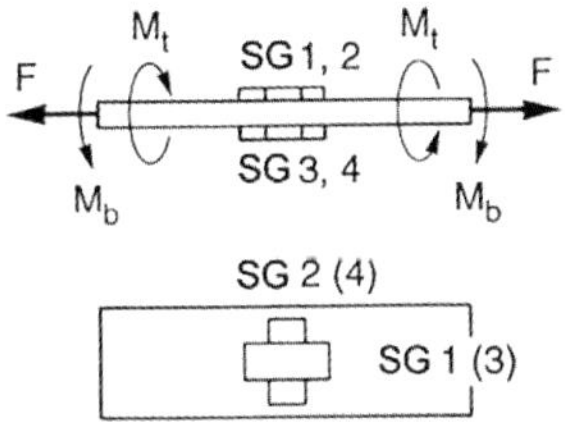

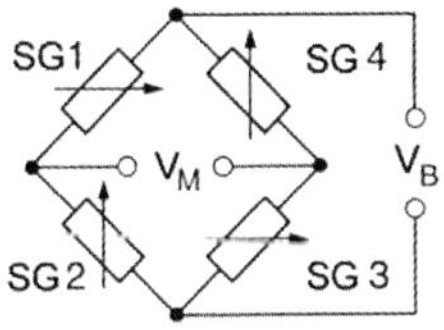

This full bridge circuit reacts to the longitudinal force in the beam and is often used for force measuring elements or simple force transducers that are subject to longitudinal force. The circuit is fully self-compensating for temperature. The gages bonded transversely to the beam axis measure the transverse strain due to the uniaxial longitudinal loading, giving ν times the longitudinal strain with a sign opposite to the longitudinal strain. The circuit does not react to bending loads, since their effects cancel each other out in the circuit. It is insensitive to twisting. The circuit output signal V_M/V_B corresponds to double the sum of the magnitudes of the longitudinal strain ε_F and the transverse strain $\nu\varepsilon_F$:

$$\frac{V_M}{V_B} = \frac{k}{4} 2(\varepsilon_F + \nu\varepsilon_F) = \frac{2k(1+\nu)}{4}\varepsilon_F \tag{4.17}$$

IX. Two pairs of strain gages arranged on a round beam at 45° to the beam axis and perpendicular to one another. The strain gages are wired to form a full bridge circuit

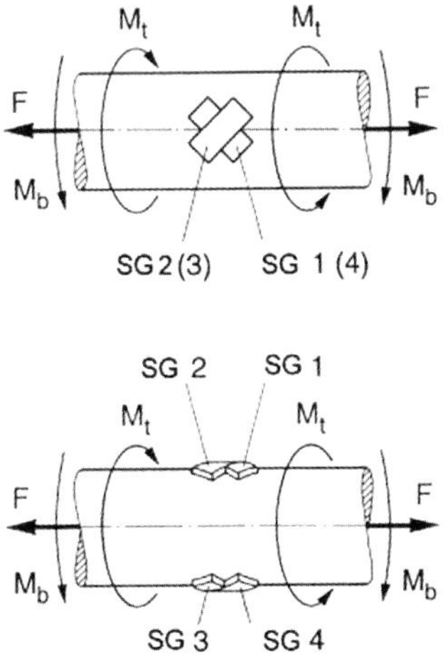

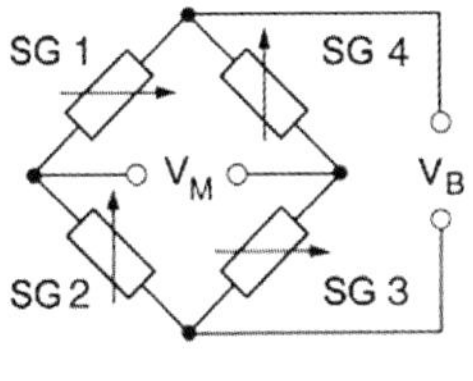

This full bridge circuit is the ideal arrangement for measuring torque. When the round beam is loaded by torque, the two principal normal stresses which have the same magnitude occur with different signs at 45° to the beam axis. The strains produced by these stresses, which are also equal in magnitude and have different signs, are measured by the strain gages. Mutually perpendicular strain gages measure strains of different signs and are wired in neighbouring arms of the bridge. The bridge circuit adds the strain magnitudes $|\varepsilon_{Mt}|$ measured by the four strain gages.

Therefore, the circuit output signal V_M/V_B corresponds to four times the strain ε_{Mt} produced by the torque in the beam surface at 45°:

$$\frac{V_M}{V_B} = \frac{k}{4}(4\varepsilon_{Mt}) \tag{4.18}$$

The full bridge circuit is fully self-compensating for temperature. It is insensitive to bending and also to longitudinal forces. With bending and longitudinal loading both gages of a pair that are located in adjacent arms are loaded to the same extent so that their effects cancel one another out.

References

[4.1] Hunter-Christie, S. (1833) Experimental determination of the laws of magneto-electric induction in different masses of the same metal, and of its intensity in different metals. *Philosphical Transactions of Royal Society,* **123**, 95–142.

[4.2] Wheatstone, Ch. (1843) An account of several new instruments and processes for determining the constants of a voltaic circuit. *Philosophical Transactions of the Royal Society, Series B,* **133**, 303–327.

[4.3] Heringhaus, E. (1982) Trägerfrequenz- und Gleichspannungsmeßverstärker für das Messen mechanischer Größen – ein Systemvergleich aus anwendungsbezogener Sicht. *MTB,* **18**, 42–49 and pp. 70–73.

[4.4] Küpfmüller, K. (1955) *Einführung in die theoretische Elektrotechnik,* 5th edn, Springer Verlag, pp. 456.

[4.5] Kreuzer, M. (1980) Ein programmierbares Präzisions-Meßgerät der Genauigkeitsklasse 0,0005 und seine Anwendungen. *MTB,* **16**, 41–46.

5 Adjustment and compensation circuits

5.1 General

The simple measures for temperature compensation described in the previous chapter are generally sufficient to fulfil the accuracy requirements in bridge circuits used for the direct measurement of strain as in experimental stress analysis. During the construction of transducers for indirect measurement of mechanical quantities, such as force, pressure or torque, substantially lower measurement uncertainties can be reached than is possible with simple temperature compensation. During the construction of transducers the output signal will be calibrated in units of the quantity to be measured. The strains in the spring body occurring at the installation points of the strain gages are mostly inconsequential. Transducers with strain gages for various levels of claimed accuracy are in use today. Transducers with strain gages in full-bridge circuits can meet the highest quality standards as shown below.

The transformation of the measured quantity, force, into an electrical signal proportional to the force will be carried out in several steps. The first step takes place in the spring body of the transducer; it consists of the conversion of the force into an elastic deformation of the spring body. The second step converts the spring body's deformation into resistance changes of strain gages mounted on its surface. In the third step the changes in resistance are used in an electrical circuit to create changes in the output voltage of the circuit. The strain gages are connected to form a Wheatstone bridge circuit to achieve this. The bridge circuit is supplied with a voltage from outside, and the output of the bridge represents the output signal of the transducer.

Resistance changes with opposite signs occurring in neighbouring arms of the bridge circuit have a favourable effect on the output signal. Thus, the aim is to construct spring bodies with positive and negative strains at their surface occurring both at the same time when load is applied. These strains create, in the strain gages installed at the spring body, positive and negative changes in resistance, depending on the location of installation. The wiring is then carried out such that positive and negative changes occur in adjacent arms of the bridge. In spring bodies formed as bending beams or shear beams the demand for strains with opposite sign and equal quantity can be almost ideally met. In cylindrical spring bodies, the strain occurring transverse and with opposed sign to the strain in loading direction can be used. This transverse strain depends on the value of Poisson's ratio, and amounts about 30% of the strain in loading direction.

Complete full-bridge circuits should be applied exclusively in force transducers with strain gages. A half-bridge circuit would require a second half-bridge outside the transducer as supplement to complete the circuit to a full-bridge. The connection leads then required between the two half-bridges would be part of the measuring components, and changes of resistance in these leads would be error sources.

Characteristic parameters of industrial manufactured full-bridge transducers will be set following particular standards and adjusted to stipulated tolerances by the

Technology and Practical Use of Strain Gages: With Particular Consideration of Stress Analysis Using Strain Gages. First Edition. Stefan Keil.

producer. This is required in order to be able to exchange transducers in the case of defects or when multiple transducers are used simultaneously in parallel. To meet these requirements it is necessary to insert additional resistors into the bridge circuit in order to adjust the output signal of the transducer and to compensate the effect of changes in temperature.

The output signal of an unloaded transducer is desired to be zero. And it should remain zero even with changes in temperature. Also desired is the adjustment of the nominal output and its independence of temperature changes within a given temperature range.

The setting of standardized values for nominal output (span), zero signal and bridge resistance, each within specified tolerances, requires careful implementation of specific procedures. In a defined temperature range the measurement properties of a transducer have to be independent of changes in temperature, so the output signal of the bridge circuit installed at the transducer also has to be temperature-independent. That means that in the manufacture, during and after installation of the bridge circuit, special measures must be taken in order to achieve complete temperature compensation for sensitivity and zero signal. Regarding the effect of temperature changes, two effects are taken into account: effects on the electrical bridge circuit and impact on property changes of materials involved. Both effects can be compensated by the circuitry.

To minimize uncertainty of a transducer, the used strain gages need to have a specific amount of creep. During manufacture the creep properties of modern foil strain gages can be adapted to the creep properties of the material of the transducer's spring body. Adiabatic changes of the modulus of elasticity can be compensated by temperature-dependent resistors.

The construction of transducers provides two possibilities for the positioning of the additional resistors: the direct installation at the spring body or installation outside the hermetically sealed interior in the externally accessible junction box of the transducer. Resistors for compensation of creep and temperature effects have to be mounted directly on the spring body, but for fine tuning they can still be affected by shunts in the junction box if the interior has already been hermetically sealed. After execution of all compensation and calibration tasks the junction box is usually sealed by the manufacturer to protect it against external influences, using a plastic compound.

During the development of powerful transducers with very high quality, a highly sophisticated strain gage installation technique was also developed. As well as a number of influencing variables, the installation of strain gages, soldering terminals, compensation and adjustment resistors and the wiring of the complete bridge circuit are of particular importance for the quality of the transducer.

Other key aspects in this context are the preparation of the surface of the spring body before installation of the gages, the adhesive used and the bonding technique used (one- or two-component cement, hot- or cold-curing adhesive, bonding under pressure, etc.).

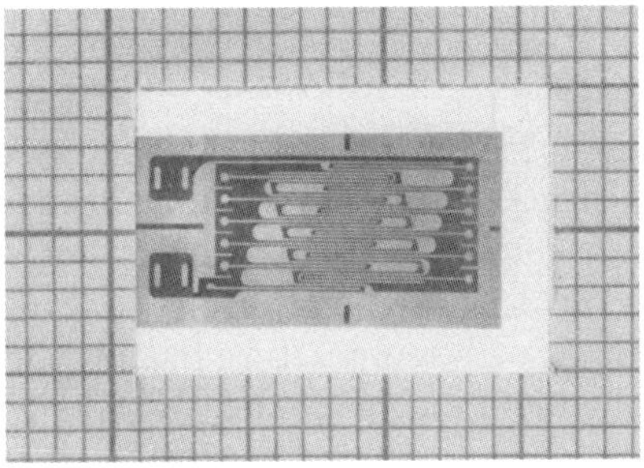

Fig. 5.1 Special strain gage with separable grid parts to change the position of the active grid parts after installation for the construction of transducers

Precise positioning and alignment of the strain gages at the appropriate sites on the spring body are also important. Eccentricities in the spring body and the strain gage arrangement can cause errors in the measurement signal and should therefore be avoided. However, errors caused by eccentricity can be compensated by the arrangement of several useful interconnected strain gages. The geometry of the spring body and the arrangement of the strain gages must be designed to be completely symmetrical to the load introduction. Absolute symmetry is not feasible in practice because the strain gages cannot be positioned absolutely accurately and because the spring body does not behave perfectly symmetrically, because of inhomogeneity in the material. Practically occurring asymmetries can be corrected by including strain gages in the circuit that open the possibility of the geometric displacement of the installation site, after sticking, by an appropriate design of the measurement grid. The centre of the measuring grid can be shifted by several connection points standing for election and by ravel of certain grid parts in order to activate additional tracks [5.1]. A strain gage developed according to the patent in [5.1] is shown in Figure 5.1.

Of particular importance with regard to the long term-stability of the measurement properties of the transducer is to protect the installed measuring system against external influences, especially against moisture. Experience shows that the hermetic sealing of the area with the strain gage installation and a carefully executed cable terminal are essential for the long-term stability of the protection against moisture. The hermetic encapsulation must not lead to mechanical shunts. Particularly important for transducers for small forces is to ensure, that air pressure fluctuations do not distort the measurement result [5.2].

Information on the performance of a transducer is given by the characteristics values given in its specification. These characteristics are standardized with respect to the terms and definitions used [5.3].

5.2 Compensation of zero shift with temperature change and zero adjustment

Even with the greatest care, installed bridge circuits have in the unloaded state a temperature-dependent output signal. Small electrical imbalances in the bridge circuit are inevitable, despite diligent efforts to use identical design for the four arms of the full-bridge by using identical strain gages and cables from identical lengths and materials. This results in different resistance changes with temperature in the four bridge arms and this generates a temperature-dependent output signal. Additional causes of zero-signal drift can be thermal expansion of the spring body material, any changes in insulation resistance, expansion differences between the transducer housing and the spring body and changes in air pressure in a hermetically sealed transducer housing.

The resultant zero signal drift caused by temperature changes can be measured and can be, almost completely eliminated by compensation measures, within a predetermined temperature range. The temperature drift of the circuit can be checked relatively easily. By applying the excitation voltage, the change in the output signal is measured as a function of temperature. This zero shift with temperature change can be eliminated by incorporating an appropriate temperature-dependent compensation resistor in the bridge circuit. There are two possibilities for carrying out the compensation: by soldering into one bridge arm a wire whose resistivity has a suitable temperature dependence (typically copper or nickel) or to use a special resistance pattern offered by the strain gage manufacturer for zero-shift compensation. In the following, the practical procedure for both options is briefly explained with an example.

Example:
On a full-bridge strain gage circuit with a resistance of 350 Ω and the gage factor of 2.0 a drift of the zero signal of 0.0320 mV/V is measured for a temperature variation between −10 °C and +40 °C. This drift corresponds to an apparent strain of $\varepsilon_S = 64 \times 10^{-6}$ (for GF = 2.0 this is 1 mV/V = 2000 µm/m). The resistance change associated with this drift can be calculated easily

$$\Delta R = k\varepsilon_S R_0 = 0.047\,\Omega \tag{5.1}$$

where $R_0 = 350\,\Omega$ is the nominal resistance of the strain gages used. With the given numerical values Eq. (5.1) gives $\Delta R = 0.047\,\Omega$, which means that at a temperature change of $\Delta T = 50\,\text{K}$ the resistance will change on one side of the installed bridge circuit to 0.047 Ω more than on the other side. This can be compensated for by inserting a temperature-dependent resistor into the side with the lower change in resistance, which varies by 0.047 Ω under a change in temperature of 50 K. This can for example be a nickel wire. Nickel has a temperature coefficient of resistivity of $TK\rho = 5 \times 10^{-3}/\text{K}$, the resistivity is $\rho = 0.08\,\Omega\text{mm}^2/\text{m}$ (note: ρ for nickel can be between $0.08\,\Omega\text{mm}^2/\text{m}$ and $0.11\,\Omega\text{mm}^2/\text{m}$, $TK\rho$ between $3.7 \times 10^{-3}/\text{K}$ and $6.0 \times 10^{-3}/\text{K}$).

The value of the nickel resistance to be inserted is given by the equation

$$\Delta R = TK_{\rho} R \Delta T \tag{5.2}$$

Solving Eq. (5.2) for R and inserting the values gives

$$\Delta R = \frac{0.47}{50 \cdot 5 \cdot 10^{-3}} = 1.88\,\Omega \tag{5.3}$$

The required wire length depends on the wire cross-section A

$$l = \frac{R \cdot A}{\rho} \tag{5.4}$$

And using the numerical values

$$l = \frac{1.88 \cdot 0.00785}{0.08} = 0.184\,\text{m} \tag{5.5}$$

The zero signal drift due to the change in temperature is compensated by soldering a nickel wire of 0.1 mm Ø and 18.4 cm length into the right arm of the bridge circuit. The arm with the lower thermal output is in this case the right arm of the bridge. This is indicated by the sign of the offset of the bridge.

Another simple way to compensate the temperature drift of the zero signal is by the use of an appropriate bondable adjustable resistor pattern. Different strain gage manufacturers offer such adjustable resistor patterns, which are cemented like strain gages to the measuring object. This possibility is illustrated here by the use of an etched nickel pattern foil with a resistance of twice 1.05 Ω, each divided into three resistance steps of 0.15 Ω, 0.3 Ω and 0.6 Ω

Figure 5.2 shows as an example such a nickel resistance pattern. The temperature coefficient of resistivity of the pattern is $TK\rho = 5 \times 10^{-3}$/K. Certain resistance values can be selected by separation of compounds in the resistance pattern.

Adjustable resistor patterns, as shown in Figure 5.2, can be installed in the same manner as a normal strain gage. The pattern can be installed during installation of the bridge circuit, as shown in Figure 5.3.

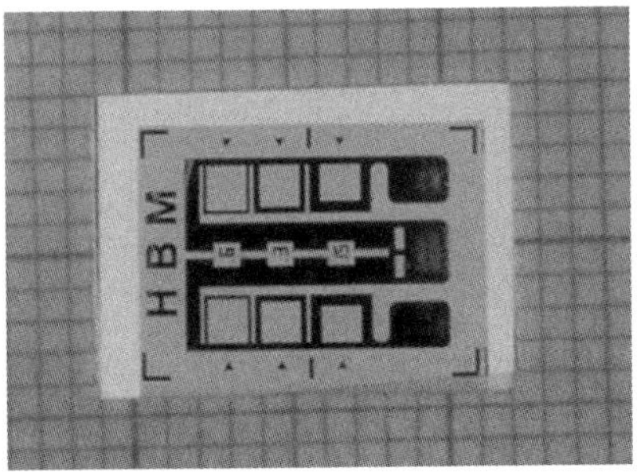

Fig. 5.2 Adjustable resistance pattern to compensate the temperature drift of zero signal (Type TN1, HBM)

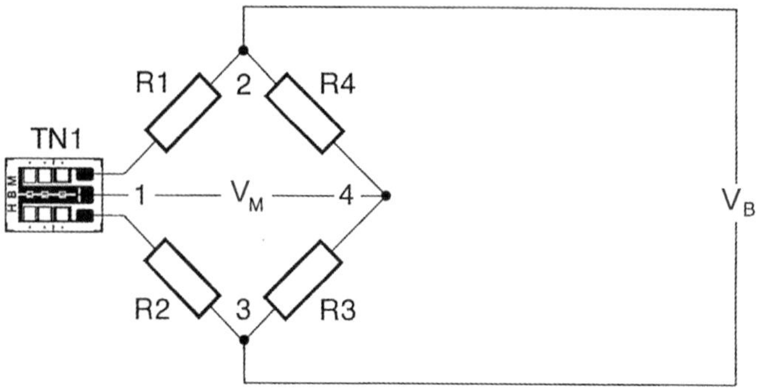

Fig. 5.3 Adjustable resistance pattern installed into a full-bridge circuit for compensation of temperature shift of the zero signal

After installation and measurement of the thermal output of the bridge circuit the suitable value of the compensation resistance must be selected by separation of the corresponding conductor line. For the previous example, a resistance of 0.188 Ω was required with the $TK\rho = 5 \times 10^{-3}$/K. Using the pattern as given in Figure 5.2 the resistance value of 0.15 Ω should be activated by separation.

In the industrial production of transducers the compensation of the zero shift can be carried out in two steps, first a coarse adjustment with a resistor installed at the spring body, and second a fine adjustment with a second compensation resistor in the junction box. When checking the compensation in a climate chamber at each temperature level it is essential to wait until a steady state of temperature distribution in the transducer is present.

After compensation of the temperature drift of the zero signal, the output signal of the unloaded transducer under applied supply voltage will be non-zero, mainly because the inserted compensation resistor has brought additional asymmetry into the circuit. It is a characteristic of the strain gage technology that very small changes in the resistance in one bridge arm lead to a significant bridge offset. This asymmetry can be removed by inserting an appropriately sized resistor in the bridge arm with the lower resistance. This must, however, be a temperature-stable resistor with no effect on the temperature compensation of the zero signal. For this resistor constantan is preferred. Here is to be noted that small asymmetries in the bridge circuit do not have an unfavourable influence on the measurement accuracy. In addition, the zero signal can be compensated by the adjustment elements of the subsequent measurement device so that during measurement the display shows zero when the transducer is unloaded.

During manufacture of a transducer attention should be paid to using strain gages with resistances as nearly identical as possible. The selection can be made using an ohmmeter. Despite careful selection and strain gage installation there will still be a small bridge asymmetry which can be eliminated by inserting a suitable resistor into the circuit. The practical implementation of such a zero balance will be explained with an example.

Example:
A full-bridge circuit of $R_0 = 350\,\Omega$ installed at the spring body of a transducer produces at the unloaded transducer an output signal of 0.332 mV/V. This signal corresponds (1 mV/V = 2000 µm/m) to an apparent strain of 664 µm/m. With Eq. (5.1) the corresponding change in resistance is

$$\begin{aligned}\Delta R &= R_0 k \varepsilon_S \\ &= 350 \times 2 \times 10^{-6} \\ &= 0.456\,\Omega\end{aligned} \tag{5.6}$$

that is in the bridge arm with the lower resistance an additional resistance of 0.465 Ω must be inserted to bring the full-bridge into balance. This can be done by inserting a temperature insensitive wire. Using, for example, constantan wire with a diameter of 0.1 mm ($A = 0.00785\,\text{mm}^2$) and $\rho_{\text{constantan}} = 0.5\,\Omega\text{mm}^2/\text{m}$

$$l = \frac{\Delta \text{R} \cdot \text{A}}{\rho} = \frac{0.465 \cdot 0.00785}{0.5} = 0.0073\,\text{m} \tag{5.7}$$

A wire of 7.3 mm length is required.

The adjustment of the zero signal of the circuit under consideration can also be carried with an adaptable resistor pattern offered by the strain gage manufacturers, made from a temperature insensitive material. The best for this purpose are resistor patterns made from constantan. Such a network is shown in Figure 5.4.

The pattern shown in Figure 5.4 has a total resistance of twice 4.73 Ω, and each can be separated in resistance steps of 2.4 Ω; 1.2 Ω; 0.6 Ω; 0.3 Ω, 0.15 Ω and 0.08 Ω or sums of these individual values. The carrier foil from polyimide can be bonded to the measuring object like a strain gage.

When using a resistor pattern that is switched to the full-bridge as shown in Figure 5.5 the conductor lines have to be separated so that with Eq. (5.6) the calculated resistance is set as closely as possible. The connector ridges marked by

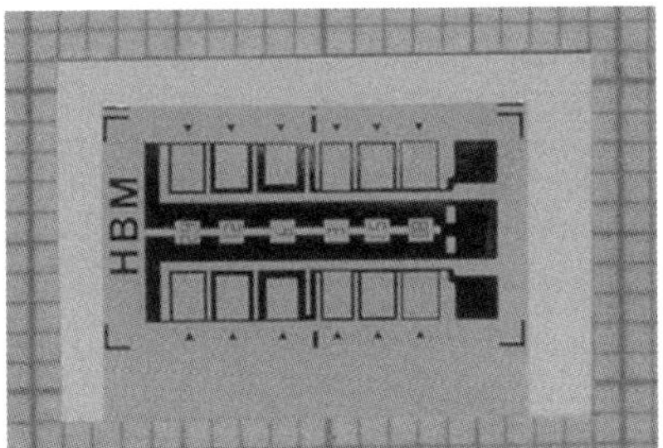

Fig. 5.4 Enlarged reproduction of the adaptable resistance pattern for adjustment of the zero signal of a full-bridge strain gage circuit (Type NA1, HBM)

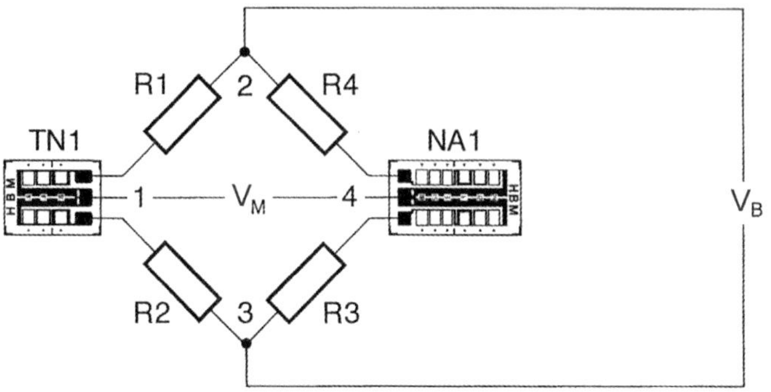

Fig. 5.5 In addition to the temperature compensation pattern in the circuit is inserted an adaptable resistance pattern for adjustment of the zero signal of the unloaded transducer

triangles at the 0.15 Ω step and at the 0.3 Ω step have to be separated at the pattern NA1, so that an additional resistance of 0.45 Ω is inserted in the bridge arm to be corrected.

Figure 5.6 shows a schematic representation of the positioning of the resistor patterns for the compensation of the temperature drift and the adjustment of the zero signal in the full-bridge circuit. In the industrial transducer production the necessary correction of the zero signal can be made in the junction box. This can be problematic with regard to the occurrence of temperature gradients inside precision transducers. As the characteristic quantity of the asymmetry of the bridge circuit the 'relative deviation of the zero signal' is stated. It is defined as the deviation of the zero signal from the value zero, related to the nominal output of the transducer.

The zero signal of a transducer has a practically useful meaning, too. If a transducer is overloaded over its mechanical load limit it will almost always result in a change of

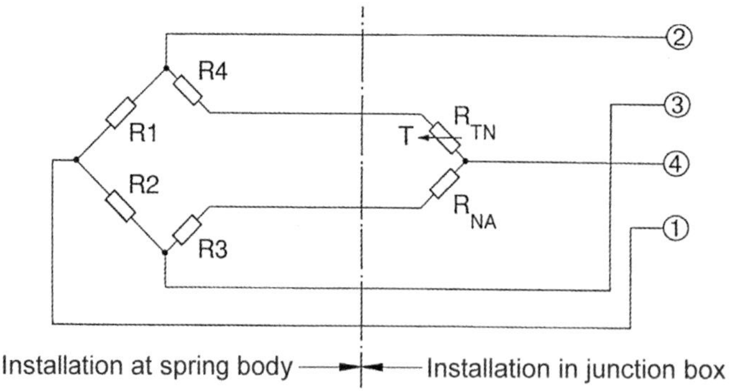

Fig. 5.6 Arrangement of the resistors: R_{TN} for temperature compensation and R_{NA} for adjustment of the zero signal

the zero signal. Thus the check of the zero signal can provide information regarding the overload of the transducer. But it is essential that the zero signal of the undamaged transducer is known. The zero signal of transducers with high quality hardly changes over a long period with correct treatment. One example was that the change of the zero signal of a force transducer with a strain gages after 13 years of industrial use was less than 0.01% [5.4].

5.3 Compensation of temperature effects on the sensitivity and linearity measures

In a strain gage circuit sensitivity means the size of the change in the output signal in response to a change in the quantity to be measured. The sensitivity of an uncompensated strain gage bridge circuit is temperature dependent. This temperature dependence is quite large, compared with the currently achievable uncertainties of precision force transducers. The reason for this is the temperature dependence of both the gage factors of the strain gages used and the modulus of elasticity of the material on which the strain gages are installed.

The gage factor of the installed strain gages usually increases slightly with increasing temperature. When using constantan foil strain gages as a standard for the temperature-dependent change of the gage factor 0.008%/K can be expected.

The modulus of elasticity of the material of the measuring object is reduced with increasing temperature, so that a constant loading of the material at low temperature causes less deformation than at a higher temperature. Thus a correspondingly lower or higher resistance change is effected in the strain gages installed on the material. The temperature dependence of the modulus of elasticity is dependent on the material used, and for alloys on their composition. The influence of temperature on the modulus of elasticity, depending on the material, is between 0.025%/K and 0.070%/K. Stainless steel and CuBe are at the lower end, aluminium alloys in the middle and titanium alloys in the upper part of that range. The values given here are to be considered as a guide only, because considerable deviations are possible caused by various alloying constituents.

The temperature effects that affect the gage factor and the modulus of elasticity result in an increase of the sensitivity in the measuring circuit as the temperature rises, so that their sum must be taken in account for compensating measures.

Table 5.1 shows the effects of temperature changes on the modulus of elasticity for four materials commonly used for transducer spring bodies.

Figure 5.7 shows the effect of temperature changes on the gage factors of measuring grids made from the two important materials constantan (58% Cu; 41% Ni; 1% Mn) and Karma (74% Ni; 20% Cr; 3% Al; 3% Fe). Note: The diagram in Figure 5.7 includes the entire measuring point, including the adhesive layer between strain gage and the measured object.

Table 5.1 Temperature effect on the modulus of elasticity of materials used for spring bodies of transducers

stainless steel	−0.03 . . . −0.035%/K
AlCuMg1	−0.04 . . . −0.05%/K
TiAl6V4	−0.06 . . . −0.07%/K
CuBe2	−0.03 . . . −0.033%/K

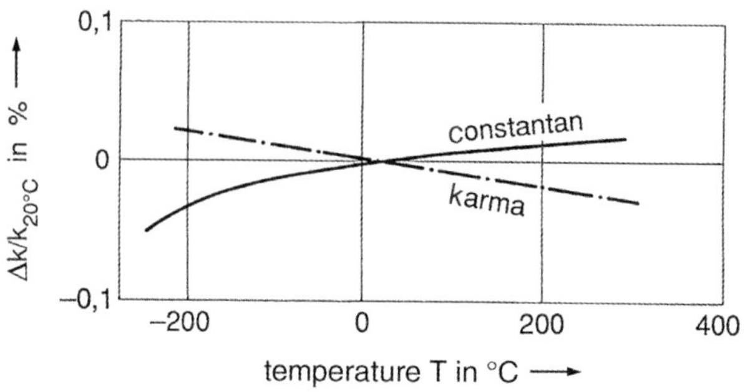

Fig. 5.7 Effect of temperature on the gage factor of constantan and karma

The influence of temperature variations on the sensitivity can be compensated by the insertion of temperature-dependent compensation resistors in the bridge excitation supply lines. These resistors change the temperature-dependence of the supply voltage, in such a way that the temperature-induced increase in sensitivity by gage factor and modulus of elasticity in the output signal is no longer noticeable.

Thus, for example with increasing temperature the supply voltage decreases due to the temperature-dependent resistors, so that the temperature-related increase in sensitivity of the bridge circuit is compensated for. These resistors, often semiconductor resistors are used, are usually mounted with a highly thermally conductive adhesive close to the strain gages directly on the spring body to record its temperature changes without delay.

With high-quality transducers, fine adjustments can be made with compensation resistors in the junction box connected in parallel to the compensation resistors at the spring body, where after closing the hermetically sealed housing readjustments can still be made. The practical implementation of a sensitivity compensation is relatively complex since the transducer must be loaded (e.g. in a material testing machine) and exposed to different temperatures (e.g. in a temperature chamber). Here, the transducer signal must be recorded with a precision measuring instrument. For precision force transducer and high precision weighing cells deadweight standard machines are used, which operate through the direct action of masses in the gravity field.

Fig. 5.8 Under floor part of a 240 kN deadweight force standard machine

Figure 5.8 shows the underfloor part of a deadweight machine for loads up to 240 kN giving an impression of such a facility. The uncertainty of the machine shown in the photo amounts of 2×10^{-5}.

The value of the temperature sensitive resistor which has to be inserted into the voltage supply leads, to compensate the temperature influence on the sensitivity, can be estimated relatively easily. The modulus of elasticity of the Cr-Ni steel used causes a sensitivity increase of 0.035%/K and the strain gage installation causes a sensitivity increase of 0.008%/K. This results in an estimated increase in the sensitivity of 0.043%/K. That means that in order to compensate this change in sensitivity a temperature dependent resistor by the same percentage has to be inserted in the supply lead.

Example:
In the example, a full-bridge circuit with 350 Ω strain gages is to be temperature-compensated with respect to the sensitivity. From experience a compensation resistance of 30 Ω is inserted in the supply lines of the full-bridge. Thus the nominal resistance of the excitation circuit of the bridge amounts 380 Ω. This resistance must change by 0.043%/K, which means a change of 0.163 Ω/K.

The compensation of the temperature effect on the sensitivity may be carried out with nickel resistors inserted into the supply leads. The temperature coefficient of resistivity of nickel is $TK_\rho = 5 \times 10^{-3}$/K. In order to produce a change in resistance of $\Delta R = 0.163\,\Omega$/K in the feed line a nickel resistor of $R = 32.6\,\Omega$ is required:

$$\Delta R = R \times TK_\rho \tag{5.6}$$

$$R = \frac{\Delta R}{TK_\rho} = \frac{0.163}{5 \cdot 10^{-3}} = 32.6\,\Omega \tag{5.7}$$

In order to produce a resistance of 32.6 Ω with nickel wire of 0.05 mm diameter, a wire length of 80 cm is needed. The use of long wires for compensation is not recommended, as they must have the same temperature as the spring body, which is difficult to achieve. These days, the use of semiconductor resistors for temperature compensation is possible; semiconductors with reliable temperature coefficients of resistivity are freely available. The correlation between temperature and change in resistance is not linear, what has an adverse effect. Therefore, the manufacturers of precision transducers prefer to use specially designed foil resistors which are manufactured photo-chemically like foil strain gages. In contrast to strain gages they consist of material with temperature-dependent resistance such as nickel. When inserting these elements into the transducer, resistors something greater than estimated are used and inserted into the feed lines which are later adjusted by shunts during the calibration procedure.

However, this calibration procedure requires relatively high expense and a lot of experience. The required temperature-dependent change of resistance in the excitation lines can be achieved by inserting temperature-sensitive resistance patterns as offered by the strain gage manufacturers into the excitation line of the transducer circuit. Figure 5.9 shows a resistance pattern for this purpose etched from nickel foil. The total resistance of 60 Ω is downgraded to part resistances of 4 Ω, 8 Ω, 16 Ω and 32 Ω. By separating conductor lines, the part resistance can be activated and connected in series.

The resistance of 32.6 Ω required for the example is divided into two parts switched in the two excitation leads because of the bridge balance, as can be seen in Figure 5.10. Pro wire 16.3 Ω is required. The patterns of each provide a part resistor of 16 Ω, and it will be activated by separating the corresponding conductor line.

The influence of temperature on the sensitivity is described in the specification of the transducer as 'temperature effect on the characteristic value per 10 K'. The characteristic value is defined as the output signal of the force transducer under nominal load.

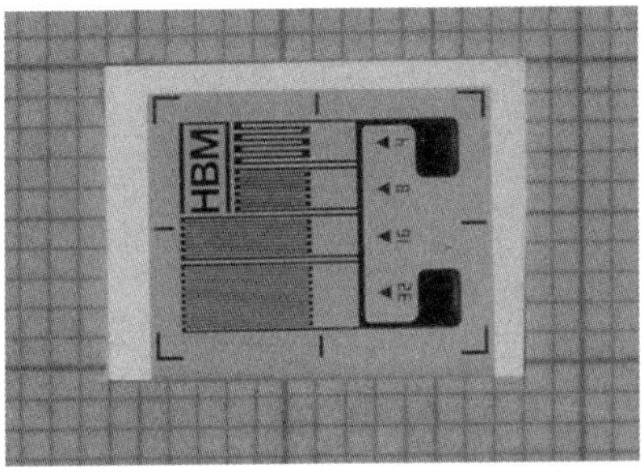

Fig. 5.9 Enlarged representation of an adjustable pattern from nickel foil for temperature compensation of the sensitivity (Type TC1, HBM)

Figure 5.11 shows a photograph of a strain gage installation on a tensile rod designed for precise force measurement with four equally spaced full-bridge circuits on its perimeter. It shows the installed calibration and compensation patterns disposed in the strain-free direction of the tensile rod, which is orientated for $\nu = 0.3$ at an angle of 61° to the axis of the rod.

The construction of the transducers is aimed at proportionality between the output signal and the measured quantity, so a linear relationship is required. To compensate for any nonlinearities strain-dependent resistors are used, which are connected into the feed lines in series. These resistors are cemented to the spring body and influence, by their strain dependence, the supply voltage in order to linearize the output signal. For reasons of symmetry, two such resistors are used, one in each of the two supply lines. Depending on the grid material, on the place of installation and on the orientation of the conductor tracks of the resistors there is a wide variability to compensate constant nonlinearities [5.5]. For high quality transducers, on each of these two resistors is a second resistor connected in parallel and arranged in the cable box.

The deviation from linearity is a specific characteristic of a transducer and is defined as the maximum relative deviation, under increasing load, of the recorded transducer

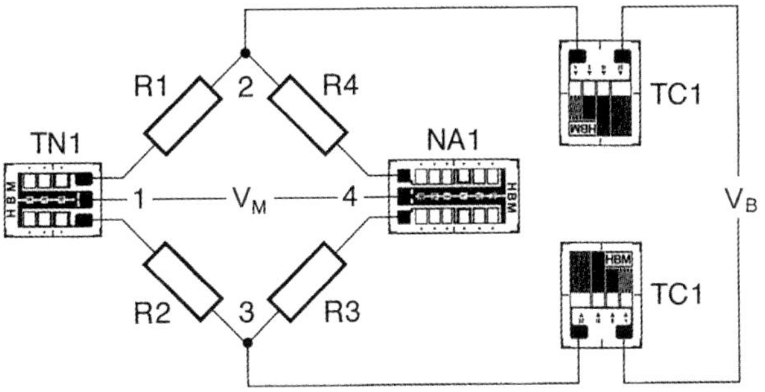

Fig. 5.10 Two resistance patterns TC1 inserted in the feed lines of a transducer for compensation of the temperature effect on the sensitivity

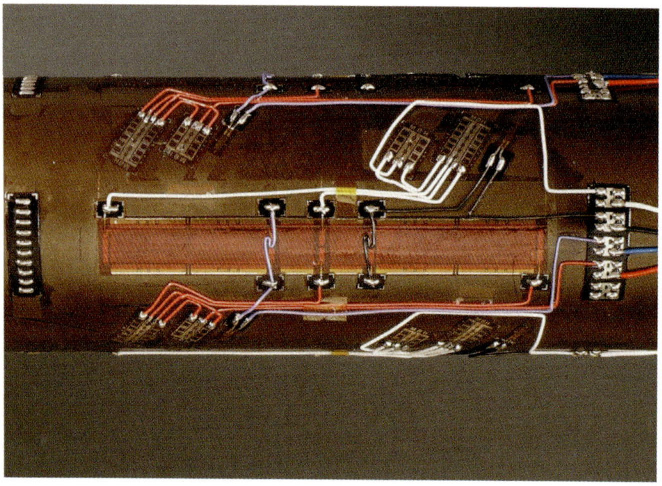

Fig. 5.11 Practical implementation of compensated and adjusted full-bridge circuits on a tensile rod

characteristic from the best-fitting straight line. The reference value for the relative deviation is the nominal output at nominal load. The best-fit straight line passes through the characteristic curve so that the amounts of maximum positive and negative deviations are of the same size. With regard to precision measurements the linearity of a transducer plays only a minor role. More important is the reproducibility of the calibration curve (characteristic) which is specified as the relative span, and is a measure of the reproducibility of the characteristic with repeated measurements. Measurement instruments (using amplifiers) with microprocessors are capable of linearizing transducer characteristics with high reproducibility.

Figure 5.12 shows the diagram of a transducer circuit with all built-in resistors for compensation, adjustment and correction purposes. It shows the resistors R_{TC} and R_{TCP} for compensation of the temperature effect on sensitivity, the resistors R_{DL} and R_{DLP} for linearization and the shunt resistor R_{RE} for adjustment of the input impedance.

5.4 Adjustment of the characteristic value

The characteristic value describes the output signal at rated force.

During the manufacture of the transducer, the aim is to set the characteristic value of the transducer to a specific standardized amount, which is called the rated characteristic value (nominal sensitivity). Widely used in Europe for force transducers with strain gages is a rated characteristic value of 2 mV/V. The characteristic value marks the end point of the characteristic curve of the transducer (the characteristic curve describes the dependence of the output on force). The change of the characteristic value is linked with a change in the sensitivity (slope of the characteristic curve). The characteristic value is represented as a quantitative relationship, which is the ratio of output voltage of the bridge to the excitation voltage. Thus, inserting resistances in

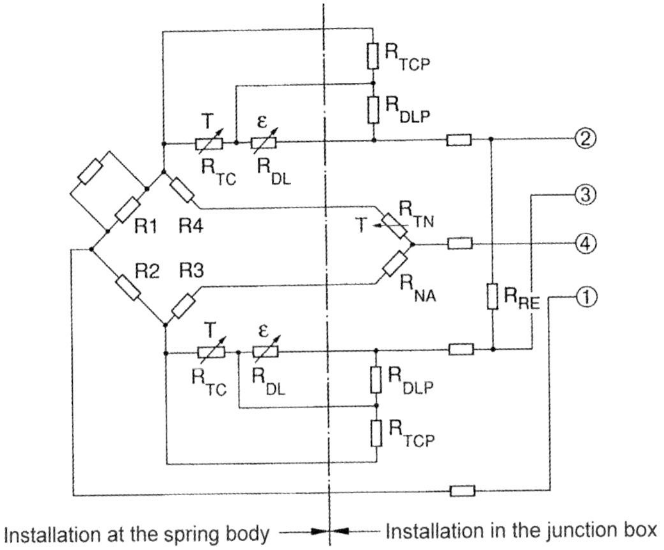

Fig. 5.12 Scheme of the full-bridge circuit for a precision transducer with all additional adjustment and compensation patterns;
R1 . . . R4: Strain gages of the full-bridge circuit,
R_{TN}: Temperature compensation of zero signal,
R_{NA}: Adjustment of zero signal,
R_{TC}: Temperature adjustment of sensitivity,
R_{TCP}: Shunt to R_{TC},
R_{DL}: Adjustment of linearity,
R_{DLP}: Shunt to R_{DL},
R_{RE}: Adjustment of the input resistance of the circuit

the feed lines changes the denominator of the quantity relation, thereby changing the characteristic value.

For mass-produced transducers, it is not possible to set each piece at exactly the same characteristic value of 2 mV/V. A measurable scattering of characteristic values will occur, depending on effort, which is specified as the tolerance of the characteristic value. This tolerance of the characteristic value has no negative effect on the measurement uncertainty, since each sensor has its individual characteristic value. This characteristic value can be measured depending on the measurement uncertainties of the available test equipment. The sensitivity tolerance is only relevant if the transducer has to be replaced in a measurement chain without the possibility of a new mechanical calibration of the measuring chain. In this case, the sensitivity tolerance is included in the uncertainty of the measurement chain. Even if several transducers need to be connected electrically in parallel for averaging, having the same characteristic value for all of the transducers is advantageous.

In comparison measurements the exact value of the characteristic value is of minor importance, rather it is the reproducibility of the characteristic value that is important. So, for example [5.6], shows, in 1986, an international force comparison standard that was introduced using precision force transducer C12 (HBM, nominal force 55 kN); after repeated loadings and unloadings it had a scatter width of the characteristic value of only 0.0001% (1×10^{-6}!). But even a load cell with strain gages for industrial applications can have excellent long-term stability of the characteristic values. So one force transducer exhibited, after 19 years of industrial use, a sensitivity deviation of <0.1% [5.4].

After completion of the compensation and adjustment measures, often a resistor is placed between the two excitation lines, in order to adjust the input resistance of the entire circuit to a rated value. When one strain gage is arranged in each bridge arm, the input resistance of the bridge circuit is the same as the nominal resistance of the strain gages used. The total resistance of the bridge circuit increases due to the additional inserted balancing and compensating resistors, but it can be brought back to the desired value by an appropriately sized shunt. This shunt is shown as R_{RE} in Figure 5.12.

5.5 Creep compensation

Creep is the time-dependent change of the output signal of a strain gage circuit installed on the spring body of a transducer after load change. Due to the complex causes it has not yet been possible to describe the creep behaviour of a strain gage circuit with reference to a general algorithm quantitatively.

Force variation time, weighting time and rise or fall time are three arbitrary selectable time criteria formulated from which the characteristic values of load creep, creep when load is relieved and initial force signal rest can be found [5.3]. It can be seen that a characteristic value for creep behaviour description must be assessed always associated with an observation period range. The main causes of creep of the strain gages and the measuring object material was explained in the previous chapters in the description of the time-dependent behaviour of strain gages [5.7]. Both show an opposite time response. This raises the possibility of mutual compensation, which is used in the construction of transducers. Today foil strain gages can be produced that are specially adapted to the creep of the measuring body's material.

Particularly with respect to cost-effective production, the requirement for an extremely low creep is difficult to satisfy using the conventional technique. The laminating technique which was developed especially for the mass production of transducers offers the possibility of installing complete bridge circuits in form of full-bridge units with conductor structures within the installation serving for creep compensation.

Figure 5.13 shows such a full-bridge manufactured with laminating technique for application in force transducers. In the figure, the points provided for separation are marked with arrows. According to a patent [5.8] these correction grids may be staggered binary in their creep values in order to enable sensitive adaptation. Furthermore, the lattice configuration is designed such that the balance of the full-bridge circuit is not disturbed by the separation of the traces for creep correction [5.9]. This technique allows the serial production of force sensors with a creep of less than 0.005% relative to the nominal sensitivity.

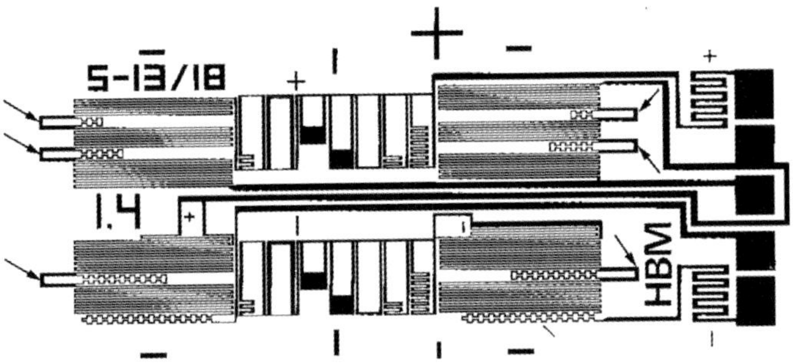

Fig. 5.13 Full-bridge circuit developed for the production of force sensors with low creep in laminating technique. Creep can be influenced by separating the conductive traces at the points marked with arrows

But even in the conventional technique the residual creep can be further reduced if one does not mind the effort required to achieve this. It is possible to arrange two bridge circuits at the measuring body, wherein the bridges differ in their geometry, so that in the one bridge negative creep is generated and in the other positive creep. Switching both bridges together results in a mean creep, which in the ideal case is equal to zero. This ideal case is practically difficult to achieve. Any remaining error, however, can be minimized in the finished transducer by 'attenuating' one of the bridges by means of resistors. For example, this was realized on the precision force transducer for international comparison measurements on force standard machines, for which creep values for loading and unloading are below 0.001%. In this transducer these resistors are arranged in the junction box accessibly from outside. Thus, the possibility of a subsequent creep correction on the finished transducer was also provided [5.6].

5.6 Full-bridge circuits connected in parallel

Every now and then practical applications require the parallel connection of several full-bridge circuits. This is usually the case when more than one bearing of a mechanical system is equipped with load cells and, as a result, the total load is required. In the case of parallel connected full-bridge circuits, at constant supply voltage, an output is obtained for the average of the signals of the individual full-bridges. However, this electric average then only truly represents the average of the applied forces, if all the bridges connected in parallel have identical characteristics (the same sensitivity) and the same output resistance. They influence each other in proportion to their output resistances. This condition is, however, given to a good approximation only at very close tolerances for nominal characteristic and output resistance.

But even transducers with large tolerances in characteristics and output resistance can be connected in parallel if they are adjusted to each other with the help of additional resistors. This adjustment is carried out through the output lines of the transducer so as not to disturb the already performed adjustments and compensations. In addition, in the output lines typically only very small currents flow. The proportioning of the resistors which are required for this adjustment is relatively simple, but that requires the knowledge of the characteristics and output resistances of the transducers involved [5.10]. The resistor needed for the correction should be divided into two halves and each half inserted in one of the two output lines of the respective transducer. For this adjustment, the actual characteristic values C and the output resistances R_A must be known for each of the transducers that are connected in parallel. The actual characteristic of a precision transducer can be found in the test report drawn up by the manufacturer. The correct determination of the output resistance with a resolution to 0.01 Ω requires some effort.

Determination of the resistors to be inserted in the output lines is briefly explained here [5.10]. The effective characteristic value C^* of a transducer which is connected in parallel with other transducers is given by

$$C_1^* = \frac{C_1}{1 + R_{A1}\left(\frac{1}{R_{A2}} + \frac{1}{R_{A3}} + \ldots + \frac{1}{R_{An}}\right)} \tag{5.8}$$

where the index 1 designates the transducer in consideration and the indices 2 to n the other transducers included in the parallel connection.

The characteristic value C^* of the total parallel circuit is determined by the sum of the effective characteristic values of the transducers involved to

$$C_{tot} = C_1^* + C_2^* + \ldots + C_n^* \tag{5.9}$$

In order that the forces acting on the interconnected transducers are received in equal proportions in the overall result, the effective parameters Ci^* of the transducers involved must be the same. This accord is accomplished by inserting resistors R_{AZ} in the output lines. Decisive for the adjustment here is the transducer with the largest quotient Q of output resistance R_A and characteristic C. To this transducer the other transducers are adapted by adding resistors R_{AZ}. The necessary output resistances R_{Ai}^* for the other transducers come from the relation

$$R_{Ai}^* = C_i Q_{\max} \tag{5.10}$$

The respective additional resistances arise from the difference

$$R_{AZi} = R_{Ai}^* - R_{Ai} = C_i \times Q_{\max} - R_{Ai} \tag{5.11}$$

Each is used with half the value in the two output lines of the transducer.

Figure 5.14 schematically shows the circuit diagram of three transducers for averaging connected in parallel with correction resistors in the output lines of two sensors.

The transducer without correction resistors in its output lines is the transducer with the largest ratio of output resistance and characteristic value, and the other two transducers are adjusted to it. In [5.10] is given a numerical example of the case

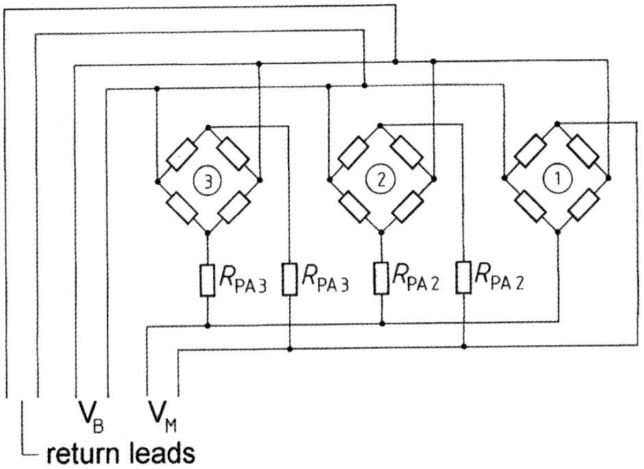

Fig. 5.14 Scheme of three transducers parallel-connected for averaging with correction resistors in the output lines

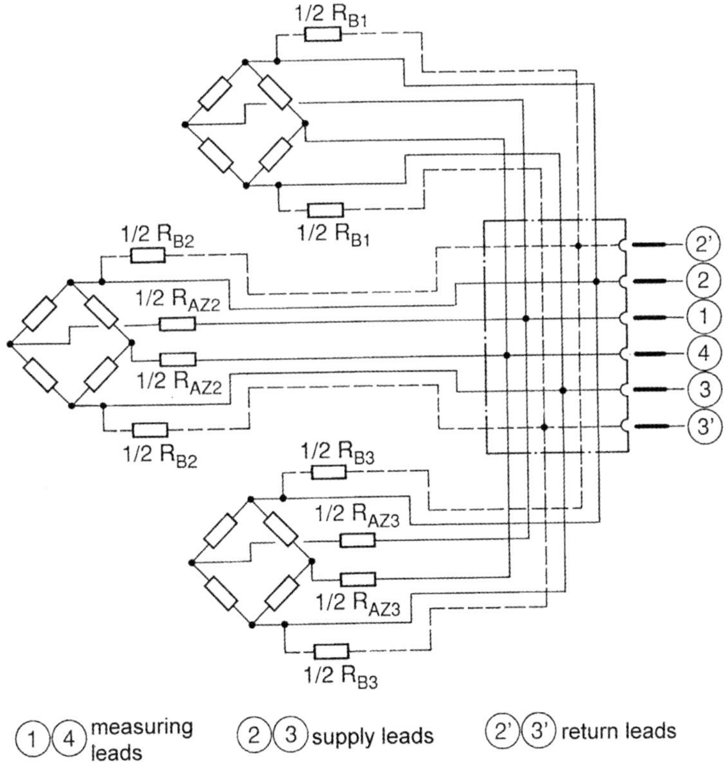

Fig. 5.15 Scheme of three precision force transducers parallel-connected for averaging with correction resistors in the output lines using six-wire circuit and two-channel principle

shown in Figure 5.14 of permanent monitoring of the weight of a vessel supported by three force transducers connected in parallel.

The basic structure of a parallel connection of three precision force transducers using six-wire circuit and dual-channel principle is shown in Figure 5.15, which shows the additional resistances R_{AZ2} and R_{AZ3} in the output lines of transducers 2 and 3 to adjust the characteristic values, as well as the resistors located in the return lines for realization of the dual-channel principle.

References

[5.1] Paetow, J. (1982) Verfahren zur Justierung von Meßwertumformern mit Dehnungsmeßstreifen und Dehnungsmeßstreifen zur Durchführung des Verfahrens, German Patent DE 29 33 676 C2.

[5.2] Paetow, J. (1962) Kraftmeßeinrichtung; German Patent 1 125 205.

[5.3] VDI-VDE-Richtlinie 2638 (April 1989) Kenngrößen für Kraftaufnehmer, Begriffe und Definitionen.

[5.4] Stein, P.K. (1984) *Some Little-Known Literature and Effects in Strain-Gage-Based Force Transducers*, Lf/MSE Publication No. 80, Stein Engineering Services, Inc., Phoenix Arizona.

[5.5] Starr, J.E. (1962) Compensation of strain gauge transducer nonlinearity, US Patent No. 3,034,346.

[5.6] Hellwig, R. (1987) Precision force transducer for international comparison measurements on force standard machines. *RAM*, **3** (1), 17–22.

[5.7] Keil, S. (1993) Chapter "Meßkörperwerkstoff", in *Handbuch der physikalisch-technischen Kraftmessung* (ed. W. Weiler), Vieweg Verlag, pp. 79–83.

[5.8] Ort, W. (1987) Meßumformer mit einer Feder und einer darauf applizierten Dehnungsmeßstreifeneinrichtung, German Patent DE 29 16 427 C3.

[5.9] Ort, W. (1983) *New Developments in Foil Strain Gages for Transducers*, Technical Note published by Hottinger Baldwin Messtechnik GmbH, Darmstadt.

[5.10] Paetow, J. (1987) Parallelschaltung von DMS-Aufnehmern. *Wägen und Dosieren*, **18**, 230–234.

6 Cable between the strain gage bridge circuit and measuring instrument

6.1 Basic

The test cables installed between the strain gages and the measuring instrument have an essential importance especially for precision measurements. The connection cable may cause significant measurement errors if certain basic rules are ignored. Generally, the cable has to fulfil two tasks: the transport of the bridge excitation voltage from the supply voltage source located in the measuring instrument to the bridge circuit; and the transmission of the measuring voltage from the bridge output to the instrument. In fundamental considerations using circuit diagrams the conductive connections are often viewed as having zero resistance. It is idealized in assuming that the cable cores are able to transport supply and measured voltage without loss and with the exclusion of external interference. This simplistic approach is allowed for measurements with low accuracy under ordinary conditions when the cable resistance is very small compared to the resistance of the strain gage bridge circuit.

In general, four wires are required within the cable for operating a strain gage full-bridge circuit, two wires for the bridge supply voltage and two wires for the bridge output voltage. Both cable pairs should be provided with their own screen.

The measuring signal delivered by a strain gage bridge circuit is usually represented as the ratio of the bridge output voltage V_M to bridge supply voltage V_B. It is often tacitly assumed that the supply voltage V_B provided by the supply voltage source arrives unchanged at the bridge circuit and the bridge output voltage V_M with its full value reaches the input of the measuring instrument (measuring amplifier or compensator). However, the intermediate cable affects this through the increasing length of the ohmic resistance of its cores, and for application of the carrier frequency method, the capacitances occurring between the wires affects the transmitted voltage and its phase angle. The effect of errors due to the cable capacitance increases with increasing carrier frequency, with increasing resistance of the connected bridge strain gage circuit and with increasing cable length.

6.2 Ohmic resistance of the cable

6.2.1 Preliminary remarks

All wires for connecting the strain gage bridge arrangement and all the wires of the connection cable between the strain gage bridge circuit and the instrument have ohmic resistors that affect the metrological performance of the entire circuitry. It is essential for the nature of the influence, to know whether the line is considered as part of the Wheatstone bridge circuit, which is an integral part thereof, or if it is disposed between the full-bridge circuit and the measuring instrument connection cable.

6.2.2 Ohmic resistance of the cable in full-bridge circuits

In full-bridge circuits it can be assumed that the internal wiring of the bridge is performed with short strands of equal length, whose resistances are low and their influences in the

Technology and Practical Use of Strain Gages: With Particular Consideration of Stress Analysis Using Strain Gages. First Edition. Stefan Keil.

four arms of the bridge cancel each other out. So here we consider the connecting cable between the full-bridge circuit and the instrument. In this cable, the effect of voltage drops caused by the ohmic resistance of the wires is important. Since the input resistance of the measuring instrument to which the strain gage bridge circuit is connected, as a rule, is almost infinite, the ohmic resistance of the two measuring lines does not play a relevant role in the calculation. However, the ohmic resistances of the feed wires affect the supply voltage, and since the sensitivity of the strain gage circuit is directly proportional to the supply voltage, voltage drops in the supply cables cause corresponding loss of sensitivity. This loss of sensitivity is dependent on the ohmic resistance and can quite easily be calculated. If we denote by E the sensitivity of the full-bridge without connecting cable and $E*$ the sensitivity of the measuring system with cable, then we have

$$\frac{E^*}{E} = \frac{R_B}{R_B + 2R_C} \tag{6.1}$$

where R_B is the input resistance of the bridge circuit and R_C is the ohmic resistance of a cable core of the feed line.

The evaluation of this equation is for copper cable (specific resistance $\rho = 1.8 \times 10^{-6}\,\Omega\text{cm}$) presented in Figure 6.1 with two different cross-sections and for two common strain gage resistances as a function of cable length.

It is seen that the cable resistances in bridge circuits with a smaller resistor R_B exert more influence on the sensitivity than with a larger R_B. Thus, a 10 m long connecting

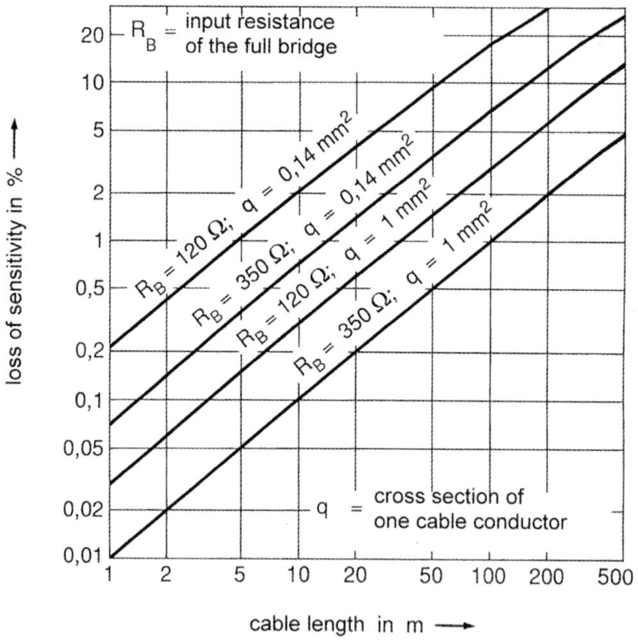

Fig. 6.1 Loss of sensitivity due to the ohmic resistance of the feed lines for the bridge resistors 120 Ω and 350 Ω for two cable cross-sections as a function of cable length

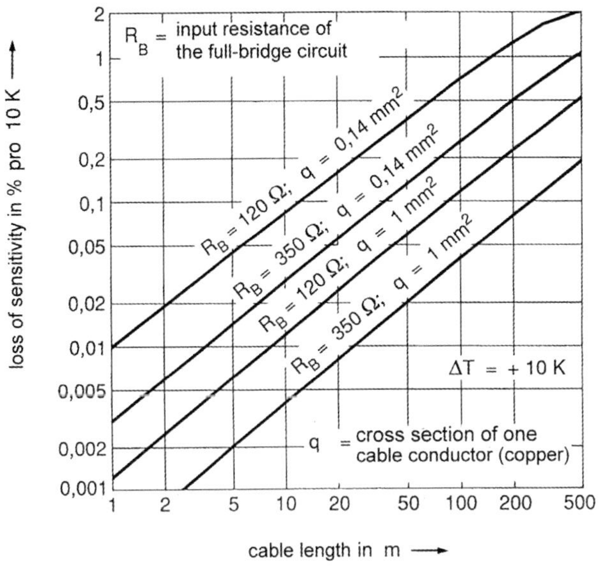

Fig. 6.2 Sensitivity shift due to temperature change of 10 K in the feed lines for the bridge resistors 120 Ω and 350 Ω for two line cross-sections as a function of cable length

cable with the core cross-section $q = 0.14\,\text{mm}^2$ in a 120 Ω bridge circuit has a sensitivity loss of 2% (which, however, can be rebalanced when calibrating the entire assembly).

When the cable is subjected to changes in temperature during the measurements, the temperature dependence of the resistivity of the conductor material (for copper the temperature coefficient of resistivity is $TK_{\text{Cu}} = 4 \times 10^{-3}/\text{K}$) causes changes in resistance of the cable. In which the voltage drop is influenced and thus the sensitivity. The relative change in sensitivity $\Delta E/E$ is then

$$\frac{\Delta E}{E} = \frac{-2\Delta R_K}{R_B + 2R_K}, \tag{6.2}$$

in which $2\Delta R_K$ means the temperature-induced change in resistance in the two feed conductors. The evaluation of Eq. (6.2) leads to Figure 6.2, which reproduces the change in sensitivity caused by a temperature change of 10 K for the same parameters as in Figure 6.1.

6.2.3 Ohmic resistance of the cable in half and quarter-bridge circuits

In half- and quarter-bridge circuits the interconnection cables between the active strain gages and the dummies in the instrument which, as supplements, complete the full-bridge circuit are components of the bridge circuit, as can be seen in Figure 6.3. So that any change in resistance in a bridge arm is influencing the measurement result in the same manner as the change in resistance of a strain gage. Furthermore,

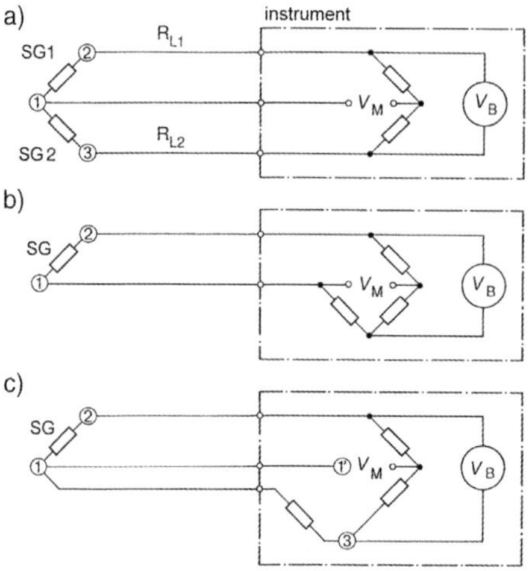

Fig. 6.3 Interconnections between the outer half- or quarter-bridge and the supplementary resistors in the instrument; a) half-bridge circuit, b) quarter-bridge with two-wire circuit, c) quarter-bridge with three-wire circuit

the resistance of the connection cable lying in series with a strain gage increases the resistance of that bridge arm and so apparently changes the k-factor of the strain gage in this arm. The k-factor is defined by the equation

$$\left(\frac{\Delta R}{R}\right)_{SG} = k\epsilon \tag{6.3}$$

in which ΔR is the change in resistance of the strain gage and R the nominal resistance of the strain gage.

If now there is a line resistance R_L in series with the strain gage, then this change in resistance of the strain gage ΔR_{SG} must be related to the total ohmic resistance of the considered bridge arm and the Eq. (6.3) reads

$$\frac{\Delta R_{SG}}{R_{SG} + R_L} = k^* \epsilon. \tag{6.4}$$

This shows that the gage factor k* for the entire considered bridge arm is smaller than k, and thus the sensitivity in the considered bridge arm is correspondingly smaller and the measured signal indicates a smaller strain than is really occurring in the strain gage.

In the half-bridge circuit as shown in Figure 6.3a the feed line R_{L1} is in series with SG1 and the feed line R_{L2} is in series with SGS2 and these influence the sensitivity as described by Eq. (6.4). The temperature-induced changes in the feed line resistances occur in two half-bridge arms in the same way and when these lines are laid together and twisted, they compensate each other.

This temperature compensation not takes place, however, in the quarter-bridge circuit as illustrated in Figure 6.3b, in which the feed and return line of the strain gage are located in one bridge arm. Here, temperature-induced changes of conductor resistance act as fully apparent strain. If, for example, a 3 m long connecting cable made of copper with 0.8 mm^2 cross-section is affected by temperature change of 40 K, the result in the two connecting wires is a change in resistance of 0.0216 Ω ($TK_{Cu} = 4 \times 10^{-3}$/K). This generates in a strain gage with $k = 2$ and 120 Ω an apparent strain of 90 μm/m and in a strain gage with resistance of 350 Ω an apparent strain of 31 μm/m. From this example, the influence of the strain gage resistance on the effect of changes of the cable resistance becomes clear.

The remedy with regard to the effects of temperature changes during measurement for quarter-bridge circuits comes in the three-wire connection as depicted in Figure 6.3c. In 1951, Ruge used this three-wire system, which was known from practical measurements with resistance thermometers, for the measurement with self-temperature-compensated single strain gages [6.1]. In this circuit, the two connection lines of the strain gage are no longer in one bridge arm, but in adjacent branches, which compensate for temperature-related effects in these two lines against each other. However, complete compensation is required for that the two wires are physically identical and over their entire length exposed to the same temperature changes. The loss of sensitivity caused by the voltage drop in the supply lines according to Eq. (6.4) is not prevented by the three-wire circuit, but it can be completely eliminated by the application of the Extended Kreuzer circuit (see Section 6.7). The third line is connected as part of the measuring diagonal to the amplifier input with almost infinite input resistance, thus their resistance and its changes have no influence on the measurement signal. Another disadvantage of the three-wire circuit is its capacitive asymmetry caused by the shorting of the two lines connected at point 1 (Figure 6.3c). This asymmetry will have greatest effect when working at high carrier frequency, but it can largely be eliminated by inserting a capacitor (between points 1′ and 3 in Figure 6.3c).

6.3 Influence of cable capacitance

6.3.1 Preliminary remarks

Measurement cables generally contain several conductors, and one or more shields, which are separated by insulation appropriate to the application cases. The individual wires form capacitances with each other and with the shielding which can affect the measurement signal to be transmitted. When working with DC measurement amplifiers the cable capacitances due to the transmitted DC signals have no disturbing influence, except that due to the dynamic behaviour of the measured quantity higher measuring signal frequencies have to be transmitted. Then the same capacitive influences are in principle effective as below for CF measurement amplifiers described. Using CF measurement amplifiers then at carrier frequencies of >1 kHz and large cable lengths, and in particular the use of unsuitable cables, the effect of cable capacitances can cause significant measurement errors if no appropriate action is taken.

Generally, the cable capacitances cause, with carrier frequency operation, three different interference effects:

- Capacitive asymmetry can create reactive voltage in the bridge which modulates the measuring range of the amplifier according to their amount.
- Capacitance causes a phase shift of the oscillation of the carrier frequency from the phase position on a supply voltage generator.
- Capacitance causes a drop in amplitude of the measured voltage.

6.3.2 Capacitive asymmetry

When working with full-bridge circuits and modern low-capacitance measurement cables with paired shielding of supply and measuring wires, then the apparent capacitive asymmetry is almost always negligible. In many applications, especially in experimental stress analysis using strain gages, often simply shielded cables are used with half-and quarter-bridge circuits. In these cases, marked capacitive asymmetries may occur. The full-bridge circuit as shown in Figure 6.4 is selected to explain the capacitive asymmetry with simple shielded measurement cables.

This shows the four ohmic resistors of the strain gage bridge circuit and the capacitances between the four parallel cable cores. A balanced capacitive symmetry requires that the capacitive effects in the individual bridge branches cancel each other out similar to the way they do with the ohmic resistances, so that with the notation of Figure 6.4 the two measuring wires are denoted 1 and 4 and the two feed wires 2 and 3, the condition $C_1/C_2 = C_4/C_3$ must be satisfied. If this condition is not met, then the state is referred as capacitively asymmetric.

If, as recommended, a cable is used in which the measurement wires and the feed wires are each provided with their own screen, the present considerations are unnecessary, since capacitive asymmetries occurring in these cables are negligible.

If the bridge is unbalanced due to the capacitive asymmetry, the resulting reactive voltage leads the signal caused by the ohmic upset by 90°. It can be separated from the actual measured signal in the demodulation of the instrumentation amplifier with a correctly set reference phase and thus will not falsify the measuring signal. The correct

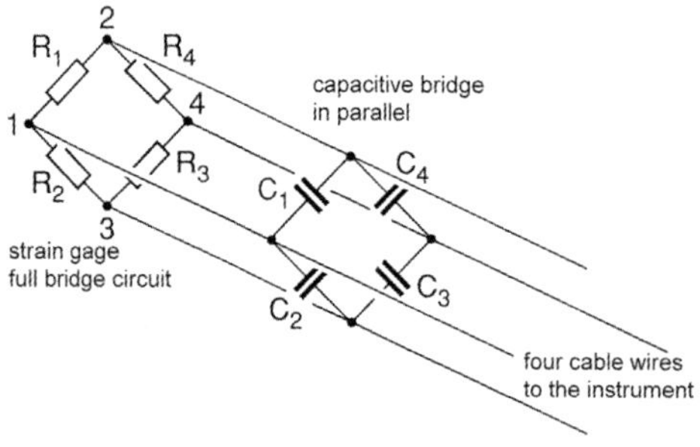

Fig. 6.4 The capacitances between the four cable cores lying parallel to the full-bridge strain gages circuit with a simple shielded cable

demodulation ensures that only the signal originating from the resistive bridge imbalance is amplified but not the capacitive imbalance. However, the demodulation is performed after the amplification of the signal supplied to the amplifier, which contained the resistive plus capacitive imbalance. Thus, with strong capacitive asymmetry the measuring amplifier may already be driven to a considerable extent, so that it is overdriven when the actual measured voltage caused by the resistive upset occurs. When overdriving occurs, the output voltage of the amplifier is no longer proportional to the resistive bridge output. However, the resulting overload can be avoided if the described capacitive asymmetry is eliminated prior to amplification.

This is done by the so-called capacitance balance (C-balance), which is carried out by means of an adjustable capacitor in the measuring amplifier. The tuning indicator of the measuring amplifier serves to control this balance, which is brought to a displayed minimum by adjusting the C-balance. The adjustment of the C-balance is performed when the display indicates minimum. If this is reached, the capacitance is balanced.

The C-balance method symmetrizes the capacitances but does not eliminate them. Capacitive asymmetry only has to be removed if, in summation with the actual measured signal, it is overdriving the amplifier. When this is not the case, then it has no distorting effect. This fact gives a measure of their tolerability. It is also not important when working with CF amplifiers with lower carrier frequency, such as 225 Hz or 600 Hz (this instrumentation amplifier for this reason does not have a C-balance).

If the adjustment range of the C-balance of the amplifier under heavy capacitive asymmetry is not sufficient to eliminate the asymmetry, then you can reduce asymmetry by coarse balancing with a suitable capacitor. This capacitor should be soldered in front of the amplifier input between the measurement and feed line. Such an additional capacitor can be soldered either at the measuring point (at the solder terminals of the installed strain gages) or at the measuring amplifier between the cable wires.

6.3.3 Phase shift

In addition to the capacitances described above which are responsible for capacitive asymmetry more capacitances occur between the cable conductors and the shields in a measuring cable with shielding. The capacitance concerned with the supply wires has no influence on the measured signal.

The capacitances between the measuring wires or between the measuring wires and the shield form, with the ohmic resistance of the strain gages, an RC network, which causes a phase shift of the oscillating measuring voltage in comparison to the supply voltage. This phase shift must be taken into account in the demodulation, that is in the separation of the measurement signal from the carrier frequency, otherwise there may be marked measurement errors. This consideration is done by the reference phase setting by which the phase angle of the control voltage of the demodulator will be adapted to the phase angle of the measured signal. The demodulator converts the AC signal voltage to a pulsating DC voltage and only then, when the control voltage

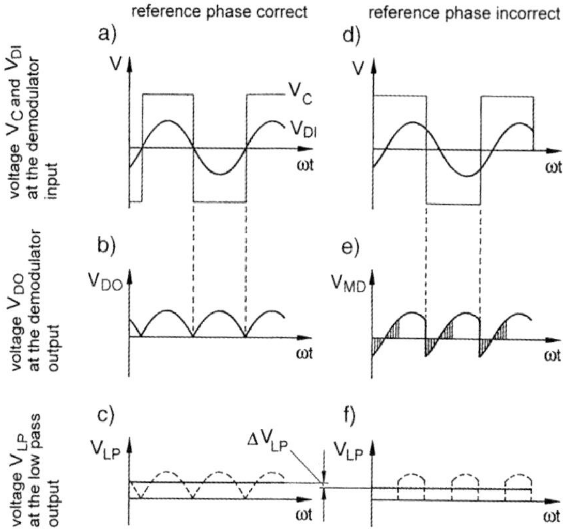

Fig. 6.5 Error of measuring signal ΔV_{LP} at the low pass output caused by incorrect setting of reference phase after demodulation of the amplified measurement signal V_{DO}

of the demodulator has the same phase angle as the measurement voltage can it be demodulated correctly without measurement error.

Figure 6.5 shows a schematic comparison of correctly and incorrectly set reference phases, and the emergence of the measurement error due to incorrect setting. Figure 6.5a shows the control voltage V_C and the amplifier output voltage V_{DI} in relation to the zero crossings of the properly set reference phase. Figure 6.5b shows the pulsating DC voltage at the output of the demodulator V_{DO} again, from the integration performed by the downstream low-pass filter giving the DC voltage signal shown in Figure 6.5c V_{LP}.

The corresponding steps in the incorrect set reference phase are shown in the 6.5d to f. The zero crossings of V_{DI} and V_C do not match in Figure 6.5d, as the demodulator therefore produces a signal for positive signals at its input and also negative portions of the pulsating voltage at its output (Figure 6.5e) that are offset by an equal share of the positive voltage in the low-pass filter and thus reduce the measured signal V_{TP}. The way the shares cancel each other in the low pass components of the voltage at the demodulator are shown hatched in Figure 6.5e.

Figure 6.5 in parts a), b), c) shows the voltage profiles under the assumption of a correctly executed C-balance. However, when reference phase is correctly adjusted with only little or no C-balance it has no distorting influence on the measured signal provided the amplifier is not overdriven. This can be shown using Figure 6.6 that reproduces the time-voltage curves with correct (a) and incorrect (b) reference phase setting [6.2].

From Figure 6.6a) it can be seen that the reactive part V_{RV} of the measurement voltage $V_{AO}{}^{*}$ which is influenced by the capacitive asymmetry runs within a half cycle

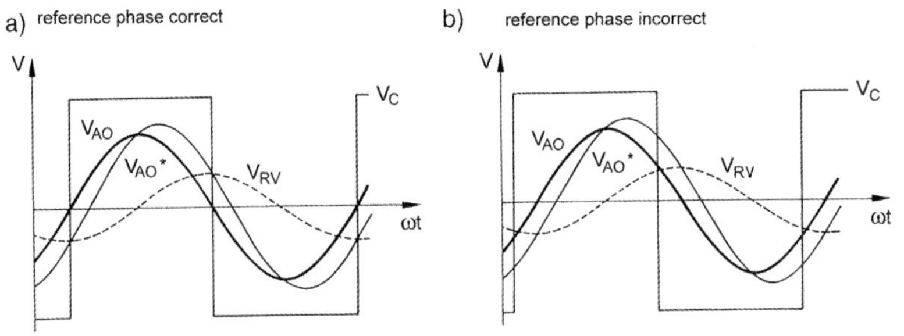

Fig. 6.6 Voltage waveforms at the demodulator input in the absence of C-balance in a) correct reference phase setting, b) incorrect reference phase; setting V_{RV}: reactive voltage due to a lack of C-balance, V_C: control voltage of the demodulator, V_{AO}: signal voltage at the amplifier output with a correct C-balance, $V_{AO}{}^*$: signal voltage at the amplifier output in the absence of C-balance

of the demodulator control voltage V_C symmetrically to the zero line and thus vanishes by the integration in the low-pass filter. A correct C-balance would convert $V_{AO}{}^*$ into V_{AO}, and the reactive voltage would then not exist. For incorrect reference phase setting, such as image 6.6b shows, the reactive voltage V_{RV} runs within a half cycle of the control voltage V_C not symmetrical to the zero line and the integration in the low-pass filter results in an additional (positive or negative) amount of signal (which is added to what, in Figure 6.5, is designated as the ΔV_{LP} error).

A correct reference phase adjustment can be made according to two different methods known as minimum and maximum method. The minimum method is applied with an unloaded transducer. The reference phase is alternately adjusted with repeated R and C adjustment, until such time that an alternating of the C-balance has no influence on the measurement signal (here zero signal). The explanation can be seen from Figure 6.6, which shows that only with an incorrect setting of reference phase does the reactive voltage provide a signal component. The maximum method is performed with a fully or partially loaded transducer (with real mechanical loaded transducer or with electrical simulation of the transducer output signal including the complete cable). Then, the display of the measured signal is brought to a maximum by means of the reference phase setting. The setting assigned to the maximum is the right one. The explanation relates to Figure 6.5, which indicates clearly that with proper reference phase setting ΔV_{LP} does not exist.

For some time carrier frequency measurement amplifiers have been available which, when using microprocessor engineering, avoid the need for reference phase adjustment and C-balance. This approach will be described in Section 6.6.

6.3.4 Amplitude drop

The output resistance of the bridge circuit connected to the two wires of the measurement line forms, together with the capacitance between these two measuring wires, a voltage divider that causes a reduction in the amplitude of the measurement voltage in addition to the already described phase shift. The magnitude of this amplitude decrease depends on the value of the carrier frequency, on the output resistance of the bridge circuit, on the ohmic resistance of the cable and on the capacitance between the two measuring wires, on the cable length and on the technological structure of the cable. Under specified conditions (e.g. only mutual capacitive coupling of the two shielded measuring conductors) one can calculate the resulting capacitive cable losses.

Figure 6.7 shows the calculated amplitude reductions for different carrier frequencies, depending on the cable length when connecting a strain gage full-bridge circuit with 350 Ω output resistance for a low capacitance measuring cable with 0.14 mm^2 copper cross section and a capacitance of 60 pF/m between the two test leads [6.3].

Figure 6.7 shows that the capacitive induced amplitude drops are quite low, even with a long cable at the low carrier frequency of 225 Hz. However, with a carrier frequency of 4.8 kHz, the losses are, at longer cable lengths, quite considerable and increase significantly with increasing strain gage resistance (output resistance of the connected bridge circuit).

However, the capacitive cable-related losses can't be calculated easily for each cable, since the capacitive couplings between measuring leads and the other conductors in the cable are not readily detectable and can't be easily quantitatively introduced into the calculation. The capacitive cable losses can, however, be relatively easily determined experimentally by carrying out measurements one after the other with and without the connection cable. From this signal difference we subtract the ohmic resistance of the cable, which is simple to calculate. The remainder is formed capacitive.

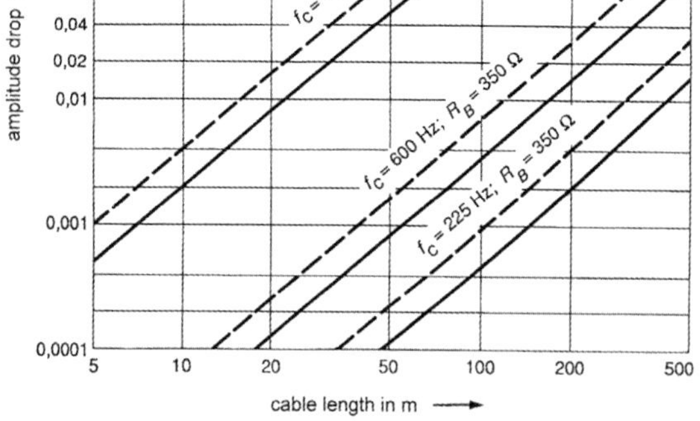

Fig. 6.7 Cable losses by capacitance in operating a 350 Ω transducer with a low-capacitance cable in a four-wire connection for different carrier frequencies as a function of the cable length [6.3]

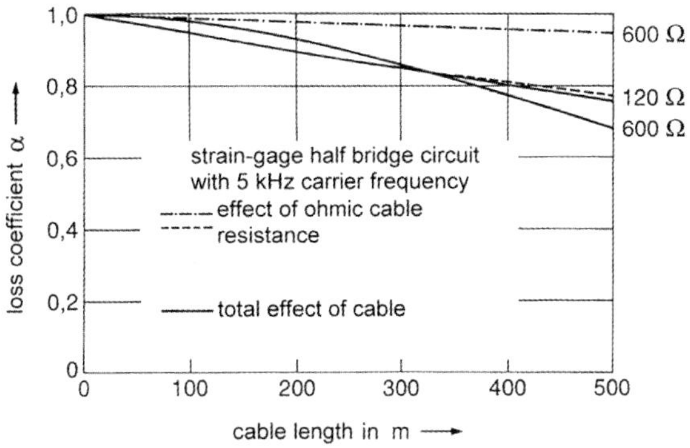

Fig. 6.8 Experimentally determined cable losses from measurements with and without connection cable for two strain gages half-bridge circuits with 120 Ω and 600 Ω strain gages and 5 kHz carrier frequency excitation

Figure 6.8 shows an example of cable losses measured against the cable length for strain gage half-bridges of both 120 Ω and 600 Ω gages on a 5 kHz carrier frequency measuring amplifier. On the ordinate is the cable loss coefficient α as the ratio of the measured value with cable to the measured value without the cable. The solid curves indicate the experimentally determined loss coefficient, the broken curves indicate the calculated loss coefficient, taking into account only the ohmic resistance. The influence of the strain gage resistance is clear. At 120 Ω nominal resistance, the effect of ohmic cable resistance is much greater than in the strain gage with 600 Ω nominal resistance, while the capacitive losses in the 600 Ω strain gage are considerably larger than the 120 Ω strain gage. The former is due to the ohmic resistance of the cable being higher than the strain gage resistance, and the latter is due to the increasing impact of the capacitive shunt with increasing terminating resistance.

These losses caused by the cable can, however, be kept away from measurement result by calibrating the entire measuring device including cable, using a suitable calibration method. One can connect a calibrator instead of the strain gage bridge circuit at the test site at the measuring cable and thus simulate a defined bridge output signal. By means of adjusting the amplification factor on the measuring amplifier (after R and C balance and reference phase setting), the cable losses can then be compensated.

6.4 Full-bridge connection in four-wire technology

Figure 6.9 shows the full-bridge circuit connected via four-wire technique to the measuring instrument (amplifier). The cable has separately shielded supply and measuring wires. Cable resistances and the effective capacitance between the measuring wires are marked.

The concept of the equipment (e.g. amplifier) designed for the measurement with strain gage transducers is already designed to minimize the cable influences. The internal resistance of the supply voltage source is approximated $R_i = 0$, and thus the

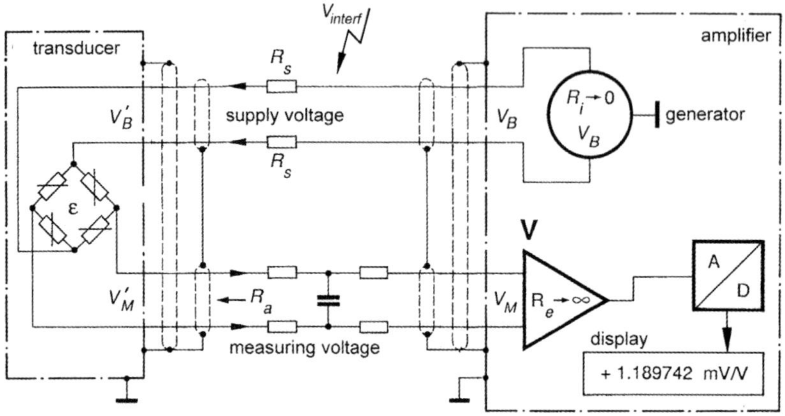

Fig. 6.9 Full-bridge circuit connected with four-wire technique

bridge supply voltage is independent of the transducer resistance. The supply voltage is balanced to ground and therefore counteracts the formation of high common-mode voltages in the measuring lines. The input resistance R_e of the amplifier goes to infinity and thus provides no load for the signal voltage. Supply wires and measuring leads are separately shielded to avoid interference due to low insulation resistances and cable capacitances. This shielding is particularly important because of the high voltage differences between supply and measuring wires. With regard to the electromagnetic compatibility (EMC), the entire cable should be protected by using a closed screen connected to the measuring instrument (amplifier), in the manner of a Faraday cage. Issues relating to the EMC are explained in Section 6.8.

Industrially manufactured measurement transducers with four-wire connection usually have a connection cable of a certain length (e.g. 3 m) firmly connected to the transducer, which is included in the calibration procedure carried out as part of the transducer manufacture. Therefore, any variation of the cable length (shortening or lengthening) influences the sensor sensitivity. The transducer cable is included in the temperature chamber, along with the transducer, for the temperature calibrations for zero signal and sensitivity. This means that for later measurements transducers and cables should be at the same temperature to avoid temperature-related errors. One should not overestimate errors that occur when there are temperature differences between cable and transducer, although they remain small in the relatively short transducer cable. So a conventional transducer cable of 3 m length with copper wires of $0.14\,\mathrm{mm}^2$ cross-section produces for a temperature change of $+1\,\mathrm{K}$ a change in resistance of $\approx 3.1\,\mathrm{m}\Omega$, which causes for a transducer resistance of $350\,\Omega$ a decrease in sensitivity of $\approx 8.8 \times 10^{-6}$, that is to $\approx 0.00\,088\%$.

The cable effects are, however, significantly larger when the length of the cable calibrated with the transducer is not sufficient and cable extensions are required. Then the sensitivity is not only affected by changes in resistance caused by temperature changes, but by the total resistance of the extension cable itself. The sensitivity change caused by resistance changes in the supply wires is generally determined by the relationship given in Eq. (6.1).

If one extends, for example, the sensor cable to 10 m, so the used copper cable with 0.14 mm^2 cross-section affects the sensitivity by the remarkable value of ≈0.7%. If, for example, a cable is used that has a lower resistance with 1 mm^2 cross-section, the loss of sensitivity is reduced to ≈0.1%. This sensitivity changes caused by cable extensions can be taken into account by calibrating the total chain consisting of transducer, measuring cable and amplifier and then the cable will not have a distorting influence on the measurement result. For this calibration to be carried out mostly on site, however, the necessary aids must be available, that is a known mechanical quantity as direct load on the transducer or an electrical calibration standard to be connected instead of the transducer and to simulate a certain transducer output signal. The limits of the estimated achievable calibration uncertainty are given by the uncertainty of the available calibration value or when using the electrical calibration standard by the uncertainty and sensitivity tolerance of the transducer. However, such calibration does not take into account the changes in resistance occurring after calibration due to temperature changes in the lines, which could have a non-negligible influence on the use of a copper cable of $\approx 4 \times 10^{-3}$/K, which is ≈0.4%/K.

In the cable wires which lead the output signal of the bridge circuit, the line resistance is of minor importance due to the almost infinitely large input resistance of the amplifier. Here the already explained capacitive effects at carrier frequency operation are significant.

When operating with carrier frequency, the transducer resistance and the capacitance of the cable form a voltage divider that produces a drop in amplitude and a phase shift of the measuring voltage. The supply voltage and the measuring voltage are alternating at the same frequency but the modulated signal is shifted by the phase angle φ. The losses caused by phase shift are less critical, since in the instrument the demodulation can be adjusted for correct separation of the measurement signal from the carrier frequency. The control voltage of the demodulator is adjusted in accordance with the phase shift occurring in the measuring line (phase reference level), and thus the influence of the phase shift can be entirely eliminated.

The amplitude drop is therefore not compensated. The amplitude drop increases with increasing carrier frequency, with increasing transducer resistance, with the ohmic resistance of the cable, with the cable length, and is also strongly influenced by technological structure of the cable. So, for example, the thinnest possible test leads should have their own shield, as shown in Figure 6.9. For the circuit reproduced in Figure 6.9, the ohmic cable losses are given in Figure 6.1 and the capacitive losses are reproduced in Figure 6.7.

In terms of cable effects there is in principle no significant difference between an industrially produced transducer and a self-installed full-bridge circuit connected via cable to the instrument. With regard to the electromagnetic compatibility (EMC), however, differences are possible. So the housing of the transducer may need to be included in a Faraday's cage formed by the shielding. This is only possible with a self-installed bridge circuit if you can attach a metallic cover, which is conductive, connected (possibly by soldering) to the measured object. However, total shielding is required only in the relatively rare cases of measurements in the field of strong electromagnetic radiation. However, it requires the object to be measured to be conductive, and so with measurement objects of non-conductive materials such as plastic, wood, concrete, and so on it is not possible.

6.5 Six-wire circuit

The sensitivity losses caused by voltage drops in the supply lines can be avoided by using the six-wire circuit [6.4]. It has been well-known since around 1970, but has only recently came to the fore.

The principle of this circuit is quite simple: Two sensing leads pickup the supply voltage as it is present at the transducer input and return it to the comparator in the instrument with the supply voltage source. Due to the almost infinite input impedance R_i of the comparator there will be no voltage drop across the sensing leads because no current flows through them. The comparator compares the sensing voltage with the reference voltage, which is equal to the required bridge supply voltage. The adjustable generator is controlled by the comparator and increases the supply voltage until the sensing signal is equal to the reference voltage. The difference between the supply voltage generated by the generator and the effective power supply voltage at the transducer is then just as large as the voltage drop in the supply lines.

This controlling of supply voltage takes into account resistance changes in the supply lines and resistance changes due to temperature changes. Figure 6.10 illustrates the basic structure of a six-wire circuit.

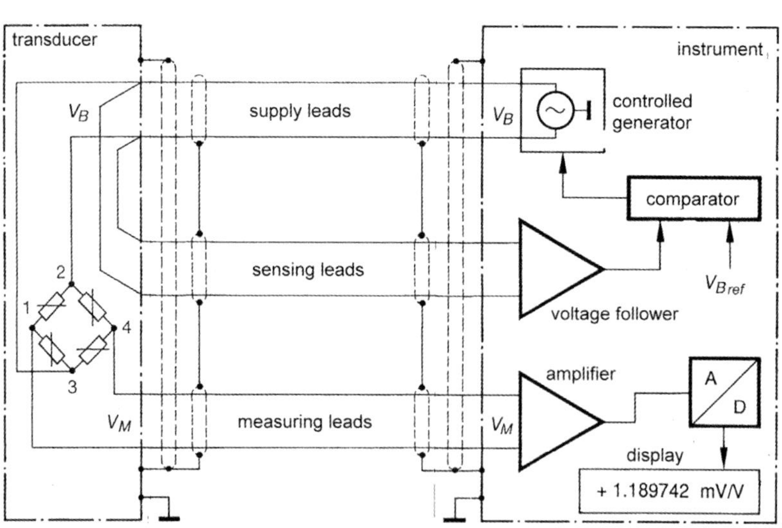

Fig. 6.10 Strain gage full-bridge connected in six-wire technology

The successful application of the six-wire circuit requires appropriate measuring instruments. Transducers designed for the six-wire connection are calibrated by the manufacturers, usually without cable, because the six-wire cable at least in terms of the supply voltage has no distorting influence. The capacitance-related losses due to amplitude drop and phase shift in the measurement cable are not compensated by the six-wire circuit. They are, however, such as Figure 6.7 shows not critical for transducers with 350 Ω strain gages and lower carrier frequency at normal cable lengths. However, they can also be eliminated by applying the dual-channel principle described in Section 6.6.

When the voltage drops in the two supply wires are identical, the six-wire circuit can be reduced to a five-wire circuit. In the five-wire circuit, which requires symmetrical transducer excitation, the supply voltage is tapped only at one sensor input point and the return wire is used for adjustment of the supply voltage [6.5].

Special attention should be paid to cases where transducers designed in six-wire circuit calibrated without cable are operated at an instrument designed for a four-wire circuit, because in this case the resistance of the supply lines acts to decrease the sensitivity. In this case, a six-wire cable will be used with each feed line connected in parallel to the associated return line, thus reducing the resistance of the supply cable. The operation of a transducer in four-wire technique calibrated with the cable to an instrument designed for six-wire technique is almost always uncritical, since these devices can switch usually between six-wire to four-wire mode. When cable extension is required, this can be performed with the six-wire technique, but the tapping points for the feedback voltage must then arranged at the end of the transducer cable. The four-wire cable which is calibrated with the transducer must never be separated from the transducer, as this changes the characteristic values that are set by calibration (increased sensor sensitivity).

The six-wire circuit is applied in the same way as for industrially manufactured transducers for self-installed bridge circuits. Here, the connection of the return lines is generally performed at the soldering terminals of the strain gage circuit to which the power supply lines are connected.

6.6 Dual-channel principle

The six-wire technique in its form as described up to now can't avoid the capacitive generated losses caused by the carrier frequency at the measuring lines. An additional measure can make it so that amplitude drop and phase shift have no influence on the measurement result [6.3,6.5]. To this end, the circuit for the voltage feedback is designed identical to the measuring circuit and thereby amplitude drop and phase shift in the feedback leads occur in the same way as in the measuring circuit.

This requires using, as a sensing cable, a cable identical to the measuring cable. A separately shielded pair cable is used for voltage feedback, which is not directly connected, but each wire has a resistor to the return points on the transducer. At the instrument side, the input impedances in the sensing circuit and in the measuring circuit are identical.

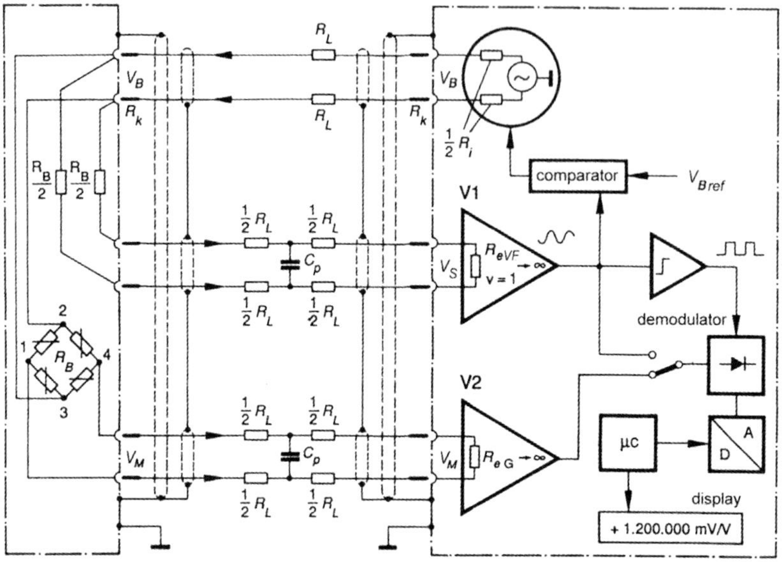

Fig. 6.11 Dual-channel principle for the transducer connection to avoid resistive and capacitive effects in cable carrier frequency mode

Each of the two resistors used at the transducer in the sensing lines has a value of half the transducer resistance (½R_B) and this ensures that the recirculated voltage experiences the same amplitude drop and the same phase shift as the measurement signal. Figure 6.11 shows the diagram of the dual-channel principle with the two resistors ½R_B in the sensing lines.

The recirculated voltage has the same phase shift as the measurement voltage and can be used to adjust the reference phase of the demodulator. Thus, the control voltage of the demodulator always has the optimum phase relationship to the measured voltage. The recirculated voltage, as described in explaining the six-wire circuit, is compared with the reference voltage, and the supply voltage is adjusted accordingly. This correction is done such that the amplitude drop is corrected by voltage increase and the phase shift by setting the phase of the power supply voltage.

In [6.3] sensitivity losses due to cable effects were determined for the four-wire circuit and the six-wire circuit with and without dual-channel principle and compared with each other on the basis of measurements carried out. As measuring device a 350 Ω full-bridge circuit was used, operated with a 4.8 kHz CF measurement amplifier with different cable assemblies. The measured results are shown in Figure 6.12.

One can see the high loss that occurs when using the four-wire connection. Even with the six-wire connection, even at 400 m cable length there is a 5% drop in amplitude. The result of the dual-channel principle is excellent: at 500 m cable length it has only 0.3% measured amplitude drop. Thereby the low residual error is

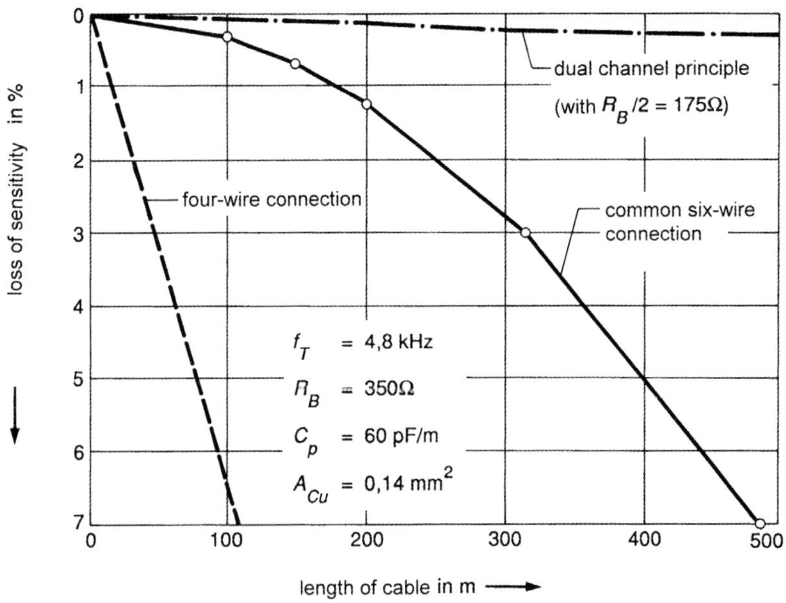

Fig. 6.12 Measured sensitivity reduction as a function of cable length for 350 Ω transducer with a 4.8 kHz carrier frequency amplifier with four-wire connection and six-wire connection with and without dual-channel principle [6.3]

caused by tolerances of the inserted resistors, which don't quite match the values $\frac{1}{2}R_B = 175\,\Omega$, and that the two pairs of conductors for the return lines and the measuring voltages in their capacitances and ohmic resistors have slight differences from each other.

Prerequisite for the practical application of the dual-channel principle is the availability of an appropriately designed measuring instrument. For a full-bridge circuit connected in six-wire technique the two resistors have to be inserted in the two return lines. Figure 6.13 shows a sketch of a proposal for the positioning of the two resistors for the application of the dual-channel principle in the case of a

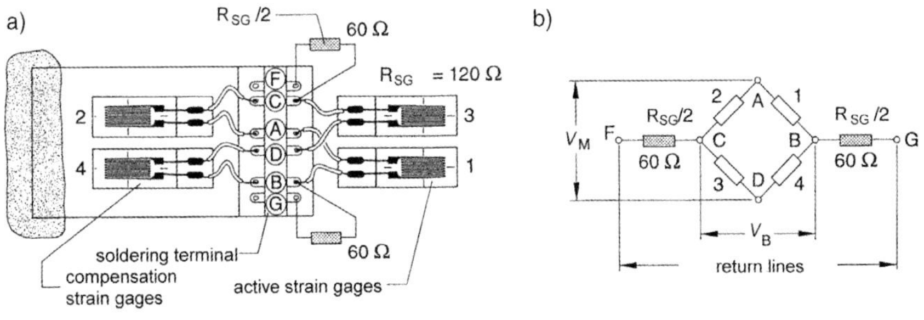

Fig. 6.13 Application of the dual-channel principle in a self-installed full-bridge; a) positioning of the two resistors ½R_{SG}; b) Schematic of the bridge circuit with wire designation for standard pin assignment

Fig. 6.14 In the plug of the connecting cable in four-wire technique on the side of the transducer soldered resistors for the application of the dual-channel principle

full-bridge circuit, which is operated at a relatively long cable with 5 kHz carrier frequency. The two resistors $\frac{1}{2}R_{SG}$ are soldered at the soldering terminal between the connection points of the return lines and the feed wires. When using a standardized connector, the feed lines are connected to the terminal points B and C, the measurement lines at the points A and D (in the half-bridge, the terminal A is used, and D is open), and the return lines are located on F and G, where G is assigned to A, and F is assigned to C.

If it is not possible to attach the resistors directly to the bridge circuit at the test object as shown in Figure 6.13 or at the transducer because of a hermetically sealed case or similar, then the resistors can be mounted at the end of the connection cable in a plug between the feed line and the associated return line. Before applying the dual-channel principle it should be considered whether the circumstances related to the measurement task (line length, high carrier frequency, required accuracy) make the associated costs necessary.

If for a transducer calibrated in four-wire technique an additional extension cable is required, it is important not to change the calibrated transducer cable. Here one can use for cable extension a six-wire cable with two resistors in the plug at the cable end as described above, soldered between each supply line and associated return line.

Figure 6.14 shows the photo of a plug as described. A resistor has been soldered between each supply wire and return line. The plug shown satisfies the EC directives governing the EMC requirements, as its housing continues the tubular shield without interruption.

6.7 Connection of half and quarter bridges

When measuring with half-bridge circuits or single strain gages the connection cables are of particular importance because they are components of the bridge circuit. In such cases the changes in resistance of the cable cause bridge imbalance as do changes of the strain gages and thus they directly influence the measured signal. As a full-bridge circuit is basically required when measuring with strain gages, installed half-bridge or single strain gages always need to be complemented by completion resistors to full-bridge and the connecting lines between the installed strain gages and the completion resistors are inevitable internal components of the full-bridge and can in the worst case, if their influence is not avoided, cause strong distortions of the measurement results. In half-bridge circuits the two feed wires of equal length should be twisted, so that they are exposed to the same temperature changes. The effects of temperature changes on the two line resistors then compensate each other, as the two feed lines are components of adjacent bridge arms. Any difference in the changes in resistance of the two feed lines will go fully as apparent strain into the measurement result.

When connecting single strain gages one applies, for the above-mentioned reasons, the three-wire circuit as already shown in Figure 6.3c or the advanced Kreuzer circuit as reproduced in Figure 6.17. In the three-wire circuit, the two connecting wires for the single strain gage are placed in adjacent bridge arms; the line for the measurement signal is not critical because of the infinitely large input resistance of the measuring amplifier. In the advanced Kreuzer circuit the strain gage is connected to four cable wires.

Figure 6.15, for example, shows the connection of a half-bridge circuit in six-wire technique. Here five wires are needed, since the half-bridge circuit requires only one measuring line (the second measuring line is located in the instrument on the supplementary half-bridge). The two resistors ½R_{SG} soldered for the use of the dual-channel principle in the sensing lines, however, have relatively short lines and low carrier frequency. It is advisable to place the two feed lines and the two return lines each in twisted pairs (each optionally screened).

The screen placed around all leads in Figure 6.15 when it is extensively connected at its entire circumference to the instrument's housing and possibly used EMC-compliant plugs, corresponds to the EC directives for the internal market of the EU. For self-installed strain gages it will not be possible in the majority of cases (and not technically required), to shield the entire strain gage installation by a Faraday cage.

The disturbing effect of changes in resistance in the connecting lines of the Wheatstone bridge circuit is also a relevant factor for the multipoint measurement

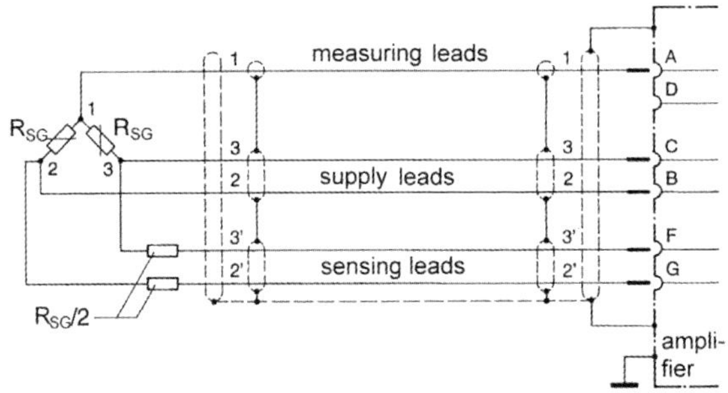

Fig. 6.15 Connecting a strain gage half-bridge circuit in six-wire technique and use of the dual-channel principle

technique. In this technique, by sequential connection of the measuring points involved in the shortest possible time a number of half-bridge or single strain gages have to be connected to the input of the instrumentation amplifier. The switching technique applied here has long been characterized by relatively slow relays, since their reproducible resistances were implemented in the switch contacts. In 1976, however, the Kreuzer circuit was presented which allowed the use of FET switches (FET = field effect transistor) despite their irreproducible transfer resistances in multipoint measuring instruments [6.6]. Thus the advantages of FET switches such as speed, freedom from wear and bounce-free switching for multipoint measurements can be used. Figure 6.16 gives the basic design of this circuit as an example of three active single strain gages (in this case three measuring grids of a rosette) with shared compensation strain gage [6.7].

In the Kreuzer circuit, similar to the previously discussed six-wire technique, the supply voltage of the measuring strain gage will be picked up at the strain gage and returned to its source and there adjusted according to the voltage drops occurring in the leads and the FET switch, similar to the six-wire technology.

The two control amplifiers A2 and A3 regulate their output voltages as required, until the returned voltage matches the reference supply voltage. In this way the voltage drops in the supply lines 2 and 3 and in the supply voltage switches S2 are compensated. The output voltages V_2 and V_3 of the control amplifiers are greater by the amounts ΔV_2 and ΔV_3 than the reference supply voltages $+\frac{1}{2}V_B$ and $-\frac{1}{2}V_B$. In the return lines 2′ and 3′ no current can flow due to the infinite input impedance of the voltage follower, thus in them and in their switch S2′ no voltage drops can occur.

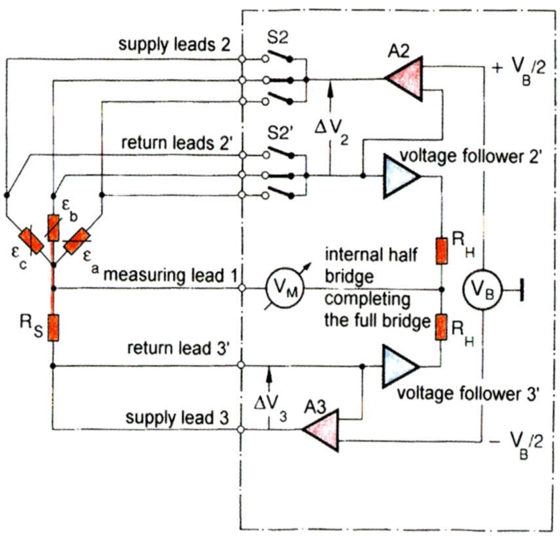

Fig. 6.16 Kreuzer circuit for the connection of single-strain gages with supplementary resistor R_S close to the measurement site

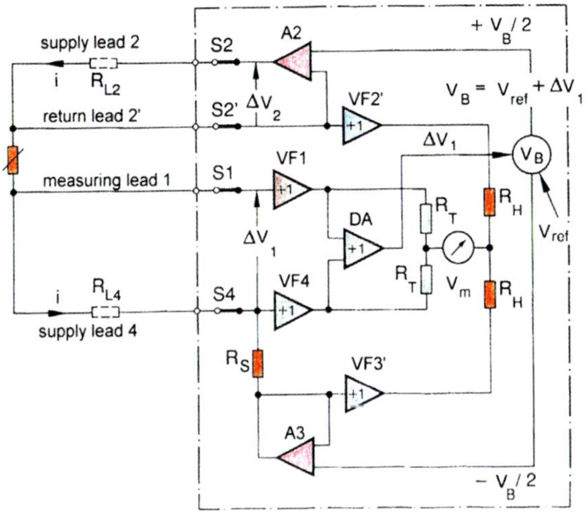

Fig. 6.17 Advanced Kreuzer circuit for switching single strain gages arranged away from the supplementary resistor R_S arranged in measuring instrument

The outer half-bridges are complemented for the full-bridge by an internal half-bridge with the resistors R_H. This internal half-bridge is fed by the voltage followers with the returned voltage, so that the external and internal half-bridge are at the same voltage.

The Kreuzer circuit eliminates almost perfectly the zero and sensitivity errors. It has the advantage that it is relatively easy to extend for a multitude of additional measuring points. Each additional measuring point only requires two additional lines (feed line and sensor wire) and two switches for their connection and disconnection. However, the Kreuzer circuit in the form shown in Figure 6.16 does not consider the voltage drops that occur in the line between the supplementary or compensatory strain gage R_S and the active strain gage. Hence it is important to arrange the supplementary strain gage as close as possible to the active strain gage. Furthermore, the direct galvanic connection between all the strain gages means that the change in the insulation resistance of one strain gage can effect affect the measurement results of all the strain gages.

To avoid these disadvantages, the Kreuzer circuit was expanded by its inventor [6.8]. With this advanced Kreuzer circuit active single strain gages over long distances up to 1000 m can be connected and still be sure that both the active and the supplementary strain gage always have the desired reference voltage V_{ref}.

This ensures that the voltage drops ΔV_1 and ΔV_2 have no effect on the sensitivity. The advanced Kreuzer circuit has the extra advantage of full galvanic isolation of the active strain gages from each other and thus the avoidance of mutual interference in modified insulation resistance or short to ground.

The advanced Kreuzer circuit has a wide range of variation. Full, half and quarter bridges can all be connected side by side at an appropriately equipped multipoint measuring instrument. The quarter-bridge circuit can use a common compensation strain gage or an externally or internally arranged supplementary resistor. The advanced Kreuzer circuit is applicable regardless of the operating mode of the amplifier, including DC or carrier frequency operation.

Especially in large-scale tests with many strain gages installed on a measurement object for stress analysis, the advanced Kreuzer circuit can provide substantial benefits when suitably designed multipoint measuring instruments are available. Each individual strain gage has to be connected to two pairs of wires. The circuit fully regulates the influence of the leads even with long cable lengths. In temperature-stable environment the large numbers of temperature compensation gages can be replaced by one completion resistor acting for all the individual strain gages. Also when connecting half-bridge circuits the Kreuzer circuit eliminates all resistance changes occurring in the feed lines.

6.8 Protection against disturbing influences

6.8.1 Protection against electrical and magnetic fields

Both electric and magnetic fields can cause noise voltages in the measuring lines due to interference [6.9]. The signal voltages occurring in the measuring lines are relatively small when measuring with strain gage transducers (of the order micro-volts), so appropriate protection measures must be provided to ensure that the interference voltages do not exceed the limit of resolution of the measurement voltage. These protective measures have to be considered separately, because of the different physical nature of electric and magnetic fields for these two field types.

All conductors that carry electrical voltages are surrounded by an electric field whose intensity is proportional to the voltage and decreases as the square of the distance from the conductor. In conductors lying within the range of electric field interference voltages are generated by capacitive coupling of charges. If the electric field is not constant but continuously changing, such as around lines with AC voltage, the interference voltage will change accordingly. Electrical interference can cause common-mode interference, that is they produce, in both wires of the measuring line, a voltage of the same polarity. Thus, they are not particularly critical, since the modern amplifier technology ignores common mode noise largely. Electrical interference can be greatly reduced by using a shielded cable. The shield absorbs the electric fields and conducts the induced charges to ground. This means that for the shield to operate effectively, it requires an effective ground connection. To avoid so-called ground loops, the shield should be grounded against electrical fields only at one selected point. Grounding the shield at two or more points can lead to galvanic coupling of interference when the ground points have different potential. In the industrial environment this is more often the case than might be expected.

However, magnetic interferences is a more serious problem, against which the normally used conductivity screens such as copper braid fail. Magnetic fields are generated by current-carrying conductors and permanent magnets. We find these, for example, in the vicinity of motors, generators, transformers and power lines.

Magnetic fields then induce noise voltages in a conductor when there is relative motion between the conductor and the magnetic field lines. This is the case when the conductor is moved through the field, when the magnetic field moves relatively to the conductor or the magnetic field itself is subject to change.

Since the normal conductivity shields are ineffective against magnetic fields, one must use shielding materials that have high permeability to accommodate the magnetic flux and keep the magnetic field away from the measurement cable. It is preferable to use, as a shield, armoured steel pipes for the leads. Particularly effective are shields from Mu-metal, a nickel–iron alloy of high permeability. This material is wound in the form of foils or thin strips around the cable. The deformation of the winding causes a reduction in permeability, especially for small cable diameters, Due to the crystalline structure of the Mu-metal its permeability has set limits [6.10].

The best shield against magnetic interference is achieved with amorphous alloys, which are designed specifically for these applications. The shielding factor of these amorphous alloys is higher by a factor of about 10 than that of crystalline shield materials with high permeability [6.11, 6.12]. Of great importance for the overall shielding effect is the execution of the shield connection at the cable ends. An effective cable shield against electromagnetic fields can be realized only through direct bilateral round contact of the device housings [6.13]. All other versions such as braids, 'pigtails' and drain wires and the connection of the shield to the electronics ground and not attached to chassis lead to loss of shielding effect, as proved in [6.13] from measured values. These questions of the shield are treated in more detail in the following section with regard to electromagnetic compatibility (EMC).

Magnetic interference generates signals in the two wires of the measuring line interference voltages of opposite polarity. This push-pull noise voltage controls the two amplifier inputs in phase-opposition and has a particular detriment especially when using DC amplifiers. Better than DC voltage measurement amplifier against push-pull noise voltage behaviour is the carrier frequency measuring amplifier. The CF amplifier transfers only signals in a narrow frequency band around the carrier frequency, thus interfering voltages only have an effect when their frequency is in the transmission bandwidth of the measuring amplifier.

Shields against strong magnetic fields can be very expensive. A recommended safety measure for reducing the effect of magnetic fields is given by the possibility of twisting the cable cores for partial self-compensation of the induced noise voltages. The aim is that the noise voltage induced in a winding is completely compensated by noise voltage of the opposite sign induced in the adjacent winding.

For the best possible self-compensation of electromagnetic interference, a shielded cable with two conductors is available in which each conductor consist of two parallel wires [6.14]. The four cable wires are not twisted, but as shown in Figure 6.18 so interwoven so that spiral cable loops are avoided. Parallel are the two wires in planes

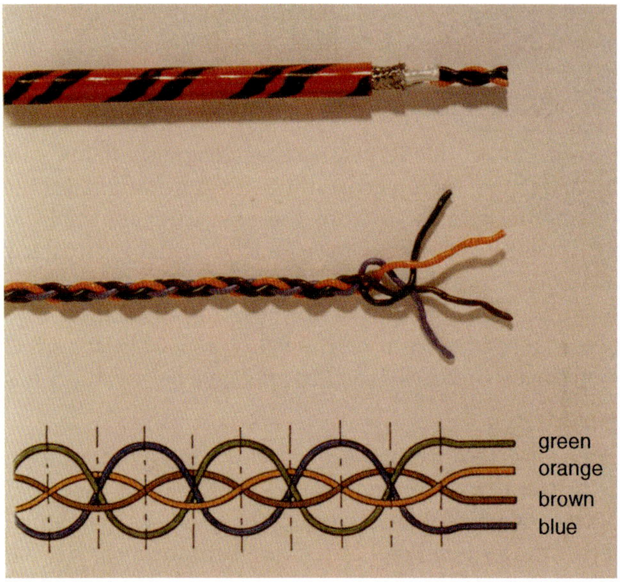

Fig. 6.18 Woven cable with two parallel pairs of conductors to compensate for the voltage generated by magnetic interference voltages in the parallel cable pairs

staggered by 90° to each other, so the conductors with the colours orange and blue and the conductors of green and brown in Figure 6.18, respectively.

6.8.2 Electromagnetic compatibility

The constantly increasing number of electronic systems used in industry and leisure applications such as personal computers and mobile phone systems leads to a continuous increase of the noise level in our environment. In order to keep the increase of these disturbances within limits, in 1996 the concept of 'electromagnetic compatibility (EMC)' was developed for the European internal market to apply to all industrial products. The concept on the one hand limits the outgoing equipment interference (EMI) implemented by appropriate protective measures and evaluates on the other hand the electromagnetic noise immunity (EMS) of devices. This approach is limited in terms of the measuring technique not on individual devices alone, but includes the entire measuring chain, including all connection cables. In order to achieve a comprehensive protection here, the entire measurement chain must be continuously surrounded by a shield, thus sealed on all sides (almost like a Faraday cage). In this case, the transition points between the device housings and the cable shields require special attention. The shield must be connected all around 'close to' the housing. The same applies to connectors: the housing is also part of the shielding around. The combination drilling of shielding braid to a solderable tip and the connection of the shield to the electronics ground and not attached to chassis can lead to loss of shielding effect and of not meeting the EMC standards. Especially with measuring arrangements with long cables the shielding has at several places connections to the earth, the EMC-compliant wiring for shielding and signal ground requires two separate leads. In some modern devices an electrical connection can be provided on demand via a switch between shield and signal ground, which may result in special cases in a reduction of hum. New EMC-compliant concepts include the transducer housing in the shield. The transducer housing is then usually connected to the protective earth, so that there may be widely spaced grounding points of the shield with long measurement leads. If there are large differences in ground potentials, an equipotential bonding conductor (copper and of the largest possible cross-section) may be necessary connecting the widely spaced grounding points. Conventional measurement chains with shield connected to signal ground must be grounded only at one single point (at the amplifier input) to avoid ground loops.

For measurement chains with strain gage transducers the questions of immunity EMS are more prominent than that of EMI interference. Especially when strain gages are directly installed at the measurement object full EMC shielding for practical reasons is not feasible in many applications. Each bridge circuit ought to be provided with a metallic cover which is conductively connected (soldered or similar) to the measurement object and to which the cable shield is connected annular to the cable passage opening. In most cases one has to leave open the cable shield to the soldering strip of the strain gage circuit. In the evaluation of existing measurement arrangements and in the design of new facilities the question should be in the foreground as to whether the EMC-compliant design is required at all. If after conventional conceptually design, measuring systems have worked smoothly and to the user's satisfaction, it is to be expected that they continue to do so, unless new sources of interference arise in the immediate environment.

6.8.3 Protection against overvoltage

Particularly critical faults may arise through overvoltage occurring in the measuring leads between the sensor and measuring instrument or in the power supply of the instrument. Overvoltage can cause severe damage or the total destruction of the entire measuring device. In modern measuring instruments equipped with semiconductor circuits only relatively low-energy voltage pulses are enough to cause damage or malfunction. Reasons of overvoltage are atmospheric charges and discharges (lightning), static, voltage caused by welding or similar arbitrary influences and the NEMP. The NEMP (nuclear electromagnetic pulse) is defined as the electromagnetic pulse resulting from a nuclear explosion. The strong gamma radiation liberated ionizes the air molecules by knocking out the electrons (Compton effect) and thereby produces within a few nanoseconds a strong electric field that radiates through the rapid change of the NEMP.

The protective measures briefly described here are based on atmospheric discharges as the cause of the overvoltage, and these protection measures are also largely effective against all other overvoltage sources. With an atmospheric discharge relevant to a measuring system subsequent phenomena are connected substantially.

The resulting impulse current during a flash of lightning has a rise time of about 1 μs between 1 kA and 100 kA [6.15]. The best protection is offered by the classic lightning rod which captures the flash, and the discharge current is caught and channelled away. By the static electricity of the air and charged thunderclouds also lines located in the surrounding area are charged.

The discharge from a flash generates a short-term increase of the earth potential as the lightning current enters the ground. The resulting overvoltage can damage devices present in the area. Part of the voltage drop across the grounding resistance of a lightning rod (magnitude 100 kV) can also occur between the nearby transducer and its surroundings, causing a flashover. So as well as the signal line, the input of the connected instrument is then endangered, although the flash has not even been taken into the signal line. It follows that transducers in the field, if possible, should well insulated when installed.

The current of a few thousand amperes associated with the discharge pulse produces in the atmosphere a strong electromagnetic field which induces high voltages and currents in the cables covering this area. These lead to overvoltage in the connected electrical devices. In this way an overvoltage can reach the measuring instruments via the cable of the mains supply.

Damage can be caused not only by direct lightning strike, but also by the electrical phenomena associated with the static charge and discharge, against which a measuring system has to be protected.

Protection circuits to divert the overvoltage to ground are industrially manufactured components available that have to be installed at selected locations in a measuring system. Good overvoltage protectors have short response times and large discharge capacitance. In the idle state they do not affect the facilities to be protected, but they can change the resistance of the line in which they are fitted. Gas-filled overvoltage arresters proved to be good for protecting circuits in telecommunications and measuring systems. In principle these consist of ceramic tubes filled with inert gas, and at the end faces of each tube a metal electrode is arranged. Gases are non-conductive media under normal pressure and temperature conditions. If the voltage increases between the electrodes, as occurs when an ignition voltage is exceeded, an avalanche-like rise of charge carriers caused by impact ionization takes place. Between the electrodes a gas discharge path is formed, which ignites independently a gas discharge. At the moment of ignition, the voltage collapses and adjusts to the value of the arc voltage of about 10–20 V. At the same time leakage impact current begins to flow. Between the maximum operating voltage and the static response voltage (ignition) a minimum safety factor of 1.5 should be taken into account.

Decades of experience in telecommunications show that a discharge capacity of 25 kA is sufficient for surge arresters in virtually all cases, including for the protection of measuring systems with strain gage transducers [6.15]. Marketable overvoltage protection circuits, which have proven themselves for the protection of measuring systems [6.16], consist of a combination of a gas arrester as a coarse protection element with a fine protective element (protection diode or capacitor).

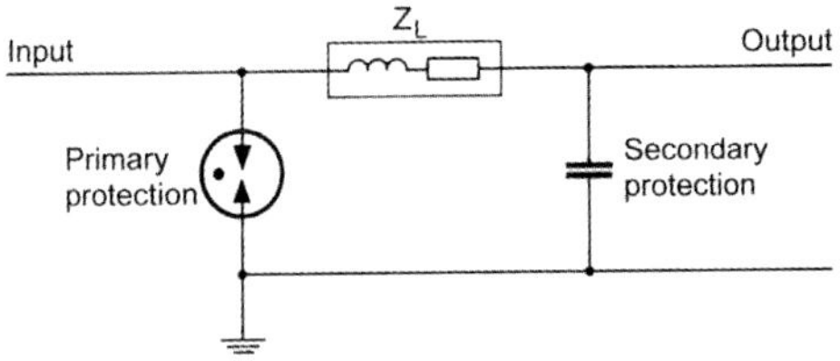

Fig. 6.19 Basic structure of a protection circuit with gas arresters and decoupling element between coarse and fine protection [6.15]

Figure 6.19 illustrates the basic structure of an overvoltage protection circuit. Between coarse (primary) and fine (secondary) protection is arranged an intermediate series element Z_L, which limits the discharge current through the secondary protection element and, if a gas-filled overvoltage protector is used as primary protection element, permits a rise of the input voltage up to the spark-over point. Z_L can take the form of either an ohmic resistor or an inductance with an ohmic component. However, a protection circuit with a pure ohmic resistance is very problematic. Since any secondary protection element, regardless of whether it is a Zener diode or a capacitor, has an inductance of at least 14 nH according to its physical length (at least 10 mm), high frequency voltage drops must be expected here which are greater by several factors than the limiting value envisaged.

Figure 6.20 shows a photograph of the realization of an overvoltage protector in Europe card format, as used in measurement technology with strain gage transducers. On a common ground plate are arranged eight paths with gas discharge tubes. Europe cards of this type can quite easily be placed in standard housings or placed in special lightning protection racks. For transducer connections in the six-wire technique it is recommended, to use one Europe card per transducer and assign to each wire a protection path.

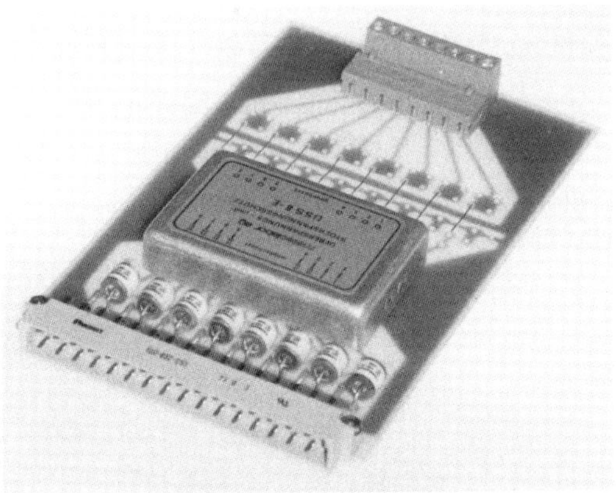

Fig. 6.20 Photo of an overvoltage protector with eight paths

References

[6.1] Ruge, A.C. (1951) Elimination of lead wire errors in the SR-4 temperature-compensated strain gage. *Testing Topics, BLH-Corp.*, **6** (3), 7–8.

[6.2] Riedhof, D. (1974) Zur Referenzphaseneinstellung bei Trägerfrequenz-Meßverstärkern. *MTB*, **10** (1), 8–10.

[6.3] Kreuzer, M. (1990) Die Schnittstelle zwischen Aufnehmer und Meßverstärker bei Gleichspannungs- und Trägerfrequenzbetrieb. *MTB*, **26**, 42–48.

[6.4] Paetow, J. (1988) Die Sechsleiterschaltung für DMS-Aufnehmer. *Wägen und Dosieren*, **19**, 12–16.

[6.5] Kreuzer, M. (1989) Anpasser, in *Handbuch für experimentelle Spannungsanalyse*, 1st edn (ed. Chr. Rohrbach), VDI- Verlag, Düsseldorf, pp. 666–705.

[6.6] Kreuzer, M. (1976) Eine Vielstellenmeßanlage mit FET-Schaltern. *MTB*, **12**, 4–9 and pp. 35–41.

[6.7] Kreuzer, M. (1983) Vergleichende Betrachtung verschiedener Schaltungsarten für das Messen mit Dehnungsmeßstreifen. *MTB*, **19** (3), 63–68.

[6.8] Kreuzer, M. (1983) Elektrische Vielstellen-Meßeinrichtung, German Patent DE 3 238 482.

[6.9] Heringhaus, E. (1982) Trägerfrequenz- und Gleichspannungsmeßverstärker für das elektrische Messen mechanischer Größen, part 2: Verhalten gegenüber externen Störeinflüssen und Auswahlhilfen. *MTB*, **18**, 70–73.

[6.10] N.N. (1991) Schutz vor magnetischen Feldern. *Markt und Technik*, no. 4 (18.1.1991), p. 18.

[6.11] Borek, L. (1982) Magnetische Kabelabschirmung mit amorphen Metallen. *Elektronik*, (4), 43–46.

[6.12] Queck, U. (1986) Das Übertragungsmedium "Kabel", part 1 and part 2. *Elektronik*, (3), 51–56; (4), 99–105.

[6.13] Mooser, J. Überschätzte Schirmwirkung. *Elektronik Journal*, **26** (23), 42–44.

[6.14] Catalog WC-2A "Inter-8 Weave Cable" (1989) Magnetic Shield Corp., Perfection Mica Co., Bensenville, 60106 Ill., USA.

[6.15] Büchler, W. and Bosshard, W. (1986) *Blitzschutz Elektronischer Geräte und Anlagen*, published by Meteolabor AG, Wetzikon.

[6.16] de Witte, H.-J. and von Lerber, A. (1989) Reconstruction of a storm-damaged bridge on the approach to the Gotthard Tunnel in Switzerland. *RAM*, **5** (2), 45–51.

7 Signal processing

7.1 Introductory viewing

In the middle of the nineteenth century, in his resistance measurements, Charles Wheatstone used a compensating measurement method. An imbalance created in one arm of his bridge was compensated by an imbalance in the adjacent bridge arm (see Figure 1.9). Complete compensation was achieved, when the output signal of the bridge circuit became zero. The amount of imbalance required for complete compensation represents the measured value.

For the supply of his bridge circuit, Wheatstone had to use an unstable DC voltage source (voltaic element), which was all that was available in the mid-19th century, and he checked the balance of the circuit with a mirror galvanometer. Thus, in principle, the minimum instrumentation that is required for resistance measurements using the Wheatstone bridge circuit is just a supply voltage source and a sensitive measuring instrument for detecting the voltage on the measuring diagonal.

So in his resistance measurements, Wheatstone used a comparative method. He compensated the resistance change to be measured by detuning of the adjacent bridge arm. This technique, known as the compensation or even zero measuring principle is still widely used today, but with the help of sophisticated modern technology. Precision compensators with the smallest possible measurement uncertainties are available for the strain gage technique, such as the extremely small measurement uncertainty of <0.0005% for resolution of a signal of 2 mV/V into 10^6 digital steps [7.1].

More often than the compensation principle, the amplification method is used in the field of practical measurement where the change in the bridge output voltage due to resistance change is measured by a suitable device. Since the output voltage and its changes are very small, the output signal for its technical exploitation must be amplified. The amplified output signal is then able to operate output devices such as displays, or perform other functions. The amplified signal can also be transmitted over longer distances, whereas the compensation method excludes this possibility. In addition, the amplification method allows the recording of time-varying quantities, whereas the compensation method is generally confined to the measurement of static or quasi-static variables. With microcomputer controlled digital compensators we can also achieve relatively high measurement rates, as will be shown later.

The development of the strain gage in parallel with that of the associated instrumentation, has meant that the compensation and the amplification principles have coexisted from the beginning up to now. Both methods have found their specific application areas. The different areas of application have led to special equipment developments that optimally meet the needs of a given application. The main criteria here are the temporal behaviour of the measured quantity, the number of measurement points to be covered and the further exploitation of the signals. With the measuring of static variables often high precision and long-term stability are of highest importance, and the multipoint measuring technique can be carried out

Technology and Practical Use of Strain Gages: With Particular Consideration of Stress Analysis Using Strain Gages. First Edition. Stefan Keil.

serially due to the time-constant signals. For dynamic variables, the transmission behaviour of the measuring instrument is critical, and it must provide a faithful reproduction of the output signal to give a good presentation of the time-varying quantity. With multipoint measurements of time-variable quantities, parallel signal acquisition and processing, depending on the frequency of the measured quantity, is often required, especially if the measured values need to be synchronously associated with each other, for example in the detection of load-dependent deformations or in the evaluation of measurements with strain gage rosettes with time-varying load on the measurement object.

The question of the further use of the measured signal for the execution of specific functions may require additional conditioning, such as when including measuring equipment in processes where the measurement signals are used to monitor or control tasks. Especially in the recent past, the strain gage measurement technique with the development of fully computer controllable digital measuring amplifiers and miniaturized digital transducer electronics has been fully integrated into the data processing environment which is these days dominated by electronics. The direct connection of measuring devices to computers opens up the possibility of letting the immediately available results of the online processed measurement signal influence the measurement process or a process control effect retroactively. Digital processing of dynamic test signals can require large storage capacities, especially if measured values need to be detected and evaluated that are of higher-frequency processes, perhaps even multi-channel and synchronously on-line. The function of time of the measured quantities and their frequency spectrum are key criteria for the selection of the measuring equipment for signal conditioning and analysis.

Instruments for strain gage technology are generally designed so that you can also connect full-bridge circuits and half-bridges and single strain gages. The complementary resistors required for their addition to full-bridge circuits when connecting half- or quarter-bridges are located in the instrument and can be switched on as required. Modern devices are prepared for transducer connection with a six-wire technique (for control of the voltage drops in the supply lines of the pickup circuit, see Section 6.5), or by the two-channel principle (for elimination of resistive and capacitive effects in cable using carrier frequency mode, see Section 6.6).

7.2 Analogue instrumentation amplifier

7.2.1 Structure of analogue signal processing

The main tasks of a measuring amplifier are to provide the excitation voltage required for the supply of the Wheatstone bridge circuit connected to the input and to amplify the analogue output signal of the bridge circuit from the millivolt range to a level in volts. Since bridge circuits with strain gages give an analogue output signal, the amplifier process has to take an analogue signal at its input stage. Until the early 1990s, the signal processing in measurement amplifiers was almost entirely analogue, and the basic structure of the analogue amplifiers, regardless of the currently available electronic components (valves, transistors, operational amplifiers) remained almost unchanged for many decades [7.2]. The basic functional

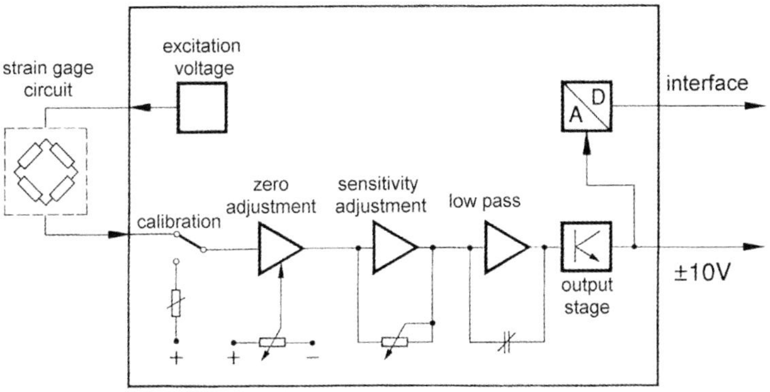

Fig. 7.1 Functional units of an analogue amplifier for strain gage measurement

diagram of the signal processing in an analogue measuring amplifier is shown in simplified form in Figure 7.1. Apart from the supply voltage source it shows the essential signal processing function units.

The main units are the generator for the excitation voltage, which supplies the strain gage circuit with the appropriate voltage; the calibration switch which enables a defined signal to be applied at the input for adjusting the gain of the amplifier; the adjustment of zero which enables the signal to be set to zero when the strain gage circuit is unloaded because, especially for measurements with strain gages, it is desirable to have a zeroing range at the measuring amplifier as large as possible, since installed strain gages, including their connections, often lead to strong asymmetries in the bridge circuit. Here, the scale value represents the amplifier input signal leading to the full scale output of the amplifier. If the measuring range on the example is set to 0.5 mV/V, an output signal of the strain gage circuit of 0.5 mV/V (which corresponds with gage factor $k = 2$ to a strain of 1000 µm/m) leads to the maximum level of the amplifier output. Normally, the measurement range of the amplifier can be set via switches in rough steps and is continuously adjustable via a potentiometer in between the fixed steps. The low-pass filter, usually with selectable cut-off frequencies, is used to filter the measured signal which is superimposed on higher-frequency disturbances. The final stage provides an analogue output signal of sufficient power. Normally there is an analogue-to-digital converter to show the readings on a digital display or transfer them to a computer.

For analogue measurement amplifiers, the coarse and fine adjustments of individual functional elements will be carried out by a combination of switches and potentiometers, which can represent a significant share of the total cost of the amplifier circuit, depending on the ranges to be covered and the required resolution.

Wheatstone bridge circuits are called passive sensor circuits, since they are not able to output a measurement signal without a power supply. This contrasts with the active sensors, in which the measured quantity itself such as electrodynamics effects generates the measured signal. For the supply of Wheatstone bridge circuits two different principles of excitation are available: the supply with DC voltage and the carrier frequency method. With the DC method the Wheatstone bridge circuit is supplied with a highly stable DC voltage, and the measurement signal is then available in the form of the DC voltage output at its measuring diagonal.

The carrier frequency method is carried out with an alternating voltage whose amplitude and frequency are highly stabilized. Accordingly, the output voltage of the bridge circuit is an AC voltage whose amplitude is used as the signal parameter. Each of the two processes requires a specially designed power source and a corresponding amplifier input stage, so that the type of power supply is also found in the description of the corresponding measurement amplifier, which is referred to as a DC voltage or CF (carrier frequency) measurement amplifier. Due to the different modes of operation of DC and CF measurement amplifiers they have specific advantages and disadvantages for the solution of different measurement tasks. These two methods, as well as their relevant differences for practical application, will now be briefly explained.

7.2.2 Bridge supply with carrier frequency

The question about the pros and cons of the use of DC or CF measurement amplifiers for measuring with strain gages can be answered clearly as follows. Whenever the frequency of the measured variable allows, it is best to use the CF method. If the frequency of the measured variable exceeds the cut-off frequency of the amplifier, which depends on the carrier frequency of the amplifier, the DC method has to be used. Strains to be measured relating to mechanical processes mainly take place in the static or low-frequency range, so that the carrier frequency measuring amplifier is predominantly used in strain gage technology. An advantage of the CF method is that only a narrow band, lying in the range of the carrier frequency, is used for transmission, and thus excellent noise suppression is achieved. Also the thermoelectric voltage generated by the solder joints or similar DC voltage disturbances are totally eliminated. CF amplifiers can achieve much higher resolution than DC measurement amplifiers, especially for small signal bandwidth (up to factor 10) because with CF measurement amplifiers the thermal noise is much lower than with DC amplifiers.

Figure 7.2 illustrates the basic structure of a carrier frequency amplifier for strain gage measurements [7.3]. An AC generator produces the bridge supply voltage with highly stabilized amplitude and frequency. Today commonly used frequencies are 225 Hz, 600 Hz, 1 kHz, 4.8 kHz and 5 kHz. We call these frequencies carrier frequencies, since the amplitude of the AC voltage the measurement signal is modulated. The output of the Wheatstone bridge circuit is an AC voltage in the millivolt range which is proportional to the measured quantity and is raised by the input stage and the downstream stages of the measuring amplifier to a level in volts. Here, a pure AC voltage is amplified in a very narrow frequency bandwidth range and all interfering signals outside the range of the carrier frequency and the intrinsic zero signal drift will be suppressed.

The input stage amplified AC signal is phase-critically rectified in the demodulator. Phase-critical indicates that the sign of the rectified voltage corresponds to the sign of the variable. At the sign change of the measured quantity (i.e. for example the transition from a positive to a negative strain in cyclic loading of the measurement object) a phase shift of 180° (phase jump) occurs in the output of the strain gage circuit.

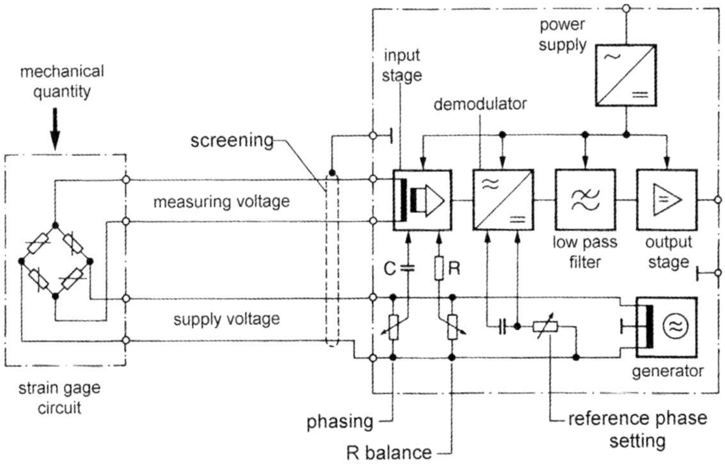

Fig. 7.2 Basic circuit diagram of a carrier frequency (CF) amplifier with connected strain gage bridge circuit

This phase jump causes a change of sign of the rectified signal in the demodulator. The rectification of the measurement signal in the demodulator converts the sinusoidal to the zero position swinging carrier frequency signal to a half-wave signal whose envelope corresponds in its course over time with the measured variable. Positive half-waves occur during positive measuring values and negative half-waves for negative values. The demodulator requires an oscillating carrier frequency control voltage which is provided from the supply voltage generator for the rectification. For lossless signal rectification in the demodulator, the zero crossings of the demodulator input signal voltage and the control voltage (reference voltage) must be identical. The identity of the zero crossings can be brought about by controlling the phase angle of the control voltage. This process is referred to as a reference phase adjustment, and is described in more detail in Section 6.3.3.

After the demodulator there is a low-pass filter, which converts the amplitudes of the carrier frequency signal into smoothed voltage signals conforming to the mechanical signal being measured. At the same time the low pass filter removes most of the carrier frequency, the small residual carrier being further attenuated in additional low pass filters until it is very small compared to the measured signals. The ripple of the amplifier output signal is thus a quality criterion for a carrier frequency amplifier. Typical values of the carrier leak voltage are less than 0.1% of the nominal output voltage. The following output stage converts the output voltage of the low pass filter into a signal of −10 V to +10 V. Figure 7.3 illustrates the described sequence of signal processing in a carrier frequency measuring amplifier using the example of a time-varying measurement signal with change of sign.

The CF amplifier should preferably be used if the frequency of the quantity to be measured lies below the high cut-off frequency of the amplifier. If the frequency of the quantity to be measured exceeds the high cut-off frequency of the CF amplifier a DC amplifier should be used.

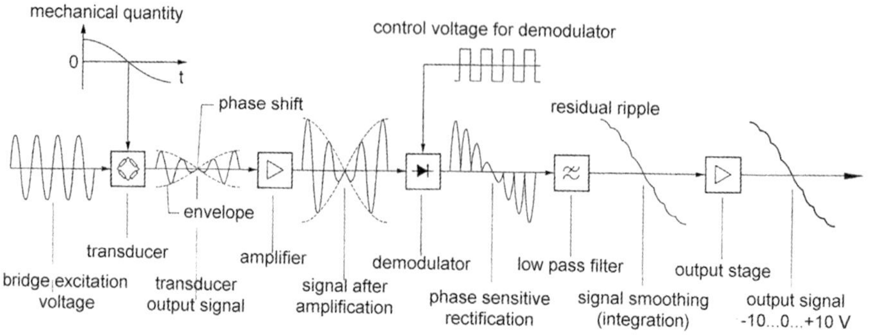

Fig. 7.3 Operation and structure of signal of the different stages of the carrier frequency principle

An important characteristic for a carrier frequency measuring amplifier is its high frequency cut-off. It is defined as the frequency at which the amplitude ratio drops by 1 dB. The cut-off frequency represents the upper limit of the frequency which can be processed by the amplifier. This frequency depends on the carrier frequency of the amplifier under consideration and is defined as the frequency at which the amplitude of a sinusoidal signal drops by 1 dB compared with zero frequency amplitude ratio which represents the static amplification factor.

The high cut-off frequency of an amplifier can be found by recording the amplitude response of the instrument: a sinusoidal test signal is fed to the amplifier's input and the response at the output is recorded. The output signal will be a sinusoidal of the same frequency but of changed amplitude and displaced phase. This shows that different frequencies at the input can cause different values of amplitude drops and different values of phase shift. Data on the time response of amplifiers to periodic signals are always related to the signal frequency.

The variation of the amplitude of the output signal in relation to the frequency is called the amplitude response of the amplifier. An example of it for a 5 kHz carrier frequency amplifier is given in Figure 7.4. It can be seen that above a certain signal frequency any further increase of the frequency causes a reduction in the amplitude of the output signal. The fall in amplitude D (damping) is usually quoted in decibels (dB). If A_1 is the reference amplitude (100%) and A_2 the reduced amplitude, the drop D in amplitude is given by

$$D = 20\lg\frac{A_1}{A_2} \text{ in dB} \tag{7.1}$$

For example, 1 dB gives an amplitude ratio $A_1/A_2 = 1.12$ (drop to 89.3%) and 3 dB a ratio of $A_1/A_2 = 1.4$ (drop to 71.4%). The frequencies given for amplitude drops of 1 dB and 3 dB are often quoted as the upper cut-off frequencies for signals that can be processed by measuring amplifiers. They are also identified on the example in Figure 7.4. The upper cut-off frequency also governs the transfer bandwidth which, however, can be further reduced by means of built-in or external filters, for example low pass filters.

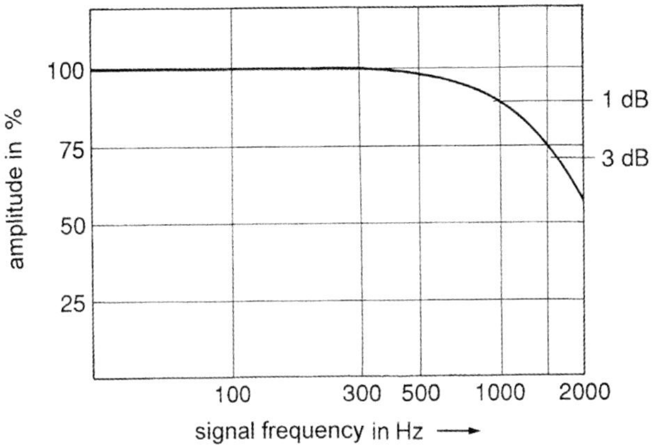

Fig. 7.4 Amplitude response of a 5 kHz carrier frequency amplifier

If there is the possibility of a dynamic calibration of the measuring device, it is possible to measure signals with constant frequency almost up to the upper limit frequency, without the error caused by the drop of amplitude. Signal frequencies, higher than the upper cut-off frequency of the amplifier must be processed either with an amplifier with higher carrier frequency or with a DC measuring amplifier. With DC measuring amplifiers signal frequencies >10 kHz can be processed without problems.

As mentioned before, different frequencies cause different values of phase shift. Plotting the phase displacement of the output signal against the signal frequency gives the phase response of the amplifier. A typical example of the phase response of a 5 kHz carrier frequency amplifier is given in Figure 7.5.

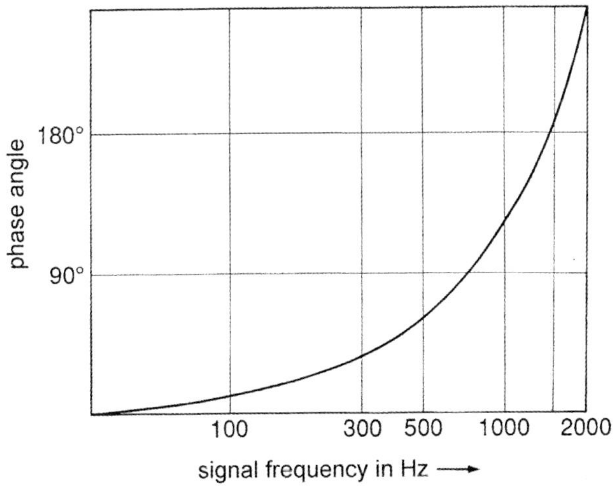

Fig. 7.5 Phase response of a 5 kHz carrier frequency amplifier

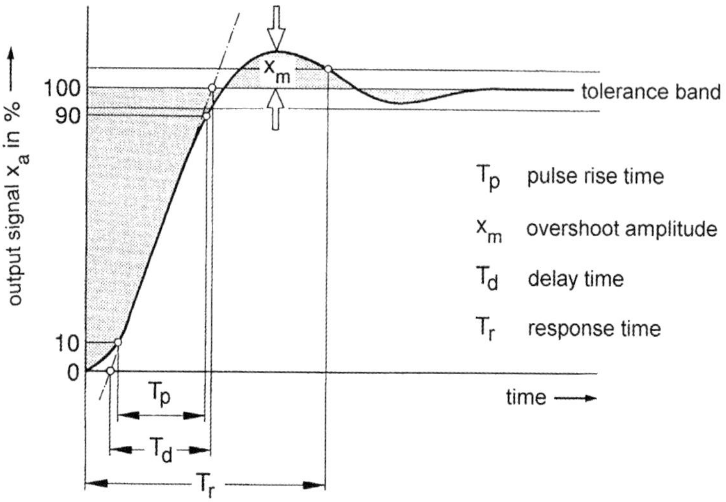

Fig. 7.6 Step-change response of a measuring instrument and definition of various parameters

An additional criterion for the assessment of the dynamic characteristics of amplifiers is the step response. The output signal of the measuring unit will be compared with the input signal. The output signal must reproduce with high accuracy the variation of the input signal with respect to time.

Using this behaviour the characteristic parameters of an amplifier can be observed without major problems by using a frequency generator: the test function (step function) is applied to the amplifier input and the amplifier output signal (response function) is viewed on an oscilloscope.

For step changes of the input signal, the output signal follows with a slower rise time and requires a certain time interval to reach the completely steady-state condition. The rise time is closely connected to the limit frequency of an amplifier and is defined by the pulse rise time. The pulse rise time is the time required by the output signal to rise from 10% to 90% of a step change at the input of 100%. Figure 7.6 shows a diagram of the step-change response of an amplifier with the definition of the main parameters.

The parameters of the response function have a special significance for the measurement of pulse-form mechanical functions. The curve of step response as can be seen in Figure 7.6 is also called transient response. The step response can be aperiodic or oscillatory, as shown in the example in Figure 7.6. The amplitude of the largest over swing (usually the first) is called the overshoot amplitude. The time taken for the output signal to return permanently to within the predetermined tolerance band after a step-change in the input signal is called the response time. The points of intersection at zero and 100% output signal of the tangent drawn to the curve at the inflection point give the delay time. In actual practice with amplifiers, the term "pulse rise time" has become popular; it means the time required for the output signal to increase from 10% to 90% of the full level when a step function occurs at the input.

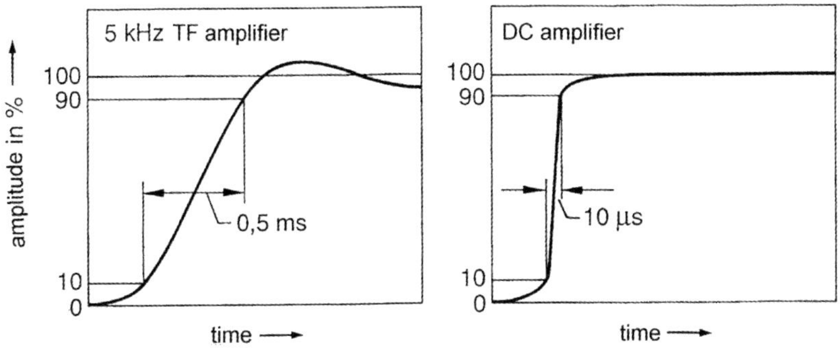

Fig. 7.7 Comparison of the pulse rise times of a CF and a DC amplifier [7.3]

The rise time is closely connected with the limit frequency of an amplifier. The total response time is influenced by the characteristic of the filters inherently required for carrier frequency amplifiers. Figure 7.7 [7.3] shows the step response times of a 5 kHz CF and a DC amplifier of a comparable price range. The DC amplifier shows the shortest step response time, which means it has the better dynamic behaviour. CF measurement amplifiers with low carrier frequency (e.g. 225 Hz or 600 Hz) have significantly longer pulse rise times. However, this is not a disadvantage, because these measurement amplifiers are not designed for dynamic measurement tasks. They have other significant advantages such as long-term stability, low temperature effect on both the zero signal and the sensitivity, good linearity and almost no capacitive cable influence. If we include a 225 Hz CF amplifier in the comparison, we will see that its rise time will amount to 20 ms.

The longer rise time of the CF method compared to the DC method is without significant importance for practical measurements in most cases, as is measured with strain gages on mechanical systems, and their inertia limits the mechanical vibrations. In the investigation of sudden stresses such as in fracture mechanics, or shock waves from explosive processes the rise times of the amplifiers are significant. For detecting these fast processes, the strain gages should be installed directly on the object to be measured, because even transducers possess pulse rise times and cause damping which, compared with those of the measuring amplifiers, could be considerable.

7.2.3 Bridge supply with direct current

Figure 7.8 shows the block diagram of a DC voltage measurement amplifier with connected full-bridge circuit. The output signal of the strain gage bridge circuit is supplied to the input stage of the measuring instrument designed as a differential amplifier. This has two inputs, and only processes the difference of the two input voltages, referred to a reference voltage lying symmetrically to the inputs of the difference amplifier.

The output signal of the strain gage bridge circuit consists of two partial voltages which are symmetrically orientated relative to the reference voltage. This signal is completely amplified by the two dual transistors arranged in different circuits. The resulting drift voltages in the transistors are, however, almost completely suppressed since they are of

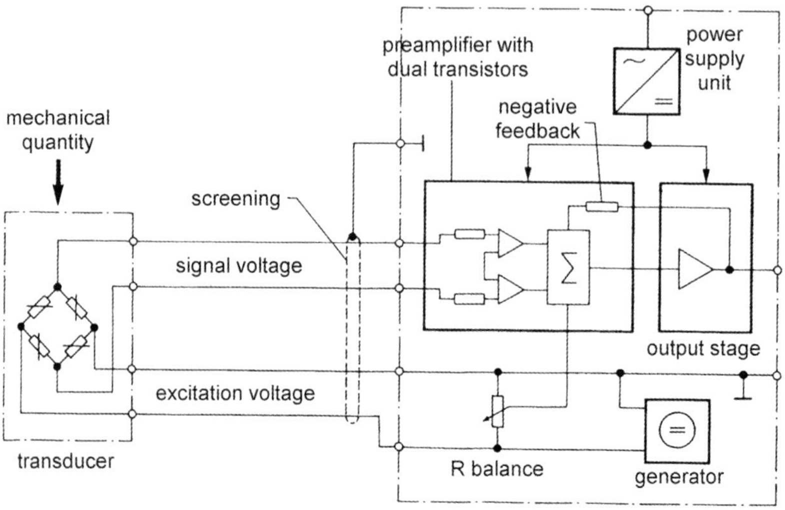

Fig. 7.8 Basic circuit diagram of a DC amplifier with connected strain gage bridge circuit

the same sign and have approximately the same value. This leads to fundamental behavioural characteristics of the DC method against interference voltages acting on the test leads. These are in phase under the action of electric fields, so they are suppressed (common mode rejection ratio, CMRR). Push-pull interference such as generated by magnetic fields, however, can be fully effectively as disturbances.

The input stage of the amplifier is followed by other amplifier stages, the last of which is referred to as the final stage. An important parameter for the user of analogue DC voltage amplifiers is the zero signal drift and its temperature dependence. When measuring amplifiers with selectable measuring ranges are used the effects of temperature on the zero signal are also dependent on the measuring range. The temperature coefficient of the zero signal is often mentioned as a characteristic of analogue DC amplifiers. It is composed of the gain-independent part of the final stage and of a part dependent on the measuring range. The absolute zero signal shifts are made larger by temperature changes at high gain. To assess the influence of temperature on the zero signal all possible measuring ranges should be checked, but at least the largest and the smallest measuring range. Generally DC instrumentation amplifiers have much higher temperature-induced zero signal drift than CF measuring amplifiers.

As mentioned in the description of the carrier frequency method, the special advantages of the DC process are in the very high upper cut-off frequency, which is far above the reach of carrier frequency amplifiers values. Another advantage is the much better transient response of the DC measuring amplifier, which was also discussed in the previous section.

The main disadvantages of the DC principle are caused by its wide transmission bandwidth. As can be seen in Figure 7.9, mains impulses as well as impulses created by power cables can cause strong disturbances. Thermoelectric voltage and galvanic voltage can also interfere with the measurement signal. Thermal voltages occur, for example, with different temperatures at the soldered points between the pins and

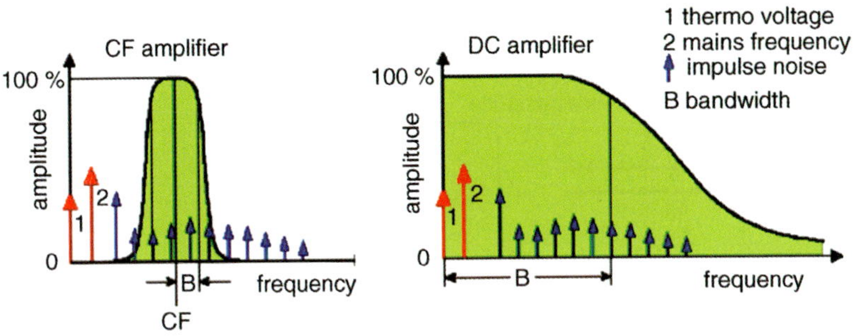

Fig. 7.9 Comparison of the liability to interferences of CF and DC amplifiers; from [7.4]

measuring lines. Galvanic voltages are caused by the formation of galvanic cells in the measuring circuit, for example, through connection between two different metals in contact with the measuring circuit. That is, for example, possible if insufficient covered strain gage installations come in contact with an electrolyte such as salt water. There the resultant DC voltages are substantially greater than the typical thermal voltages. Galvanic voltages in the strain gage measuring circuits have been observed of several hundreds of millivolts [7.3].

In the DC method electrolysis may be caused in damp, corrosive atmospheres. The constant excitation voltage leads to corrosion of the terminals or of the strain gages itself. This corrosion does not generate interference voltages in the measuring circuit, but can lead to failure of the measuring point after a relatively short time. Such electrolysis doesn't occur with the CF method.

Figures 7.10 and 7.11 show qualitatively the effects of interferences when using the DC voltage procedure as well as the CF frequency principle.

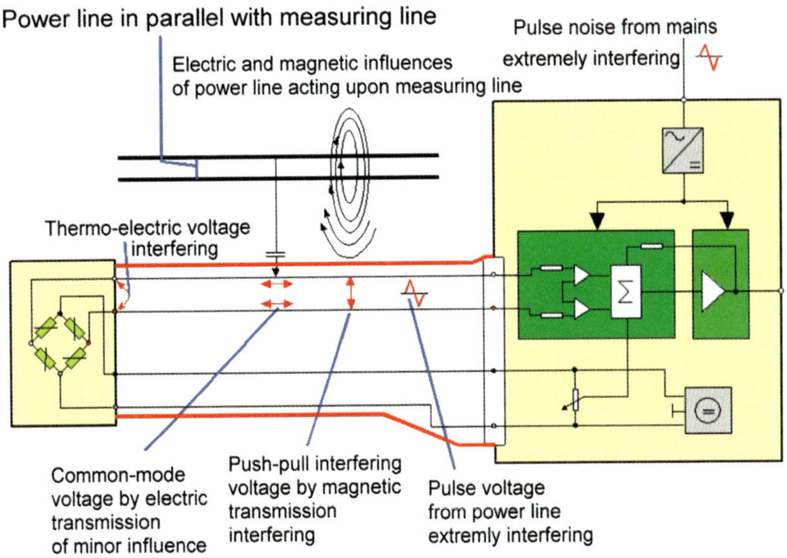

Fig. 7.10 Influence of interferences acting on measuring systems with DC amplification

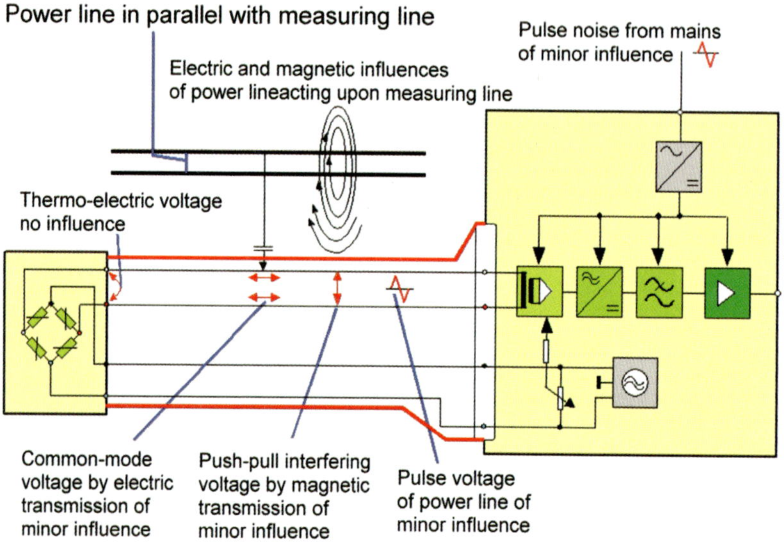

Fig. 7.11 Influence of interferences acting on measuring systems with CF amplifiers

7.3 Digital amplifier concepts

7.3.1 Computer controlled analogue amplifier

Our present-day industrial environment is mainly dominated by digital data processing and computer technology. Therefore there is an urgent desire to link up strain gage measurements effectively into this sphere. Using analogue amplifiers the direct link between the measurement instrument and the computer provides the opportunity of immediately making available not only the measurements, but also the evaluated results. Because of the analogue signal processing in the amplifier it is not possible to directly control the measurement equipment functions using a computer.

In order to satisfy the requirement for computer controllable measurements, attempts have been made, as shown in Figure 7.12, to convert digital commands issued by the computer into analogue commands within the instrument, so that at least the most important amplifier settings can be carried out by the computer

However, this method requires suitable digital-to-analogue converters in the instrument. The setting parameters, digitally specified by the computer for the separate amplifier functions, are each converted to analogue values in the functional units to be adjusted so that analogue control signals can bring about the desired changes in the analogue functional units. This means that a digital-to-analogue converter must be present for each amplifier functional unit.

A problem arises with the required high resolution of the digital-to-analogue converter. For example, the zero point balance must be able to be carried out correctly within the measurement ranges with high sensitivity. The use of digital-to-

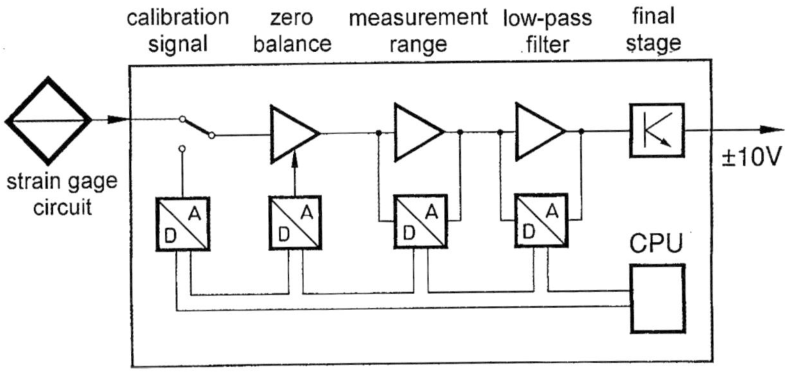

Fig. 7.12 Basic design of a computer controllable analogue measurement amplifier

analogue converters with sufficiently high resolution and with the required high stability on the analogue output signal cannot be justified for economic reasons. If lower resolution can be tolerated, then a correct zero point balance will not be able to be carried out when working on the sensitive amplifier ranges, because zero-balance skipping will occur. Therefore it can be said that amplifiers based on the structure shown in Figure 7.12 represent a compromise between measurement and computer technology, but almost always to the detriment of the measurement.

The logically obvious solution is to digitize the transducer signal before it is processed in the measuring amplifier and then to carry out a purely digital signal processing, but this has long been unsuccessful for the lack of a suitable analogue-to-digital conversion method with which the analogue transducer signal could be implemented without any loss of information, resolution and bandwidth in a digital data stream of high data rate. Only in the late 1980s were appropriate conversion methods available, which opened the way to a completely new concept of measuring instruments. The conventional amplifier structure composed of analogue circuits for signal processing is replaced by a new range of measuring instruments consisting mainly of digital function units.

7.3.2 New methods of analogue-to-digital conversion

The optimal solution of the problem can be achieved by pure digital processing in the measuring instrument. This requires the digitization of the output signal of the strain gage bridge circuit before signal processing in the amplifier. The digitization has to be carried out over the complete measuring range of the connected bridge circuit at high speed and also with high resolution.

A new digitizing technique was the key to building digital measuring amplifiers, with which it was first possible to digitize analogue transducer signals without loss of information [7.5,7.6]. Without loss of information means here that a high signal resolution is required, and high measurement rates have to be realized. These two requirements have so far been contradictory, that is the usual methods of digitizing

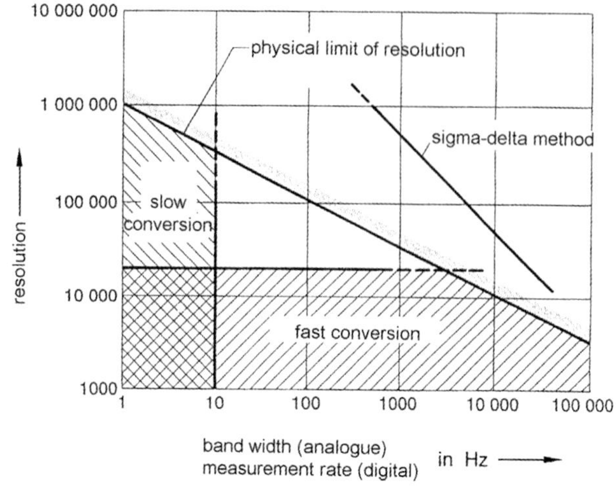

Fig. 7.13 Resolution and measuring rates of available conversion principles

were either quick and allowed only limited resolutions or gave high resolution, but only slowly.

Digitization techniques as available nowadays (sigma delta method or MPDM principle) that enable the design of fully computer controlled measurement instruments. The analogue signal in the input section of the measuring instrument can be digitized over the complete range without loss of resolution.

Figure 7.13 shows in a diagram resolution and sample rate of conventional conversion methods. The drawn physical resolution limit is due to the thermal noise of the transducer circuits. The fast conventional converters are generally constructed in the form of divider chains, in which the signal voltage to be digitized is compared in discrete steps with a reference voltage. The reference voltage is generated with the aid of divider chains (e.g. resistor networks) whose switching state of each digitized signal represents the voltage. Divider chains reach very high conversion rates as shown in the diagram, but the resolution is limited to 12 bits or 16 bits depending on the economic viability of the effort required. High resolution can be achieved with an integral counting method, in which the ratio of the analogue detection voltage to the reference voltage is represented by the ratio of two times, which are determined by simple counting. With increasing resolution achieved by lengthening the beats the measurement rates are smaller.

A reasonable resolution is limited by the signal-to-noise ratio, as already explained in Section 4.4. As shown there, in the example, a $350\,\Omega$ strain gage performs at room temperature a thermal noise voltage of $10\,\text{nV/B}^{-2}$, with an analogue measurement signal of $10\,\text{mV}$ this results to a signal-to-noise ratio of $10^6/\text{B}^{-2}$ (B here means the bandwidth in Hz). You can find the limit of useful resolution for this case from this value quite easily by substituting B for the desired frequency in Hz. For example, with $10\,\text{Hz}$ as frequency, we obtain $\approx 316\,000$ parts as the upper limit of resolution.

In principle, the upper limit of resolution is irrelevant for the new conversion method from the viewpoint of technical realization. We consider the multiple-pulse duration modulation (MPDM) conversion method, which combines the integral counting method with the divider chain and uses the associated benefits. The background to this description is the use of two pulse-width modulated voltages in the compensation elements. As a controlling element, a combination of pulse-width modulated signals is used instead of the divider networks. For relatively low technical outlay very high resolution is achieved at very high measurement rates. The method allows, for a sample rate of 38.4 kHz (repetition frequency of both the pulse-width modulated voltages), a signal bandwidth of 4 kHz. Thus the digital data stream after the conversion contains all the information contained in the analogue signal from the largest signal amplitude to the thermal noise voltages, from static to rapidly changing measurement signals in the kHz frequency range [7.7].

7.3.3 Structure of digital measuring amplifier

With the help of the fast high-resolution analogue-to-digital conversion fully computer controlled measuring amplifiers can be built relatively easily. The digital signal processing that is now possible in the amplifier brings, as well as computer control, more significant advantages over the analogue signal processing. Digital signals are long-term and temperature stabile because there is no drift or hysteresis. This is also important in respect to setting limits or storage of signal values. The processing of digital signals can be made by very precise calculation algorithms with appropriate software for computational algorithms which are not affected by temperature and long-term drifts and component tolerances.

This opens up completely new possibilities for calibration and filter functions. Digital filtering is only based on computational algorithms, so variable analogue components with tolerances and flutter can be excluded. The possible diversity of viable filter functions is described in detail in [7.8]. Thus, for example, the signal bandwidth between 1 Hz and frequencies above 1000 Hz can be changed in small increments, just by changing the filter coefficients. One advantage of digital filtering lies in its crystal controlled precision: the running times and amplitude transitions of the signals in all channels are identical in the case of multi-channel operation (with the same filter settings in the channels). online evaluations of measurements with strain gage rosettes are easily feasible even with time-varying loads. Adjustment of zero signals and taring can be performed by simple subtractions, and variations of the measuring range are carried out by multiplications. The digital arithmetic functions can be realized with a special gate array, so additional features are relatively easily available as space needing options for analogue amplifiers such as limit switch, alarm level monitoring or peak value memory can be included in the confined space of a digital amplifier. Here, a digitally stored measured value remains constant, independent of time and temperature changes. In addition, the digital amplifiers can be used quite easily with the aid of corresponding software. In Figure 7.14 the structure of a digital measuring amplifier is schematically shown.

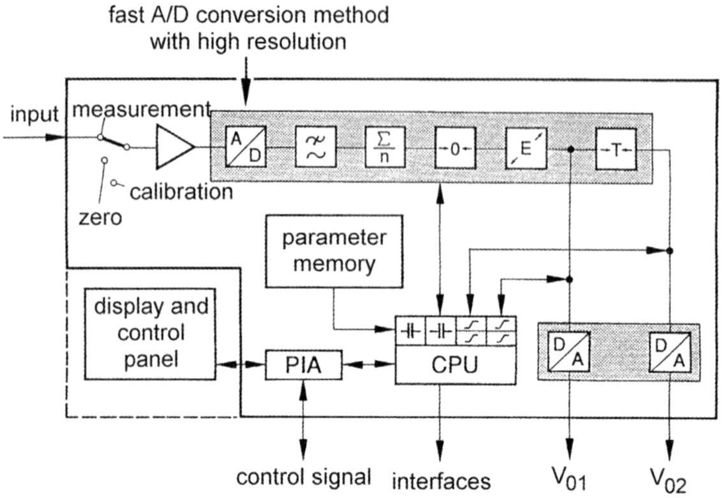

Fig. 7.14 Basic structure of a digital measuring amplifier

With the increasing range of functions due to the digital processing technology and the shrinking hardware components thanks to modern electronic components, the front panels of the instruments become very small and thus the instruments are hard to handle. It is no longer possible, especially for multi-channel devices, to locate all the controls for all the functions in easily manually operated version on the geometrically very limited equipment surfaces. But fully computer controlled equipment does not require manual operation in any case. However, to equip a fully computer-controllable device with the additional option of manual operation, it is necessary to physically separate the manual controls from the actual amplifier. This leads to the formation of separate display and control units, from which single- and multi-channel devices can be centrally operated manually. These control units can be completely separated from the actual hardware device in a further step to which they are connected only when needed via cable.

All settings on the amplifier can be performed from a PC via a computer interface, which is located in the amplifier between the display and the functional units. This design allows for deactivating the control panel when the unit is simply being controlled by the computer. But it also allows the operation of the amplifier by manual control without use of operation by the computer. If a device is to be exclusively controlled by the computer, you can completely dispense with its idea of a manual control panel.

The digital structure offers the further option of saving the device settings to perform a measurement task once undertaken, including the transducer adaption with the associated calibration values in the allocated memory of the considered measurement channel and it retrieve when needed. Such a parameter memory may be housed not only in the amplifier, but also in the transducer [7.9]. Then, the transducer type, bridge excitation voltage, calibration values, low pass filters, limit values, range setting and amplifier mode are stored in this module, and on connecting the transducer to the amplifier, the affected amplifier channel is immediately restored to the operating state matched to the transducer. This self-identification of the transducer by the parameter memory saves time and avoids subjective setting errors, facilitates the reproduction of advance settings and simplifies especially measurements with many sensors, and gives error-free channel adaptation.

Digital measuring amplifiers have on their input side an analogue part. The strain gage bridge circuit provides an analogue signal which is raised in the analogue part of the amplifier to a level suitable for further processing. The measuring range of the analogue amplifier covers the entire range of the strain gage bridge signal and therefore does not require range switching. It also has no zero adjustment elements, because the balancing takes place in the digital part. In the analogue amplifier part, the supply voltage for the transducer is created, and there are also the function units necessary for the six-wire circuit or the dual-channel principle. When operating with carrier frequency the analogue part performs the demodulation of the signal voltage and in a low-pass filter the smoothing. Everything else is then performed in the digital part, after the analogue transducer signal has been digitized without loss of information. The digitization records all analogue signal amplitudes including small thermal noise voltages from the transducer resistances as well as the rapidly changing signals up to the cut-off frequency of the analogue amplifier part.

With the aid of the digital function units a permanent calibration of the analogue preamplifier can be realized, and thus a substantial reduction in the uncertainty of the overall amplifier system can be achieved. At certain intervals, a zero signal and a calibration signal will be switched to the input of the analogue part and will readjust the amplifier characteristic. Any temperature or long-term drifts that occur in the analogue part are thereby fully compensated. Thus, very high precision can be achieved almost effortlessly, which with pure analogue amplifiers could only be realized with considerable effort.

The output signal of the digital amplifier is available as a digital gross and net signal. This signal can also be converted to analogue again and provided as an analogue signal at the amplifier output, so that then such an amplifier appears outwardly to be an analogue device but internally works digitally. For this conversion the digital net and gross output are fed to another gate array, which performs the digital-to-analogue conversion and two gross or net signal proportional analogue output voltages (−10 V to +10 V) are generated.

Transmission of the digital output signals of the amplifier to the computer can be performed either serially or in parallel by means of suitable interfaces [7.10]. In general, the serial interface is slower than the parallel interface, because of the sequentially performed transmission. With the parallel interface each set of eight bits is transmitted simultaneously. However, the serial interface can cover much greater distances than the parallel. Before the operation with a serial interface various parameters such as baud rate (transmission speed), number of data bits and number and parity of stop bits must be set by the operator. With serial interfaces, transmission distances of more than 1000 m can be bridged, and additional bus operation with multiple participants is possible. As a prerequisite for this bus mode, the measuring equipment must be equipped by the manufacturer with software that allows for specific response of individual devices.

7.4 Compensation method

In the middle of the 19thcentury Wheatstone used a comparing measurement method in his resistance measurements. An imbalance created in one arm of his bridge was compensated by an imbalance in the adjacent bridge arm. Complete compensation was achieved, when the output signal of the bridge circuit became zero. The amount of imbalance required for complete compensation represents the measurement value.

Compensators were being used for the first static strain gage measurements shortly after the invention of the strain gage because of the simple practical application of the compensation principle and the simple design of compensators for the measurement of the imbalance of a Wheatstone bridge circuit as caused by a change in resistance.

In contrast to the principle of amplification, which delivers the measurement signal directly as a difference related to a reference value, the compensation method compensates the measurement signal using zero adjustment.

The compensation method requires a display to control the zero adjustment (Wheatstone used a galvanometer to accomplish control) and a compensation circuit to carry out zero adjustment.

Figure 7.15 shows in simplified form the block diagram of an analogue compensator. The compensator receives the output signal V_M of the strain gage bridge. This voltage will be superimposed on the compensation voltage V_C created by the compensation circuit. The signal difference $V_M - V_C$ will be displayed and adjusted to zero using the zero adjustment at the compensation circuit.

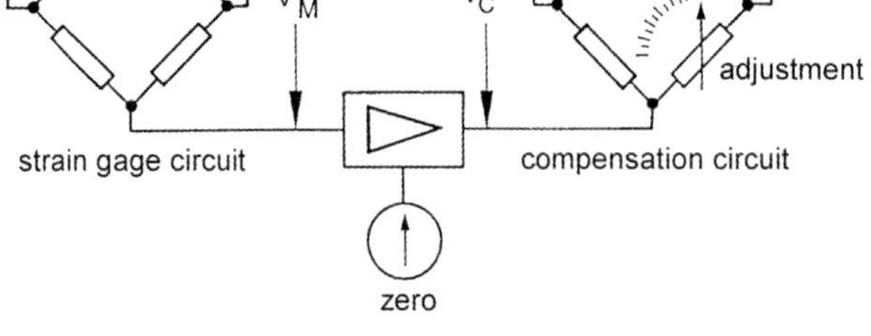

Fig. 7.15 Principle of the analogue compensator

The method is independent of the magnitude of the supply voltage because the supply voltage for the strain gage circuit and the compensating circuit is taken from the same source. The internal resistance of the power supply source is approximately zero due to the type of circuit employed, so the resistance of the strain gage bridge has no effect on the results. The difference former is only used as a null indicator so that the method is independent of its amplification factor. This means that the accuracy of the system is only dependent on passive elements (resistors) which can be manufactured with very high precision.

These characteristics allow absolute calibration of the compensator by the manufacturer in units of the measured quantity. No more calibration is needed in normal use.

The use of compensators is restricted to the measurement of static or quasi-static processes. Compared with amplifiers in the same price range, compensators have the advantages of higher accuracy as well as of better long-term stability.

The compensation bridge circuit with its adjustment element is arranged in series with the measurement bridge circuit. Superimposition of the signal V_C of the compensation bridge on the transducer's output signal V_M generates the differential signal $\Delta V = V_M - V_C$. This differential signal will be displayed and adjusted to zero using the zero adjustment unit at the compensation circuit.

The zero adjustment unit is tightly coupled to the display which shows the amount of the required compensation value. All elements used for zero adjustment are potentiometers calibrated with high accuracy, so that the amount of the required adjustment represents the measurement value with very little uncertainty.

Any possible uncertainty of the characteristic of the zero amplifiers has no influence and doesn't matter. Excitation of both bridge circuits (transducer and compensation circuit) occurs using carrier frequency voltage supplied by a bridge transformer with galvanic separated secondary windings. So variations of the excitation voltage are affect both circuits in the same way and have no repercussions on the measurement result.

Using the principle shown in Figure 7.15, instruments for strain gage measurement have been designed with the adjustment facility calibrated in the strain unit μm/m. These instruments are running with low carrier frequency and allow measurement of static quantities with high resolution. Battery-powered manual compensators have been very popular because of their ease of handling, the large balance range and the field reliability.

Modern precision compensators combine the compensation principle with elements of the amplification technique in order to use the prevailing advantages of both principles.

The measurement signal V_M is fed to the input of the differential preamplifier MA. The further signal processing is carried out as described before. This concept uses the symmetrical input stage of the amplifier which causes high noise suppression (common mode voltage suppression). Asymmetry caused by the adjustment equipment in the compensation circuit will be avoided with the amplifier.

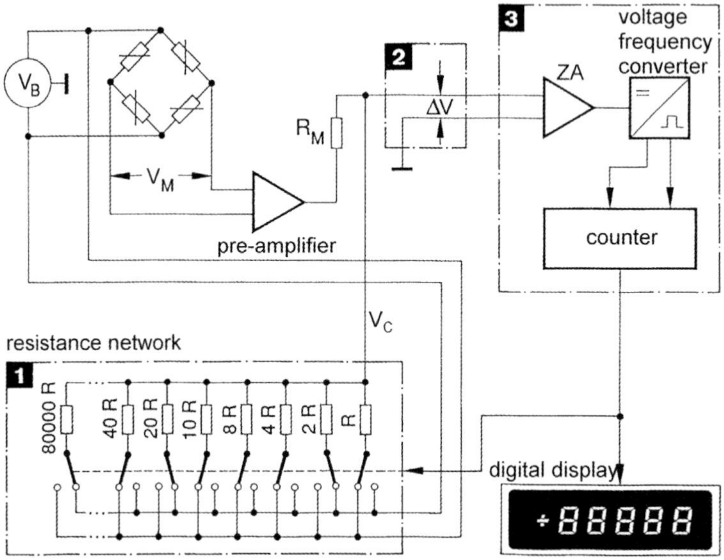

Fig. 7.16 Combination of the compensation principle with an amplifier [7.11]

Figure 7.16 shows the block diagram of the combination principle. The diagram shows the three basic elements of the compensator: The zero adjustment unit (1), which creates the compensation voltage V_C and represents the measurement value; the difference former (2), which forms the voltage difference $\Delta V = V_M - V_C$; the compensation circuit (3) which controls the adjustment unit. The resistor network of the adjustment unit is directly connected to the excitation voltage V_B of the strain gage circuit.

Instruments as shown in Figure 7.16 have displaced the pure analogue compensators and are widely used as battery-powered instruments in practical applications. They can be used for measurements with half- and full-bridge strain gage circuits and offer a wide range of connectable resistances.

With the introduction of microcomputer technology newly designed adjustment units in compensators have been created [7.11–7.13]. In 1972 the proposed idea [7.14] to perform multi-stage pulse duration modulation led to a microcomputer-controlled circuit of highest performance [7.15]. Figure 7.17 shows the block diagram of a compensator unit using the principle of the two-stage pulse duration modulation. This principle allows high resolution and fast measuring rates.

A short explanation of the analogue-to-digital conversion method as used in the compensator can be given with the aid of Figure 7.17, which shows the circuit principle in a two-stage version. Each pulse duration modulated stages can be modulated in 1000 steps (time slices), enabling a total resolution of 1 000 000 steps. The compensation current i_C is derived from the two pulse-duration modulated rectangular voltages, one rectangular voltage giving the coarse compensation and the other the fine compensation.

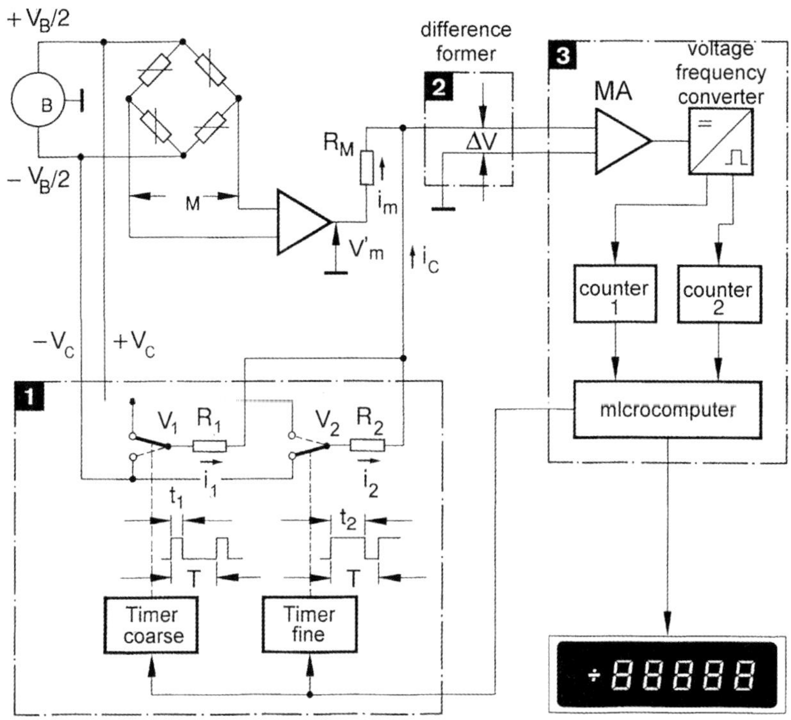

Fig. 7.17 Microcomputer-controlled compensator using pulse duration modulation

When the fine compensation reaches 1000 digits, a carry occurs, which means the coarse pulse duration is increased by one coarse time slice and the fine pulse duration is reset to zero. Only low demands are made on the ADC in the balancing circuit, because its output value is always zero when balanced, and its zero drift is corrected by the integrator.

If the sum of compensation current i_C plus measurement current i_M is not equal to zero, the voltage frequency converter will produce pulses which the counter adds up, and input them to the microcomputer. The microcomputer changes the pulse durations in the timers so that the sum of i_C plus i_M becomes equal to zero. Since the compensation is made continually and since both currents must average to zero over any period of time, the value of the compensation current must swing about the correct measurement value, because it can only change incrementally. This means that the resolution of a single measurement which is also very high can therefore be further increased by the addition of consecutive measurement values.

With a period of the sampling frequency in both the timers of 1 ms and a counting frequency of 1 MHz the time slices (pulse durations) created by the timers are variable in 1/1000 steps. With a fine compensation resistance 10^3 times as big as the coarse compensation resistance, 1000 time steps of the fine compensation correspond to one step of the coarse compensation. In this case the resolution amounts to 10^6 (1000 steps in coarse compensation with 1000 fine steps each).

Two-stage compensation circuits are used today in connection with carrier frequency preamplifiers as particularly powerful measuring instruments with high resolutions and concurrently relatively high measurement rates [7.16]. Due to the digital signal structure, very effective filter circuits can also be realized, even if large dynamic signal components are superposed, and so the static or quasi-static measurement values can be determined with high precision [7.17]. Preferred areas of application are measurements with strain gage transducers in weighing and material testing.

7.5 Multipoint measurement

7.5.1 Preliminary remarks

When measuring with strain gages, particularly in the experimental test technique, measuring objects are often provided with multiple measuring points at a variety of locations in order to obtain information about the strain occurring at the different measuring points during the load test. This can be the case with relatively small objects with only a few measuring points, but also with large structural components with more than 100 measuring points. In order to perform such measurements with commercially reasonable effort, specially developed multipoint measuring instruments are used. A multipoint measurement instrument is a measuring device to which a certain number of measuring points can be connected simultaneously. Depending on whether the measured variable is static or whether it is subject to more or less rapid changes over time, one must use the appropriate equipment. With static loads it is possible to connect the measuring points one after the other, using one signal processing channel, such as with one measuring amplifier. For time-varying variables that serial connection of the measuring points is not feasible, unless the temporal change of the measured value is so slow compared to the flow velocity of a scanning rate (time for a single recording of all connected measuring points) that one can still speak of a quasi-simultaneous scan of all measuring points. But this is relatively rare. In most cases of the recording of dynamic signals the signals of the individual measuring points must be processed in parallel channels, for example in order to be able to represent action and reaction signals in a clear temporal relationship. Synchronous data acquisition and processing is especially necessary when simultaneously detected signals have to be synchronized linked together, such as in the evaluation of the strain gage rosette measurements under dynamic stress.

Serial and parallel data acquisition devices require different concepts. Serial detection requires switches, because one measuring point after the other is switched to a specific channel amplifier. We therefore refer to instruments of this type as switching units or scanners.

In parallel signal processing an amplifier is provided for each measuring point so that the processing of the measuring signals is carried out simultaneously in all channels, and then the measured values can be displayed in their temporal assignment to one another. This equipment is therefore referred to as a multi-channel device or system. Multi-channel devices are generally more expensive than scanners.

7.5.2 Serial multipoint measurement

Serial multipoint measurement means to scan a lot of measuring points one after the other in rapid succession. The switches required to scan the measuring points represent connecting resistors in the measurement circuit, which are not reproducible with the required accuracy. The measurement circuitry must be designed so that voltage drops across the switching elements and measuring leads cannot affect the readings. This requirement is especially difficult to accomplish when applying strain gages in half- and quarter-bridge circuits because in measurements using strain gages the changes in measured resistance are very small, so that changes in resistance in the measuring circuit can cause large errors. For instance, if strain gages are switched by means of field-effect transistors having a forward resistance of about 20 Ω and a temperature coefficient of 0.4% per K, zero shift is greater than 160 000 μm/m and zero temperature dependence is greater than 640 μm/m per K. This error is even larger than the measured signal itself.

This example shows quite clearly that with the normal Wheatstone bridge circuit it is very easy to get into trouble, especially if there are high lead resistances and the switching devices do not have an extremely low reproducibility error.

Modern high-feedback electronic circuits permit the design of circuits that can prevent the undesirable influences. In 1976, a circuit was introduced by Kreuzer, and some years later improved by him with an extension [7.18,7.19]. This circuit is based on the Wheatstone bridge but avoids its error-producing disadvantages almost completely. The great advantage of the advanced Kreuzer circuit is that it also takes into account voltage drops across the connecting leads between the active strain gages and the supplementary resistor.

This technique not only relates to resistance changes caused by the switches, but regulates all kinds of voltage changes in the connecting lines, such as those caused by changes in temperature. Also, voltage drops in very long measuring cables are corrected easily by the Kreuzer circuit. The advantage of this circuit for the multipoint measurement technique is that semiconductor switching (FET switches, FET = field effect transistor) can be used for the switching technique. FET switches could not previously be used in scanners because of their non-reproducible and strongly temperature-dependent insertion resistances. Now the advantages of FET switches can be used with the elimination of the disadvantages. The advantages of FET switches are: semiconductor switches allow quick, bounce-free switching with no wear; they are reliable; and the power required for their activation is negligibly small. In Section 6.7, the Kreuzer circuit was explained in its original and advanced form with the Figures 6.16 and 6.17.

Today, on the basis of this circuit technique, very convenient scanning instruments are available for strain gage technology that also allow the connection of inductive transducers, thermocouples, resistance thermometers, and both voltage and current

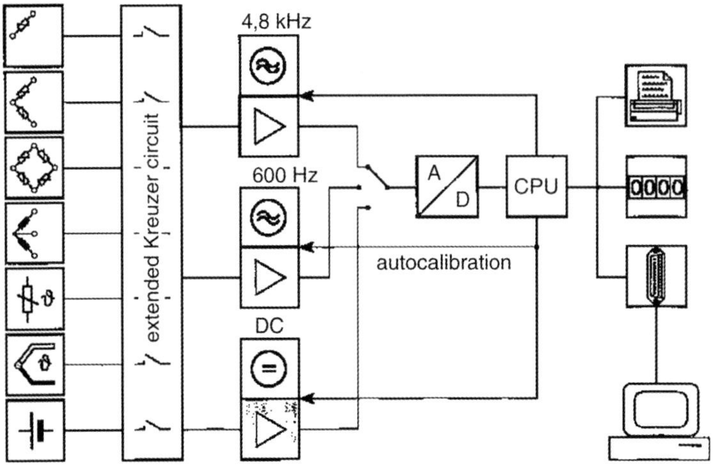

Fig. 7.18 Block diagram of the structure of a serial multipoint measuring instrument using the Kreuzer circuit

signals. The devices have different amplifier channels (e.g. 600 Hz and 4.8 kHz carrier frequency and DC voltage measurement amplifier) that can be assigned to the individual measuring points by programming the instrument. The achievable sampling rates are relatively high; for example with 600 Hz carrier frequency a maximum rate of 200 measurement points per second is possible [7.20]. At higher carrier frequencies or in DC operation, the measurement rates are correspondingly higher. Here the scanning technique is what sets the upper limit of the measuring rate rather than the selected measuring amplifier with its transient response.

The advanced Kreuzer circuit produces error free measurement up to very long cable distances of about 1000 m whereby the energizing voltage at the measuring points is monitored by feedback leads and adjusted to balance out voltage drops caused by cable and transfer resistances. Figure 7.18 shows a block diagram of the basic structure of a scanning instrument for serial multipoint measurements.

In such devices – when operating with carrier frequency – the switching of the measuring points can be precisely synchronized with the phase of the carrier frequency, thus the heating effects in the strain gages during the zeroing measurement and during the actual measurement are identical and compensate each other [7.21]. These devices provide a wide variation and a high degree of reliability through their diverse connection possibilities and self-test functions (cable break, short to ground, measuring the volume resistance), especially for the connection of single strain gages in multipoint measurements for experimentation work [7.20]. Also autocalibrations of the measuring amplifier and the connected measuring points are possible.

Important criteria for the use of scanners when measuring with strain gages are the offered connection possibilities. Apart from half-bridge circuits, single strain gages with a common internal completion resistor or external compensation gages can be connected using the advanced Kreuzer circuit for experimentation work. The possibility of assigning compensation gages to certain groups of active gages may have favourable implications for the cabling. A line circuit with galvanic separation of all measurement and compensation gages can bring advantages, especially in harsh environments; the ground connection of a single strain gage does not affect the measurement results of the other strain gages. Nowadays modern scanners have capabilities for alarm level settings that can be used for monitoring tasks [7.22]. Further criteria for use in monitoring tasks are selectable time intervals between the automatic run of measurement cycles and the interim storage of measured values.

Often the question of query rate or sample rate is ignored, so it will be dealt with here. In general, high speeds are not required when measuring static or quasi-static signals; one can achieve better noise suppression here with a little more patience. If, for example integrating converters are used for the analogue-to-digital conversion in the scanning device, the selection of a slightly longer integration time can be advantageous. In most integrating converters, the option is given of choosing a desired integration time within specified limits (e.g. 1 ms to 100 ms). Thus, for example, 50 Hz interference will be completely suppressed at an integration time of 20 ms. With the non-integrating method sample-and-hold elements are used in the step before the converters, which keep the measuring signal constant for the implementation (sampling) period of time required. This technique requires the arrangement of relatively complex filtering, and limits the useful bandwidth to less than 10% of the sampling frequency.

Figure 7.19 represents the timing chart of a scanning procedure. Using FET switches the switch-on time is very short. The ensuing rise time of the measurement amplifier depends on its bandwidth, rising time decreases with the subsequent transient time of the amplifier, which depends on its bandwidth; it decreases in CF amplifiers with increasing carrier frequency, and reaches a minimum with DC amplifiers. As an average, one can assume for 225 Hz carrier frequency a rise

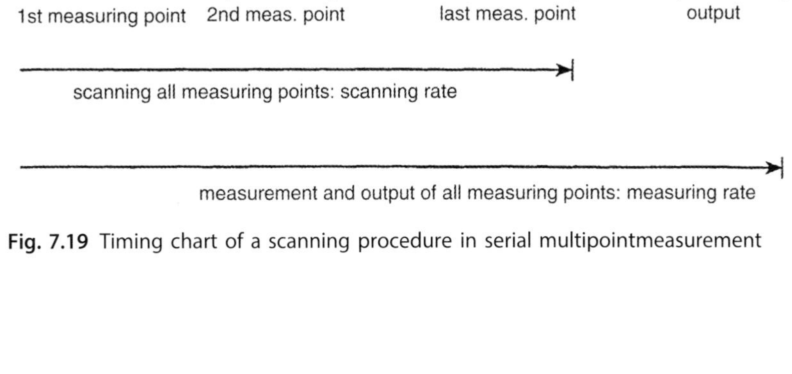

Fig. 7.19 Timing chart of a scanning procedure in serial multipointmeasurement

time of about 25 ms and for a DC amplifier about 0.8 ms. The time required for the analogue-to-digital conversion, which has been already mentioned briefly, can be set using an integrating method and can be assumed to be 20 ms. Storing the measured values typically requires no time. Certainly the time required for printing the results has to be taken in account. The sum of the scanning times of all measuring points is called scanning rate: this is the number of measuring points recorded per second. After recording all measuring points, certain calculations, autocalibrations and zero-point corrections will be carried out. The time required for output depends on the interface used.

Assuming all favourable conditions for rapid scanning, serial scanning instruments still remain limited to static or quasi-static variables. It follows that multi-channel systems with the number of parallel measurement channels corresponding to the number of measuring points have to be used instead of switching units for multipoint measurements of dynamic variables.

7.5.3 Parallel multipoint measurement

When operating with parallel multipoint systems, the measurement points are not switched on individually, but each point is connected to the entire measuring system with a associated amplifier channel so that all connected measuring points are scanned at the same time (synchronously) and the measured signals are available in parallel at the outputs of the channels. In this technique, no switches are needed; each measuring point is permanently connected. The measuring amplifiers remain in the steady state, so that the response time is to be considered only with respect to rapid changes in signal, but not with respect to any switching-on operation.

The objective of measurements with parallel multipoint systems is always the synchronous display of the measured signals. Either as time functions with a clear temporal relationship of the measured values or in association with each other, such as during the recording of one or more load-dependent strains, the measured strains are displayed in dependence on the measured load at exactly the same time in an xy-graph.

In a purely analogue multipoint system, which may consist of several parallel operating amplifiers in the simplest case, at the output are available parallel analogue signals that can be recorded with analogue equipment. Instruments of this type are to some extent still in operation today, despite the advance of digital technology.

The request to evaluate the measurement results obtained by parallel operating analogue amplifiers in computers is connected to the digitizing of the measuring signals. This requires an analogue-to-digital converter in each channel with a suitable interface for connection to a bus through which the measuring signals of all the parallel-connected channels are fed to the computer. The number of connected measuring points, the applicable cable lengths and the speed of signal transmission determine the type of bus used. Such systems are very complex and have the disadvantage that no adjustments can be performed by the computer on the amplifiers, which serves only to evaluate the measured values.

Another solution to the parallel processing of detected signals is given with the use of a multiplexer. In this case, a common analogue-to-digital converter is used for all measurement channels, but the output signals from the converter are supplied to the multiplexer. The multiplexer in principle causes a time lag between the measurement signals recorded in parallel in one scanning cycle. For this reason, in each channel a sample-and-hold circuit is arranged in front of the multiplexer in order to transfer measuring signals temporal associated with each other to the multiplexer. However, because of the time-discrete scanning of the measurement signal by the multiplexer, this requires that, before the sample-and-hold circuit has to be switched, is placed an analogue low-pass filter as an antialiasing filter. Furthermore, it is not helpful to use an integrating analogue-to-digital conversion method behind the multiplexer.

Meanwhile, low-cost high-performance analogue-to-digital converters that allow easy implementation in each measurement channel have largely displaced the multiplexer from the modern multipoint measurement technique. The great advantage here is that the digital amplifier technology, as described in Section 7.3, provides parallel digital output signals for all channels, with exact time equality, for further processing by the computer, in which each channel has its own analogue-to-digital converter. A device operating with the MPDM conversion method is described in [7.23,7.24], and Figure 7.20 is the block diagram of such a multi-channel device.

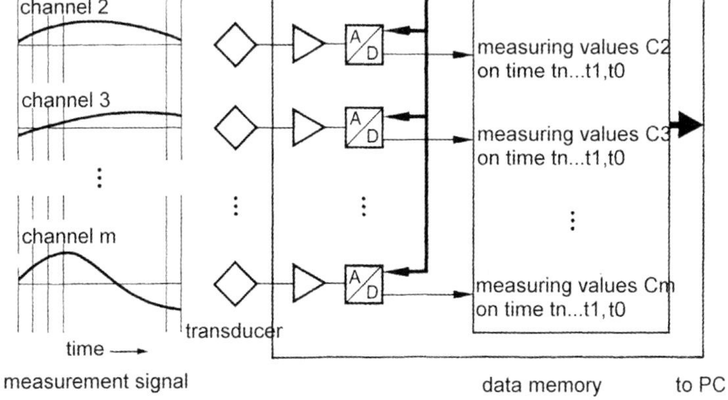

Fig. 7.20 The measurements are digitized in all channels exactly synchronous in time and the time relationship is retained for computer evaluation

An instrument configured according to Figure 7.20 fitted with 10 channels can capture, for example, a maximum of 9600 readings per channel each second with 16-bit resolution. That means, at the maximum sample rate of 9600 samples per second, the measuring signals of three rosettes can be processed up to a frequency of the measured quantity of 1–2 kHz without loss of data in continuous operation with a computer. The sample rate is not fixed at 9600, but can be set for each device between 1 and 9600.

The individual channels can be equipped with different amplifiers, so that in addition to strain gages, other transducers can be connected, such as inductive transducers, thermocouples and resistance thermometers. The measuring channels are synchronized by a central quartz oscillator, which guarantees an exact temporal allocation for the measured signals of all channels. The central processor unit collects the measured values of all channels and stores them in a buffer memory, from where they are transmitted via suitable interfaces to the connected computer. As digital measuring amplifiers are fully computer controlled, all settings can be made from a computer in such parallel multi-channel measuring instruments. They are ideally suited for online measurements in which the measured values are evaluated immediately by appropriate programs, and the evaluation results are still available for access during the running measurement.

This digital parallel multipoint measurement technique allows the online evaluation of measurements with strain gage rosettes under dynamic stress. This option is used especially for stress analysis in dynamic service load tests [7.25]. The device technology ensures that each item of data supplied by the three rosette grids and linked in the evaluation of measured values is evaluated precisely in its time allocation.

With today's available software it is also possible to display the time functions of the two principal stresses and their orientation angle for each rosette, in addition to the time characteristics of the three measured strains. Since the memory of the multi-channel digital multipoint measuring instrument for temporary storage of the measured values before passing them on to the computer can be extended [7.24], at higher frequencies of the measured variable, the data evaluation can be carried out without problems even if the evaluation is slower than the acquisition.

With the possibility of online evaluation of dynamic rosette measurements a long-cherished wish of experimental mechanics comes true.

Experimental stress analysis with online measurement devices is performed on safety-related components of motor vehicles, aircraft and in many fields of mechanical engineering. In special test devices, for example, components of motor vehicles are dynamically loaded with realistic simulated multiaxial force combinations. Strain gage rosettes are installed at selected points on the object to be investigated, whose signals are fed to the digital measuring system in order to detect the resulting stresses. The fully computer controlled measuring system links synchronously recorded strains in multiple channels for evaluation in the connected PC. On the screen next to the measured strains, the time courses of the principal stresses and their orientation angle can be tracked.

References

[7.1] Kreuzer, M. (1980) Ein programmierbares Präzisions-Meßgerät der Genauigkeitsklasse 0,0005 und seine Anwendung. *MTB,* **16** (2), 41–46.

[7.2] Kehrer, R. (1991) Synthese aus Analog- und Digitaltechnik ändert Verstärkerstrukturen. **VFI** (2), 64–70.

[7.3] Heringhaus, E. (1982) Trägerfrequenz- und Gleichspannungsmeßverstärker für das Messen mechanischer Größen – ein Systemvergleich aus anwendungstechnischer Sicht. *MTB,* **18** (2), 42–49 and No. 3, pp. 70–73.

[7.4] Laible, M. (2014) Kriterien zur Auswahl von Messverstärkern; in Seminarmanuskript Nr. 32357.00.01 der Technischen Akademie Esslingen.

[7.5] Kreuzer, M. (1987) Meßverstärker mit schneller hochauflösender Digitalisierung, part 1 and part 2. *Elektronik,* **14**, pp. 77–80 and **15**, pp. 68–73.

[7.6] Kreuzer, M. (1991) Neuartige Universal-Meßverstärker. *MTB,* **27**, 18–27.

[7.7] Kreuzer, M. (1992) *Neue Strukturen bei Meßverstärkern,* Techn, Note MD 9202, HBM GmbH Darmstadt.

[7.8] Kreuzer, M. (1986) Digitale Filterfunktionen für das elektrische Messen mechanischer Größen. *MTB,* **22** (2), 53–61.

[7.9] Kehrer, R. (1992) Selbstidentifikation des Aufnehmers erhöht Bediensicherheit. **VFI** (5), 34–36.

[7.10] Kitzing, H. (1992) *Mechanische Größen digital messen,* MSR-Magazin, No. 9/10, pp. 54–57.

[7.11] Kreuzer, M. (1985) Moderne mikrocomputergesteuerte digitale Kompensatoren, part 1: Schaltungen für Digital-Kompensatoren. *MTB,* **21** (2), 35–38.

[7.12] Kreuzer, M. (1980) Anwendung des Verstärker- und Kompensationsprinzips bei Meßgeräten für die Dehnungsmeßtechnik, part 2: Digital-Kompensator DK 37A. *MTB,* **16** (1), 13–15.

[7.13] Kreuzer, M. (1979) Anwendung des Verstärker- und Kompensationsprinzips bei Meßgeräten für die Dehnungsmeßtechnik, part 1, Digitaler Dehnungsmesser DMD 20. *MTB,* **15** (3), 64–69.

[7.14] Horn, K. Schaltungsanordnung zur digitalen Messung der meßgrößenabhängigen Verstimmung einer elektrischen Brückenschaltung, German Patent PS 2 260 439.

[7.15] Kreuzer, M. (1982) Verfahren und Vorrichtung zum Umsetzen einer Meßspannung in einen digitalen Wert, German Patent DE 2 952 311.

[7.16] Kreuzer, M. (1985) Moderne mikrocomputergesteuerte digitale Kompensatoren, part 2: Wägeelektronik für selbsstätige Waagen zum schnellen Abwägen von schüttfähigem Wägegut. *MTB,* **21** (3), 76–72.

[7.17] Kreuzer, M. (1986) Moderne mikrocomputergesteuerte digitale Kompensatoren, part 3: Präzisions-Digitalkompensator mit sehr effektiver digitaler Filterung. *MTB,* **22** (1), 16–20.

[7.18] Kreuzer, M. (1976) Eine Vielstellenmeßanlage mit FET-Schaltern, part 1. Schaltungskonzeption. *MTB,* **12** (1), 4–9.

[7.19] Kreuzer, M. (1983) Vergleichende Betrachtung verschiedener Schaltungsarten für das Messen mit Dehnungsmeßstreifen. *MTB,* **19** (3), 63–68.

[7.20] Kreuzer, M. (1989) Umschaltanlagen für die Versuchstechnik. *VDI-Berichte,* **731**, 67–83.

[7.21] Kreuzer, M. (1986) How to avoid errors caused by heat effects in strain gage measurements when using scanning units, *Proc. 8th Int. Conf. Stress Analysis, Amsterdam,* pp. 447–454.

[7.22] Aschaber, M. (1991) Meßtechnische Überwachung der Setzungsbewegung des Flußpfeilers der Inntal-Autobahnbrücke bei Kufstein. *MTB,* **29** (1), 1–7.

[7.23] Kreuzer, M. (1992) Multichannel system for high precision static and fast dynamic measurements. *Measurement,* **10** (1), 8–13.

[7.24] Kitzing, H. (9–10 1992) *Mechanische Größen digital messen,* MSR Magazin, pp. 54–57.

[7.25] Müller, A. and Grubisic, V. (1993) Simulation und Erfassung komlexer mehrachsiger dynamischer Beanspruchungszustände an Fahrzeug-Achskomponenten. *MTB,* **29** (2), 21–26.

8 Calibration of devices for measuring with strain gages

8.1 Introduction

The strain gage bridge and its circuit form the recorded mechanical strain into an electrical signal containing information about the value of the measured variable. The strain gage bridge circuit is connected to electrical measuring instruments which process the electrical signal and output the value of the measured quantity in direct or encrypted form. These devices, together with the bridge circuit, form a series of devices called a measurement chain. At the beginning of this chain can be arranged a strain gage installation intended for strain measurement or even a transducer equipped with strain gages. To execute a measurement, the information output at the end of the chain must convey the amount, the sign and the unit of the quantity to be measured in an intelligible form. This requires the production of a clear correlation between the output information from the measuring device and the value of the measured variable. The creation of this relationship is referred to as calibration.

The calibration is an essential prerequisite for the proper carrying out of a measurement. Depending on the measurement task, the effort of the calibration procedure can be very different. In some cases, you can, after zero adjustment, calibrate with an available known mechanical quantity, such as by fitting of weights on a load cell and register the thereby emitted signals. From the assignment of the signals to the loading comes the result: the calibration or calibration curve. This type of calibration can be relatively easily implemented and can require little effort.

In other cases, the actions to be taken for calibration can be very expensive, such as when complicated devices are required to load a measurement site with defined loads. An impressive example of this is given in [8.1], where the calibration of the strain gage arrangement installed on a bulkhead in the bow of the research vessel *Polarstern* is described, in order to detect the forces acting on the frame while passing through the ice floes. For this calibration, a later removed loading trestle was welded to the outside of the ship in the dry dock. Force controlled by a force transducer was introduced via hydraulic cylinder at the measurement site. In such cases, the expense required for a calibration can be extremely high. Often, the ingenuity of the engineer is required for performing a calibration. One always strives to obtain the relationship between variable and the signal within a predetermined limit of uncertainty of measurement.

8.2 Measurement chain

Devices for measurement with strain gages are basically constructed in the form of measurement chains. The first link of a measurement chain is an installed strain gage bridge circuit, either installed directly at the measurement object or part of a measurement transducer. The bridge circuit outputs an analogue electrical signal which represents the value of the process variable. The end of the measurement chain is formed by an issuer who outputs, for example, the measured values in the

Technology and Practical Use of Strain Gages: With Particular Consideration of Stress Analysis Using Strain Gages. First Edition. Stefan Keil.

desired form as an analogue or digital display, as printed out or in converted signal form for further processing in a computer. Between the first chain link and the issuer, there are one or more signal-processing devices with the mission to provide the output of the sensing circuit in a form understandable to the issuer. In the following the first link of the chain is called a transducer, which applies to both free installed strain gages as well as transducers equipped with strain gages.

For each measurement task an optimally suitable measurement chain is necessary. In the design of this chain the type of measured variable, the expected maximum value, the allowable measurement uncertainty and the form of presentation of the results all influence the selection of individual links of the measurement chain.

For dedicated measuring tasks, unchanged measurement chains are applied for a long time, for example in industrial production or in the monitoring of processes. In many technical areas, however, measurements chains are put together, which are temporarily assembled for solving temporary measurement tasks from single devices and then separated again. These cases of spontaneous assembling of measurement chains are especially common when measuring with self-installed strain gages. This creates the practical requirement of measuring devices that can be combined into specific chains for use in different measurement tasks. This requirement is met by the creation of hardware systems, which consist of a variety of interchangeable and coordinated transducers, conditioning units, output units and accessories. The interchangeability of the individual links of the chain in order to create specific and tailored chains, which are adapted to the respective measuring tasks requires the standardization of the links used in terms of the characteristic values of their input and output variables. The output characteristics of a device located in the chain should match the input characteristics of the following link or may be made to match by appropriate measures.

The quantity to be measured acts on the transducer, and its output represents the value of this quantity in the form of an analogue electrical signal, which is supplied to the subsequent measuring device, in most cases, an amplifier. It is advantageous for the design of measurement chains with regard to the interchangeability of transducers if transducers are used that have standardized output signals. This normalization leads to the production of transducer-type series within which transducers for various useful staggered nominal loads (measuring ranges) have the same standardized output signal.

Transducer circuits with strain gages are passive sensors that require electrical power for their operation. This supply is preferably performed at constant voltage (also constant AC voltage), because the transducer output signal is represented in relation to the amount of the supply voltage in the unit in mV/V.

For a measuring amplifier it is important that its measuring range be adapted to the range of the transducer signal. This measurement range, for example, can be set at universal amplifiers in coarse steps and between these steps with a continuous fine adjustment to desired values. With this adjustment identity can be established between the measurement range of the amplifier and the nominal output signal of the transducer, but the range can also be adjusted as desired, for example, what is of importance for part-loaded transducers or for calibration purposes. This is regardless of whether the amplifier carries out the signal processing in analogue or digital form.

The final link of a measurement chain provides the needed information on the value of the measured quantity. The nature of this final element, and thus the form of information output can be very different. The output options cover a broadly spread range from simple analogue pointer instrument to convenient graphics on the monitor in digital chains with online connected computer.

8.3 Characteristic and sensitivity

The sensitivity of a measuring instrument is defined as the relation between the change of the output signal and the change of the input signal causing the change in the output signal. Figure 8.1 gives the definition of sensitivity in graphical form.

In this graph, the output signal O of a measuring arrangement is shown as a function of the input signal I. The curve representing this function is called the characteristic curve of the amplifier.

The sensitivity is obtained by definition of the differential quotient dO/dI of the characteristic curve and is thus identical to the slope of the curve at this point. This means that measuring instruments with nonlinear characteristics may have at every point of their characteristic a different sensitivity. This may be due to the applied physical methods, such as in non-contact inductive displacement transducers, or be deliberately performed.

The modern microprocessor technology allows the realization of amplifiers that automatically adapt their characteristic to the characteristic of the transducer, to establish a linear relationship between the measured variable and the output signal. In general, however, the characteristics of measuring instruments for the strain gage measurement technique are linear within very tight tolerances and therefore have, within their ranges, constant sensitivities.

Any amplifier can be used for a particular range or for multiple switchable measuring ranges. The range limits are defined by specifying the nominal values of the input

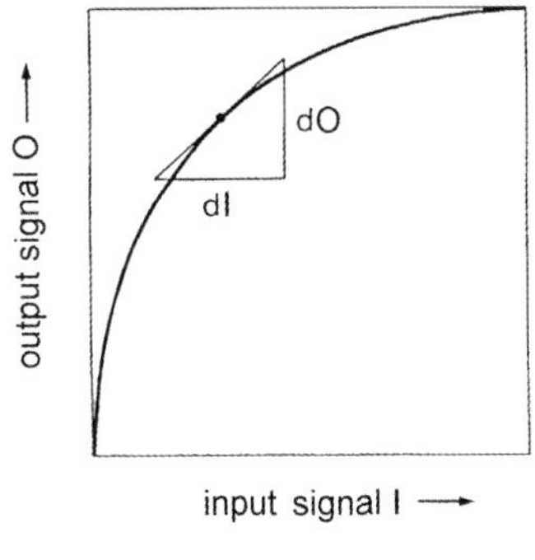

Fig. 8.1 General definition of sensitivity

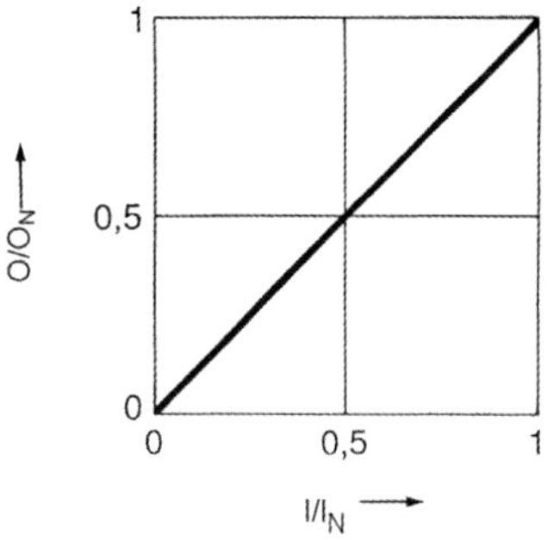

Fig. 8.2 Measurement range of an amplifier in related representation; O = output signal, O_N = nominal output signal, I = input signal, I_N = nominal input signal

signal. Thus, for example, for transducers with strain gages with the indication of the nominal value of the measured mechanical quantity (nominal power, nominal pressure, nominal torque, etc.) and any required direction indication (e.g. tensile or compressive force), the range of the transducer is set. In signal processing measuring instruments for the identification of possible measurement ranges the values of the input signal are given, which produce the nominal output of the device. The nominal input signal of a measuring device is assigned the nominal output signal by the characteristic.

Figure 8.2 shows the range-related representation of an amplifier, assuming a straight curve. Within the measurement range specified for an instrument the uncertainties given for the individual parameters of the instrument are not exceeded.

Characteristics of measuring instruments may be changed undesirably by influences such as changes in temperature, but also by intentionally caused changes, such as being changed by circuitry interventions. In terms of their impact on the characteristic, there are multiplicative and additive influences. The multiplicative influences affect the slope of the characteristic curve and thus the sensitivity of the device concerned. They cause, as shown schematically in Figure 8.3a, a rotation of the curve

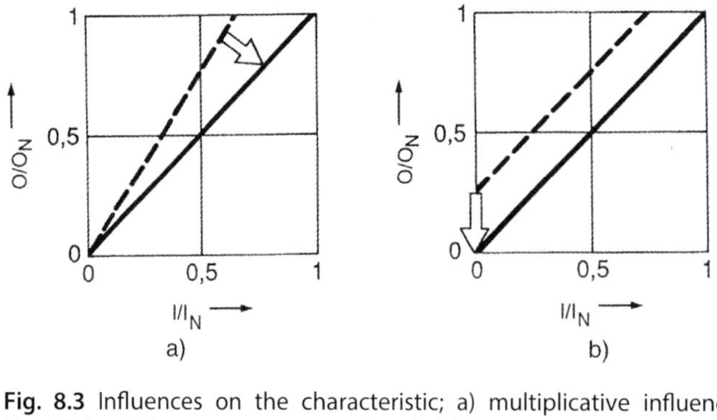

Fig. 8.3 Influences on the characteristic; a) multiplicative influences cause change in sensitivity, b) additive influences cause zero signal shift

about its origin. The additive influences that are in absolute terms regardless of the amount of the measured signal cause a parallel shift of the curve, as Figure 8.3b shows. This means that zero signal shifts are associated with additive effects.

The suggestibility of the characteristic curve is used during calibration by using the existing controls on the measuring instrument (e.g. zeroing the parallel shift; measurement range to slope change) which brings the characteristic of a device to a desired position. It can be set by parallel displacement through the zero point, for which the input signal is zero and the output is assigned to zero. By varying the slope the instrument's sensitivity is varied and a defined end point is set (measuring range limit).

8.4 Calibration of the entire measurement chain as a measuring device

If a measurement chain remains unchanged for a longer duration for a particular measurement task then it can be advantageous to calibrate the entire measurement chain as an autonomous system. The best way to perform this is to load the transducer with a defined mechanical quantity, that is a force transducer with a known value of force, or a pressure transducer with a pressure, and to compare the indicated measuring value with the applied load. Applying changes in load step by step one finds corresponding values of load and indication. It is usually possible during the calibration procedure to carry out adjustments on to the amplifier or the indicator, so that the measured value can be displayed in units of the quantity to be measured or an appropriate multiple of it. This also permits both accurate zero adjustment and the decimal interpolation of the scale.

The method used to calibrate the entire measurement chain makes use of the adjustment facilities for the characteristic curve of the amplifier provided by its operating controls. The characteristic curve of an amplifier represents the relation between the output signal and the amplifier input. The characteristic curve of a measuring amplifier is normally represented by a straight line.

Figure 8.4 shows a diagrammatic example of the two principal methods of varying the characteristic curve of a measuring amplifier with the aid of zero balance

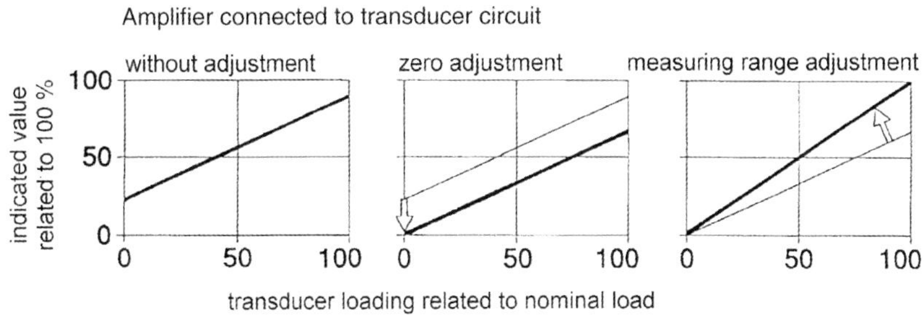

Fig. 8.4 Modifying the characteristic curve of an amplifier through adjustment of zero signal and sensitivity

control and the measuring range selector. The zero balance provides parallel shift of the characteristic curve without affecting its slope. Adjusting the sensitivity, that is the measuring range, changes the slope of the characteristic curve without interfering with its point of intersection with the abscissa. On measuring amplifiers it is usually possible to select the measuring range stepwise with the coarse-adjustment control and continuous between the steps with the fine-adjustment control.

If the measurement is carried out without altering the calibrated sensitivity, the relationship between the measured quantity at the transducer and the measured value indicated at the end of the measurement chain will be that established during calibration.

When calibrating the measuring chain as a complete system it is not essential to know the values of the signals passed between the individual elements of the chain. The calibration using known values of the mechanical quantity gives the most accurate results. The uncertainty according to this calibration method lies either on the accuracy of the known mechanical quantity itself, the reading accuracy of analogue displays, the resolution of digital displays, or the quality of interpolation between the calibrated load steps. It must be assumed, that the amplifier provides the required accuracy in its characteristic properties.

Frequently, however, the user of the measurement equipment will not be able to calibrate to known values of the particular mechanical quantity, either because no suitable devices are available (very high loads), or for example when using strain gages, known values of strain cannot be produced for calibration. In such cases other methods of calibration must be employed.

8.5 Compensators

Compensators based on a simple measurement principle permit high accuracies. The principle is that the actual measurement signal is compensated to zero by means of a zero adjustment unit. The position of the adjustment unit represents the measured value.

With analogue compensators the positioning of the zero adjustment unit takes place either manually or automatically by means of a servomotor. Modern compensators realize the compensation principle digitally. Thereby a zero adjusting circuit, which is composed of a zero amplifier, a voltage–frequency converter and a bidirectional counter, changes the final control element until the difference between the measurement signal and the compensation voltage is zero. The position of the adjustment is used for the digital display of the measured value.

The basic concept of linking the zero adjustment circuit, positioning unit and display allows compensators to be calibrated during manufacture. Thus, the relationship between measured signal is set at the compensator input and display. High quality precision compensators are able to resolve a measurement signal of 2 mV/V in up to 10^6 digital steps [8.2].

A compensator forms, together with a calibrated transducer, a measurement device that is immediately ready for measurement without any calibration procedures needing to be performed by the user. The uncertainty here depends on the tolerance of the nominal sensitivity of the transducer and the accuracy class of the compensator.

Because of their internal calibration, compensators are particularly suitable for measuring with strain gages, since a defined relationship exists between the input signal in mV/V and the display in μm/m. The calibration of these compensators is based on the existing relationship (for gage factor $k = 2$) between strain and bridge output signal of 2000 μm/m = 1 mV/V, that is it indicates under the applied input signal of 1 mV/V at its display 2000 scale divisions. This relationship can be adapted to the actual gage factor of the strain gages used with the help of a gage factor dial. This eliminates the need for mathematical correction of the measured value for the cases in which $k \neq 2$.

8.6 Transducers

8.6.1 Transducers calibrated by the manufacturer

Nowadays transducers for electrical measurement of mechanical quantities, calibrated by the manufacturer, are used in many applications. Thus for the respective transducer a valid relationship between the value of the force acting on the transducer and the value of the transducer output signal is given. Relevant information can be obtained from the transducer data sheets. The important ratio here is the tolerance of the rated characteristic value (nominal sensitivity), which indicates the uncertainty of the actual output signal at nominal load with respect to the nominal sensitivity (rated output).

The designation of a transducer corresponds to the measured mechanical quantity. One thus speaks of force transducers, load cells, torque transducers, pressure transducers, and so on. The range of a transducer is limited by the marking of the rated load, so depending on transducers of nominal power, the nominal torque, the nominal pressure, and so on. The designation of the transducer output comes from specifying the output signal at rated load. The sensitivity of the transducer is defined as the change in its output signal related to the input signal causing this output. If the relationship between values of transducer loading and emitted signal (curve) is linear, the transducer sensitivity for the entire range is defined by the specification of the output at nominal load. The output signal at nominal load, that is the nominal output signal is based on the published VDI Guideline 2637 'Characteristics of load cells' also referred to as the rated characteristic value (nominal sensitivity).

The calibration of transducers by the manufacturer also demands, especially with high accuracy requirements, a large amount of effort for calibration. Sensors that work with metallic strain gages as transforming elements, have met the highest accuracy demands of today's advanced manufacturing technologies. Thus, for example, an international force comparison standard for comparison measurements at various national metrological institutions is carried out with a strain gage force transducer with 48 strain gages in its bridge circuit with highest repeatability of its rated output [8.3].

The calibration of transducers requires very reliable calibration standards. Calibration standards embody with very low uncertainty the value of a mechanical quantity and are used by the manufacturer to load the transducer in order to calibrate the transducer. Simultaneously the transducer output signal is measured with a precision device during this loading. For example, a calibration force must be available for the amount of the transducer's nominal load for the calibration. The calibration force must be available with an uncertainty which is an order of magnitude less than that required for the transducer uncertainty. Such calibration standards are available in the calibration laboratories of the DKD (German Calibration Service) and are connected to the standards of the PTB (Germany's national metrology institute).

One characteristic of the quality of the calibration of a transducer is the sensitivity tolerance that specifies the tolerance of the rated characteristic value (nominal sensitivity).

8.6.2 Calibrating transducers self-made by the user

Users who can deal professionally with strain gages are able to produce themselves transducers with strain gages. These sensing elements can easily pickup for example force, pressure or torque. They can be made using as measurement objects the original components of structures to be examined which are equipped with strain gages, thus fulfilling in addition to their actual function still a measurement task. Examples of this are pasted with strain gages links for measuring tensile forces in chains or applied to shafts strain gages to detect torques. Special measuring elements are made frequently by users who adapt in their structure specially designed members fitted with strain gages. This is necessary if suitable measuring elements are not available on the market. Examples are special torque measuring wheels or measuring platforms.

Users have to calibrate their self-made transducers themselves and have to take into account the expected influencing factors for the intended measurements, such as temperature changes and so on. The simplest method is again the previously described method of calibration of the entire measurement chain with a known mechanical quantity. This method also has, in addition to its reliability, the advantage that the sensitivity of the conditioning unit (amplifier) can be adjusted during calibration so that the measured value will appear in the unit of quantity to be measured on an indicator.

When applying the known mechanical quantity for calibration of the measuring system, it must be ensured that the direction of the effect (e.g. the force vector) coincides with the direction of the measured variable and that overlaps and interactions, as for example can arise from lateral forces and torque, are avoided. The influence of temperature changes that can occur during the measurements should be determined during calibration. If an interfering influence of temperature is detected, a corresponding correction of measured values should be performed, either by improving the temperature compensation in the gage circuit or by simultaneous temperature measurement in order to perform a correction of the readings. If the self-made transducer cannot be calibrated together with the entire measuring chain, then the characteristic of the transducer has to be determined. As a function of defined values of the loading, the output signal of the transducer is recorded. In this case, by repeating these measurements it is possible to obtain information about hysteresis and reproducibility of the characteristic. Variations in the calibration temperature provide information about effects of temperature changes on the characteristic curve. Key characteristics in the assessment of the self-made measuring element during calibration are the detected constancy of the zero signal, the temperature effect on the zero signal and the sensitivity and reproducibility of the recorded curve. With good reproducibility the linearity of the characteristic curve is of secondary importance if the plotted curve shape is taken into account in the evaluation.

8.7 Calibration measures

8.7.1 Calibration unit

In general, industrial manufactured transducers are calibrated units; they deliver a defined electrical output when loaded with a defined mechanical quantity. The nominal output of a transducer represents the transducer output signal created by loading the transducer with nominal load. Therefore, a defined output signal of a calibrated transducer represents a defined value of the mechanical quantity to be measured.

However, it is also possible to produce a defined electrical signal by a calibration unit simulating the transducer output signal. The calibration unit is connected to the measuring chain in place of the transducer during the calibration procedure. The output signal of the calibration unit is then used to carry out the adjustment of the characteristic curve of the amplifier.

A calibration unit usually contains a Wheatstone bridge circuit and produces output signals by connecting shunt resistors to the internal bridge arms, causing balance in the circuit. Figure 8.5 shows a simplified circuit diagram of a calibration unit which

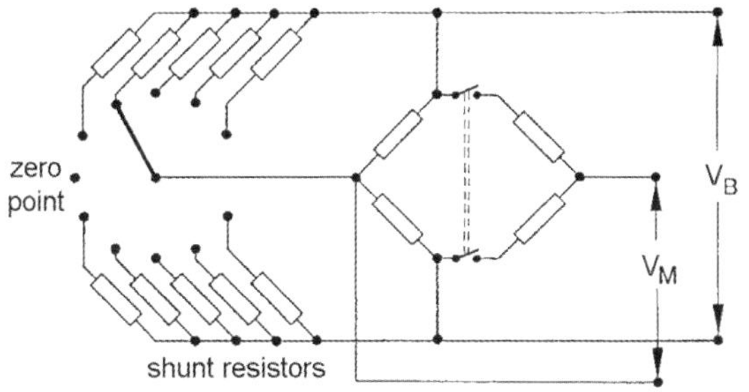

Fig. 8.5 Simplified circuit diagram of a calibration unit

enables various shunt resistors to be connected and so can generate various values of output signal.

In order to ensure interchangeability of transducers and calibration units the nominal values of the input resistance of the calibration unit and the transducer must be identical. The reason is that the bridge supply voltage is affected by the input resistance and the signal voltage is influenced by the output resistance.

Calibration units are not bound to any particular measurement chain but can be used for all measuring devices having a resistance comparable to strain gage measuring transducers. Another advantage of the independent calibration unit is that it can be used at the site of the transducer. Any changes in sensitivity caused by possibly long lengths of cable between transducer and amplifier are therefore accounted for in the calibration.

When performing a calibration, the calibration unit is connected in place of the transducer. After zero adjustment, the nominal output of the transducer is simulated by the output signal of the calibration unit (e.g. 2 mV/V). This signal is used to adjust the sensitivity of the amplifier.

In addition to these calibration units there are bridge normal that fulfil even higher accuracy requirements. They are used to calibrate equipment described herein and other precision devices to be calibrated. These are generally based on resistance dividers and are each set to only one signal value in order to avoid the influence of switching errors. A further reduction in the calibration uncertainty is achieved through the use of inductive dividers in calibration standards, due to the function of the inductive effect principle only for application of the carrier frequency method. In [8.4], such a bridge standard is described by the PTB. Its stable long-term uncertainty is certified as less than 5×10^{-6} for a voltage ratio of 2 mV/V at 225 Hz carrier frequency [8.5]. Such devices, however, are only important for users of measuring instruments with regard to their use in the device manufacturing. For normal measurement tasks they are usually not needed.

8.7.2 Measurement amplifier with built-in calibration device

Measuring amplifiers offer with their measurement range adjustment the possibility of adjusting the transmission factor continuously over wide ranges. In addition, measurement amplifiers offer the option of choosing various bridge supply voltages for the transducer, whereby the upcoming measurement signal with the unit mV/V is changed accordingly to the amplifier input as a voltage ratio. It is therefore extremely advantageous if a measuring amplifier has a calibration device which permits to establish a defined relationship between the signals at the amplifier input and the amplifier output.

For this purpose measurement amplifiers have a calibration device which produces a defined voltage ratio in mV/V which can be connected to the amplifier input. This signal is irrespective of the set amplifier range. The calibration of the measuring amplifier is usually carried out with a connected transducer, but depending on the amplifier design, two different procedures are possible. The calibration signal is superimposed on the currently pending transducer signal, or at turn-off of the transducer signal the calibration signal is pending alone at the amplifier input.

The calibration of the adjustment of the amplifier using the internal calibration signal, however, has in relation to the above-described methods the disadvantage that the measuring cables disposed between the transducer and the amplifier are not included in the calibration. This disadvantage does not occur when industrially manufactured transducers are used with fixed connection cables, since the calibration of the transducers was performed during manufacture, including conductor cables. This disadvantage can be considerable with long cables or cable extensions in four-wire technology. When applying the six-wire technology, this disadvantage is reduced considerably, since the voltage drops in the cables are compensated and thus cannot falsify the calibration.

With regard to the generation of the calibration signal to be switched to the amplifier input, different methods are possible. They differ in terms of their usefulness for the used amplifier type (e.g. CF or DC voltage measurement amplifier), as well as cost relating to the uncertainty of the calibration. Some instrumentation amplifiers have a so-called internal reference signal that can be switched with high reproducibility to the amplifier input. This reference signal can be used for the reproduction of a certain adjustment of the measuring range. If you want later on to restore a specific setting, you switch on the reference signal before you change the original setting and record the display. Later on you can restore the gain setting by switching on the reference signal and adjusting the measuring range to the recorded display.

8.7.3 Shunt calibration

Resistors can be switched in parallel to one or more branches of the strain gage bridge circuit, so with appropriate dimensioning of these shunt resistors a defined offset in the bridge can be achieved. The shunt resistance reduced the total resistance of that bridge section, so the resulting bridge imbalance simulates a reduction of resistance (compression of the strain gage) in the parallel bridge arm. Figure 8.6 shows schematically the arrangement of a shunt resistor for calibration.

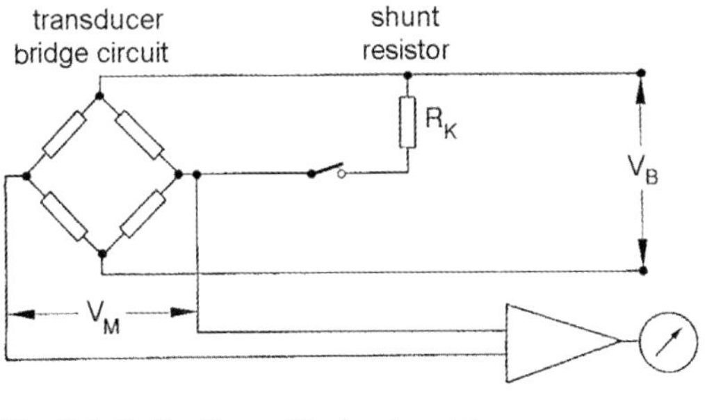

Fig. 8.6 Calibration with shunt resistor

With regard to the arrangement of the shunt resistor and the location of its switch there are different possibilities. The shunt resistor can be arranged either in the transducer or in the measuring amplifier, as are its switching point.

The question of the arrangement of the high-resistance shunt resistor and its switch point is important due to the often different locations of sensor and amplifier. In cases of unfavourable arrangement at high carrier frequencies the connection cables have important effects on the value of the calibration signal. When operated with DC voltage the arrangement of shunt resistance and its switch point is of minor importance, since the scarcely detectable ohmic cable resistance occurs. Thus, in operation with DC voltage or the often used low carrier frequency of 225 Hz, the shunt resistance in the transducer and its switch point can be arranged in the measuring amplifier without uncontrollable effects on the calibration occurring through the cable. However, if such arrangements are to operate with the equally frequently used carrier frequency of 5 kHz, strong distortions of the calibration signals set up for DC can occur. These distortions are caused by cable capacitance imbalances occurring at high carrier frequencies. This problem can be solved by placing the switch point near the shunt resistor in the transducer.

If the shunt resistor and its switch point are arranged in the amplifier, the effective cable asymmetry is significantly reduced (low-impedance termination in the transducer). This results in the possibility of using a symmetrical arrangement of several shunt resistors in the measuring amplifier, to largely avoid the effects of capacitive cable asymmetry. A shunt signal produced in this way does not capture the cable influences and is therefore only interpreted as a reference signal. Owing to the parallel connection, the reference signal is superimposed on the measured signal coming from the transducer. The reference signal is reproducible and can be used for the proper reproduction of a calibration otherwise made.

The best arrangement for the reliability of the calibration with a shunt resistance is achieved by positioning the shunt resistance and the switch at the measuring point. This solution has the disadvantage that the switch must be actuated at the measuring point by hand or via a relay. This requires some effort, but in comparison to the benefits this may be acceptable. You can avoid this effort by using a calibration unit that is connected at the amplifier instead of the transducer. However, the use of a calibrator is often more expensive than the remote shunt for longer distances between the measuring point and the amplifier or it may be difficult or impossible to reach test points.

On the basis of the circuit diagram in Figure 8.7 in the following it is shown that it is easy to calculate the required shunt resistance R_S to produce a defined calibration signal. In a balanced bridge in a bridge arm, a shunt resistor R_S is connected in parallel to the resistor R. Thus, the resistance of the bridge arm affected reduces to:

$$R_C = \frac{R_S \cdot R}{R_S + R} \tag{8.1}$$

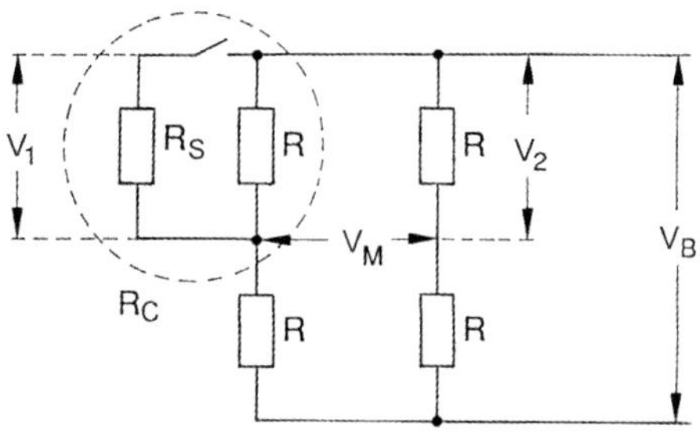

Fig. 8.7 Principle of switching a shunt in parallel to one bridge arm

For the circuit shown in Figure 8.7, the following relationships apply to V_B as a bridge excitation voltage and V_M as the output voltage:

$$\frac{V_1}{V_B} = \frac{R_C}{R + R_C} \tag{8.2}$$

$$V_M = V_2 - V_1 \tag{8.3}$$

$$V_2 = \frac{1}{2} V_B \tag{8.4}$$

This is

$$V_M = \frac{\frac{1}{2} V_B \cdot R}{2R_S + R} \tag{8.5}$$

For $R_S \gg R$ Eq. (8.5) can be approximately written as

$$\frac{V_M}{V_B} = \frac{R}{4R_S} \tag{8.6}$$

The resolution of this equation with respect to R_S yields the desired relation for the shunt resistor

$$R_S = \frac{V_B \cdot R}{4V_M} \tag{8.7}$$

Example: A 120 Ω strain gage bridge circuit is to be detuned with the aid of a shunt to produce a signal of 0.5 mV/V. According to Eq. (8.7) a shunt resistor of 60 kΩ is required.

The shunt calibration can be especially advantageously applied, if the measuring points are located in inaccessible or more remotely located measuring sites. In parallel to one of the active strain gages of the installed bridge circuit a shunt is arranged with a relay for switching on or off. When the shunt is turned on a defined bridge, imbalance is produced corresponding to a calibration signal. Especially when strain gages are installed in hard to reach complex structures this technology opens, in addition to calibration, the possibility of monitoring the soundness of the measuring point. The shunt is switched on at the measuring point by means of a relay as necessary. This requires two additional cable conductors for controlling the relay (for example, 12 volts DC). It should be used with a multicore cable, in which the supply for the relay is incorporated.

Furthermore it is advisable in such cases, to carry out the signal conditioning as close as possible at the test site, in order to keep the failure-prone transducer cable short. The trouble-free amplified measurement signal is then passed via the connection cable to the measuring centre, which can then be far away.

Example of the application of a shunt calibration [8.6]: The successful use of this type of measuring points fitted with shunt and relay occurred in 1987 on board the Soviet ice-breaker *Mudjug*. The ship had been fitted with a new type of bow. During extensive testing in arctic conditions, 54 full-bridge strain gage measuring points mounted on beams in the deck and bottom structure were simultaneously interrogated. These were checked each time, using shunt calibration after ice-breaking had taken place.

Using a relay, the shunt is directly connected in parallel at the point of measurement to one of the active strain gages in the full Wheatstone bridge. A multicore cable is used, providing two cores for the 12 V DC relay supply. Figure 8.8 shows the arrangement of a full strain gage bridge circuit, including two active and two temperature compensating strain gages. The relay and shunt are located on the

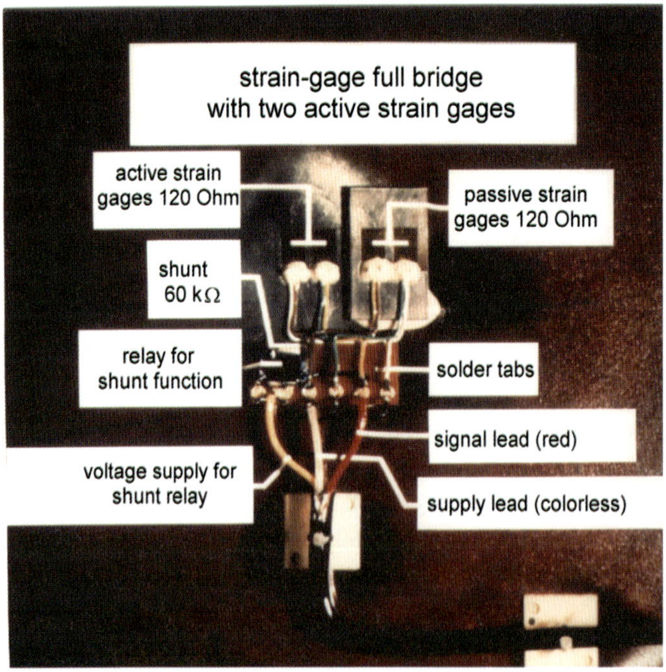

Fig. 8.8 Mounted strain gages with shunt and relay

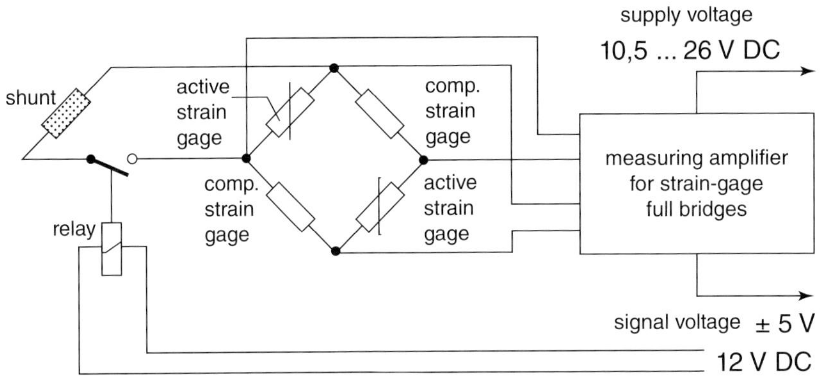

Fig. 8.9 Type of circuit used on board ship

solder tab strip and are assembled before installation. Figure 8.9 shows the circuit arrangement that was used.

It is important that the measuring point is made as flat as possible so that a waterproof covering layer can be applied and so that the risk of damage is minimized. When fitted on board ships, this type of measuring point may then be no longer accessible without a great deal of dismantling. Cable lengths of up to 100 m often occur with the cable in exposed positions. Repeated strain gage calibration involving the complete cabling is an important requirement for the reliability of automatically interrogated signals and for proving correct operation over extended periods.

The two compensation strain gages are located on a small mechanically unloaded steel block made from ship's material secured with adhesive, in addition to the active gages on the vessel wall. To switch on the shunt suitable relays are easy obtainable commercially for little money. Drift, plastic deformation at the measurement location, poor installation, cable damage and amplifier problems can be detected by switching in the shunt during equipment inspection in dock or during test measurements at sea.

8.8 Calibration of measuring arrangements with self-installed strain gages

8.8.1 Wheatstone bridge circuit

For measuring with Wheatstone bridge circuits the key to the calibration relationship between the measured strain ε and the bridge output signal V_M/V_B is given with the equation

$$\frac{V_M}{V_B} = \frac{1}{4} k\,\varepsilon \tag{8.8}$$

Solving Eq. (8.8) for the measured mechanical strain ε, we obtain

$$\varepsilon = \frac{4\,V_M}{k\,V_B} \tag{8.9}$$

the relationship between the value of the measured variable and the output signal of the bridge circuit. If one looks for the strain causing a bridge output of 1 mV/V (1 mV/V = 10^{-3}), we obtain from Eq. (8.9)

$$\varepsilon = \frac{4}{k} 10^{-3} \tag{8.10}$$

with $k = 2$

$$\varepsilon = 2 \times 10^{-3} = 2000\,\mu\text{m/m} \tag{8.11}$$

Thus, for $k = 2$ with an active strain gage in the bridge circuit we have

$$1\,\text{mV/V} = 2000\,\mu\text{m/m} \tag{8.12}$$

Using Eq. (8.12), the required relationship for a calibration is found.

For example, if a strain gage is used with $k = 2.1$, it is clear at a bridge imbalance of $V_M/V_B = 10^{-3}$ corresponding to Eq. (8.10) that

$$\varepsilon = \frac{4}{2.1} 10^{-3} = 1.904 \times 10^{-3} \tag{8.13}$$

Because the gage factor k is greater than 2, here a smaller strain than 2000 µm/m is sufficient to produce a signal of 1 mV/V. According to Eq. (8.13) a measured bridge output of $V_M/V_B = 1$ mV/V corresponds to mechanical strain $\varepsilon = 1904$ µm/m at the active strain gage with the gage factor $k = 2.1$.

8.8.2 Example of the calibration of a measuring device for measuring with self-installed strain gages

If strain has to be measured in the surface of a member using self-installed strain gages in half-bridge circuit, the active gage attached to the measurement object is connected to a second gage serving as temperature compensation gage and completes the half-bridge circuit. The measuring instrument is an amplifier with digital display. The gage factor of the strain gages used amounts to $k = 2.12$. The calibration of the measurement chain should be done with a calibrator. The bridge resistance of the calibrator corresponds to the nominal resistance of the strain gages used and amounts to 120 Ω.

For the calibration of the measuring system after the successful wiring the following steps have to be performed:

- Connect the calibrator to the input of the measuring amplifier. Both the calibrator and the input of the amplifier must be set to 'half-bridge.'
- Set the calibrator to zero signal and perform zero adjustment at the measurement amplifier.
- Set the calibrator to an appropriate calibration signal, for example $V_M/V_B =$ 0.5 mV/V. A calibration is suitable if its value corresponds to about the maximum expected measurement signal.

- Adjust the measuring range of the amplifier. The basis for this yields Eq. (8.8):

$$\varepsilon = \frac{4}{k} \cdot \frac{V_M}{V_B} \tag{8.14}$$

In this equation the used calibration signal and the gage factor k of the strain gage are introduced:

$$\varepsilon = \frac{4}{2.12} \cdot 0.5 \times 10^{-3} = 943 \times 10^{-6} \tag{8.15}$$

Now with an applied calibration signal of 0.5 mV/V the measurement range of the amplifier is adjusted so that at the indicator the value 943 appears. Thus, the transmission factor of the measuring amplifier and indicator is set correctly for the intended strain measurement: one display unit corresponds to 1 μm/m. (Note: It is recommended after span adjustment to again perform a zero check and to correct the zero point shift caused by any coarse measurement range adjustment. After this zero-point correction it is then good to recalibrate the range value with the fine adjustment):

- Disconnect the calibration device from the amplifier input.
- Connect the strain gage bridge circuit prepared for measuring to the amplifier input.
- Zero adjustment the amplifier with connected strain gage bridge circuit without changing the previously adjusted measuring range.
- Carry out the measurement.

Many currently available amplifiers for measuring with strain gages have an internal gage factor adjustment via a switch on the front panel, or with digital amplifiers the gage factor can be set with the operation software. By using these facilities you can perform the calibration as before, as if the strain gage used for measuring has the gage factor $k=2$, but you have to set the actual k-factor (e.g. $k=2.12$) on the amplifier. This changes the previous example in such a way that with applied calibration signal you set the gage factor to $k=2.1$ and the measuring range is adjusted instead to 943, so that at the display the value 1000 appears.

Note: If the calibration is carried out without using a calibration unit but rather with the internal calibration signal of the amplifier, then one proceeds in the same manner as described with the calibration unit. However, this means that the influence of the measuring cable between the strain gage installation and the amplifier is not taken into account.

References

[8.1] Westram, A. (1987) The measurement instrumentation of MV *Polarstern* for an icebreaking expedition. *RAM*, **3** (1), 23–29.

[8.2] Kreuzer, M. (1980) Ein programmierbares Präzisionsmeßgerät der Genauigkeitsklasse 0,0005 und seine Anwendungen. *MTB*, **16** (2), 41–46.

[8.3] Hellwig, H. (1987) Precision force transducer for international comparison measurements on force standard machines. *RAM,* **3** (1), 17–22.

[8.4] Kreuzer, M. (1981) Kalibrieren des digitalen Präzisions-Meßgeräts DMP 39 mit einem speziellen Brückennormal. *MTB,* **17** (3), 67–73.

[8.5] Ramm, G. (1990) Calibration of bridge standards for use in strain-gage measurements. *RAM,* **6** (1), 26–30.

[8.6] Westram, A. (1990) Shunt calibration of strain gage measuring points for automatic long-term measurements. *RAM,* **6** (2), 53.

9 Determination of mechanical stresses from strains measured with strain gages

9.1 Introduction

The application of strain gages to the monitoring and safety verification of structural members has increased over the past several years. Despite significant advancement in calculation methods – mainly due to the development of the appropriate software – it is still necessary to employ experimental stress analysis. In addition to the cases where strength verification is required for safety reasons, experimental stress analysis is also essential for the determination of stress distribution on complex structural members, or on structures for which calculation techniques are too difficult or involve unacceptable cost.

One of the primary advantages of strain gage technology is its easy application on structural members in their normal operating environment. In many cases strain gages allow stress analysis under the operating conditions that affect the measuring object. With knowledge of the necessary material laws, the stress condition at the measuring point can be determined from the measured strains. Moreover, the capital investment for instrumentation involved in strain gage measurement is small. These are some of the reasons why strain gages are the preferred technique for experimental stress analysis.

Stress analysis using strain gages consists of two steps:

- measurement of the strain created by stress in the surface of the specimen,
- calculation of the stress from the measured strain.

Here it must be emphasized that it is impossible to measure stress with strain gages. We are able to measure the strain with high accuracy, but the estimation of stress requires calculation methods based on appropriate material laws. Now we will consider stress calculation from the measured strain using strain gages. This assumes the ability to carry out satisfactory strain measurements with strain gages.

Stress analysis as considered here is limited to the elastic deformation range of materials, where strain is proportional to stress and is reversible. After load relief in the elastic deformation range the material returns to its original form. Most analyses using strain gages are limited to the elastic deformation range; in these cases the deformation of the material under study does not reach the yield point.

Stress analysis using strain gages is also feasible in the elastoplastic deformation range of the material. In this range reversible (elastic) deformation as well as irreversible (plastic) deformation occurs. Therefore the stress is no longer proportional to the strain. This has to be considered by the applied evaluation technique.

Technology and Practical Use of Strain Gages: With Particular Consideration of Stress Analysis Using Strain Gages. First Edition. Stefan Keil.

9.2 Terms of stress and strain

In order to provide understanding for the stress analysis to be performed, the relationship between stress and strain the terms stress and strain are briefly explained here.

The term *strain* describes a translational change in the shape of a material. Generally, strain means the relative change of the distance between two defined points on the surface or within a body specified by the straight line connecting these two points [9.1]. It describes the change in length Δl related to the base length l_0 of the straight line between these points. The known initial distance prior to the occurrence of the strain is mainly used as the reference length for the strain definition, so that the technical strain results to

$$\varepsilon = \frac{\Delta l}{l_0} \tag{9.1}$$

In the measurement technique, the effective strain ε_{eff} is becoming increasingly important, and so, for the base length for the strain definition, it is not the initial length l_0, but the respective length l that is used [9.2]. As a result the effective strain is

$$\varepsilon_{eff} = \int_{l_0}^{l_0+\Delta l} \frac{\mathrm{d}l}{l} = \ln\frac{l_0 + \Delta l}{l_0} = \ln(1 + \varepsilon) \tag{9.2}$$

Because of the logarithmic relationship between the technical strain ε and the effective strain, the effective strain is also referred to as logarithmic or natural strain. For small strains, the difference between ε and ε_{eff} is very small, as one can easily understand by substituting real values, so that for technical strain measurements in the range $\varepsilon < 1\%$ the difference between ε and ε_{eff} can almost always be neglected. For $\varepsilon = 1\%$ $\varepsilon_{eff}/\varepsilon = 0.995$ or $\varepsilon - \varepsilon_{eff} = 50\,\mu\text{m/m}$. As for the evaluation of the measurement results, the difference between technical and effective strain is interesting: this difference is given as a function of technical strain in Figure 9.1 for the range $1000 \times 10^{-6} \leq \varepsilon \leq 100\,000 \times 10^{-6}$ in diagrammatic form.

However, for strain measurements of large material deformations in the elastoplastic deformation range – which are increasingly required, especially on pipes and vessels – the difference is gaining importance. In this context, the question is whether the specified gage factor of the used strain gages is valid in the elastoplastic deformation range. The specified gage factors k are determined by the manufacturer in accordance with the valid standards in the elastic deformation range between $-1000 \times 10^{-6} \leq \varepsilon \leq 1000 \times 10^{-6}$. In this area, the difference between technical and effective strain is at most $0.5\,\mu\text{m/m}$. For measurements in the elastoplastic range at large strains corrections of the specified k-factor are required.

The term *stress* is defined as the force per given area. Stress arises, for example, from internal forces in the material at equilibrium with externally applied forces. If one

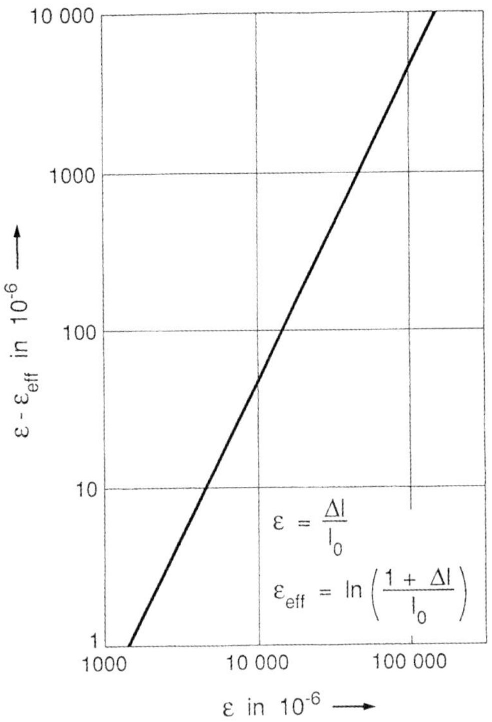

Fig. 9.1 Difference between technical and effective strain as a function of the technical strain

puts through any point P of a body a planar section and considers in this plane at the point P a sufficiently small surface element dA, the related force dF to this surface represents the stress vector $\wp$

$$\wp = \frac{dF}{dA} \tag{9.3}$$

By rotation of the plane around the point P, the stress vector $\wp$ is changed.

The totality of all $\wp$ describes the state of stress at point P. A stress vector can be decomposed as a spatial vector associated to the three directions of its effective surface orthogonal coordinate system. The stress perpendicular to the surface element dA is called the normal stress component, and the two components located in the area are referred to as the shear stress.

The amounts of these stresses are the scalar quantities σ_1, τ_1 and τ_2. Using the unit vectors n, t_1 and t_2 gives the relationship shown in Figure 9.2:

$$\wp = n\sigma_n + t_1\tau_1 + t_2\tau_2 \tag{9.4}$$

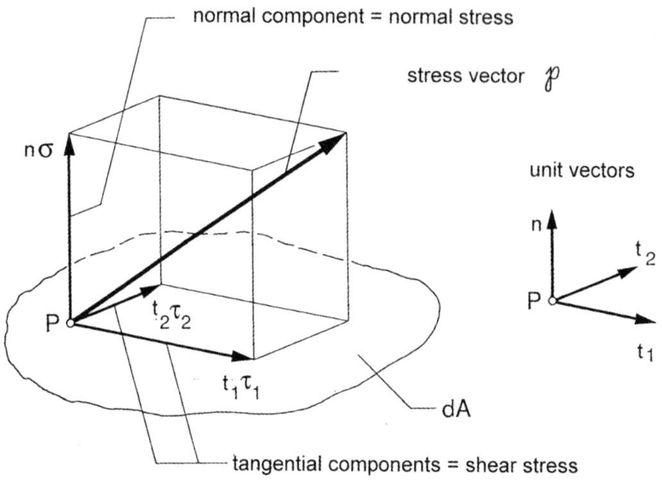

Fig. 9.2 Components of the stress vector at a surface element dA

9.3 Elastic deformation and stress of a tensile rod under uniaxial tensile loading

Stress calculations from measured strains are based on the material laws describing the relationship between stress and strain. In the elastic deformation range of homogeneous materials Hooke's law is valid and forms the basis for the stress calculation from measured strains. In order to demonstrate this law we will consider a cylindrical rod of circular cross-section subject to a tensile force, as illustrated in Figure 9.3. The unloaded rod, indicated by the dotted line, is of length l_0, diameter d_0 and cross-sectional area A_0. A force F along the axis of the rod creates a load expressed by the stress σ, that is the force per unit of initial cross-sectional area.

The load on the rod results in deformation, illustrated by the solid line in Figure 9.3. The rod has been stretched by the tensile force F and experiences a longitudinal strain ε_l. At the same time, the diameter of the rod has decreased. The transverse strain has an opposite sign compared to the longitudinal strain and is therefore called lateral contraction. The absolute value of the quotient of the lateral strain and longitudinal strain is called Poisson's ratio ν. Poisson's ratio plays a particularly important role in stress analysis using strain gages.

The deformation that occurs in the rod under tensile load depends on the force F, thus on the stress, and on the mechanical properties of the rod material. In order to determine the relationship between load and deformation for a particular material, standardized tensile tests are performed on standard specimens and the measured strain values are plotted against the stress. This produces the familiar stress–strain diagram of the material under study.

Figure 9.4 shows a stress–strain diagram for a mild carbon steel obtained from a standardized tensile test. We can see that in a certain range below the upper yield point the strain of the material is proportional to the stress.

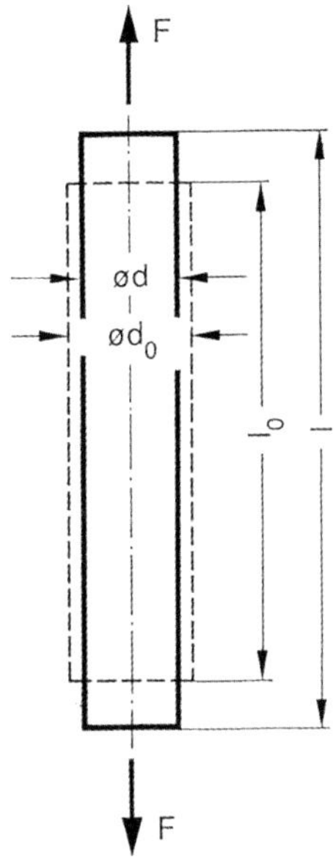

$$A_0 = \frac{d_0^2}{4}\pi \quad \text{initial cross section}$$

$$\sigma = \frac{F}{A_0} \quad \text{nominal stress}$$

$$\varepsilon_l = \frac{l - l_0}{l_0} \quad \text{longitudinal strain}$$

$$\varepsilon_q = \frac{d - d_0}{d_0} \quad \text{transverse strain}$$

$$\nu = \left| \frac{\varepsilon_q}{\varepsilon_l} \right| \quad \text{Poisson's Ratio}$$

Fig. 9.3 A cylindrical rod of circular cross-section under tensile load, showing longitudinal and transverse deformations with uniaxial stress, and a definition of Poisson's ratio as a material characteristic

In the range between zero and the limit of proportionality there is a linear relationship between stress and strain. This linear relationship appears in the diagram as a straight line which is represented by the equation

$$\sigma = E\varepsilon_l \tag{9.5}$$

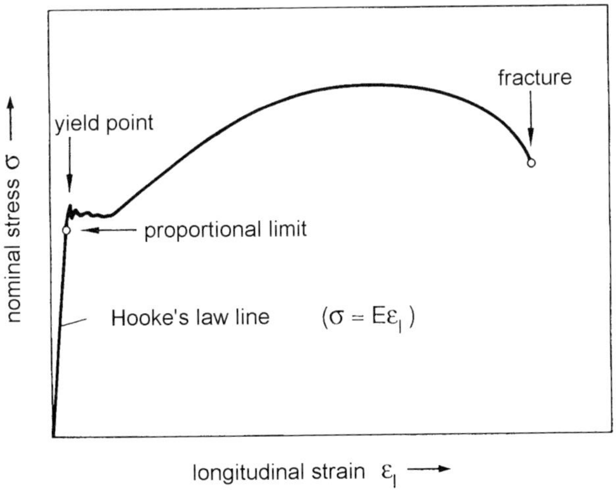

Fig. 9.4 Stress–strain diagram of a mild carbon steel

This equation is known as Hooke's law for the uniaxial stress state [9.3]. The proportionality factor between stress and longitudinal strain, which varies from material to material, is called the modulus of elasticity E (Young's modulus). E represents the slope of the straight line in the diagram. Within the range of Hooke's law materials behave elastically; this means that the deformation is reversible and that there will be no permanent deformation upon complete release of the load. If the stress exceeds the yield point, plastic (permanent) deformation will occur, in addition to the elastic deformation.

The range of simultaneous elastic and plastic deformation is called the elastoplastic range. In the elastoplastic deformation range Hooke's law is not valid and therefore not applicable for the calculation of the stress from the measured strain. In this range neither the linear relationship between stress and strain described by the modulus of elasticity nor Poisson's ratio as defined for elastic deformation are valid. But stress analysis with strain gages is also possible in the elastoplastic deformation range. Suitable strain gages and evaluation methods are available [9.4]. For this reason the following observations are restricted to the elastic range of deformation.

The factor of proportionality between stress and strain – the modulus of elasticity E – is the parameter that describes the elastic deformation behaviour of the material. Figure 9.5 shows the straight-line portions of the stress–strain curves for various metals with different slopes for different moduli of elasticity. With decreasing modulus of elasticity the elastic strain increases for the same stress in the material.

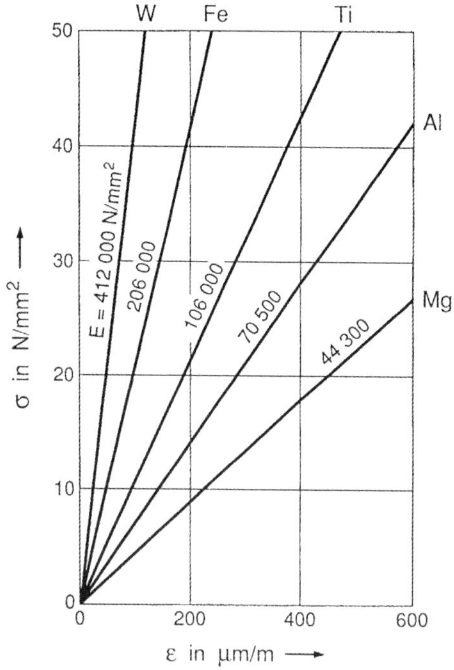

Fig. 9.5 The slope of Hooke's law line represents the modulus of elasticity of various metals [9.5]

Figure 9.3 showing a cylindrical rod subjected to a uniaxial tensile load, supplies the answer to the question about the magnitudes of the strains which would be measured with strain gages installed in the longitudinal and transverse directions. The longitudinal strain ε_l, occurring in the direction in which force F acts, is derived from Eq. (9.5):

$$\varepsilon_l = \frac{\sigma}{E}. \tag{9.6}$$

and by introducing Poisson's ratio ν, the transverse strain becomes

$$\varepsilon_q = -\nu\varepsilon_l \tag{9.7}$$

Here it is particularly interesting that a negative strain occurs transverse to the bar axis, although no stress acts in this direction. Equations (9.6) and (9.7) describe the strains arising in the direction of the bar axis and at an angle of $\varphi = 90°$ to it. Between these values the magnitudes of the strains are not known that arise at any other intermediate value of φ and which would be measured by strain gages installed at this angle. The magnitude of the stress occurring in this direction is also unknown. These quantities can, however, be relatively simply derived, based on Figure 9.6, which shows the force and stress components acting in a sectional plane at an angle φ through the tensile bar. Figure 9.6a shows the force components F_n and F_t defined at the angle φ normal and tangential to the sectional plane:

$$F_n = F\cos\varphi; \quad F_t = F\sin\varphi \tag{9.8}$$

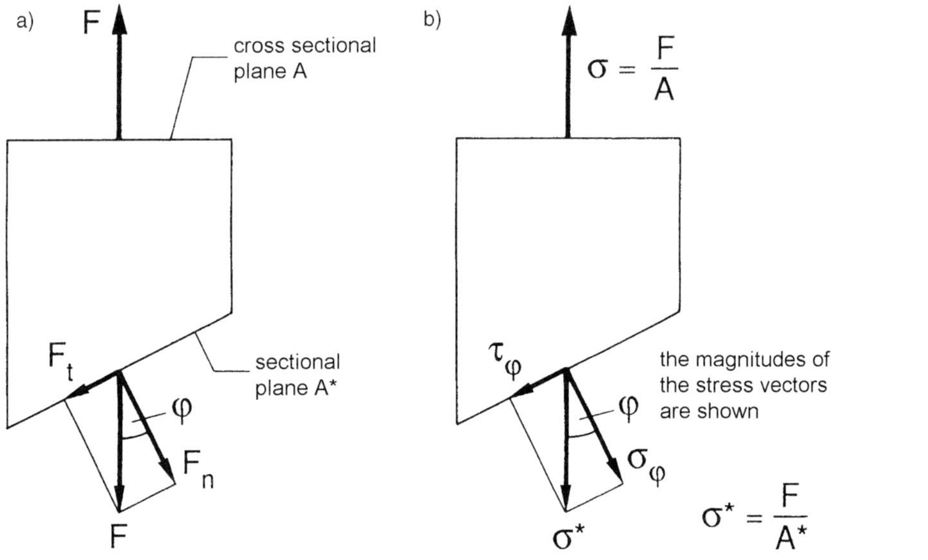

Fig. 9.6 a, b Force components and stresses acting in a sectional plane at an angle φ through the tensile bar

If these force components are referred to the area of the sectional plane A^*, which is

$$A^* = \frac{A}{\cos\varphi} \tag{9.9}$$

then the normal stress acting in the direction φ is obtained

$$\sigma_\varphi = \frac{F_n}{A^*} = \cos^2\varphi\frac{F}{A} \tag{9.10}$$

with $\sigma = \frac{F}{A}$ this becomes

$$\sigma_\varphi = \sigma\cos^2\varphi \tag{9.11}$$

The shear stress τ_φ in the sectional plane is given by

$$\tau_\varphi = \frac{F_t}{A^*} = \sin\varphi\cos\varphi\frac{F}{A} \tag{9.12}$$

and with $\sigma = \frac{F}{A}$

$$\tau_\varphi = \sigma\cdot\sin\varphi\cdot\cos\varphi \tag{9.13}$$

Therefore, this determines by calculation the normal and shear stresses in the sectional plane at the angle φ. They are reproduced in a polar diagram in Figure 9.7 for all possible angles φ. Here, since the stress distribution is symmetrical, the

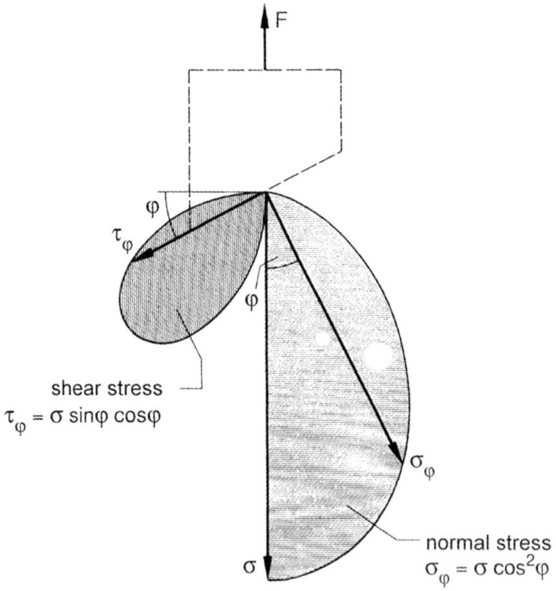

Fig. 9.7 Normal and shear stresses on a planar area cut from the uniaxial loaded tensile bar at an angle φ (polar diagram)

diagram is only shown for $0° \leq \varphi \leq 90°$. The right part of the illustration shows the distribution of the normal stress with the maximum stress in the longitudinal direction ($\varphi = 0°$) and the minimum stress in the transverse direction ($\varphi = 90°$). From the distribution of shear stress on the left, it can be clearly seen that the shear stress maximum occurs at $\varphi = 45°$.

Figure 9.7 shows that in a uniaxial tensile test normal stresses do not just occur in the loaded material in the longitudinal direction of the bar, but also in any sectional plane at an angle φ. It is only for $\varphi = 90°$, that is the transverse direction, that the normal stress is zero. If an infinitesimal elemental area is taken from the bar surface at an angle φ, then the normal stresses σ_φ and $\sigma_{\varphi+90°}$ act on the cut surfaces of this element as shown in Figure 9.8. This means that in this elemental area a biaxial (planar) stress state exists. This statement is important for the determination of the strain occurring in the direction φ, because it is influenced by the transverse strain which is produced in the direction φ by the stress acting in the direction $\varphi + 90°$.

The normal stresses on the elemental surface shown in Figure 9.8 are

$$\sigma_\varphi = \sigma \cdot \cos^2\varphi \tag{9.14}$$

$$\sigma_{\varphi+90°} = \sigma \cdot \sin^2\varphi \tag{9.15}$$

Taking into consideration the influence of the transverse strain, the strain in the direction φ is then

$$\varepsilon_\varphi = \frac{1}{E}\left(\sigma_\varphi - \nu\sigma_{\varphi+90°}\right) \tag{9.16}$$

Substituting Eqs. (9.14) and (9.15) into Eq. (9.16) gives

$$\varepsilon_\varphi = \frac{\sigma}{E}\left(\cos^2\varphi - \nu\sin^2\varphi\right) \tag{9.17}$$

and

$$\varepsilon_\varphi = \varepsilon_l\left(\cos^2\varphi - \nu\sin^2\varphi\right) \tag{9.18}$$

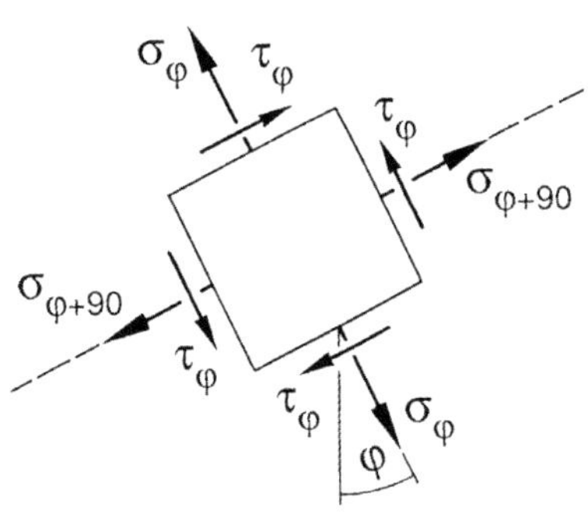

Fig. 9.8 Normal stresses and shear stresses in a sectional plane through a uniaxially loaded tensile bar at an angle φ

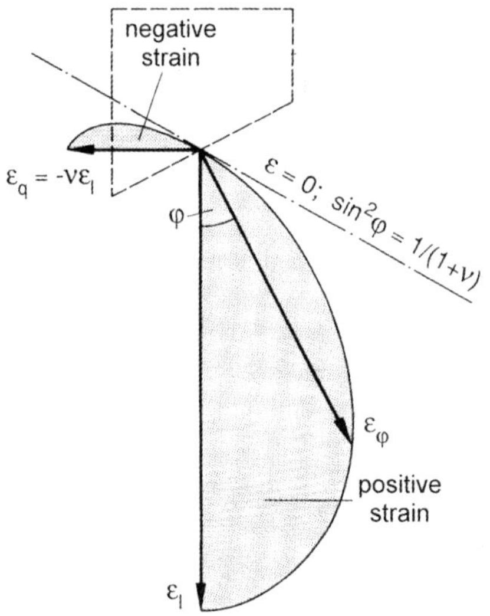

Fig. 9.9 Strains in a uniaxial loaded tensile bar in relationship to the angle φ (polar diagram)

Equation (9.18) can be further simplified to

$$\varepsilon_{\varphi} = \varepsilon_l \left[1 - (1 + \nu)\sin^2\varphi\right] \tag{9.19}$$

In Eq. (9.19) the strain occurring in the bar at an angle φ is defined as a function of the longitudinal strain ε_l for all possible angles φ. A strain gage mounted on the tensile bar with its measuring grid in the direction φ measures this strain.

Figure 9.9 represents Eq. (9.19) for the angular range $0° \leq \varphi \leq 90°$. The mutually perpendicular strains arising in the direction of the bar axis and in the transverse direction have opposite signs. The direction in which this change of sign occurs is free of strain. The angle φ of this strain-free direction can be easily found by setting ε_{φ} to zero in Eq. (9.19). It is a function of Poisson's ratio:

$$\varepsilon_{\varphi} = 0 \Rightarrow \sin\varphi = \sqrt{\frac{1}{1+\nu}} \tag{9.20}$$

and $\nu = 0.3$ produces an angle $\varphi = 61.3°$.

If a strain gage is bonded to a tensile bar of material with a Poisson's ratio of $\nu = 0.3$ and with its measuring grid in the direction 61.3°, then despite the tensile loading in the bar, no strain is measured. The normal stress existing in this strain-free direction is given in Eq. (9.14) and is depicted graphically in Figure 9.9.

From the above explanations it will be apparent to anyone concerned with practical applications that even with really trivial uniaxial tensile tests, the relationship

$$\sigma = \varepsilon \cdot E \tag{9.21}$$

only applies in the longitudinal direction. In all other directions the application of this equation leads to errors that can have catastrophic effects. Thought has only to be given to those cases where strain gages are mounted in the strain-free direction or transverse to the bar axis. In the former case one would conclude that, due to the lack of a signal despite the high load, the measurement direction was stress free. In the latter case the conclusion would be that compressive stresses exist in a stress-free direction!

For the element shown in Figure 9.8, cut from the uniaxially loaded tensile bar, there is one direction in which the normal stress σ_φ has its maximum value. For a uniaxially loaded tensile bar, this is at $\varphi = 0°$ (direction of the bar axis). For this direction the shear stresses at the edges of the cut surface (Figure 9.8) disappear and the normal stress in the direction perpendicular to $\varphi = 0°$, that is in the direction transverse to the bar axis, becomes zero. This phenomenon explains why a stress state is always termed uniaxial when, on rotating the section under consideration, a direction is found in which no normal stress exists. The maximum normal stress of the uniaxial stress state under consideration arises in a direction perpendicular to it. The direction with the maximum normal stress is termed the principal direction of the uniaxial stress state.

9.4 The biaxial stress state

So far, only the uniaxial stress state has been considered. Hooke's law as expressed for the uniaxial stress state is only applicable for this stress state, and only when the strain ε_l occurs in the axis of the uniaxial stress state. It has to be expected that, at the surface of loaded members, biaxial stress fields will occur. Therefore we need to expand Hooke's law for general application. Figure 9.10 serves to illustrate the necessary considerations.

Figure 9.10a shows the longitudinal elongation and the lateral contraction resulting from principal stress σ_1 acting on a thin plate. When the plate is subjected to a second

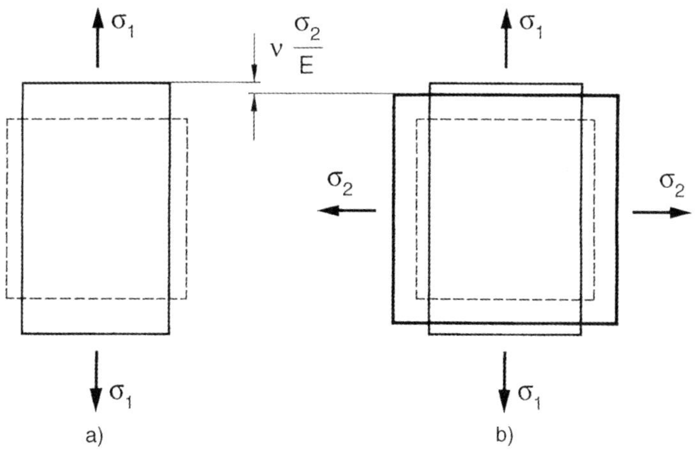

Fig. 9.10 Deformation of a plate due to principal stresses of a biaxial stress state; a) longitudinal elongation and lateral contraction resulting from a principal stress σ_1, b) action of principal stress σ_2 causes not only an elongation in direction 2, but also a reduction of the elongation in direction 1

stress σ_2 acting perpendicular to σ_1, as illustrated in the right side of Figure 9.10, the result is not only an elongation in direction 2, but also a reduction of the elongation in direction 1. It can be seen that biaxial stress involves the superposition of transverse and longitudinal strain. Therefore, Hooke's law as formulated for uniaxial stress is not valid for multiaxial stress because it does not allow for lateral deformation. The component of deformation in direction 1 produced by the stress σ_2 is

$$\varepsilon_{q2} = -\nu \frac{\sigma_2}{E} \tag{9.22}$$

and must be subtracted from the elongation produced by σ_1 to provide the total deformation in direction 1

$$\varepsilon_1 = \frac{\sigma_1}{E} - \nu \frac{\sigma_2}{E} \tag{9.23}$$

Thus for the biaxial stress state, the strains in directions 1 and 2 (principal strains) can be written

$$\varepsilon_1 = \frac{1}{E}(\sigma_1 - \nu\sigma_2) \tag{9.24}$$

$$\varepsilon_2 = \frac{1}{E}(\sigma_2 - \nu\sigma_1) \tag{9.25}$$

The result of solving Eqs. (9.24) and (9.25) for σ_1 and σ_2, respectively, is Hooke's law for the biaxial stress state, which is the fundamental relationship for stress analysis using strain gages in the elastic deformation range:

$$\sigma_1 = \frac{E}{1-\nu^2}(\varepsilon_1 + \nu\varepsilon_2) \tag{9.26}$$

$$\sigma_2 = \frac{E}{1-\nu^2}(\varepsilon_2 + \nu\varepsilon_1) \tag{9.27}$$

Equations (9.26) and (9.27) formulate Hooke's law for the biaxial stress state. The designation goes back to the English natural scientist Robert Hooke who, in 1675, hid the essential result of his findings in the anagram '*ceiiinosssttuv*'. Three years later in 'Lectures de potential restitutiva', or 'Explaining the power of springing bodies' [9.5] he gave away the sense of his anagram: *Ut tension sic vis*; with the meaning that the force of a spring behaves as its deflection [9.6]. Hooke formulated his findings in a uniaxial form, as we would call it today. The introduction of Young's modulus 1802 [9.7], the inclusion of stress and the extension to multiaxial stress was stated by Cauchy much later, in 1822 [9.8].

This form of Hooke's law for the biaxial stress state is written using stress and strain subscripts 1 and 2 to indicate the principal stresses and strains for the stress state under consideration. The principal stresses are the extreme stresses, that is the largest and smallest stresses. These two stresses are perpendicular to one another in the principal axes 1 and 2; conventionally $\sigma_1 \geq \sigma_2$.

The goal of stress analysis is to determine the principal stresses and the directions of the principal axes. This procedure is greatly simplified if the directions of the principal axes are known before the analysis, since this leaves only the principal stresses to be determined. This can be done by orienting two strain gages so that one grid measures the principal strain ε_1 along principal axis 1 and the second grid measures the principal strain ε_2 along principal axis 2. The principal normal stress can than be calculated by substituting the principal strains ε_1 and ε_2, the modulus of elasticity E and Poisson's ratio ν into Eqs. (9.26) and (9.27).

Unfortunately, in most applications the directions of the principal axes of the stress state to be determined are unknown, so that mounting of the strain gages aligned to the principal directions is not possible, and the principal strains cannot be measured directly. There are only a few cases for which the principal directions are known. Examples for loaded members with known directions of the principal axes are standardized specimens under laboratory test conditions and thin-walled cylindrical vessels under internal pressure.

Because the directions of the principal axes are in most cases unknown, their determination forms a significant part of stress analysis. The following section should simplify the understanding of this part of stress analysis and of the equations used.

For the biaxial stress state we can also define the strain in the direction normal to the plane under consideration (principal direction 3) as

$$\varepsilon_3 = -\nu \frac{1}{E}(\sigma_1 + \sigma_2) \tag{9.28}$$

The biaxial stress state creates a triaxial state of deformation. The strain acting in the principal direction 3 cannot be measured with strain gages. But that is no disadvantage in experimental stress analysis, because knowledge of the strains ε_1 and ε_2 acting in the plane of the biaxial stress state and their directions is sufficient for the experimental analysis of a biaxial stress state.

In the planar stress state, stress does not occur only in the principal directions 1 and 2, but also in other arbitrary directions within the plane. This can be viewed as a planar stress field in which the principal stress σ_1 is a maximum value and the principal stress σ_2 is a minimum value. As an aid to describing of the arbitrarily orientated stresses in the plane, a Cartesian coordinate system is used which is rotated around the system of principal axes by the arbitrary angle φ.

If we were to cut a small element from the plane shown in Figure 9.11 and observe the resulting stresses, we would find stresses σ_x and σ_y acting normal to the edges of the cut, as well as shear stresses τ acting parallel to the cut. It is only when the cut edges are orientated coincident with the principal stresses σ_1 and σ_2 that they are free from shear stress. Applying equilibrium conditions to the element shown in Figure 9.11, the relationship between principal stresses and stresses in the cut edges can be determined as a function of the angle φ. This procedure is explained in many textbooks and therefore will not be described here in detail.

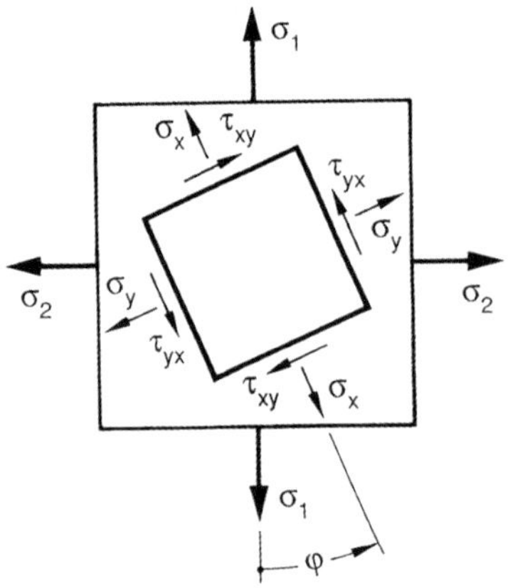

Fig. 9.11 Biaxial stress state: principal stresses σ_1 and σ_2 and the stresses in any arbitrary direction

The shear stresses τ_{xy} and τ_{yx} act parallel to the cut: the first subscript designates the coordinate normal to the cut and the second subscript designates the direction of the action of the shear stress, in which $\tau_{xy} = -\tau_{yx}$.

The equilibrium conditions lead to the following equations:

$$\sigma_x = \frac{\sigma_1 + \sigma_2}{2} + \frac{\sigma_1 - \sigma_2}{2} \cos 2\varphi \tag{9.29}$$

$$\sigma_y = \frac{\sigma_1 + \sigma_2}{2} - \frac{\sigma_1 - \sigma_2}{2} \cos 2\varphi \tag{9.30}$$

$$\tau_{xy} = -\tau_{yx} = \frac{\sigma_1 - \sigma_2}{2} \sin 2\varphi \tag{9.31}$$

In surfaces of measurement objects we find almost exclusively biaxial stress states. Only in surfaces under hydraulic or pneumatic pressure triaxial stress states occur, such as on the inner wall of a pressure vessel.

Formulating Hooke's law for the triaxial stress state we have to take into consideration the principal stress σ_3 acting normal to the measurement plane and the longitudinal and transverse strains generated by this stress:

$$\varepsilon_1 = \frac{1}{E}(\sigma_1 - \nu\sigma_2 - \nu\sigma_3) \tag{9.32}$$

$$\varepsilon_2 = \frac{1}{E}(\sigma_2 - \nu\sigma_3 - \nu\sigma_1) \tag{9.33}$$

9.5 Mohr's circle

A handy graphical aid for illustrating the relationship between the principal stresses and the stresses occurring in arbitrary directions is Mohr's circle [9.9, 9.10]. As shown

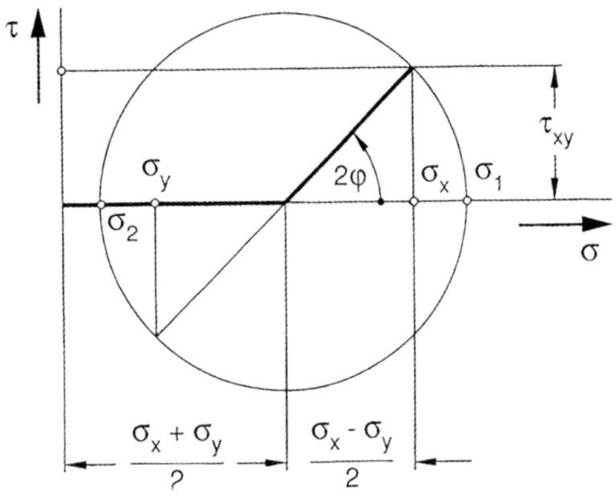

Fig. 9.12 Mohr's circle, a graphical representation of the relationship between principal stresses and stresses acting in arbitrary directions

in Figure 9.12, it is drawn on an orthogonal σ-τ-coordinate system. Its centre lies on the abscissa and it intersects the stress axis at the principal normal stresses σ_1 and σ_2.

When a straight line is drawn through the centre of Mohr's circle at an angle 2φ, the points of intersection on the circle graphically gives the values for σ_x, σ_y and τ_{xy}.

Recalling that φ is the angle at which the x-y coordinate system has been rotated around the principal axes 1 and 2, the same method can be reversed such that the principal axes and φ can be determined when the stresses σ_x, σ_y and τ_{xy} are known. From Mohr's circle it is easy to obtain the equation for the calculation of the principal stresses:

$$\sigma_{1,2} = \frac{\sigma_x + \sigma_y}{2} \pm \sqrt{\frac{\left(\sigma_x - \sigma_y\right)^2}{4} + \tau_{xy}^2} \tag{9.34}$$

as well as for the angle

$$\varphi = \frac{1}{2} arctan \frac{2\tau_{xy}}{\sigma_x - \sigma_y} \tag{9.35}$$

It is usual to set the condition that $\sigma_1 \geq \sigma_2$and $\sigma_x \geq \sigma_y$. The angle φ describes the direction of the x axis and is rotated in mathematically positive direction (counter-clockwise) starting from the principal axis 1.

It can be seen from Mohr's circle that the maximum shear stress $\tau_{\max}$ occurs when $\varphi = 45°$ $(2\varphi = 90°)$.

$$\tau_{12} = -\tau_{21} = \frac{\sigma_1 - \sigma_2}{2} \tag{9.36}$$

It is also possible to draw Mohr's circles representing triaxial stress states. In this case, as can be seen in Figure 9.13, all three principal stresses will be marked on the

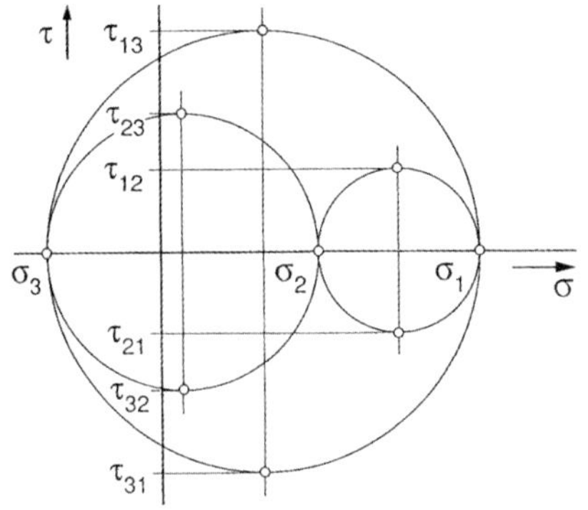

Fig. 9.13 Graphical representation of a triaxial stress state

abscissa. The differences $(\sigma_1 - \sigma_2)$, $(\sigma_2 - \sigma_3)$ and $(\sigma_3 - \sigma_1)$ define the diameters of three circles. Consequently there are three principal shear stresses:

$$\begin{aligned} \tau_{12} &= \frac{\sigma_1 - \sigma_2}{2} \quad \text{or} \quad \tau_{21} = \frac{\sigma_2 - \sigma_1}{2} \\ \tau_{23} &= \frac{\sigma_2 - \sigma_3}{2} \quad \text{or} \quad \tau_{32} = \frac{\sigma_3 - \sigma_2}{2} \\ \tau_{31} &= \frac{\sigma_3 - \sigma_1}{2} \quad \text{or} \quad \tau_{13} = \frac{\sigma_1 - \sigma_3}{2} \end{aligned} \tag{9.37}$$

The sign of the principal shear stresses is fixed by the sign of the differences of the assigned principal normal stresses. The radii of the circles represent the amounts of the principal shear stresses. The sum of all the principal shear stresses is zero.

The complete representation of a biaxial stress state by Mohr's circles consists of three circles, because $\sigma_3 = 0$ defines an additional point at the abscissa. Figure 9.14 shows the three circles of the biaxial stress state with the diameters $(\sigma_1 - \sigma_2)$, $(\sigma_1 - 0)$

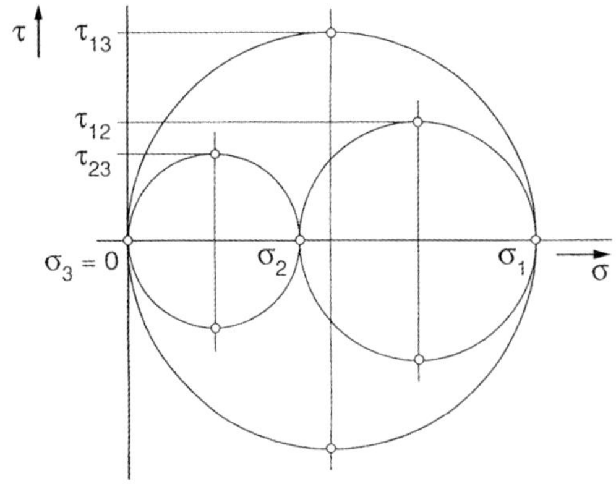

Fig. 9.14 Complete representation of the biaxial stress state by Mohr's circles

and $(\sigma_2 - 0)$. The principal shear stresses are

$$\tau_{23} = -\tau_{32} = \frac{1}{2}\sigma_2 \tag{9.38}$$

Using the representation as given in Figure 9.11 we can't discern the principal shear stresses $\tau_{13}$and τ_{23}. If σ_1 and σ_2 have the same sign, τ_{13} will become the maximum principal shear stress.

A very special case of the triaxial stress state is the hydrostatic stress state. Here all three principal normal stresses are equal, and they are identical to the hydraulic pressure. In this load case there are no shear stresses (all differences of the normal principal stresses are zero). There are only elastic deformations of the loaded body, and because there are no shear stresses there can't be any plastic deformations, as proved experimentally by Bridgman [9.11]. It is not possible to draw Mohr's circle for the hydrostatic stress state, because in the $\sigma - \tau$ diagram consists only of the point $\sigma_1 = \sigma_2 = \sigma_3 = p$ at the abscissa. The strains we find from Hooke's law for the triaxial stress state are

$$\varepsilon_1 = \varepsilon_2 = \varepsilon_3 = \frac{1}{E}p\,(2\nu - 1) \tag{9.39}$$

Some special cases of the biaxial stress state for which the principal directions are known before the measurement are given in the following. In the examples the principal strains are measured with strain gages adjusted to the principal directions, and from the results the principal stresses are calculated using the Eqs. (9.26) and (9.27).

Example 9.1 Pure tensile load, Figure 9.15

Only σ_1 acts as normal principal stress, and the normal principal stresses $\sigma_2 = \sigma_3 = 0$. The principal shear stresses τ_{12} and τ_{13} also exist, and they are equal.

If, from the mechanical circumstances, the principal direction 1 is known (the direction of the principal stress σ_1) the strain gage would be installed in this direction in order to measure the principal strain ε_1. Using Eq. (9.5), which gives Hooke's law for the uniaxial stress state, we can calculate the principal stress σ_1, provided that the modulus of elasticity E of the material is known.

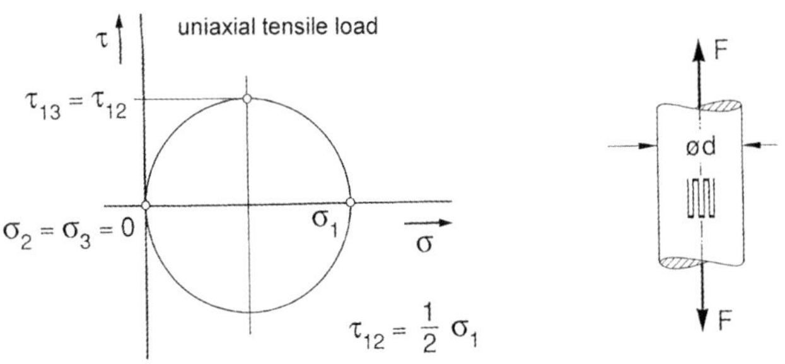

Fig. 9.15 Mohr's circle for uniaxial tensile load

Example 9.2 Pure torsion, Figure 9.16

Under pure torsion, both the principal stresses σ_1 and σ_2 have the same value but different signs. The third principal stress is $\sigma_3 = 0$. The maximum principal shear stress τ_{12} acts at the angle $\varphi = 45°$ ($2\varphi = 90°$) related to the principal direction 1. In order to measure the principal strains ε_1 and ε_2 using strain gages, the strain gages have to be installed in the directions of ε_1 and ε_2, consequently at 45° to the principal direction 1.

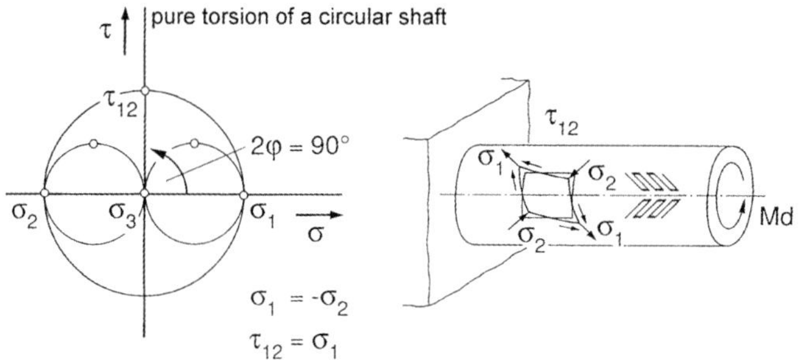

Fig. 9.16 Pure torsion and accompanying Mohr's circles

Example 9.3 Thin-walled cylindrical vessel subject to internal pressure, Figure 9.17

In this case, the principal stresses appear in the tangential and axial directions. On thin-walled vessels the stress in the tangential direction is about twice the axial stress. To carry out stress analysis on the outer wall of the vessel one strain gage will be installed in the tangential direction and a second strain gage in the axial direction. This will record the principal strains $\varepsilon_1$and ε_2. Introducing the principal strains ε_1 and ε_2 into Hooke's law (Eqs. (9.26) and (9.27)) gives the principal stresses σ_1 and σ_2, provided the modulus of elasticity and Poisson's ratio are known.

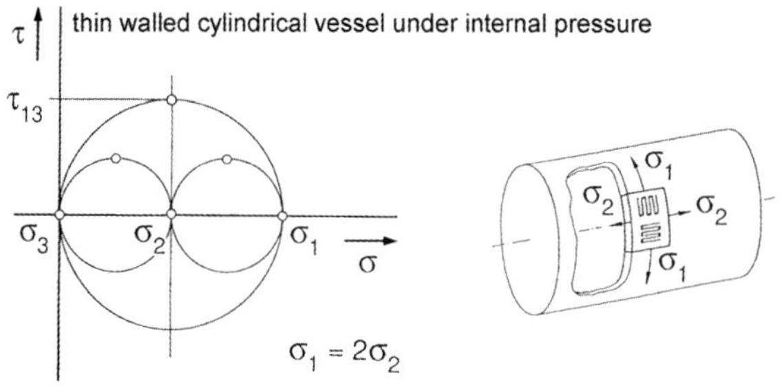

Fig. 9.17 Thin-walled cylindrical vessel under internal pressure and accompanying Mohr's circles

This allows the stress analysis of tubes. But it needs to be considered that as well as the stresses caused by the internal pressure, additional stresses which are caused by longitudinal forces, can apply torsion and bending to the tube.

Example 9.4 Combined bending and shearing stress, Figure 9.18

The special cases described in the examples 1 to 3 offer the simple possibility of installing strain gages in the principal directions, because the principal directions are well known before the measurements are taken. Therefore the principal strains ε_1 and ε_2 can be measured directly and can be used to calculate the principal stresses σ_1 and σ_2 using Hookés law (Eqs. (9.26) and (9.27)).

General loading cases with unknown principal directions are more common than the special cases shown in the earlier examples. It is impossible to measure the principal strains with strain gages if the principal directions are not known, because we don't know in what directions the strain gages have to be installed. A typical example for this general loading case is given with the combination of bending and torsion in Figure 9.18.

Here the orientation of the principal directions is set by the relationship between the bending stress and the shear stress and is not known before the measurement is carried out. Therefore the determination of the orientation angle of the principal stresses is part of the measurement task.

Mohr's circle given in Figure 9.18 shows that the orientation angle changes with alteration of the ratio of bending to shear stresses. Consequently under time dependent loading, time dependent alteration of the orientation of the principal directions can occur.

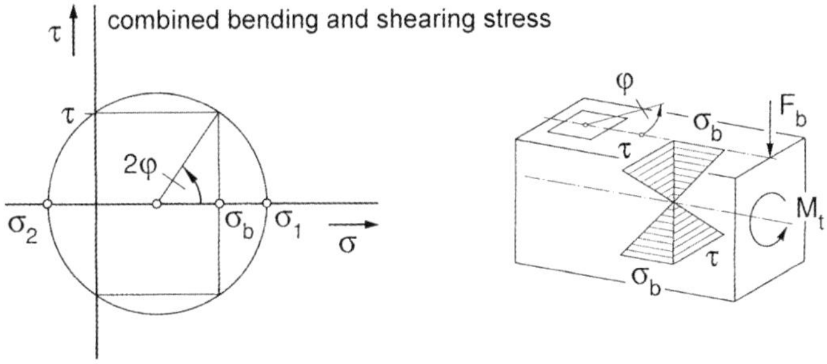

Fig. 9.18 An example of unknown principal directions before measurement is taken: combined bending and torsion

9.6 Deformation circle

Frequently, the principal directions of the stress state are not known before the measurement begins and the strain gages cannot be installed in these directions. Therefore it is impossible to measure the principal strains directly. In these cases, as

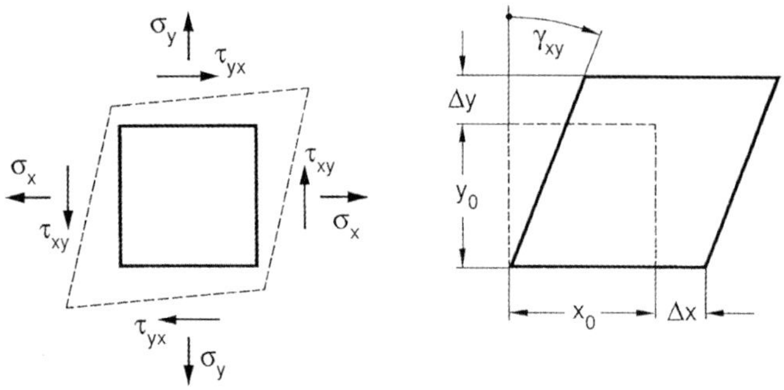

Fig. 9.19 Deformation of a small element created by normal and shear stresses

well as the principal stresses σ_1 and σ_2, there is the orientation angle as a third unknown quantity.

Although all materials under stress experience three-dimensional deformation, for stress analysis with strain gages, only the stresses existing in the material surface are of importance. Illustrated in Figure 9.19 are the possible stresses and deformations that could occur in a small element cut from the surface of a material under stress. The normal stresses σ_x and σ_y produce strains ε_x and ε_y, causing length changes in the x and y directions. The shear stresses τ_{xy} and τ_{yx} produce shear deformation, causing a change in angle γ_{xy}.

Analogous to the Mohr's circle method for graphically determining the stresses in arbitrary directions, strains and distortion angles are illustrated using a deformation circle [9.12]. As shown in Figure 9.20, this circle is represented in deformation coordinates, whereby the strain ε is used as the abscissa, and the distortion angle $\frac{1}{2}\gamma$ is used as the ordinate. The centre of the deformation circle lies on the abscissa, and its points of intersection with the abscissa determine the principal strains ε_1 and ε_2.

The deformation circle provides an easy method for determining the strains occurring in arbitrary directions along the material surface. To find an unknown

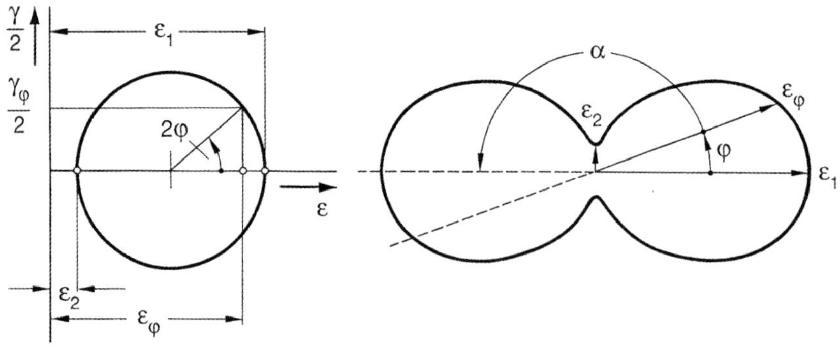

Fig. 9.20 Deformation circle and planar strain field corresponding to the deformation circle

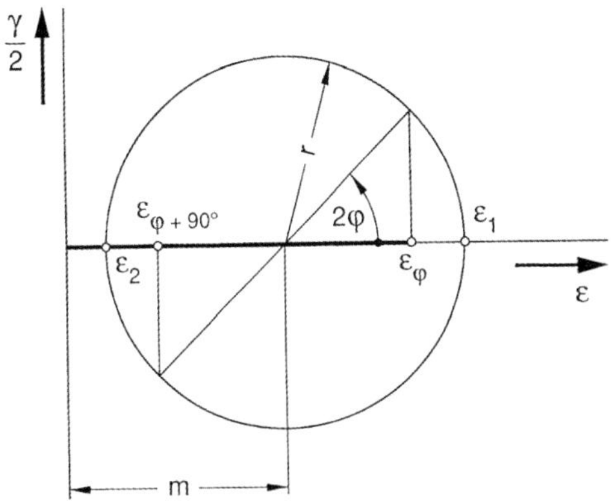

Fig. 9.21 Deformation circle with auxiliary terms r and m

strain ε_φ at an angle φ to the principal direction 1 – assuming the principal strains and directions are known – a perpendicular is dropped to the abscissa from a point on the circle where the radius at 2φ intersects it (Figure 9.20 left).

Figure 9.20 right, illustrates in polar coordinates the planar strain field corresponding to the deformation circle shown in Figure 9.20a left. The significance here is that in both illustrations the angle φ or 2φ respectively is turned in a positive rotational sense from the same reference direction.

From the deformation circle shown in Figure 9.20 the relationship between the effective strain ε_φ and the principal strains ε_1 and ε_2 can be derived:

Using auxiliary terms m and r introduced in Figure 9.21 one can find the strains

$$\varepsilon_{\phi,\phi+90} = m \pm r \cdot \cos 2\varphi \tag{9.40}$$

With

$$m = \frac{1}{2}(\varepsilon_1 + \varepsilon_2) \tag{9.41}$$

and

$$r = \frac{1}{2}(\varepsilon_1 - \varepsilon_2) \tag{9.42}$$

(9.40) becomes

$$\varepsilon_{\varphi,\varphi+90} = \frac{1}{2}(\varepsilon_1 + \varepsilon_2) \pm \frac{1}{2}(\varepsilon_1 - \varepsilon_2)\cos 2\varphi \tag{9.43}$$

Equation (9.43) provides the key to experimental stress analysis using strain gages, assuming the principal directions of stress are unknown. It contains as unknowns the principal strains ε_1 and $\varepsilon_{2,}$ as well as the angle φ, assuming that the strain in any arbitrary direction φ can be obtained by measurement. Measuring the strain in three different directions at the surface of a specimen will give three Eq. (9.43) to determine the three unknowns. This method assumes that the angles between the measuring axes are known, which is the case when strain gage rosettes are used.

9.7 Types of rosettes and grid notation

A strain gage rosette containing three measuring grids provides three separate readings which, with the aid of mathematical evaluation, can be used to determine the principal strains and their directions at the measuring point. The mathematical formulas required for this evaluation depend on the type of rosette used, as well as the notation of its measuring grids. Various methods of measuring grid notation for rosettes with the same grid arrangement can be found. This leads to different formulations of equations for the principal strains and their directions. These equations have no universal validity, but must always be viewed in connection to the method of notation employed in the derivation of the equation.

Errors are commonly encountered in determining the directions of the principal axes because of the various ways of defining the rotational sense of the angle between measuring directions and principal axes. According to Figure 9.23, the angle φ is measured starting at the principal axis 1, rotating in a positive sense to the reference axis. However, when carrying out measurements using rosettes, the inverse procedure must be followed, that is measurement directions are known and the principal directions must be found. Therefore, it is practical to introduce an angle α which is rotated from the reference axis of the rosette in a positive direction to the principal axis 1. Thus, it is necessary to specify a reference axis for the rosette. The grid notation used here is based on a standardization proposal presented in 1972 [9.13].

Two types of strain gage rosettes are preferred in experimental stress analysis: those with measuring axes at 45° to one another, and those with axes at 60° to one another. Figure 9.22 shows rosettes with the measuring axes at 45°. This grid arrangement is known as the 90° rosette, 45° rosette, and rectangular rosette. For our purposes here, we will call it the 0/45/90° rosette.

Figure 9.23 shows rosettes in which the measuring axes are at 60°. This we will call the 0/60/120° rosette. It is otherwise known as the 0/120/240° rosette, delta rosette or equiangular rosette.

In the following section, evaluation formulas are introduced for the 0/45/90° and 0/60/120° rosettes. The measuring axes for the individual grids are denoted

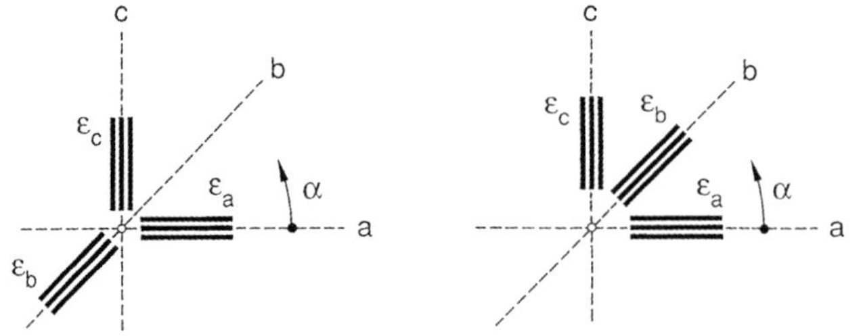

Fig. 9.22 Configuration and notation of the measuring axes of 0/45/90° strain gage rosettes

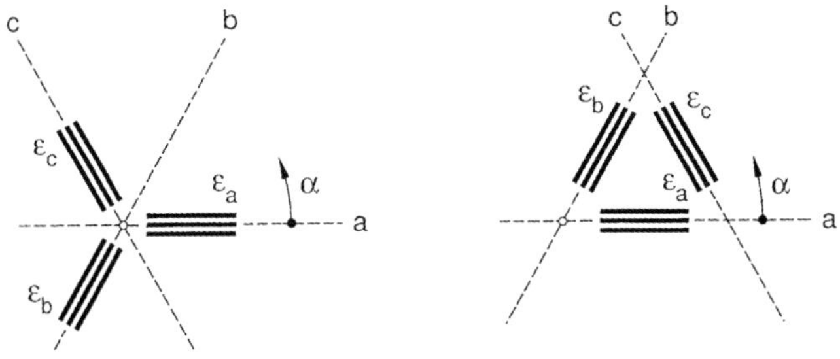

Fig. 9.23 Configuration and notation of the measuring axes of 0/60/120° strain gage rosettes

consecutively by the letters a, b, and c in the positive direction of rotation. The reference axis of the rosette is designated '*a*". Using 0/45/90° rosettes, the reference axis will be the first measurement axis encountered when moving in a positive direction of rotation. The principal axes are given the notation 1 and 2, in conformance with normal praxis in materials strength analysis.

The direction notation is employed in the form of subscripts for the different strain components, whereby the measured strains are ε_a, ε_b and ε_c, and the principal strains are ε_1 and ε_2, where $\varepsilon_1 \geq \varepsilon_2$. The orientation angle α of the rosette is measured in the positive direction of rotation from the rosette reference axis *a* to the principal axis 1.

9.8 Evaluation formulas for 0/45/90° strain gage rosettes

From the deformation circle in Figure 9.24 representing the measurement result of a 0/45/90° rosette we have

$$\varepsilon_a = m + r\cos 2\alpha \tag{9.44}$$

$$\varepsilon_b = m + r\sin 2\alpha \tag{9.45}$$

$$\varepsilon_c = m - r\cos 2\alpha \tag{9.46}$$

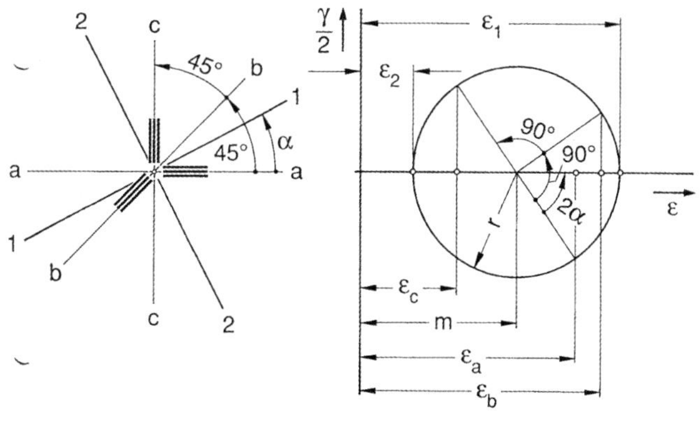

Fig. 9.24 Strains measured using a 0/45/90° rosette and the associated deformation circle

where

$$m = \frac{\varepsilon_a + \varepsilon_c}{2} = \frac{\varepsilon_1 + \varepsilon_2}{2} \tag{9.47}$$

Determination of the orientation angle α**:**
From Eq. (9.44):

$$r\cos 2\alpha = \varepsilon_a - m \tag{9.48}$$

From Eq. (9.45):

$$r\sin 2\alpha = \varepsilon_b - m \tag{9.49}$$

Division

$$\tan 2\alpha = \frac{\varepsilon_b - m}{\varepsilon_a - m} \tag{9.50}$$

(9.47) in (9.50):

$$\tan 2\alpha = \frac{2\varepsilon_b - \varepsilon_a - \varepsilon_c}{\varepsilon_a - \varepsilon_c} \tag{9.51}$$

Equation (9.51) is the defining equation for the orientation angle of the principal direction $\alpha$when using a 0/45/90° rosette. The orientation angle is obtained from a tangent function of 2α which always gives an ambiguous result because $0° \leq 2\alpha \leq 360°$. The following is a brief mathematical explanation meant to prevent errors caused by the ambiguity of the tangent function.

The tangent of an angle is a quotient. As shown in Figure 9.25, the numerator N is the part of the tangent (opposite side) subtended by the angle 2α, and the denominator D is the radius (adjacent side). In the first quadrant, numerator and denominator are positive, so the quotient is also positive. In the third quadrant, the quotient is again positive, but the numerator and denominator are negative. From this the ambiguity of the tangent function throughout the range $0° \leq 2\alpha \leq 360°$ is easily seen.

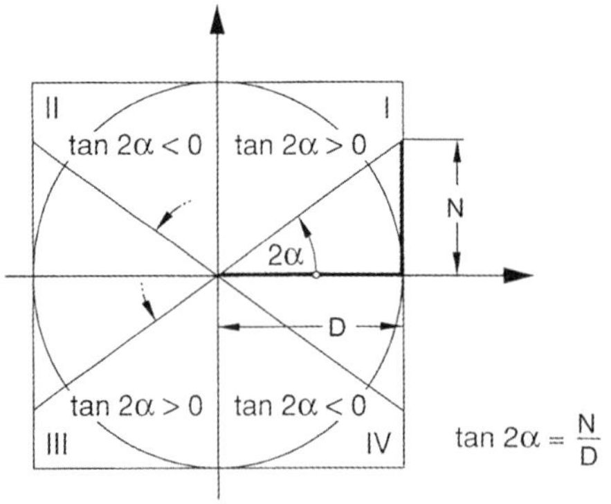

Fig. 9.25 Quadrant locator for unambiguous location of the tangent angle

Table 9.1 Relationship between the angle α and the values for the numerator N and the denominator D

N	≥ 0	>0	≤ 0	<0
D	>0	≤ 0	<0	≥ 0
2α	$\arctan \frac{N}{D}$	$\pi - \arctan \frac{N}{D}$	$\pi + \arctan \frac{N}{D}$	$2\pi - \arctan \frac{N}{D}$
α	$0° \leq \alpha < 45°$	$45° \leq \alpha < 90°$	$90° \leq \alpha < 135°$	$135° \leq \alpha < 180°$

For a clear determination of 2α, it is necessary to examine the sign of the numerator N and the denominator D. This is possible using Table 9.1, which determines unambiguously the angle α. This angle can then be measured from the rosette reference axis 'a' in a positive sense of rotation. This fixes the principal axis 1. Principal axis 2 is perpendicular to principal axis 1.

If using the 0/45/90° rosette according to Eq. (9.51) we obtain numerator N and denominator D:

$$Z = 2\varepsilon_b - \varepsilon_a - \varepsilon_c \tag{9.52}$$

$$N = \varepsilon_a - \varepsilon_c \tag{9.53}$$

The ambiguity of the tangent function is clearly evident from Figure 9.26. In the case of a positive quotient the angle 2α can lie in the quadrant I or III, with a negative quotient in quadrant II or IV.

Determination of the principal strains:
One can derive from Figure 9.24

$$\varepsilon_{1,2} = m \pm r \tag{9.54}$$

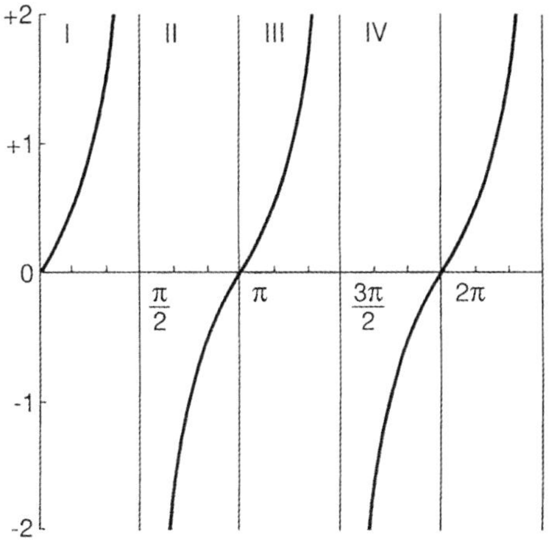

Fig. 9.26 Tangent function. Ambiguity in the field 0 to 2π

With Eq. (9.45)

$$\sin 2\alpha = \frac{\varepsilon_b - m}{r} \tag{9.55}$$

Auxiliary equation

$$\cos 2\alpha = \sqrt{1 - \sin^2\alpha} \tag{9.56}$$

(9.55) in (9.56)

$$\cos 2\alpha = \sqrt{1 - \frac{(\varepsilon_b - \varepsilon_m)^2}{r^2}} \tag{9.57}$$

(9.57) in (9.44)

$$\varepsilon_a = m + \sqrt{r^2 - (\varepsilon_b - m)^2} \tag{9.58}$$

(9.47) in (9.58)

$$r = \frac{1}{2}\sqrt{2}\sqrt{(\varepsilon_a - \varepsilon_b)^2 + (\varepsilon_c - \varepsilon_b)^2} \tag{9.59}$$

Equation (9.59) represents the radius of the deformation circle. The diameter of the deformation circle represents the maximum distortion occurring

$$2r \equiv \gamma_{12} \quad \text{thus} \quad r \equiv \frac{1}{2}\gamma_{12} \tag{9.60}$$

With (9.59) and (9.47) the definition equation of the principal strains is

$$\varepsilon_{1,2} = \frac{\varepsilon_a + \varepsilon_c}{2} \pm \frac{1}{2}\sqrt{2}\sqrt{(\varepsilon_a - \varepsilon_b)^2 + (\varepsilon_c - \varepsilon_b)^2} \tag{9.61}$$

Thus the principal strains are known in the plan under consideration. They act in directions 1 and 2, as found previously (Eq. (9.51)). Knowledge of the third principal strain acting perpendicular to the plan under consideration is not required for the analysis of the stress state in the surface of the measuring object. However, it can be found from Hooke's law with

$$\varepsilon_3 = \frac{\nu}{\nu - 1}(\varepsilon_1 + \varepsilon_2) \tag{9.62}$$

Now that the principal strains are known the principal stresses can be determined using Eqs. (9.26) and (9.27). Directions of the principal stresses and of the principal strains are identical.

Summary of important formulas for the evaluation of strains measured using a 0/45/90° strain gage rosette:

$$\varepsilon_{1,2} = \frac{\varepsilon_a + \varepsilon_c}{2} \pm \frac{1}{2}\sqrt{2}\sqrt{(\varepsilon_a - \varepsilon_b)^2 + (\varepsilon_c - \varepsilon_b)^2} \tag{9.61}$$

and

$$\alpha = \frac{1}{2} arctan \frac{2\varepsilon_b - \varepsilon_a - \varepsilon_c}{\varepsilon_a - \varepsilon_c} \tag{9.51}$$

Other notation of Eq. (9.51)

$$\tan 2\alpha = \frac{2\varepsilon_b - \varepsilon_a - \varepsilon_c}{\varepsilon_a - \varepsilon_c}$$

Note: These equations are only valid for the rosette grid notation indicated in Figure 9.24.

9.9 Evaluation formulas for 0/60/120° strain gage rosettes

For the evaluation, a general deformation state is assumed and illustrated with the deformation circle in Figure 9.27. From deformation circle in Figure 9.27 representing the measurement result of a 0/60/120° rosette one can derive:

$$\varepsilon_a = m + r \cos 2\alpha \tag{9.63}$$

$$\varepsilon_b = m + r \cos(120^\circ - 2\alpha) \tag{9.64}$$

$$\varepsilon_c = m + r \cos(120^\circ - 2\alpha) = m + r \cos(60^\circ - 2\alpha) \tag{9.65}$$

In which

$$m = \frac{\varepsilon_a + \varepsilon_b + \varepsilon_c}{3} \tag{9.66}$$

Determination of the orientation angle α:

Equations (9.64) and (9.65) can also be written

$$\varepsilon_b = m + r(\cos 120^\circ \cos 2\alpha + \sin 120^\circ \sin 2\alpha) \tag{9.67}$$

$$\varepsilon_c = m - r(\cos 60^\circ \cos 2\alpha + \sin 60^\circ \sin 2\alpha) \tag{9.68}$$

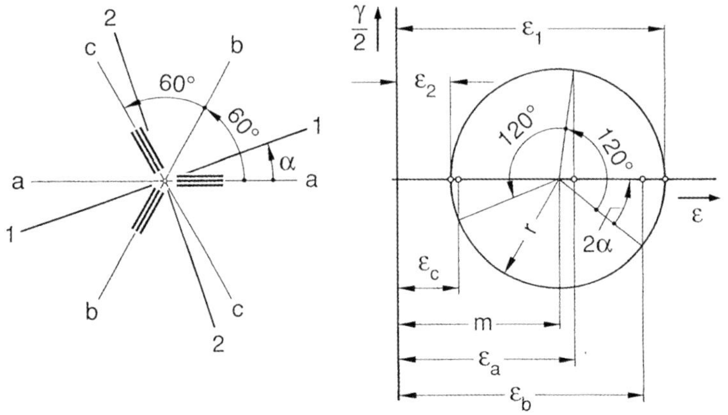

Fig. 9.27 Strains measured using a 0/60/120° rosette and the associated deformation circle

Introducing the auxiliary terms

$$w = \sqrt{1 + \tan^2 2\alpha} \tag{9.69}$$

and $\sin 2\alpha = \frac{\tan 2\alpha}{w}$ or rather

$$\cos 2\alpha = \frac{1}{w} \tag{9.70}$$

Equations (9.63)–(9.65) become

$$\varepsilon_a = m + \frac{r}{w} \tag{9.71}$$

$$\varepsilon_b = m - \frac{r}{2w}\left(1 - \sqrt{3} \cdot \tan 2\alpha\right) \tag{9.72}$$

$$\varepsilon_c = m + \frac{r}{2w}\left(1 - \sqrt{3} \cdot \tan 2\alpha\right) \tag{9.73}$$

from (9.72)

$$r = \frac{2w(m - \varepsilon_b)}{1 - \sqrt{3} \cdot \tan 2\alpha} \tag{9.74}$$

from (9.73)

$$r = \frac{2w(m - \varepsilon_c)}{1 + \sqrt{3} \cdot \tan 2\alpha} \tag{9.75}$$

equate (9.74) = (9.75)

$$\frac{2w(m - \varepsilon_b)}{1 - \sqrt{3} \cdot \tan 2\alpha} = \frac{2w(m - \varepsilon_c)}{1 + \sqrt{3} \cdot \tan 2\alpha} \tag{9.76}$$

Introducing (9.66) into (9.77) leads to the determination equation for the orientation angle α

$$\tan 2\alpha = \frac{\sqrt{3}(\varepsilon_b - \varepsilon_c)}{2\varepsilon_a - \varepsilon_b - \varepsilon_c} \tag{9.77}$$

or rather.

$$2\alpha = \arctan \frac{\sqrt{3}(\varepsilon_b - \varepsilon_c)}{2\varepsilon_a - \varepsilon_b - \varepsilon_c} \tag{9.78}$$

Equation (9.78) is the defining equation for the orientation angle of the principal direction α when using a 0/60/120° rosette. The orientation angle is obtained from a tangent function of 2α which always gives an ambiguous result because $0° \leq \alpha \leq 180°$. Concerning the orientation angle α for the 0/60/120° rosette the same applies as for the 0/45/90°rosette. To avoid ambiguity caused by the tangent function, the numerator and denominator must be considered separately.

Determination of the principal strains

With the help of Figure 9.27 the definition equations for the principal strains can be derived.

From Figure 9.27:

$$\varepsilon_{1,2} = m \pm r \tag{9.79}$$

From Eq. (9.71)

$$r = w(\varepsilon_a - m) \tag{9.80}$$

Equation (9.70) leads to

$$r^2 = (\varepsilon_a - m)^2 (1 + \tan^2 2\alpha) \tag{9.81}$$

(9.77) in (9.81)

$$r = \sqrt{(\varepsilon_a - m)^2 + \frac{1}{3}(\varepsilon_b - \varepsilon_c)^2} \tag{9.82}$$

And with (9.66)

$$r = \sqrt{\left(\frac{2\varepsilon_a - \varepsilon_b - \varepsilon_c}{3}\right) + \frac{1}{3}(\varepsilon_b - \varepsilon_c)^2} \tag{9.83}$$

Equation (9.83) represents the radius of the deformation circle, Eq. (9.79) can be written

$$\varepsilon_{1,2} = \frac{\varepsilon_a + \varepsilon_b + \varepsilon_c}{3} \pm \sqrt{\frac{2\varepsilon_a - \varepsilon_b - \varepsilon_c}{3} + \frac{1}{3}(\varepsilon_b - \varepsilon_c)^2} \tag{9.84}$$

Now that the principal strains are known, the principal stresses can be determined using Eqs. (9.26) and (9.27). The directions of the principal stresses and of the principal strains are identical.

Summary of important formulas for the evaluation of strains measured using a 0/60/120° strain gage rosette:

$$\varepsilon_{1,2} = \frac{\varepsilon_a + \varepsilon_b + \varepsilon_c}{3} \pm \sqrt{\frac{2\varepsilon_a - \varepsilon_b - \varepsilon_c}{3} + \frac{1}{3}(\varepsilon_b - \varepsilon_c)^2} \tag{9.84}$$

and

$$2\alpha = \arctan \frac{\sqrt{3}(\varepsilon_b - \varepsilon_c)}{2\varepsilon_a - \varepsilon_b - \varepsilon_c} \tag{9.78}$$

or rather

$$\tan 2\alpha = \frac{\sqrt{3}(\varepsilon_b - \varepsilon_c)}{2\varepsilon_a - \varepsilon_b - \varepsilon_c} \tag{9.77}$$

Note: These equations are only valid for the rosette grid notation indicated in Figure 9.27.

References

[9.1] DIN 13 316 (September 1980) *Mechanik ideal elastischer Körper; Begriffe, Größen, Formelzeichen*, Beuth Verlag, Berlin.

[9.2] Ludwik, P. (1909) *Elemente der Technologischen Mechanik*, Verlag J. Springer, Berlin, p. 17.

[9.3] Hooke, R. (1678) *Lectures de potentia restitutiva, or of spring explaining the power of springing bodies*, London.

[9.4] Keil, S. and Benning, O. (1979) On the evaluation of elasto-plastic strains measured with strain gages. *Experimental Mechanics*, **19**, 265–270.

[9.5] Troost, A. (1980) *Einführung in die allgemeine Werkstoffkunde metallischer Werkstoffe Bd. I*, Bibliographisches Institut AG, Zürich.

[9.6] Szabo, I. (3. Auflage 1987) *Geschichte der mechanischen Prinzipien*, Birkhäuser Verlag Basel, Boston Stuttgart, p. 356.

[9.7] Young, T., *A Course of Lectures on Natural Philosophy and the Mechanical Arts*, published 1802, cited in [6] p. 390.

[9.8] Szabo, I. (3. Auflage 1987) *Geschichte der mechanischen Prinzipien*, Birkhäuser Verlag Basel, Boston Stuttgart, pp. S393–S402.

[9.9] Mohr, O. (1900) Welche Umstände bedingen die Elastizitätsgrenze und den Bruch eines Materials? *VDI-Zeitschrift*, **44**, 1524.

[9.10] Mohr, O. (1901) Zur Festigkeitslehre. *VDI-Zeitschrift*, **445**, 740.

[9.11] Bridgman, P.W. (1945) Effect of high hydrostatic pressure on the plastic properties of metals. *Reviews of Modern Physics*, **17**, 3–14.

[9.12] Rötscher, F. and Jaschke, R. (1939) *Dehnungsmessungen und ihre Auswertung*, Springer-Verlag, Berlin.

[9.13] Keil, S. (1987) Stress calculation from measured strains in the elastic deformation range. *RAM*, **3** (2), 56–62.

10 Application examples of elastic deformation

10.1 Initial considerations

There are numerous fields of application for stress analysis using strain gages. Applications are found in the most varied specialist areas. Depending on the task in hand the measurement engineer is confronted with different requirements with regard to the measurement results and the ambient conditions. There are measurement tasks that can be solved relatively quickly with not much effort, such as the monitoring of the tensile load on a tensile beam under laboratory conditions. However the solving of some measurement problems can be very complex, such as static multipoint measurements on rotating parts at high temperature and under strong magnetic fields. Sometimes during measurement work, unexpected complications arise in what was originally thought to be a relatively simple task. Perhaps this is because various considerations have been neglected during the preparation. Apart from frequently repeated routine measurements, it is recommended that a plan is laid out, preferably based on a check list, before starting to setup the measurement point.

The most important point in the preliminary considerations is the objective of the measurement. It is possible to solve numerous widely varying measurement problems using strain gages, but strain gages are also subject to physical limits which restrict their application.

Carefully executed measurements, including their preparation, demand time. A measurement carried out under the pressure of time almost always has a detrimental effect in the course of the measurement. Therefore, the measurement engineer should not be influenced by outside pressures and should avoid reducing the amount of care taken. Experience has shown that a task carried out calmly and steadily will lead to reliable results more quickly than a measurement taken hastily under time pressure. Time is also important in measurement preparation, because a measurement can only be started when all that is required is available; this also includes the measurement object itself in a state ready for testing (perhaps disassembled), ancillary personnel and the measurement devices including the leads and all other auxiliary equipment.

Another point is to consider what to do when a measurement does not conform to what was expected. Often measurements are carried out to confirm an impression obtained from theoretical studies. The measurement engineer or his client often knows the 'approximate answer' even before the measurement. But if the measurement result does not agree with this impression, then the measurement need not always be erroneous. The reverse is often true.

If despite repeated careful measurement and after subsequent checking of the measurement equipment, the measurement still does not agree with the preconceived model, then it is the model that should be withdrawn and not the measurement result. To say that a measurement does not conform to a preconceived model is no argument against the correctness of the measurement [10.1]. This does not mean

Technology and Practical Use of Strain Gages: With Particular Consideration of Stress Analysis Using Strain Gages. First Edition. Stefan Keil.

however that plausibility checks should be ignored, because these are a valuable aid in the assessment of measurement results. In the following, two examples are given which, with a superficial assessment of the measurement results could suggest that the measurement was at fault, even though the results conformed to actual mechanical conditions. The underlying reasons in both examples are not easily recognized at first glance by the untrained eye.

Example 1: A loading frame was developed for teaching purposes in which tensile, bending and torsion loads could be introduced into clamped bodies, enabling the deformation of the bodies to be measured. A simple explanatory model used here is a bending beam with uniform rectangular cross-section, clamped at both ends. This beam is shown in Figure 10.1.

With a symmetrical bending load due to a force acting at the centre of the beam, the curve for the bending moment $M_{b,comp}$ can be easily calculated. The distribution of strain occurring in the beam surface can be computed as a function of the loading force. Using strain gages it should be possible to verify that the theoretical calculation of the strain is correct. However, the calculated strains were not able to be verified by the mechanical measurements. For each measuring point there was a difference, and this was also constant for all measuring points. Because it was easily understood, the computation could not have been incorrect, so it was thought that there must have been a measurement error, perhaps an additive error from the zero-point compensation. The measurement was checked carefully and the result could not be faulted. Certainly, any mechanical engineer with some experience will immediately recognize the cause of the problem. The two points at which the beam is clamped are not entirely rigid, but instead have some degree of flexibility. This means that the slope of the elastic line of the beam in the clamps is not equal to zero. The computer method assumes ideally rigid clamping. The curve of the actual bending moment in the beam $M_{b,act}$ and the elastic line of the beam can be found from the measurement results, as can the actual clamping conditions (slope of the elastic line in the clamps).

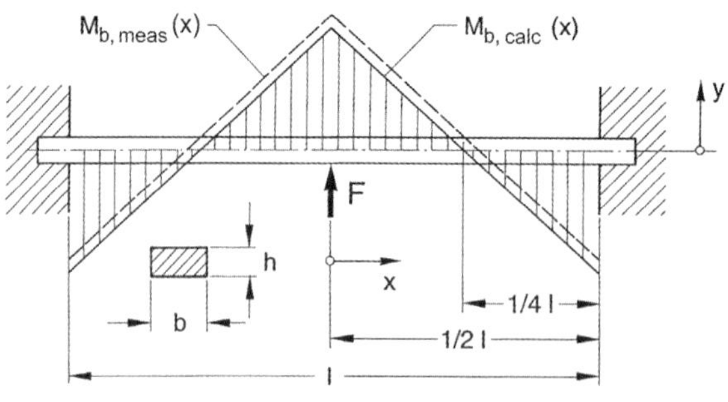

Fig. 10.1 The calculated curve of bending moment in a beam assuming rigid clamping compared with the actual curve of the bending moment assuming non-ideal rigid clamping

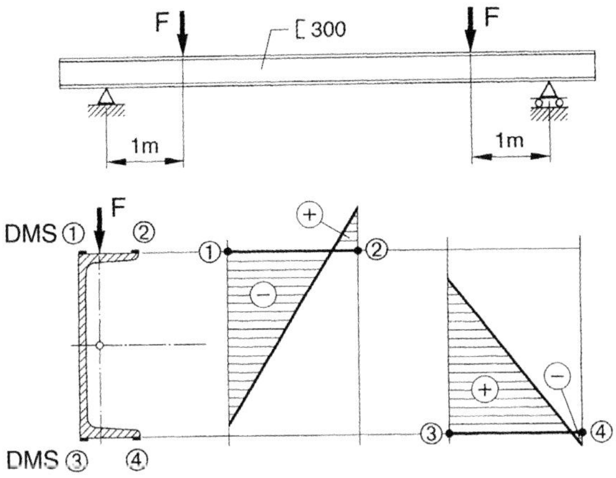

Fig. 10.2 Stress distribution in the flanges of a bending beam with unsymmetrical cross-section

Example 2: Here a steel beam (300) supported at two points is loaded in the plane of its centre of gravity (Figure 10.2) by two bending forces so that a constant bending moment is created between the two force transfer points in the centre section of the beam [10.2]. Before the beam was loaded, strain gages were bonded to the centre section in the longitudinal direction as shown in the illustration on the external edges of the upper and lower flanges.

Strangely, when the beam is loaded, the strain gages on the open side of the flange on the 'compression side' of the beam show tensile strains and those on the 'tensile side' show compressive strains. This measurement result obviously contradicts the expectation and an erroneous measurement was suspected. In fact the measurement result is in complete agreement with the mechanical theory, because in this case unsymmetrical bending is involved for which the requirements of the general theory for the bending of a beam are not fulfilled. This theory assumes that the cross-sectional shape of the loaded beam remains uniform – a feature which is only present with symmetrical cross-sectional shapes and with loads acting in the plane of symmetry. The bending theory does not apply for unsymmetrical loads as in this example, because the cross-sectional shape cannot remain uniform.

10.2 Principal directions are known

10.2.1 Examples of measurement objects loaded longitudinally

Applications in which measurements are made on longitudinally loaded civil construction or machine components occur relatively frequently. Almost always, the longitudinal direction of the usually slim measurement object is identical with principal direction 1 of the stress state. During mounting, the active strain gage or gages are then aligned with their measurement directions in this direction. The orientation of the principal direction does not change due to possible superimposed bending loads along the longitudinal axis, since the bending stresses also act in the longitudinal direction. However, supplementary torsion loads do affect the orientation of the principal directions so that superimposed torsion loads are not included in the application examples presented here.

Example: Load measurements on the mast of an ocean-going yacht [10.3]
The objective of the test was the measurement of stresses arising on the 34 m high mast of the 24.3 m long yacht during various weather conditions and sailing manoeuvers with different wind directions relative to the course.

To achieve this, a strain gage with its measurement direction along the axis of the mast was mounted at a measurement cross-section. Four gages were applied at a height of about 15 m on the circumference of the mast, one each located forward, aft and on each side. Strain gages matched to aluminium were used, appropriate to the mast material. A common compensation or dummy gage was arranged for the four active strain gages, as shown in Figure 10.3, on a small plate of the mast material.

This plate was fixed to the mast with X60 cement so that it could assume the temperature of the mast through contact with it, while remaining mechanically unloaded. All strain gages were connected using the three-wire technique. The selection of the measuring point was made via a multipoint unit that coupled the selected active gage with the dummy gage to form a half-bridge. In this manner the strains occurring at the measuring point could be measured, taking note of the gage factor k. The longitudinal stresses could then be found using Hooke's law for the uniaxial stress state, including Young's modulus for the mast material. The superimposed bending stresses were obtained from the differences of the longitudinal stresses. Figure 10.4 shows the circuit diagram for the described configuration.

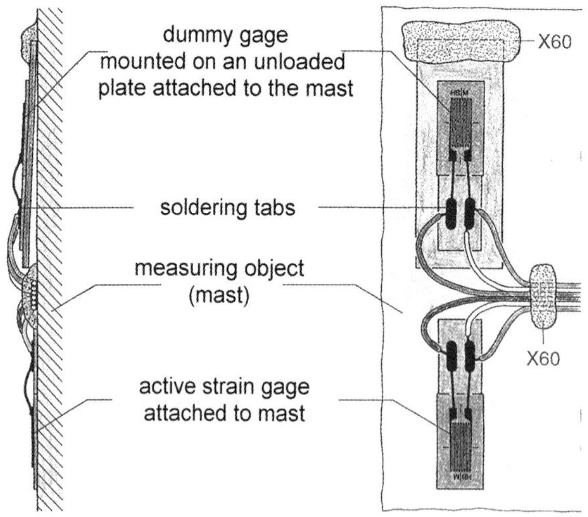

Fig. 10.3 Diagram showing arrangement of a measuring point on the mast with the active strain gage and a dummy gage mounted on an unloaded plate; the three-wire technique is used for the connection

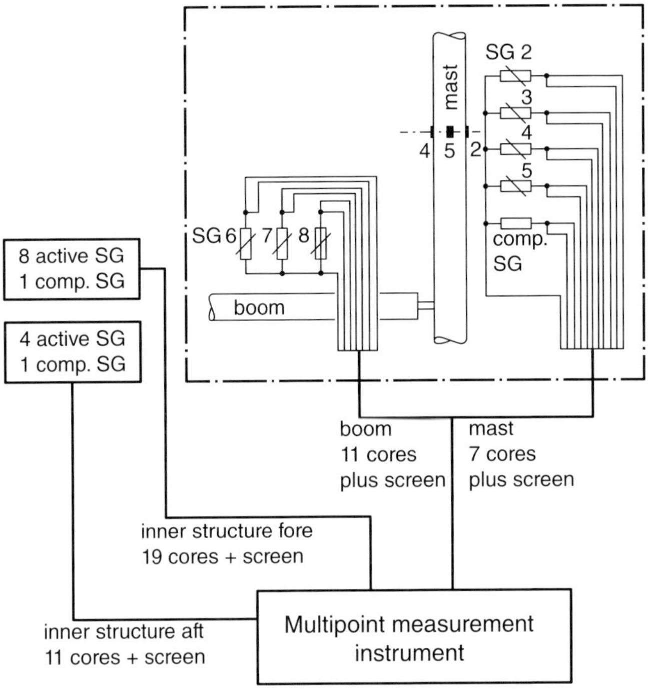

Fig. 10.4 Circuit diagram of the measurement equipment installed on board an ocean-going yacht

Table 10.1 shows an example of a loading case with the simultaneously measured strains on the mast cross-section and the stresses computed from them using a Young's modulus of 70 kN/mm^2 of the mast material (the time lag due to the switching frequency of the measurement system is not significant here due to the quasi-static mast loading).

If these stresses are compared assuming a constant compressive stress which is superimposed with a bending of the mast, then bending components are found towards the aft and the starboard sides. Measurement results from other strain gages mounted at the circumference under test would be needed for a clear indication of the longitudinal section of the mast in which the maximum bending occurs.

Table 10.1 Longitudinal strains and stresses found at the measuring cross-section on the mast for one loading case

SG position on mast	strain $\times 10^{-6}$	stress in N/mm^2
forward	−246	−17.2
port	−340	−23.8
aft	−324	−22.7
starboard	−241	−16.9

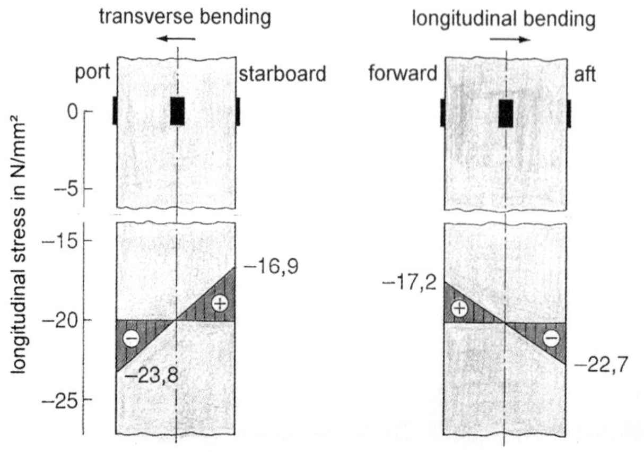

Fig. 10.5 Superimposition of longitudinal and bending stresses in the transverse and longitudinal directions as found from the measurement results given in Table 10.1

Figure 10.5 shows diagrammatically, on the transverse and longitudinal vertical sections of the mast, the superimposition of longitudinal and bending stresses found from the measurement results.

Example: Measurement of the horizontal longitudinal forces on the rollers of a belt-filter press in ore processing [10.4]

The measurement objective was the acquisition of the horizontal forces which the pressing rollers of a 22 t belt-filter press exert on their bearings and therefore on the machine frame. This data was needed to optimize the design of the machine. The required measurements had to be undertaken in actual press operation under very harsh ambient conditions (forced water, dust, electrical interference, high sound levels). To solve the problem, one of the bearings was placed as shown in Figure 10.6

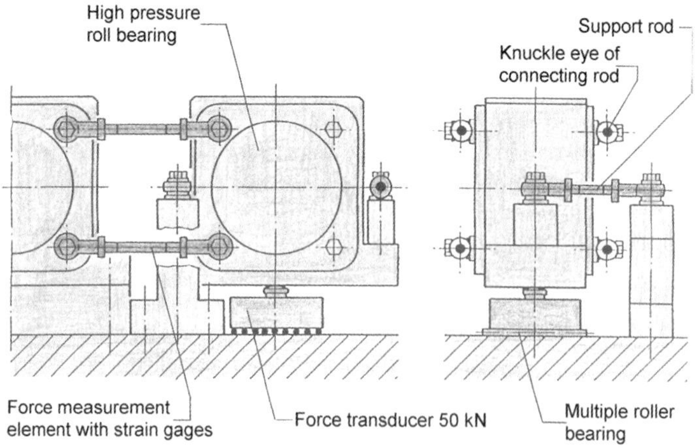

Fig. 10.6 Support bars, which act as measuring elements, on a press-roller bearing [10.4]

on a force transducer to measure the vertical force and supported using link rods with knuckle eyes at the side. The measurement of the supporting side forces took place using strain gages mounted on the support bars and it is described here in more detail. The support bars were designed for a maximum load of 10 kN.

As a practical definition, it was decided that with loading at maximum load the strain producing the measurement signal should be of the order of 1000 µm/m. This means that with modulus of elasticity for the bar material of $E = 208\ 000\,\mathrm{N/mm^2}$, a longitudinal stress in the bar of $\sigma = 208\,\mathrm{N/mm^2}$ arises. With $A = F/\sigma$ this gives a measurement cross-section of $A = 48\,\mathrm{mm^2}$, that is a diameter of $d = 7.8\,\mathrm{mm}$ for a circular cross-section. A bar diameter of 8 mm was chosen. Using this value a bar strain of 956 µm/m is calculated for a load of 10 kN.

The measurement arrangement was setup as a full-bridge circuit with four active strain gages, two mounted in the longitudinal direction and two in the transverse direction. To simplify the installation, strain gages with two measuring grids at right-angles to one another were used (HBM type XY11 3/350). The short grid length of 3 mm was suitable for the restricted space available. The longitudinally aligned gages measure the full longitudinal strain and the transverse gages acquire the transverse strain caused by the longitudinal strain on the bar. The transverse strain occurs with an opposite arithmetical sign.

Figure 10.7 shows a measuring element, a sketch of the mounted strain gage and the bridge circuit diagram. The two longitudinally aligned strain gages are wired in opposite arms of the full-bridge, as are the two transverse gages. With this

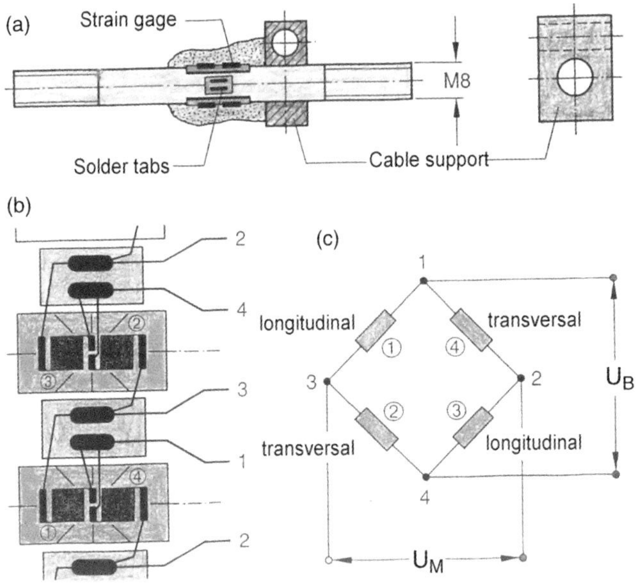

Fig. 10.7 Bar-shaped measuring element for measuring the **vertical** forces on a press-roller bearing; a) measuring element; b) mounted strain gage; c) bridge circuit

arrangement the absolute magnitudes of the strains from the four strain gages are added to form the bridge output signal. Since the negative transverse strains in neighbouring arms of the bridge combine with the strain gages measuring the longitudinal strains, their signals are included with a reversed sign, that is a positive sign, in the output signal.

The full-bridge circuit is self-temperature compensating. Due to the arrangement of the strain gages, measuring points at opposite positions on the circumference of the bar compensate for any bending stresses that arise. This is because they produce strains of the same magnitude but opposite sign in the two longitudinally aligned strain gages which are in opposing arms of the bridge.

In order to avoid bending stresses in the bars from the outset, the bars are fitted with knuckle eyes at their ends. The bar which is designed as a measuring element has a right-hand and a left-hand thread so that position tolerances can be easily compensated during assembly. Due to the harsh ambient conditions a cable connector could not be used and a permanent cable connection was provided via a cable cleat which provided strain relief for the cable. The asymmetry in the bridge circuit due to the internal wiring was able to be balanced out without any problem using the resistive balancing device in the following measuring amplifier.

With the selected strain gage configuration on the bar the bridge output signal V_M/V_B is

$$\frac{V_M}{V_B} = \frac{k}{4}(\varepsilon_l + \nu\varepsilon_l + \varepsilon_l + \nu\varepsilon_l) \tag{10.1}$$

$$= \frac{k}{4}\varepsilon_l(2 + 2\nu) \tag{10.2}$$

The material of the measuring element has a Poisson's ratio of $\nu = 0.29$ so that the circuit bridge factor is $(2 + 2\nu) = 2.58$. The strain gages that were used were taken from a production batch with a gage factor of $k = 2.1$. Eq. (10.2) then becomes

$$\frac{V_M}{V_B} = \frac{2.1}{4} \cdot \varepsilon_l \cdot 2.58 \tag{10.3}$$

$$\frac{V_M}{V_B} = 1.3545 \cdot \varepsilon_l \tag{10.4}$$

If the longitudinal strain $\varepsilon_l = 956\,\mu\text{m/m}$, then the calculated sensitivity of the measuring element is obtained with

$$\frac{V_M}{V_B} = 1.2949 \frac{\text{mV/V}}{10\,\text{kN}} \tag{10.5}$$

This contains some uncertainties which are mainly caused by the figures used for the modulus of elasticity E, Poisson's Ratio ν, the gage factor tolerance of the strain gages used and the tolerances with which the measuring bar was made.

Fig. 10.8 Measuring element exposed to harsh ambient conditions on the bearing of a belt-filter press [10.4]

To obtain reliable measurements, this type of measuring element is calibrated in a testing machine to find the actual sensitivity. If this calibration is carried out with the same measuring amplifier as is later used for the actual measurements, then the force can be displayed in the correct figures by adjusting the measurement range on the amplifier during the calibration. However, it must be ensured that the set amplifier measurement range is not changed after this calibration unless a reference signal (e.g. a control signal) is available at the required time to reproduce the setting. Figure 10.8 is a photograph of the measurement arrangement on a roller bearing taken during press operation.

10.2.2 Measurement objects subject to bending loads

10.2.2.1 Initial consideration

Direct load measurement with strain gages is often the preferred method used on components subject to bending loads, because the bending on the surfaces of the component produces both positive and negative strains of almost equal magnitude. With the formation of the signal from strain gages in a Wheatstone bridge configuration, this property is very useful since strains occurring in neighbouring arms of the bridge are included in the signal as the difference. Therefore, if strains arising in neighbouring arms of the bridge have opposite signs, then their magnitudes add in the bridge output signal. By mounting strain gages on the tension and compression sides of a bending beam and with the wiring of these gages in neighbouring bridge arms there is the additional advantage of compensation for longitudinal loads in the beam. Longitudinal loads, tensile or compressive, which are superimposed on the bending loads, cause strains of the same sign in the strain gages and have the same magnitude. Their signals mutually compensate in neighbouring arms of the bridge.

10.2.2.2 Bending beam clamped at one end

Design of a bending element for measuring the drive torque on the camshaft of a six-cylinder diesel engine [10.5]
The objective of the measurement task is the determination of the curve for drive torque on the camshaft of a six-cylinder engine. This data was needed for finding the losses due to friction. Temperature changes between 40 °C and 80 °C were to be expected during the measurements. Torques between −15 Nm and +15 Nm at speeds of between 100 rpm and 2000 rpm were also to be expected on the camshaft drive.

For the measurements the original toothed-belt pulley used for the camshaft drive was substituted by a measuring wheel with bendable spokes. The torque transfer caused twisting of the toothed ring holder relative to the hub. The twisting is proportional to the torque.

Two bending springs each fitted with two strain gages were positioned diametrically opposed on the wheel for the measurement of the displacement of the outer ring with respect to the hub. These bending elements were located as measuring elements between the hub and the toothed ring holder as shown in Fig. 10.9.

The bending springs are screwed to the hub. The outer ring carries the adjustment screws which transfer the twisting of the toothed ring relative to the hub to the springs. These screws enable the pretensioning on the springs to be set. By using different adaptor rings on the hub the free length of the spokes could be shortened and the spring stiffness changed to alter the resonant frequency of the measuring wheel. Without an adaptor ring the measuring wheel has a resonant frequency of 95.8 Hz.

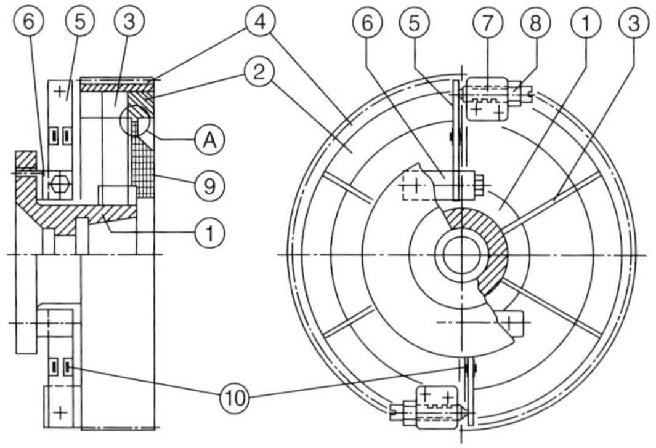

(1) hub
(2) gear carrier
(3) spoke
(4) toothed wheel rim
(5) measuring spring with SG
(6) clamping device
(7) holder for adjusting screw
(8) adjusting screw
(9) fitting flange
(10) strain gage

Fig. 10.9 Measuring wheel for finding the torque on the camshaft drive with two bending springs for the measurement of the twisting of the toothed ring relative to the hub in proportion to the torque [10.5]

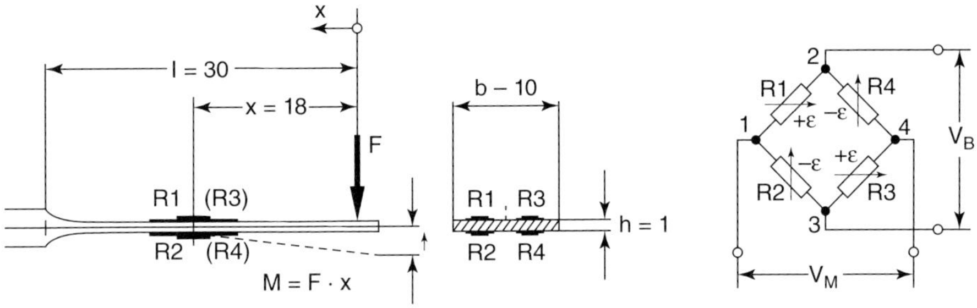

Fig. 10.10 Arrangement of a measuring spring with a strain gage full-bridge circuit

Bar material with a cross-section of 1 mm × 10 mm in 50CrV4 with an apparent yield point of $R_{p0.2} = 1175\,\text{N/mm}^2$ was available as a working material for making the measuring elements. With this cross-section and the dimensions of the measuring wheel as shown in Figure 10.10 a free spring length of 30 mm was produced.

The bending moment M_b produced at point x by the transverse force F is $M_b = F \cdot x$ where x is the lever length for the acting force. The maximum bending moment occurs at the clamping point. The bending stress σ_b in the spring surface is

$$\sigma_b = \pm\frac{M_b}{W_b} = \pm\frac{F_x}{W_b} \tag{10.6}$$

where W_b is the section modulus of the bar cross-section under bending. The arithmetical sign $\pm$ is needed because the bending stresses on one side of the bar occur as tensile stresses, but as compressive stresses on the other side. For the rectangular cross-section under consideration

$$W_b = \frac{b \cdot h^2}{6} \tag{10.7}$$

Equation (10.6) then becomes

$$\sigma_b = \pm\frac{6Fx}{bh^2} \tag{10.8}$$

If 50% of the apparent yield point is set here for σ_b as the upper limit for a spring load, the free spring length of 30 mm for x together with the other values of the spring shape to give F, then the upper limit for the permissible force F is obtained:

$$F = \pm\frac{\sigma_b \cdot bh^2}{6x} \tag{10.9}$$

$F = \pm 32.6\,\text{N}$

The deflection f at the end of the spring caused by this force is given by the relationship

$$f = \frac{Fl^3}{3EI} \tag{10.10}$$

where E is the modulus of elasticity of the spring material at 210 000 N/mm^2 and I is the moment of inertia for the rectangular cross-section, that is $I = bh^3/12$. If the figures are substituted, then at the point of action for the force F the deflection f is 1.68 mm.

Now the output signal for a full-bridge circuit formed from four active strain gages on the spring as shown in Figure 10.10 is estimated. If Hooke's law for the uniaxial stress state (which is present here with pure bending) is introduced as $\sigma_b = E\varepsilon$ into Eq. (10.6), then rearranging to give the strain ε

$$\varepsilon = \pm \frac{Fx}{EW_b} \tag{10.11}$$

If $x = 18$ mm and the other figures that have already been given are substituted, then strain ε at the measuring point is $\pm 1676\,\mu$m/m. As shown in Figure 10.10, the four strain gages in the full-bridge circuit are arranged in the longitudinal direction on the spring, so that two gages are installed on one side and the two others are mounted exactly opposite on the other side. The bridge is wired such that the gages on one side are located in opposite arms of the bridge, giving the sum of the four separate signals as the output signal for the circuit. Using the designations given in Figure 10.10 for the bridge circuit

$$\frac{V_M}{V_B} = \frac{k}{4}(\varepsilon_1 - \varepsilon_2 + \varepsilon_3 - \varepsilon_4) \tag{10.12}$$

where

$$\varepsilon_1 = -\varepsilon_2 = \varepsilon_3 = -\varepsilon_4 \tag{10.13}$$

If Eq. (10.13) is substituted into (10.12), this gives

$$\frac{V_M}{V_B} = k\varepsilon_1 \tag{10.14}$$

The strain gages that are used all originate from the same production batch and have the gage factor $k = 2.05$. If this value is substituted with the strain already calculated for the maximum load $\varepsilon = \pm 1676\,\mu$m/m into Eq. (10.14), then the effect of the maximum load on the electrical signal is obtained, $V_M/V_B = 0.003\,435$, that is $V_M/V_B = 3.445$ mV/V. When stating the sensitivity of the measuring element, the force or the deflection can be used as a reference. If the signal is required as a measure of the deflection, then the computed signal is referred to the maximum deflection $f = 1.68$ mm and the computed sensitivity of 2.05 (mV/V)/mm is obtained for the measuring element. In carrying out the test, the full-bridges of both measuring elements on the measuring wheel were wired in parallel. The mean of the two separate bridges formed the output signal of the parallel circuit.

When complete, the measuring wheel was calibrated with known torques. To do this, a spring-steel band was placed around the wheel which was mounted on a horizontally aligned cone. The band was loaded on both ends with weights so that the total force corresponded to the transverse force due to the toothed-belt

(a) (b)

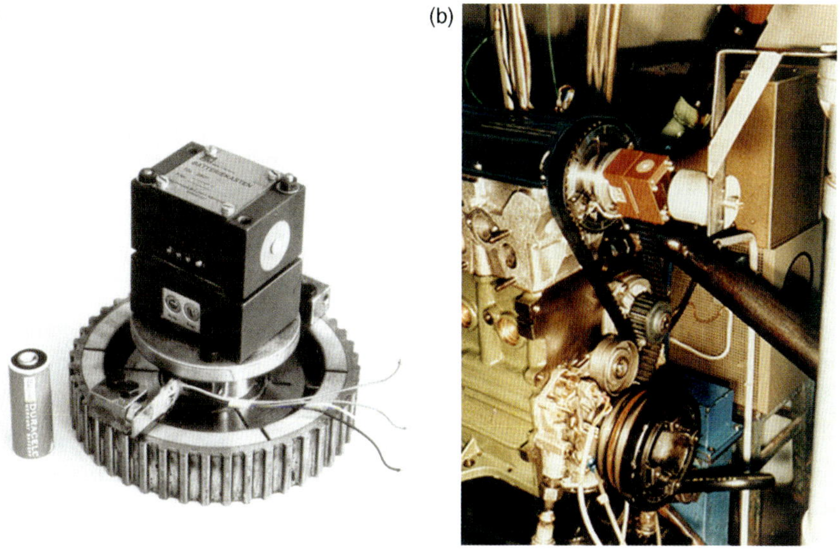

Fig. 10.11 The measuring wheel fitted with two bending springs a) wheel ready for mounting with the parts for non-contact signal telemetry joined to the hub via a flange b) wheel mounted on the engine camshaft

drive. The torques were then produced so that weights could be transferred from one side of the band to the other. This method gave a constant transverse force for different torques. The transmission of the measuring signal from the rotating wheel to the stationary part of the measurement equipment was carried out using non-contact signal telemetry. Figure 10.11 shows the measuring wheel ready for mounting with the signal telemetry parts joined to the hub via a flange and also the wheel installed on the engine camshaft. Figure 10.12 shows an example of a torque curve in dependence of the angle of rotation of the camshaft using the described set up.

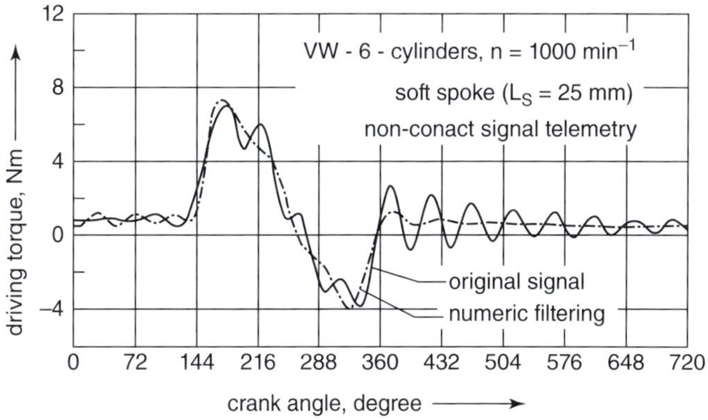

Fig. 10.12 Trace of the camshaft driving torque measured with the described measuring wheel

10.2.2.3 Bending beam clamped at both ends

Example: design of a measuring cage for torque measurement of stepper motors [10.6] During the course of a test-rig development the driving torque of a small stepper motor used in office machines was to be measured in relation to the rotated angle. To avoid signal telemetry from the rotating motor shaft and in order to be able to interchange the stepper motors under test without affecting the measurement arrangement, the torque measurement device shown in Figure 10.13 was designed in the form of a measuring cage with bending beams for measuring the reaction torque on the motor stator.

The reaction moment is transferred from the motor stator to a mounting flange by four bending beams clamped at both ends in the cage. Strain gages, which acquire the strains resulting from the bending of the beams, are mounted on the beams. These strains are proportional to the torque to be measured.

The distribution of the bending moment in a bending beam clamped at both ends is shown in Figure 10.14. Here the bending moment at the centre of the beam, and therefore also the strain, is equal to zero. To obtain a large strain signal the strain gages must be mounted as close as possible to the ends of the beam, because the bending moment and hence the strain reach their maximum values at the points of clamping.

The measuring cage should be as stiff as possible due to the required high resonant frequency and to prevent degrading the measurement of the rotated angle. But it must be compliant enough to enable sufficiently large deformations to enable signals to be produced by the strain gages. A maximum torque of 4000 N/mm was specified for the design of the measuring cage. During the dimensioning a maximum strain of 1000 μm/m was assumed at the clamping points. The radius r of the point of action for the forces F on the bending beams was defined as $r = 44$ mm to suit the shape of the stepper motors to be tested.

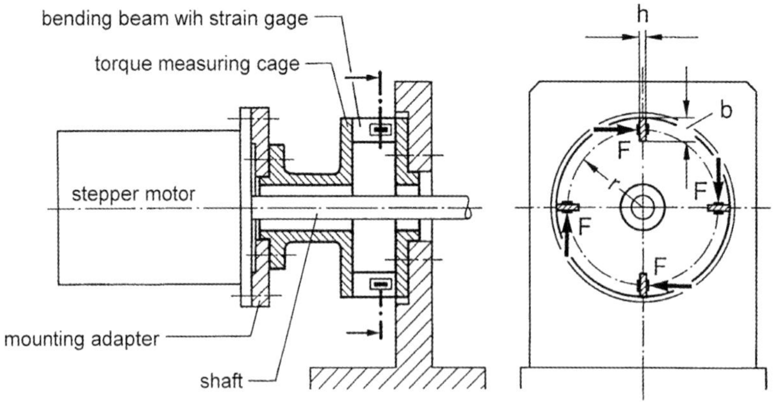

Fig. 10.13 Measurement of motor torque using measuring cage with bending beams and strain gages

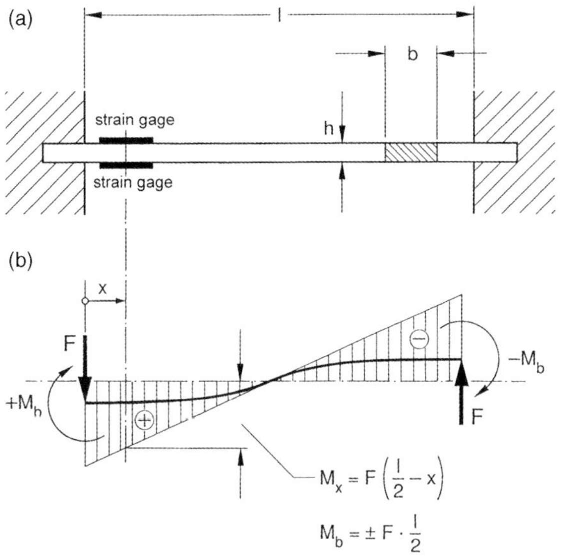

Fig. 10.14 Diagram of a bending beam in the measuring cage a) mounting arrangement with double-ended clamping b) applied loads and bending moment curve

As indicated in Figure 10.12 the quantity M_t can be resolved into four forces F, which act on the bending beams with the leverage r:

$$F = \frac{M_t}{4r} \tag{10.15}$$

The bending moment M_b at the clamping point of the beam is [10.7]

$$M_b = \pm\frac{F \cdot l}{2} = \pm\frac{M_t \cdot l}{8r} \tag{10.16}$$

The strain in the beam surface can be found via the bending strain which is given by the bending moment M_b and the section modulus under bending W_b of the beam cross-section.

$$\pm\sigma_b = \frac{M_b}{W_b} = \frac{M_b \cdot l}{8rW_b} \tag{10.17}$$

As can be seen from Figure 10.12, bending beams with rectangular cross-section, thickness h and width b are used. Their section modulus W_b is

$$W_b = \frac{bh^2}{6} \tag{10.18}$$

Therefore the bending stress is

$$\pm\sigma_b = \frac{3M_t l}{4rbh^2} \tag{10.19}$$

Using Hooke's law $\sigma_b = \varepsilon E$ for the uniaxial stress state the strain at the clamping point is

$$\pm\varepsilon = \frac{3M_t l}{4Erbh^2} \tag{10.20}$$

Here the known values are substituted and the unknown variables found. With $M_t = 4000$ Nmm, $\varepsilon = 1000\,\mu\text{m/m}$, $E = 210\,000\,\text{N/mm}^2$ (Young's modulus for C 60 steel) and $r = 44$ mm, a guide value for the bending beam dimensioning is

$$\frac{l}{bh^2} = 3.08 \tag{10.21}$$

Based on the working material available and the geometrical shape, the following were used: $h = 1$ mm and $l = 34$ mm. Solving for b then gives $b = 11.04$ mm. For dimensioning, b was specified at 12 mm. Calculating again with $b = 12$ mm, $h = 1$ mm and $l = 34$ mm gives a strain $\varepsilon = \pm 920\,\mu\text{m/m}$ for the maximum torque of $M_t = 4000$ Nmm at the clamping points.

As indicated in Figure 10.14, two strain gages (6 mm grid length, 120 Ω nominal resistance) were mounted on each bending beam. The centre of the grid was positioned at the point $x = 5$ mm. Using interpolation of the strain between the clamping point ($x = 0$; $\varepsilon = 920\,\mu\text{m/m}$) and the centre of the beam ($x = l/2$; $\varepsilon = 0$), the strain occurring at this point during maximum loading was $\varepsilon_x = 650\,\mu\text{m/m}$.

There are now two possible configurations. To configure two full-bridges from the total of eight mounted strain gages and to wire them in parallel to form the mean or to configure *one* full-bridge with two strain gages with their outputs in the same sense in each of the bridge arms. As shown in Figure 10.15, the latter method was chosen.

The strain gages located in one arm of the bridge are both subject to the same magnitude of strain and therefore also experience the same change in resistance. Twice the change of resistance ΔR occurs in each arm of the bridge. This change must be referred to the sum of the resistances in the strain gages which are wired in series in the arm of the bridge. This means that a relative change of $2\Delta R/2R$ is obtained in each arm, that is the same value as for a bridge circuit with only one strain gage per arm of the bridge. However, it must be noted here that the nominal

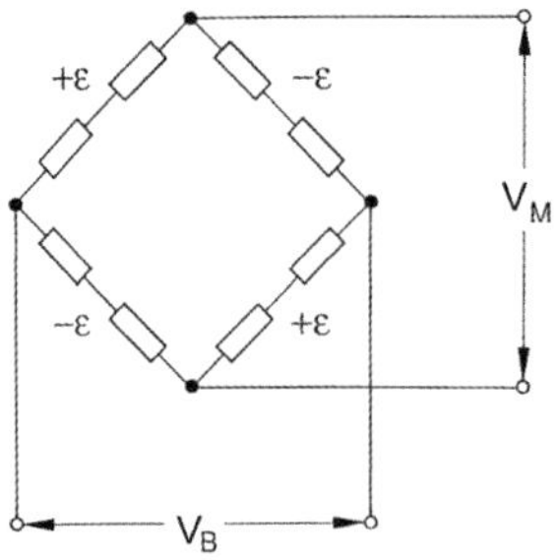

Fig. 10.15 Full bridge circuit consisting of eight identical strain gages with two active strain gages in each arm of the bridge

resistance of the bridge circuit (both the excitation circuit and the measurement circuit) is double the strain gage resistance, for example 240 Ω here.

The question of whether it is better to use the described series connection of two strain gages per arm of the bridge or two separate full-bridges, each formed from four strain gages and wired in parallel, depends on the case under consideration. The parallel configuration reduces the total resistance of the two parallel connected bridges to half the nominal resistance of the strain gages used. This can have consequences for the following signal processing, because there are finite limits on the input resistance of a measuring amplifier. However, the parallel circuit has the advantage that if defects or faults occur, the two full-bridges can each be tested individually, simplifying fault diagnosis.

In the example under consideration here the strain at the measuring points is $\varepsilon = 650\,\mu\text{m/m}$. The bridge equation for the full-bridge circuit

$$\frac{V_M}{V_B} = \frac{k}{4}(\varepsilon_1 - \varepsilon_2 + \varepsilon_3 - \varepsilon_4) \tag{10.22}$$

with $k = 2.1$ gives a bridge signal of

$$\frac{V_M}{V_B} = k \cdot \varepsilon = 1365 \cdot 10^{-6} = 1.365\frac{\text{mV}}{\text{V}} \tag{10.23}$$

On loading the measuring cage with the maximum torque of $M_t = 4000$ Nmm, the installed strain gages supply a calculated output signal of 1.365 mV/V. Calibration of the measurement device for practical applications is needed to find the actual relationship between the measurement signal and the applied torque. The calculation includes uncertainties mainly due to the mechanical tolerances in producing the cage, the tolerance on the k factor, inaccuracies in the positioning of the strain gages and the deviation of the actual Young's modulus of the bar material from the value used. Figure 10.16 shows a photograph of the torque measuring cage.

Fig. 10.16 Photograph of the torque measuring cage with bending beams for the measurement of stepper motor driving torque

10.2.2.4 Bending beam freely supported at both ends

Example: Measuring rowlock for acquiring the force/time characteristic when rowing [10.8]

When analysing the processes taking place during the sport of rowing, it is particularly important to find the time-related curve of the force with which the rower is propelling the boat. The force resulting from the pulling force exerted by the rower and the force of the water acting on the oar blade is represented by the force subjected by the oar on the rowlock. This force can be resolved into force components, one in the longitudinal direction of the oar and the other a transverse force F. The component F is especially suitable as a measurement quantity for the tests. At the points of time at which the oar axis is perpendicular to the longitudinal axis of the boat it is identical to the propelling force of the boat. It can also be relatively easily measured because it is transferred to the rowlock in a direction perpendicular to the oar axis. To achieve this, a bending beam fitted with strain gages was built into a normal production rowlock. The force of the oar F acted in the centre of the beam which was supported at both ends in the rowlock. Figure 10.17a shows a drawing of the beam which is loosely supported at one end and firmly supported at the other end. The force F to be measured is transferred to the beam through a pressure plate. This force is converted to a proportional electrical signal by a strain gage full-bridge circuit mounted on the beam. Figure 10.17b shows a photograph of a practical version of the beam.

During the dimensioning of the measuring beam a maximum rowing force of 1500 N was assumed. Due to the restricted space in the rowlock the maximum breadth and span of the bending beam that could be obtained were 20 mm and 48 mm respectively. If, taking into consideration the loading of the beam, a beam thickness of 4 mm is specified, then a bending stress of 337.5 N/mm^2 occurs at the centre of the beam, as the following calculation shows:

Maximum bending stress (10.24)

Maximum bending moment $$M_{b\,\max} = \frac{Fl}{4} \qquad (10.25)$$

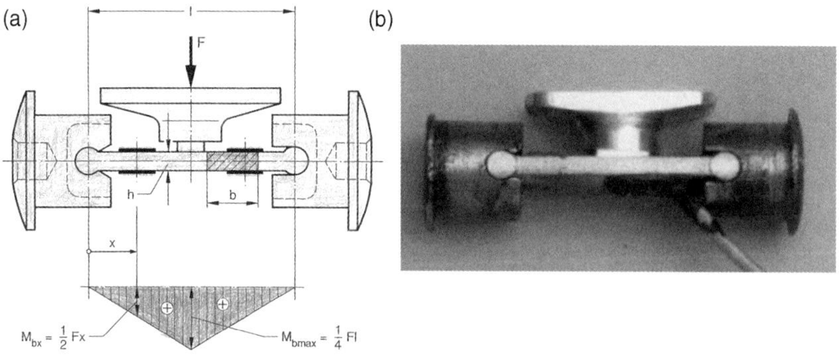

Fig. 10.17 Measuring beam for mounting in a rowlock used to measure the rowing force; a) diagram showing construction, b) photograph of measuring beam

(10.25 in 10.24)

$$\sigma_{b\,\max} = \frac{Fl}{4W_b} \tag{10.26}$$

$$\text{Section modulus} \quad W_b \frac{bh^2}{6} \tag{10.27}$$

(10.27 in 10.26)

$$\sigma_{b\,\max} = \frac{6Fl}{4bh^2} \tag{10.28}$$

Using the stated figures the maximum bending stress in the centre of the beam becomes

$$\sigma_{b\,\max} = 337.5 \frac{\text{N}}{\text{mm}^2} \tag{10.29}$$

This is a relatively high value considering that the load is an alternating one. This stress could be reduced by selecting a shorter beam span. However, shortening the span would mean that there was less space for mounting the strain gages and a thicker beam would cause problems with the design of the beam supports. Steel is the only suitable material with strength compatible with this type of load, so tempered steel or spring steel could be used for example. The measuring beam was produced using the stated figures and the strain gages were mounted so that the centres of the measuring grids were positioned at a distance x = 12 mm from the points of beam support. With Young's modulus of $E = 206\,000\,\text{N/mm}^2$ for the beam material the calculated strain on the strain gages at the point x is given by the following calculation.

The bending moment at the point x is

$$M_{bx} = \frac{2x}{l} M_{b\,\max} \tag{10.30}$$

and the corresponding stress at the point x is

$$\sigma_{bx} = \frac{2x}{l} \sigma_{b\,\max} \tag{10.31}$$

Using Hooke's law for the uniaxial stress state, $\varepsilon_x = \sigma_{bx}/E$, the strain at the point x in the following is produced after inserting the figures

$$\varepsilon_x = \frac{2x}{El} \sigma_{b\,\max} \tag{10.32}$$

$$\varepsilon_x = 819\,\mu\text{m/m} \tag{10.33}$$

If the four strain gages are connected to form a full-bridge, so that the strain gages loaded in the same sense are located in opposite arms of the bridge, then the calculated output signal for loading the beam with the specified maximum load of

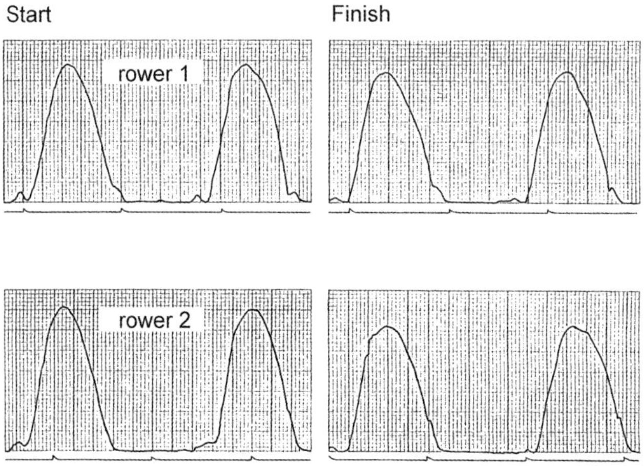

Fig. 10.18 Example of a measurement trace obtained with the test rowlock: change in the 'rower's signature' due to fatigue (two-minute rowing spell, two rowers) [10.8]

1500 N becomes

$$\frac{V_M}{V_B} = \frac{k}{4}(\varepsilon_1 - \varepsilon_2 + \varepsilon_3 - \varepsilon_4) \qquad (10.34)$$

With $k = 2.1$ this produces a figure of $V_M/V_B = 1.719\,\mathrm{mV/V}$

This gives a calculated relationship between the output signal and the measurement quantity. However, calibration of the measurement arrangement is necessary for correct evaluation of the measurement signals acquired during the tests because the calculated result includes uncertainties mainly due to the production tolerances of the measuring beam and its supports, the Young's modulus of the beam material, the k factor tolerance and the positioning accuracy of the strain gages. The calibration can be carried out by applying forces of a known magnitude, for example in a testing machine.

The cavity required for mounting the measuring element was milled out from the plastic body of the normal production rowlock. The test rowlock in this form proved excellent for the task, producing measurement traces that have been termed 'The rower's signature' [10.8] ('Handschrift des Ruderers'). From the traces of force/time recorded by a pen recorder, a specialist can assess the quality of the movement sequence and the training condition of the rower. An example of a measurement trace produced by a measurement system comprising the test rowlock, a 5 kHz cf. amplifier and a pen recorder is reproduced in Figure 10.18.

10.2.3 Measurement objects subject to torsion loads

Torsion or twisting loads are always associated with a multiaxial stress state. If Mohr's circle for a cylindrical body subjected to a pure torsion load is considered, then it will be realised that the shear stress τ occurring in the twisted cross-section produces principal stresses σ_1 and σ_2 rotated by 45° to the direction of torsion. For pure torsion, both principal stresses have the same magnitude, but opposite signs. The same applies to the principal strains, which are associated with these two principal stresses and which can be measured with strain gages. The fact that the principal strains have opposite signs is very convenient for the configuration of a bridge circuit. These two strains are measured by two strain gages rotated by 90° relative to one another and wired in neighbouring arms of a Wheatstone bridge in which the two signals are added. Strain gage manufacturers supply special strain gages for this application, having two 'Christmas tree' measuring grids. In these gages the two individual measuring grids are positioned close to one another so that with the longitudinal or bending loads that are superimposed on the torsion loads on the measuring object, it can be assumed that the corresponding longitudinal and bending strains affect both gages to the same degree. Since the gages are located in neighbouring arms of the bridge circuit, the associated resistance changes cancel one another out, producing a signal that represents only the torsion load.

Example: Measurement of torque on a hollow shaft of an agitator ball mill for paint production [10.9]
The measurement task consists of the determination of the stirring power of an agitator ball mill for paint production under various milling conditions. The power can be found by measuring the torque and the speed of the shaft. The measurement of the speed is no problem and in this case is carried out by inductive methods. The drive shaft was designed as a hollow shaft with an additional internal pipe for water cooling. This arrangement precluded the installation of a normal commercially available torque transducer.

The only feasible method was the direct installation of strain gages on the drive shaft. An accessible section of shaft approximately 25 cm in length was available for this. Strain gages specially designed for the measurement of torsion loads were bonded at diametrically opposite points on the shaft and wired in a full-bridge configuration. The signal transmission from the rotating shaft was implemented using a slip-ring transmission unit mounted on the upper end of the shaft. The measurement leads from the bridge circuit to the slip-ring unit had to be routed in a Teflon tube within the hollow shaft which was drilled at two points and fitted with suitable seals. The basic arrangement of the measuring points and the slip-ring transmission unit on the drive shaft of the mill is illustrated in Figure 10.19.

The strain gage arrangement should be designed for a maximum torque $M_t = 400$ Nm. The internal and external diameters of the hollow shaft on which the strain gages are bonded were 14 mm and 24 mm respectively. The strains to be measured by the strain gages are caused by the principal stresses resulting from the torsion. These stresses arise at 45° to the axis of rotation. With pure twisting loads (without superimposed longitudinal or bending forces) the principal stresses at 45° to the axis of rotation have the same magnitude as the shear stresses τ produced by the torque M_t:

$$\tau = \frac{M_t}{W_p} \tag{10.35}$$

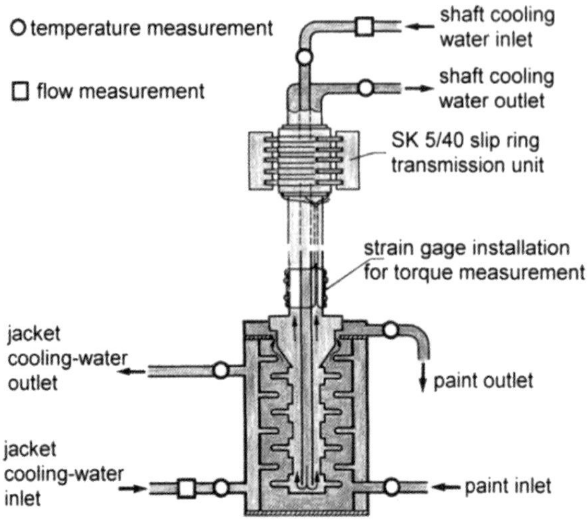

Fig. 10.19 Location of the strain gage arrangement and other measurement points on the drive shaft of a stirred ball mill

Where W_p is the section modulus under twisting for the circular cross-section under consideration:

$$W_p = \frac{\pi}{16 d_a}\left(d_a^4 - d_i^4\right) \tag{10.36}$$

With $d_a = 24$ mm and $d_i = 14$ mm this gives $W_p = 2400$ mm^3. If this sectional modulus is substituted in Eq. (10.35) with the maximum torque $M_t = 400\,000$ Nmm, the shear stress τ becomes

$$\tau = 167\ \mathrm{N/mm^2} \tag{10.37}$$

With $\tau = \sigma_1 = -\sigma_2$ and Hooke's law for the biaxial stress state the following is obtained for the strains with the modulus of elasticity: $E = 206\,000$ N/mm^2 and Poisson's ratio $\nu = 0.3$

$$\varepsilon_1 = \frac{1}{E}(\sigma_1 - \nu\sigma_2) = \frac{\sigma_1}{E}(1 + \nu) \tag{10.38}$$

$$\varepsilon_2 = -\varepsilon_1 \tag{10.39}$$

$\varepsilon_1 = -\varepsilon_2 = 1054\ \mu\mathrm{m/m}$

If the strain gages are connected to form a full-bridge, the bridge output signal is given by

$$\frac{V_M}{V_B} = \frac{k}{4}(\varepsilon_1 - \varepsilon_2 + \varepsilon_3 - \varepsilon_4) \tag{10.40}$$

and with $\varepsilon_1 = -\varepsilon_2 = \varepsilon_3 = -\varepsilon_4$

$$\frac{V_M}{V_B} = k\varepsilon_1 \tag{10.41}$$

Putting $k = 2.06$ $\frac{V_M}{V_B} = 2171 \times 10^{-6} = 2.171\ \text{mV/V}$

The calculated relationship between the measurement signal and the torque on the shaft has now been found. With the action of a torque of 400 000 Nmm the strain gage arrangement produces an output signal of 2.171 mV/V.

From the beginning of the tests it was apparent that the measuring signal depended on the cooling water temperature. The reason for this temperature effect was found to be the arrangement of the wiring and the solder pads for the connecting leads. Since two strain gages in the full-bridge circuit were located on the shaft circumference, diametrically opposite to each other, two connecting leads must be passed around half the shaft circumference. These leads each form a constituent part of an arm in the full-bridge.

To compensate for the effect of the temperature on these leads, they must be situated in neighbouring bridge arms and exactly the same length.

Figure 10.20a shows the unsuitable circuit with the connecting leads in opposing arms of the bridge. The effect of the temperature dependent resistance changes is

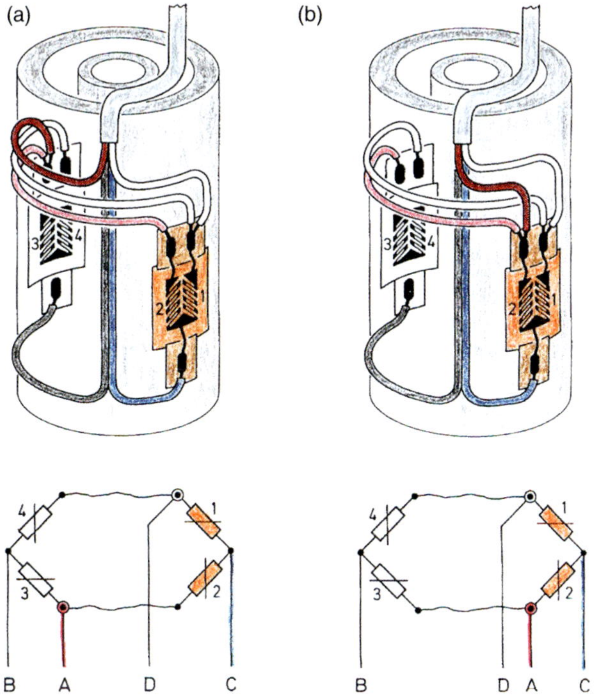

Fig. 10.20 Diagram of the arrangement of the strain gage full-bridge circuit on the drive shaft of the mill; a) unsuitable circuit with long connection leads in opposing arms of the bridge; b) better wiring of the long connecting leads in neighbouring arms of the bridge

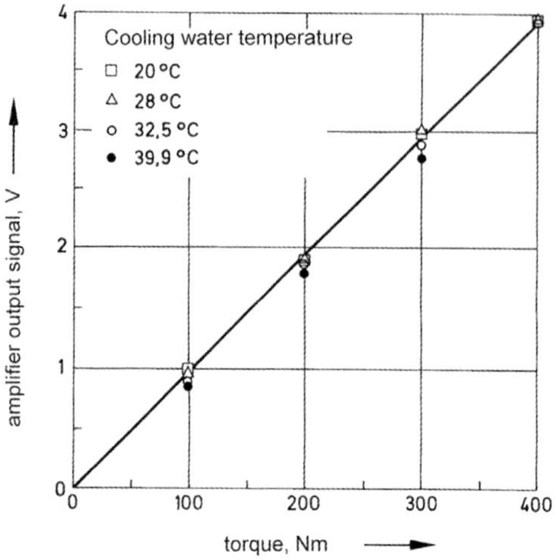

Fig. 10.21 Calibration characteristic for the relationship between the amplifier output signal and the torque on the drive shaft of the mill

added. The correct arrangement of the connecting leads in neighbouring arms of the bridge is shown in Figure 10.20b. Figure 10.21 shows that with the correct circuit the measuring signal is almost independent of the temperature for different torque loads. However, the signal required about two minutes to reach its final value after constant cooling water temperature was achieved, because the thermal transfer to the solder pads and strain gages took different periods of time.

However, for correct measurement of the torque, a calibration of the measurement arrangement with known torque levels is recommended. This calibration was carried out with the shaft removed. The horizontally positioned shaft was clamped at the lower end and loaded with defined moments, free of transverse forces on the upper end through a double lever system. To do this a weight was suspended on one side so that its weight acted downwards and an equal weight pulled the other side of the lever upwards via a deflection roller. The complete measurement system, including the amplifier, was calibrated. The amplifier measurement range selector (sensitivity) enabled the correct numerical relationship between the amplifier output signal and the torque to be established. The calibration values for different cooling water temperatures are reproduced in Figure 10.21.

10.2.4 Shear stresses in a beam

Besides the bending stresses that appear in a beam there are shear stresses which are created by the shearing force acting perpendicular to the axis of the beam. The principal normal stresses occur at 45° to the shear stress. The strains caused by the principal normal stresses are measurable by means of strain gages. Therefore the strain gages must be arranged at 45° to the direction of the shear force. The strain gages should be arranged symmetrically as close as possible to the neutral axis where bending stresses are minimum and shear stresses are maximum.

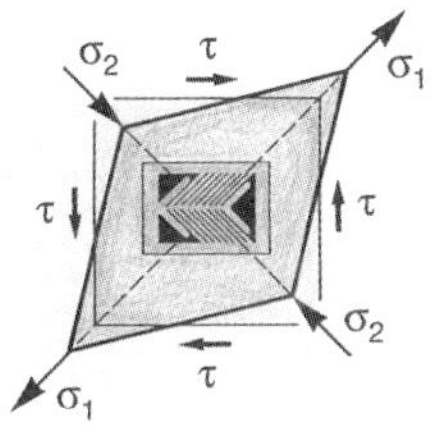

Fig. 10.22 Shear stresses and equivalent principal normal stresses in the state of pure shearing stress

Considering the square element given in Figure 10.22 due to the shear stresses one of the diagonals will be stretched and the other diagonal will be compressed, changing the square into a rhombus as shown. The element shown is in equilibrium under the shearing stresses only. Such a state of stress is called pure shear. This pure shear state is equivalent to the state of stress produced by tension in one direction and an equal compression in the perpendicular direction, as indicated in Figure 10.22.

In the state of pure shearing stress the principal axes of the stress state are at 45° to the longitudinal axis of the beam and the corresponding principal strains are of equal magnitude and opposite sign. To measure these principal strains a pair of strain gages have to be mounted with their measuring directions orientated along the two principal axes. Specially designed strain gages containing two adjusted grids perpendicular to each other are available in order to facilitate the installation of the strain gages.

This effect can be used to design shear beam load cells for force and mass measurement [10.10]. The force to be measured acts as shearing force and produces shear stress. The strains caused by the principal normal stresses are measured with strain gages. The strain gages, arranged at 45° to the direction of the shear force and wired to a Wheatstone bridge circuit, produce a signal directly proportional to the applied force. Shear beam load cells are low-profile load cells that are well suited for application in weighing bridges and platforms. They are simple to mount and they can be used to measure tensile or compressive forces. Advantages are low influence of side loads, excellent linearity, low hysteresis and no influence of loading point position.

Example: Design of a force transducer using the shear-field principle to measure forces on a support bar in an oven at 230 °C [10.11]

Hardening ovens for Rockwool products mainly consist as shown schematically in Figure 10.23 of two chain grids which circulate through an oven. The grids transport the material to be processed. Following the direction of the material flow the synthetic resin is first applied and then hardened in subsequent stages in a section heated by gas. The carrying grids consist of support bars orientated transverse to the direction of movement. The bars are driven by roller chains which circulate to the right and left in the oven.

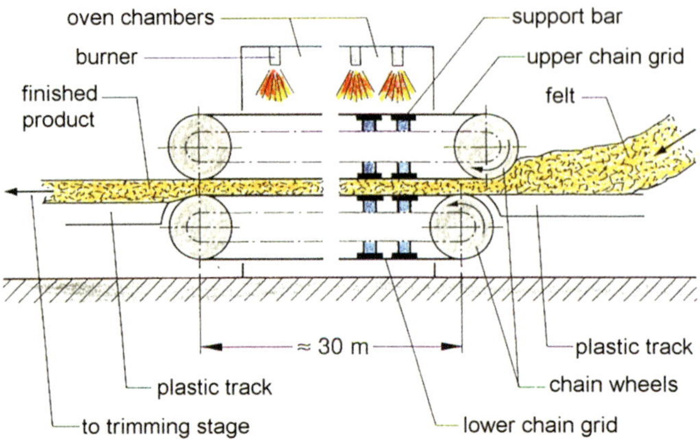

Fig. 10.23 Diagrammatic representation of a hardening oven for Rockwool

Figure 10.24 shows a photograph of the diversion station for the upper grid at the production outlet; the roller chain can be seen to the right of the picture.

The support bars are bolted at their ends to the links of the chain. The support bars form static beam models that are loaded by the product pressure as well as their own weight.

The reason for carrying out the measurements described below was that fractures of the chain links and the ends of the support bars occurred. These fractures led to substantial consequential damage, causing unwanted downtimes. The measurement task was to measure the forces on individual moving parts of the chain grid at about

Fig. 10.24 Diversion point of the upper transport grid at the product outlet; the driven roller chain can be seen on the right

230 °C. The excitation of the transducers and the transmission of the measurement signals were to be implemented by trailing leads.

The requirements were for the design of a robust force measuring element with high bending stiffness to fit the small installation space of about 150 mm × 80 mm × 50 mm. A force transducer was designed according to the shear-field principle for a nominal maximum load of 30 kN. The nominal sensitivity was selected as 1.5 mV/V for reasons of economy. The transducer has to withstand measurements made at 230 °C without suffering damage.

Based on previous experience a basic construction form was chosen as given in Figure 10.25. The transducer has to be secured at one end to a rib, which has been welded between the supports and at the other end to a load introduction block. The dimensions of the spring body can be seen in Figure 10.25. The strain gages were installed at the bottoms of the blind holes, drilled into the spring body from both sides, forming a diaphragm. The cross-section of the spring body at the measuring point shows the profile of an H-girder. A height of 46 mm was chosen.

The shear stress τ at the measuring point can be calculated using the formula [10.12]:

$$\tau = F/A_{web} \tag{10.42}$$

The permissible maximum shear stress has to be compared with the yield point of the material used for the spring body. The yield point of 42CrMo4 is given as >700 N/mm^2. It is recommended to allow a maximum stress in the spring body of about 20% of the yield stress, so we set up $\tau_{max} = 140\ \text{N/mm}^2$. With this it is possible to calculate the required cross-sectional area of the web:

$$A_{\text{web}} = \frac{F}{\tau_{\text{max}}} = \frac{30000\ \text{N}}{140\ \text{N/mm}^2} = 215\ \text{mm}^2 \tag{10.43}$$

From this area we find the width b of the web:

$$b = \frac{A_{\text{web}}}{h} = \frac{215}{46} = 4.66\ \text{mm} \tag{10.44}$$

It has been fixed to 4.6 mm.

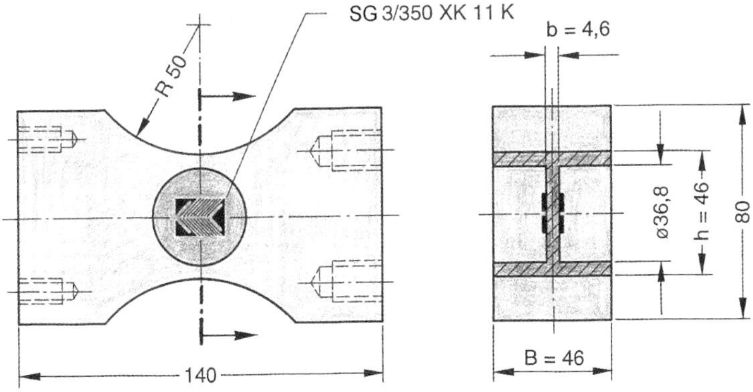

Fig. 10.25 Dimensions of the spring body

Measurements using strain gages at 230 °C require certain compensation measures. The temperature response of the sensitivity is mainly formed from the temperature dependence of the gage factor for the strain gages used and of Young's modulus for the deformation element. A change of the gage factor of about 0.008%/K and a change of Young's modulus by about 0.035% per K are assumed.

The compensation of this change in sensitivity with temperature can be accomplished by inserting a temperature-sensitive resistor in the bridge excitation supply line. When the resistor is of the correct size and temperature sensitivity, it serves to vary the bridge voltage so as to counteract the temperature-dependent changes in the bridge output.

Nickel resistors are well suited for the compensation of the change in sensitivity with temperature. From experience the nickel resistance required for compensation was estimated to about 30 Ω. From the strain gage nominal resistance of 350 Ω and the estimated value of the compensation resistance of 30 Ω a total resistance for the excitation circuit of 380 Ω is given.

The sum of the sensitivity changes from the gage factor and Young's modulus is 0.043%/K using the values mentioned. The resistance of the excitation circuit must therefore also increase by 0.043%/K in order to obtain the desired sensitivity compensation.

Nickel will be used as the material for the compensation resistance. The change in resistance which has to be compensated is 0.043%/K. Nickel has a temperature coefficient of $TK_{NI} = 5 \times 10^{-3}$/K. That means for the circuit with 380 Ω:

$$380\,\Omega \times 0.043\%/\mathrm{K} = 0.163\,\Omega/\mathrm{K} \tag{10.45}$$

In order to provide compensation of the 0.163 Ω/K a particular value of nickel resistance is required:

$$TK_{\mathrm{Ni}} = \frac{\Delta R}{R} = 5 \times 10^{-3}/\mathrm{K} \tag{10.46}$$

$$R = \frac{\Delta R}{TK_{\mathrm{Ni}}} = \frac{0.163}{5 \times 10^{-3}} = 32.6\,\Omega \tag{10.47}$$

For reasons of symmetry, two resistances of 16.3 Ω each are inserted into each of the excitation lines. In this case it was more practical to insert two adjustable nickel foils on a polyimide carrier especially offered by the manufacturers for this purpose. The resistance of these foils is in certain steps adjustable to the required value by cutting off the printed line at a particular point. Two nickel foil resistors which were adjustable to 16 Ω each have been inserted into the two excitation lines.

The required output signal including the reserve for the purpose of adjustment was set to 1.6 mV/V. The total resistance of the excitation circuit is 382 Ω. This effects a change in sensitivity of

$$\frac{E^*}{E} = \frac{R_{\mathrm{SG}}}{R_{\mathrm{SG}} + R_{\mathrm{Comp}}} = \frac{350}{382} = 0.92 \tag{10.48}$$

In Eq. (10.48) E is the sensitivity without voltage drop in the excitation line and E^* is the sensitivity in view of the fact that the resistance has increased. Taking this loss of sensitivity into account, the desired output must be:

$$\frac{1.6\,\text{mV/V}}{0.92} = 1.74\,\text{mV/V} \tag{10.49}$$

A full-bridge strain gage circuit is applied to the shear field:

$$\frac{V_M}{V_B} = \frac{k}{4}(\varepsilon_1 - \varepsilon_2 + \varepsilon_3 - \varepsilon_4) \tag{10.50}$$

For pure shear stress state:

$$\varepsilon = \varepsilon_1 = -\varepsilon_2 = \varepsilon_3 = -\varepsilon_4 \tag{10.51}$$

(10.51) in (10.50):

$$\frac{V_M}{V_B} = k\varepsilon \rightarrow \varepsilon = \frac{1}{k}\frac{V_M}{V_B} \tag{10.52}$$

Putting $k=2$ and $V_M/V_B=1.74$ gives $\varepsilon=870\times10^{-6}$

This result has been obtained by calculation. That means that under the nominal load of 30 kN a strain of $\pm870\times10^{-6}$ will occur in the shear field and will create an output signal of the full-bridge circuit of 1.74 mV/V.

The principal normal stresses $\sigma_1=-\sigma_2$ which are creating the principal strains, $\varepsilon_1=-\varepsilon_2=870\,\mu\text{m/m}$ can be found using Hooke's law for the biaxial stress state

$$\sigma_{1,2} = \frac{E}{1-\nu^2}\left(\varepsilon_{1,2} + \nu\varepsilon_{2,1}\right) \tag{10.53}$$

Introducing Young's modulus of $E=2.09\times10^5$ and Poisson's ratio of $\nu=0.3$ the principal normal stresses will be found to $\sigma_1=-\sigma_2=140\,\text{N/mm}^2$. This value corresponds to the prerequisite that the maximum stress should not exceed 20% of the yield point. With this the dimensions of the spring body given in Figure 10.25 are confirmed.

The arrangement of the strain gages on the shear diaphragm in the centre of the measurement cavity is shown diagrammatically in Figure 10.26a. In Figure 10.26b can be seen how the strain gages are wired to form a full-bridge circuit. It also shows how the temperature compensation resistances are implemented in the excitation supply lines.

For the construction of the force measurement transducer strain gages were used with glass fibre reinforced phenolic resin carriers. According to the manufacturer, these strain gages are capable of measurements at 230 °C.

For practical reasons, temperature compensation of the zero signal and zero balance of the bridge circuit could be ignored. The hot curing method and the short identical internal lead lengths ensure good bridge symmetry. The slight amount of asymmetry that occurred could be compensated without any problem by the amplifier that was used.

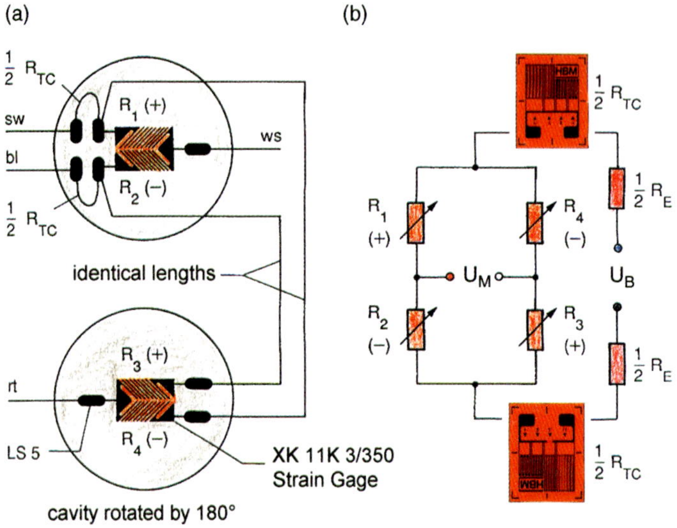

Fig. 10.26 Strain gages with glass-fibre reinforced phenolic resin carrier on the shear diaphragm of the spring body; a) strain gage arrangement, b) wiring diagram

One problem was the electrical connection using high melting point solder, requiring a temperature of 400 °C at the tip of the solder iron. The covering of the measuring points, which had to withstand 230 °C, demanded particular attention. After fitting the compensation resistances the measurement cavity was thinly covered with SG250 covering agent. When the silicone rubber (SG250) had hardened, the cavity was encapsulated with an approximately 3 mm layer of Epple 37 sealant and then hardened without crack formation in the oven in step-by-step thermal cycles up to 250 °C. Figure 10.27 shows the finished force measurement elements, extremely left with the clamping device used for calibration of the elements. The press device for hot curving is in the foreground.

A typical display for an overall test with 7000 measurements is shown in Figure 10.28 where the upper trace shows the force on the bar ends and the lower trace gives the guide-block force.

Fig. 10.27 The finished force measurement elements: extreme left with calibration clamping device, and the press device for hot curing is in the foreground

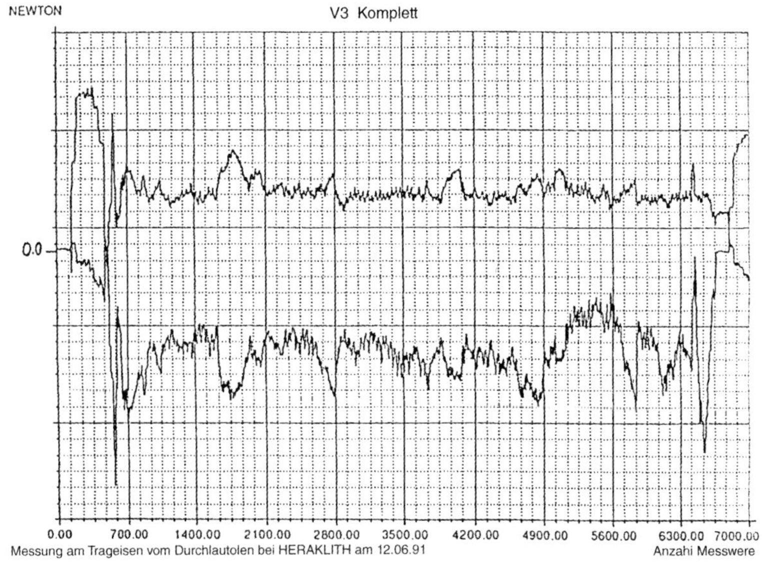

Fig. 10.28 Typical display of the measurement results for a support bar

The solution of this measurement task has not only produced knowledge about the 'inside' of a hardening oven, but also valuable practical information about the temperature compensation of user constructed force transducers in operating regions above 70 °C.

10.3 Stress analysis for unknown principal directions

10.3.1 Example of stress analysis using a 0/60/10° rosette

In order to demonstrate the application of the derived formulae, an example stress analysis will be performed using actual measurement values. The demonstration involves the mounting of a 0/60/120° rosette on a specimen to determine the planar stress state at the measuring point. The first step following the installation of the rosette to the unstressed specimen is to ascertain the reference axis of the rosette and to label the individual measurement axes according the guidelines previously laid out.

Figure 10.29 shows the rosette and relevant notation. A point of practical importance is that the wiring from the rosette should be clearly marked with the same notation as the individual grids, in order to avoid confusion when evaluating the signals from several rosettes.

Assuming a satisfactory and error-free measurement procedure, the following three readings were obtained from the rosette on the loaded specimen:

$$\begin{aligned} \varepsilon_a &= 309 \cdot 10^{-6} \\ \varepsilon_b &= 86 \cdot 10^{-6} \\ \varepsilon_c &= 205 \cdot 10^{-6} \end{aligned} \qquad (10.54)$$

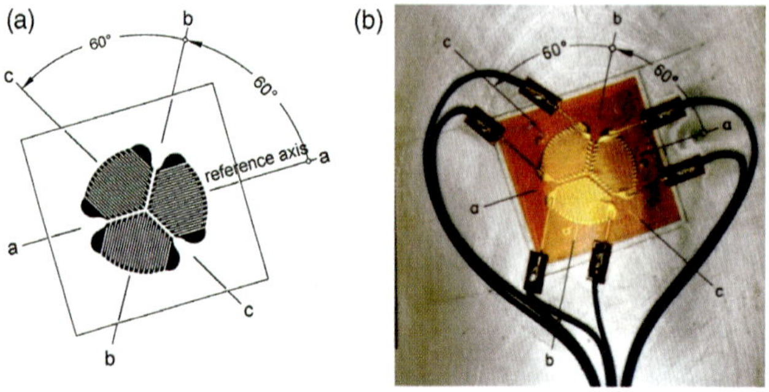

Fig. 10.29 The 0/60/120° rosette a) installed on a measurement object, b) designation of measuring directions

Calculation of the angle 2α for the 0/60/120° rosette is performed by substituting the measurement values into Eq. (9.77):

$$\tan 2\alpha = \frac{\sqrt{3}(86 - 205)}{618 - 86 - 205} = \frac{N}{D} \tag{10.55}$$

$$\frac{N}{D} = \frac{-206}{327} \tag{10.56}$$

Since $N < 0$ and $D > 0$, the angle 2α lies in the fourth quadrant. In the fourth quadrant

$$2\alpha = 2\pi - \arctan\frac{N}{D} \tag{10.57}$$

with

$$\tan\left|\frac{N}{D}\right| = 0.63 \tag{10.58}$$

and

$$\arctan\left|\frac{N}{D}\right| = 32° \tag{10.59}$$

therefore,

$$2\alpha = 360° - 32° \tag{10.60}$$

$$\alpha = 164° \tag{10.61}$$

Figure 10.30 shows a graphical aid for the determination of 2α by means of the tangent function. It is obvious that without an individual examination of numerator and denominator, the angle 2α could be mistakenly placed in the second quadrant instead of correctly in the fourth quadrant (328°).

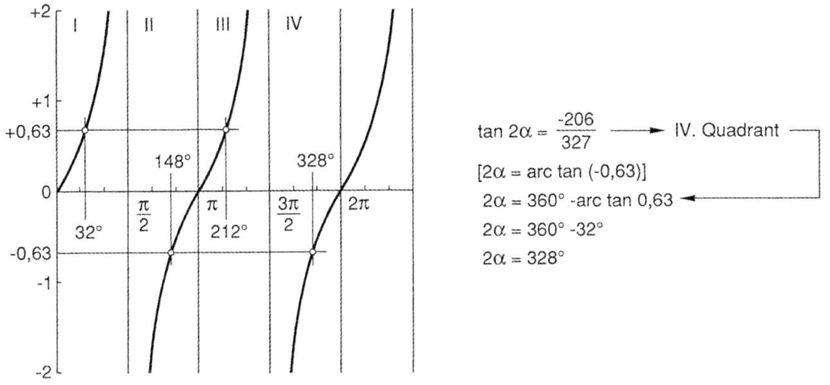

Fig. 10.30 Graphical representation of the tangent function, showing the ambiguity of determining the angle

Now that the orientation angle for the principal axis 1 has been determined, it can be seen that the principal axis 1 is at 164° to the reference axis, measured in the positive direction of rotation. Principal axis 2 is, of course, perpendicular to principal axis 1. Figure 10.31 shows the calculated orientation angle and the principal axes.

Substituting the readings from Eq. (10.54) into the defining Eq. (9.84) for the principal strains of the 0/60/120° rosette gives:

$$\varepsilon_{1,2} = \left(200 \pm \sqrt{16601}\right) \cdot 10^{-6}$$
$$\varepsilon_1 = 329 \cdot 10^{-6}$$
$$\varepsilon_2 = 71 \cdot 10^{-6}$$

With this, the principal strains have been established.

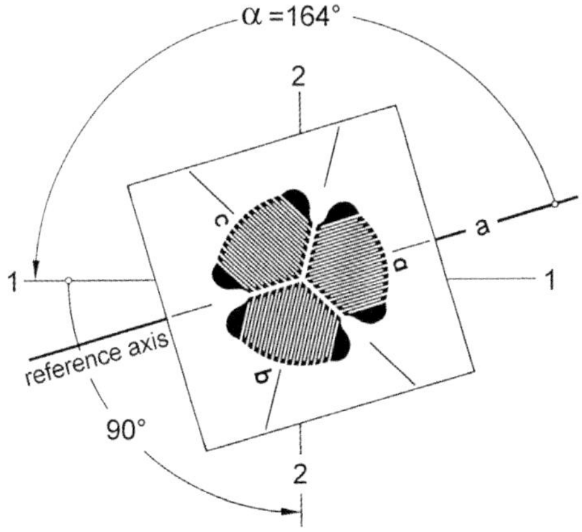

Fig. 10.31 Fixing the calculated principal axis 1 by measuring the angle in a positive rotation from the rosette reference axis 'a'

The equations representing Hooke's law for the biaxial stress state provide the necessary relationship for calculating the principal normal stresses (under the assumption of elastic deformation, which can be checked by measuring again when the specimen has been unloaded):

$$\sigma_1 = \frac{E}{1-\nu^2}(\varepsilon_1 + \nu\varepsilon_2) \tag{10.62}$$

$$\sigma_2 = \frac{\mathrm{E}}{1-\nu^2}(\varepsilon_2 + \nu\varepsilon_1) \tag{10.63}$$

The modulus of elasticity and Poisson's ratio for the material under consideration are: $E = 206\,000\ \mathrm{N/mm^2}$ and $\nu = 0.3$. Substituting the principal strains ε_1 and ε_2 calculated previously, and the material parameters E and ν in Eqs. (10.62 and 10.63) gives:

$$\sigma_1 = \frac{206000}{0.91}(329 + 21) \cdot 10^{-6} = 79\ \mathrm{N/mm^2}$$

$$\sigma_2 = \frac{206000}{0.91}(71 + 99) \cdot 10^{-6} = 39\ \mathrm{N/mm^2}$$

Thus, both the principal axes and principal stresses are now known. This provides a clear picture of the plane stress at the point of measurement and completes the demonstration of experimental stress analysis using strain gages.

Finally, a clarification should be made to show the effect of transverse strain (Poisson's ratio) on the principal strains and principal stresses at the measuring point. Note that the principal normal stress σ_1 is twice as large as σ_2. However, the principal strain ε_1 is more than 4.5 times as large as ε_2. The transformation of strain into stress without consideration of the effect of the transverse strain can therefore lead to large errors. There is a distinct possibility of making a wrong estimation of the actual stress condition if a hurried conclusion is drawn.

10.3.2 Loading of a shaft with superimposed torsion and bending

Very often cases arise in practice where shafts are simultaneously loaded with torsion and bending. Usually, the directions at which the separate components of loading act are known, but without knowing the magnitudes of the locally superimposed components no conclusion can be drawn about the orientation of the principal directions and the magnitudes of the principal stresses. With a shaft its axis almost always forms the axis of rotation for the torsion loading and the neutral plane for bending. Taking these facts into account, the bending strains can be measured with strain gages bonded longitudinally to the shaft axis, and the strains produced by the torsion loading can be acquired by strain gages bonded at 45° to the shaft axis. Using these measurements, the shear and bending stresses can be calculated, and the principal stresses and their directions found using Mohr's circle. A pre-requirement for this measurement technique is the paired arrangement of strain gages for the measurement of the torsional loading. This is needed so that the bending effect is eliminated from the result for torsional loading as a result of the simultaneous action of the bending strain on the two gages in neighbouring arms of the bridge. The same applies for the measurement of the bending strains with two gages located on the shaft circumference diametrically opposite one another. These gages are also located in neighbouring arms of the bridge so that torsion which affects both gages equally is compensated.

Example of simultaneous bending and torsion of a shaft: measurement of the loads transferred to a horizontal bar by a gymnast [10.13]
When a gymnast performs on the horizontal bar, the forces acting on him through his hands are transferred to the horizontal bar, and with relatively slow movements cause bending and torsion in the bar proportional to the forces. The forces acting are typical for a particular sequence of movements, and their time relationship therefore represents a dynamic characteristic for a certain exercise. The objective of the measurements described here was to obtain the dynamic characteristics of exercise sequences of more able and less able gymnasts using a horizontal bar fitted with strain gages. The measurements were to enable conclusions to be drawn about the state of training and the learning process when performing exercises on the horizontal bar.

In line with the objective of the tests, strain gages were bonded to the horizontal bar. These enabled the forces causing the bending of the bar in the vertical and horizontal directions to be measured. In addition strain gages were mounted for the measurement of the torque acting on the bar. The signal was recorded after amplification as a function of time using a pen recorder.

For the sake of simplicity, the bar is considered statically to be a bending beam and the forces from the gymnast's hands are taken to have a resultant acting at one point on the bar. According to the representation shown in Figure 10.32 the supporting forces F_a and F_b have different magnitudes depending on the point of action of the resultant F.

Using the terms given in Figure 10.32, the following applies in the region a on the bar

$$F_a x_a = C_B \varepsilon_a \tag{10.64}$$

in region b

$$F_b x_b = C_B \varepsilon_b \tag{10.65}$$

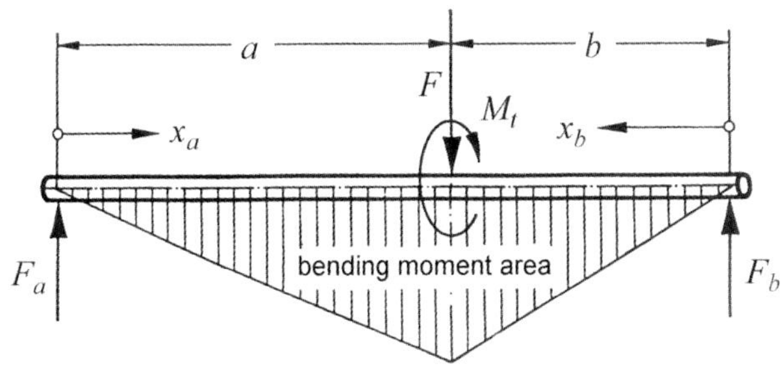

Fig. 10.32 The horizontal bar as a bending beam; simplified representation of the applied loads and definition of the terms used

where C_B is a constant relating to the elastic properties of the horizontal bar and ε_a and ε_b are the longitudinal strains measured at the points x_a and x_b respectively. If an active strain gage is applied at a distance of x_0 from each end of the bar, then the measured strains at the ends a and b are proportional to the supporting forces F_a and F_b as in the following

$$\varepsilon_a = \frac{F_a x_0}{C_B} \tag{10.66}$$

$$\varepsilon_b = \frac{F_b x_0}{C_B} \tag{10.67}$$

The sum of the two measured strains is

$$\varepsilon_a + \varepsilon_b = \frac{F \cdot x_0}{C_B} \tag{10.68}$$

This sum is therefore proportional to the total force F, irrespective of where it acts on the bar. The maximum of the bending moment and therefore also the maximum strain occurs at the point of action of this force F. Due to the gymnast's hands gripping the bar, the strain gages could not however be mounted in the centre of the bar, but only towards the ends at the points x_0.

With respect to the torque caused by the friction of the hands, the horizontal bar behaves as a shaft clamped at both ends. The shear stresses to the left and right of the load transfer point which are caused by the torque M_t produce strains with directions at an angle of 45° to the shear plane. These strains are measured by strain gages aligned at 45° to the bar axis. Depending on the position of the load transfer point, the torque is divided between the sections of the bar to the right and left. In the section of the bar labelled a the following equation can be written for the strain occurring at 45°

$$\varepsilon_{a45} = C_t M_{ta} \quad \text{with} \quad M_{ta} = M_t \frac{b}{a+b} \tag{10.69}$$

in section b

$$\varepsilon_{b45} = C_t M_{ta} \quad \text{with} \quad M_{tb} = M_t \frac{a}{a+b} \tag{10.70}$$

Here, the elastic properties of the bar under torsional loading are combined in the constant C_t. The sum of the strains occurring in both sections of the bar is given by

$$\varepsilon_{a45} + \varepsilon_{b45} = C_t M_t \tag{10.71}$$

In order to make the sensitivity as high as possible with the measurement arrangement, two full-bridges were configured for each of the components to be measured. One was mounted on the right side and the other on the left side of the bar. Therefore, for the three components, vertical and horizontal bending together with torsion, six full-bridges were mounted on the bar. The strain gage arrangement and its configuration to form three full-bridges is shown for one end of the bar in Figure 10.33.

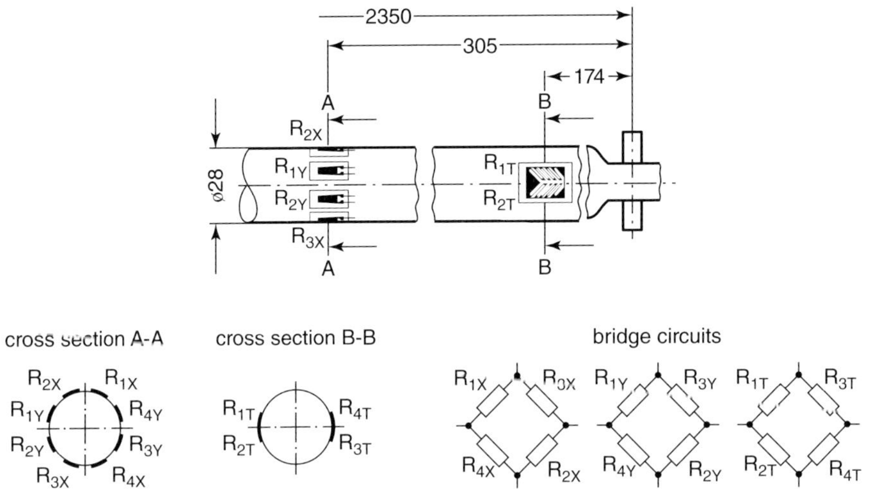

Fig. 10.33 Mounting plan of the strain gages on one end of a horizontal bar showing their configuration to full-bridges

The strain gages on the other end of the bar were positioned in the same way. In the picture the strain gages are labelled *R* and the index letter defines the measurement quantity being acquired. Here, *X* indicates the horizontal force, *Y* the vertical force and *T* the torsion.

Due to the constant level of the moment in the section of the bar under consideration, the gages for the determination of the twisting torque could be positioned close to the ends of the bar. For the gages measuring the bending, a position near the centre of the bar was advantageous due to the increasing bending moment towards the centre, but this was only partly possible due to the gymnast's hands. The normal commercially available horizontal-bar bearing was substituted by a specially designed cardan bearing to avoid different characteristics of the bar bearing in the horizontal and vertical directions as well as hysteresis effects.

After setting up, the measurement arrangement was calibrated directly in the relevant measurement quantities. The calibration was carried out using weights, which were directly transferred to the bar as a vertical force and as a horizontal force via deflection pulleys. The weights were suspended on a metal arm which loaded the horizontal bar at two points which were separated by the shoulder distance (50 cm) to take into account the distance between the gymnast's hands. To calibrate the measurement channel for torque, the weights were suspended on the bar via two levers. Figure 10.34 shows the calibration curves found in this way.

Six channels are used for the signal acquisition. The summations according to Eqs. (10.68 and 10.71) are carried out in the computer. By including the proportionality figures found during the calibration, the pen recorder immediately supplies the results during the ongoing gymnastic exercises. Figure 10.35 shows an example of

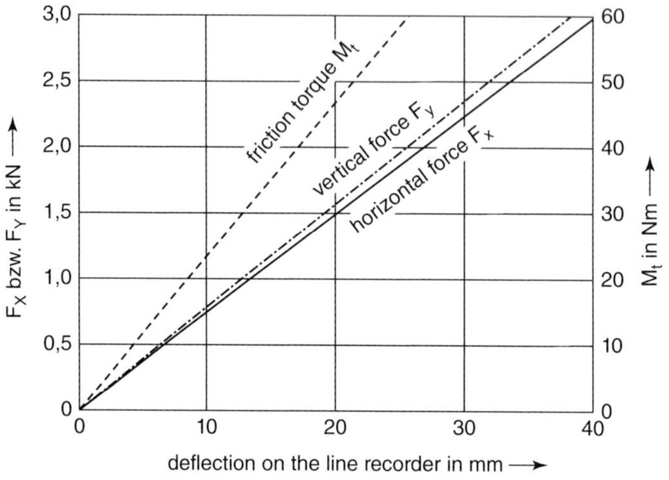

Fig. 10.34 Calibration curves of vertical and horizontal force together with the frictional moment for the measuring points on the horizontal bar found using loading weights

the traces of the vertical and horizontal forces together with the frictional moment of the gymnast's hands on the bar while performing large circles.

The resonant frequency of the horizontal bar can be taken from the measurement traces. It is 11 Hz. The time to execute a giant swing took about 2.4 s and therefore corresponds to a frequency of about 0.41 Hz. Higher oscillation frequencies of about

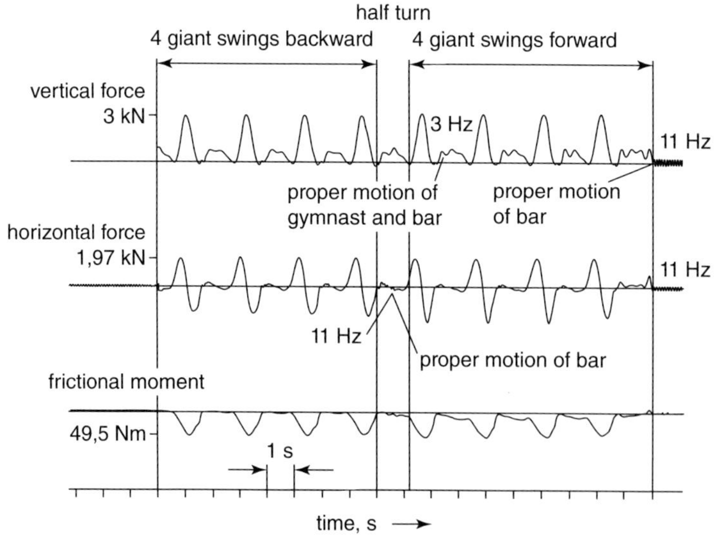

Fig. 10.35 Reduced-size traces from the pen recorder; traces of the vertical and horizontal force and of the torsion moment on the horizontal bar during the performing of large circles

3 Hz only occur when the gymnast whips in the handstand position after completing a rotation. Therefore, the processes which take place lie in the uncritical region and confirm the assumed proportionality between the force and strain.

As was expected, the measuring signals due to the frictional moments were very small in comparison to the signals from the horizontal and vertical bending of the bar. A rough calculation using the terms from Figure 10.32 provides information on the stresses resulting from the loading of the bar. The highest bending stress in the centre of the bar is

$$\sigma_b = \frac{M_b}{W_b} \tag{10.72}$$

with

$$M_b = \frac{F \cdot l}{4} \quad \text{and} \quad W_b = \frac{\pi \cdot d^2}{32} \tag{10.73}$$

Using the bar diameter $d = 28$ mm, the following is obtained

$$\sigma_b = 0.2726\,F \text{ in N/mm}^2. \tag{10.74}$$

If the force $F = 3000$ N is substituted as measured, then the relatively high bending stress $\sigma_b = 818\ \text{N/mm}^2$ is obtained. Therefore, one of the most important requirements for the steel to be used for the horizontal bar (high quality steel) has been formulated.

The shear stress τ from the frictional moment

$$\tau = \frac{M_t}{W_p} \tag{10.75}$$

with $W_p = \frac{pd^3}{16}$ and $d = 28$ mm

becomes

$$\tau = \frac{M_t}{4310} \quad \text{in} \quad \frac{\text{N}}{\text{mm}^2} \tag{10.76}$$

With $M_t = 50\,000$ Nmm as measured, then

$$\tau = 11.6 \frac{\text{N}}{\text{mm}^2} \tag{10.77}$$

10.4 Simultaneous measurement of multiple load components

Strain gages are ideal for building measuring elements with which one can simultaneously measure several different load components. The Wheatstone bridge circuit favours this by both their ease of installation and the variable design options that they offer. Figure 10.36 shows how in a simple manner various bridge circuits can be used on a beam so that several different load components can be measured: longitudinal forces, bending moments in different planes and torque.

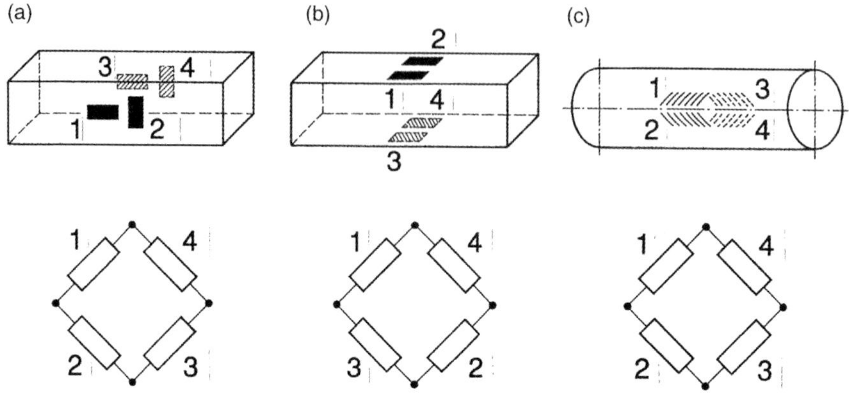

Fig. 10.36 Measurement of different load components using strain gages in different arrangements in Wheatstone bridge circuits; a) bridge circuit for measurement the longitudinal force, b): measurement of bending moment, c) arrangement for measurement of torsion

The installation in Figure 10.36a is suited to measuring longitudinal forces acting on the bar. This circuit is insensitive to bending loads because of the opposing strains on the strain gages on the opposite sides of the beam. Also, the circuit is insensitive to torque, because no strains occur in the measuring directions of the strain gages.

In Figure 10.36b the strain gages are mutually interconnected so that bending load causes addition of all four strain gage signals. There also is compensation of strains caused by longitudinal forces, the signals caused by them cancel each other out. The same applies for torque.

The measuring arrangement of Figure 10.36c is sensitive to torsion and to shearing forces. The arrangement optimally picks up the strains at 45° to the direction of the shear stresses.

These possibilities offered by the Wheatstone bridge circuit are used for the design of sensor elements which are able to record simultaneously different load components.

Example: Strain gages as sensors for controlling the manipulative robot hand OEDIPUS [10.14]
In the process of gripping and holding objects with a robot hand, the determination of the forces involved is of decisive significance. This can be achieved using strain gages applied in a suitable manner to the finger element, as can be seen in Figure 10.37.

Precise location of the points at which the force acts can be obtained by the addition of pressure measurement film on the end of the element. The number of strain gage bridge circuits and their points of mounting were given by the six unknown quantities that were to be determined. These were the three force components and the coordinates of the point of action for the resultant force. This requires six strain gage bridge circuits.

Figure 10.38 shows in the form of a development the arrangement of the strain gages on the four sides of the element and their wiring in four full-bridges: Bridge (9)

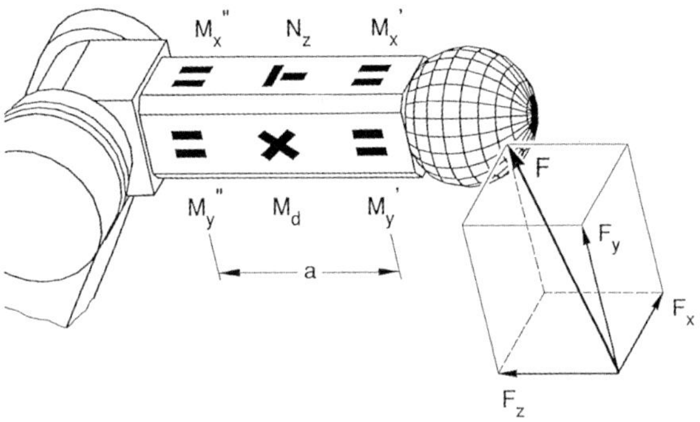

Fig. 10.37 Resolution of random force acting on the end element of the gripper into components of a Cartesian coordinate system

measures the longitudinal force, two bridge circuits (8) measure the bending moments in two different planes. Circuit (14) measures the moment of torsion. In addition, all three variants are independent of the temperature due to the temperature compensation in the full-bridges.

The assumption that each circuit only reacts to the relevant load is based on the ideal case and is only a theoretical idea. Slight deviations in the strain gage positioning and the inhomogeneity of the material used cause unpredictable interference due to

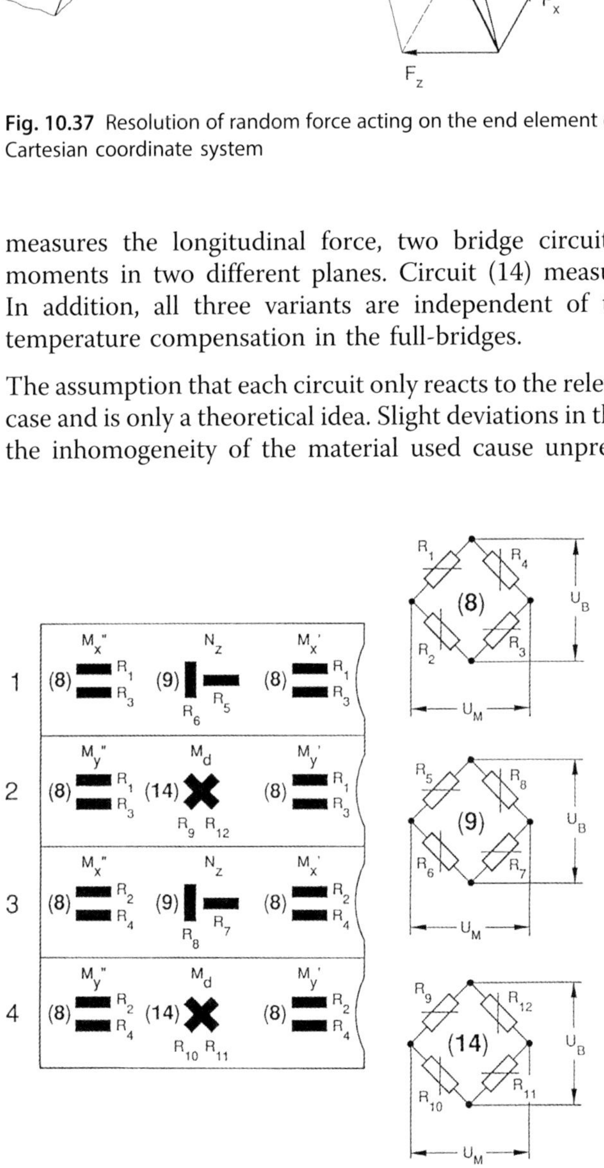

Fig. 10.38 Strain gages applied to the gripper and the connecting of the gages into full-bridges

Fig. 10.39 Photograph of the gripper element with applied strain gages

crosstalk which can, however, be measured. Correction characteristics can be derived from the results obtained and these are used as the initial quantities for a correction procedure [10.15–10.17].

This correction method, termed decoupling, takes place in a software module.

Figure 10.39 shows a photograph of the gripper element with applied strain gages as developed by the Austrian Aerospace Company. The gripper enables interesting and promising terrestrial and space applications to be considered [10.14].

Example: Triaxial measurement of the cutting forces on the teeth of a bucket wheel excavator [10.18]

By way of this example of measurement of multiple load components the possibility of decoupling in crosstalk may be explained [10.18]. The object of the example is a bucket wheel excavator. On the grounds of efficiency the optimization of cutting processes occurring at the teeth of the excavator buckets is of particular importance. Therefore measurements were carried out on a 500 t compact bucket wheel excavator during normal operation at a large opencast mine.

The main aim of the project was the measurement of the forces on the cutting teeth in three orthogonal directions, but neither the cutting geometry nor the holding of the cutting tool could be modified. The excavator possesses 12 excavator buckets on the bucket wheel with an average cutting diameter of 7.5 m. The driving power of the bucket wheel is 670 kW for a theoretical delivery rate of 3000 m^3/h.

The main problem with the measurement of a number of force components on only one test object is presented by the mutual cross-coupling between the measured quantities in the different directions. In this case, the external geometry of the test object was predetermined with the selection of the test tooth. Therefore, the only possible method of obtaining a test object geometry which was convenient for measurement was through the use of recesses.

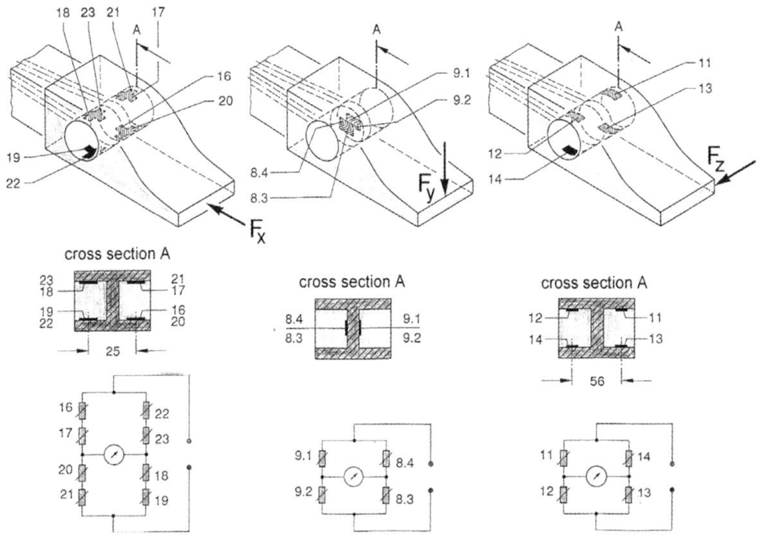

Fig. 10.40 Strain gage arrangements in the recess of the tooth for measuring the force components F_x, F_y and F_z

The strain gages that were used to sense the deformation were mounted in these recesses. In covering the measuring point, the dimensions had to be within those representing the original tooth volume, because the flow of the excavated material from the cutter past the test tooth should not be different from that with the normal tooth.

The arrangements of the strain gages in the recesses can be seen in Figure 10.40. For measuring the longitudinal force F_x, four double-grid strain gages, as shown in the figure, were mounted close to the shear diaphragm, such that measurement grids 16 to 19 located in the circumferential direction sense the longitudinal strains on the circumference of the cylindrical recesses. The measurement grids 20 to 23 located at 90° to them in the longitudinal direction of the tooth, sense the corresponding transverse strains. The force F_y is measured according to the principle of the shear field force transducer. As Figure 10.40 shows, two double grid strain gages are located in the two centres of the shear diaphragm. The measurement directions are arranged at ±45° to the x direction. Extremely short internal bridge cable lengths could be provided by drilling the shear diaphragm.

The sensing of the force F_z was carried out with four separate strain gages. As illustrated in Figure 10.40 they are arranged on the circumference of the recesses at a certain distance from the shear diaphragm, so that under the action of F_z the gages 11 and 13 are extended and the gages 12 and 14 are shortened. By selecting the distance from the centre of the diaphragm the nominal sensitivity can be adjusted between wide limits.

Figure 10.41 shows a test tooth removed from the bucket wheel after having worn down to the limit of wear.

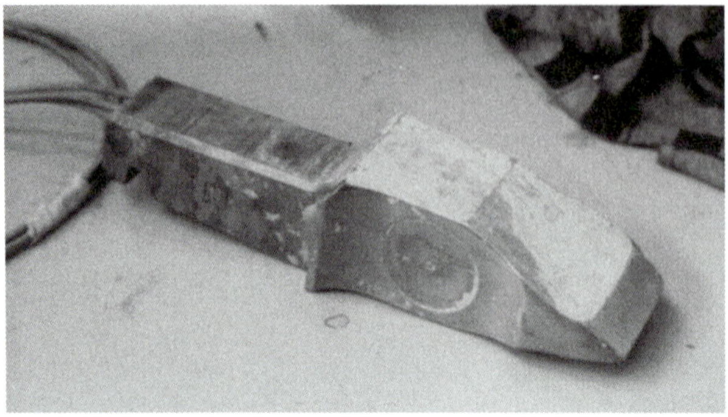

Fig. 10.41 A test tooth removed from the bucket wheel after having been worn down to the limit of wear

The transfer of the bridge supply voltage and the measuring signal between the rotating test teeth and the measuring amplifiers located in the operating cabin also posed a few problems. The measuring cable potted in drilled holes in the region of the tooth fixtures passed first through a short armoured conduit to a rotating distribution box mounted on the bucket. Here they are soldered to strong solder terminals, connecting them to their branch of the measuring cable. An armoured conduit then leads radially from here to the centre of the bucket wheel. A slip-ring head is mounted on the end of the rotating bucket wheel shaft on the wheel side. Figure 10.42 illustrates the cabling of three test teeth in a photograph.

Fig. 10.42 The measurement cable conduit extends from the test teeth to the slip-ring head on the shaft of the bucket wheel

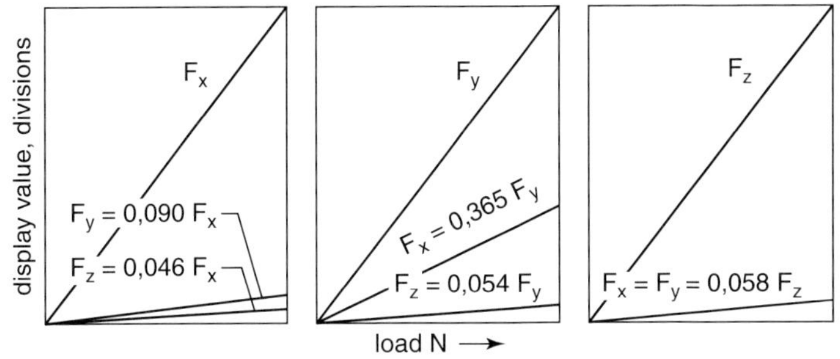

Fig. 10.43 Calibration curves showing the cross-coupling effect

For calibration, the test tooth was clamped in a fixture. It was then loaded and relieved in stages in each of the three coordinate directions x, y and z, up to a nominal load of 50 kN. Hardened steel balls were used to transfer the load. The plots from the calibration tests are reproduced in Figure 10.43. Here the interactions for the forces F_x and F_z still lie under the required 5% limit. Only F_y showed a pronounced, but defined, cross-couple effect on F_x.

The absolute amount of this cross-coupling can be eliminated from these results and can be represented in a matrix. The output signal of force F_x, designed with F_x^*, also contains signal shares generated by the forces F_y and F_z:

$$F_x^* = q_{xx}F_x + q_{xy}F_y + q_{xz}F_z \tag{10.78}$$

Without any mechanical cross-coupling effect: $q_{xx} = 1$ and $q_{xy} = q_{xz} = 0$.

In Eq. (10.78) the coefficients q_{ij} ($i = x \ldots z$ and $j = x \ldots z$) describe the shares of the assigned forces to the output signal.

Corresponding to Eq. (10.78) we can also formulate equations for the signals $F_y{}^*$ and $F_z{}^*$:

$$F_y{}^* = q_{yx}F_x + q_{yy}F_y + q_{yZ}F_Z \tag{10.79}$$

$$F_z{}^* = q_{zx}F_x + q_{zy}F_y + q_{zz}F_z \tag{10.80}$$

The relationship between the forces and the measurements can be represented in a matrix

$$D = \begin{vmatrix} q_{xx} & q_{xy} & q_{xz} \\ q_{yx} & q_{yy} & q_{yz} \\ q_{zx} & q_{zy} & q_{zz} \end{vmatrix} \tag{10.81}$$

Without cross-coupling, the coefficients q_{xx}, q_{yy} and q_{zz} will be 1 and all coefficients with mixed indices are 0; the assigned determinant becomes

$$D = \begin{vmatrix} 1 & 0 & 0 \\ 0 & 1 & 0 \\ 0 & 0 & 1 \end{vmatrix} \tag{10.82}$$

From Figure 10.43 we find the interaction coefficients and can insert these coefficients in Eqs. (10.78 and 10.80)

$$\begin{aligned} F_x^* &= 1F_x + 0.365F_y + 0.58F_z \\ F_y^* &= 0.090F_x + 1F_y + 0.058F_z \\ F_z^* &= 0.046F_x + 0.054F_y + 1F_z \end{aligned} \tag{10.83}$$

These are three equations containing three unknowns. These three unknowns F_x, F_y and F_z can easily be determined.

In [10.19] the relatively simple method of solving the equation system using determinants is described, with reference to the example of a three-component wind-tunnel scale model. Another example of a three-component transducer system with remarkably low crosstalk is given in [10.20].

10.5 Diaphragm rosettes

10.5.1 Basics

In a circular membrane that is firmly clamped at its rim, a strain distribution arises with concurrent positive and negative strains when subjected to a hydraulic pressure. This strain distribution has led to the design of special strain gage rosettes that are called diaphragm rosettes.

Diaphragm rosettes are preferably used for the manufacture of pressure transducers and, less commonly, for force transducers. A diaphragm rosette contains on the smallest possible area four measurement grids interconnectable to a full-bridge circuit, with their measurement directions adapted to the directions of strain occurring in a diaphragm in an ideal way.

The stress and strain distributions in a thin pressure-loaded membrane are shown in Figure 10.44. Strain and stress distribution are symmetrical. Principal directions are the radial and tangential direction. This provides an illustration of the stress in polar coordinates. With uniform pressure load in the centre of the membrane, tangential and radial stresses are identical, as are the tangential and radial strains.

For depiction of stress and strain distribution is chosen the radius r as spatial coordinate starting at the centre of the diaphragm. The diaphragm has radius R. With p as pressure loading of the membrane, s as membrane thickness, Young's modulus E and Poisson's ratio ν of the membrane material, the principal stresses arising in tangential and radial direction in the membrane are

$$\text{Tangential stress} \quad \sigma_t = \frac{3p}{8s^2}\left[R^2(1+\nu) - r^2(1+3\nu)\right] \tag{10.84}$$

$$\text{Radial stress} \quad \sigma_r = \frac{3p}{8s^2}\left[R^2(1+\nu) - r^2(3+\nu)\right] \tag{10.85}$$

In the centre of the board ($r = 0$) the tangential and radial stresses take the same value σ_z

$$\sigma_z = \frac{3p}{8s^2}\left[R^2(1+\nu)\right] \tag{10.86}$$

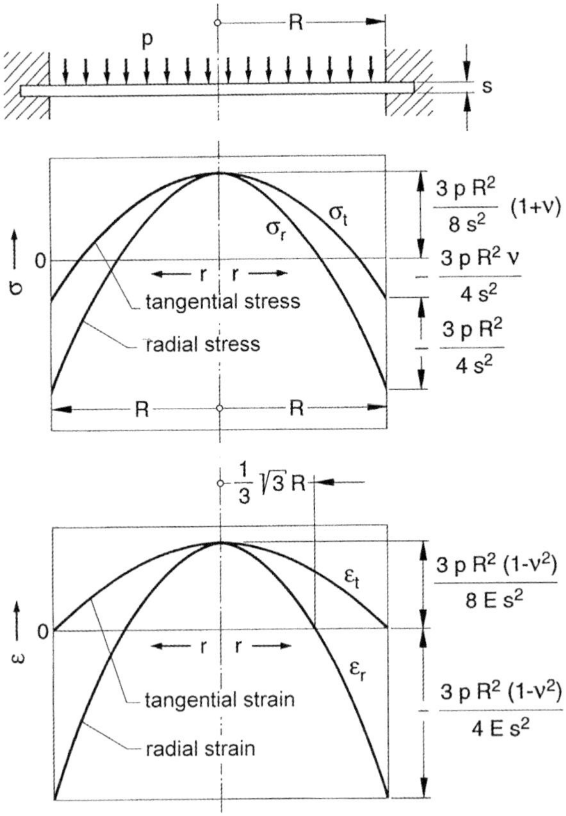

Fig. 10.44 Stress and strain distribution in a clamped diaphragm under uniform pressure

Using Hooke's law for the biaxial stress state, the principal strains can be calculated

$$\varepsilon_t = \frac{1}{E}(\sigma_t - \nu\sigma_r) \quad \text{and} \quad \varepsilon_r = \frac{1}{E}(\sigma_r - \nu\sigma_t) \tag{10.87}$$

With Eqs. (10.84 and 10.85):

$$\text{Tangential strain} \quad \varepsilon_t = \frac{3p(1-\nu^2)}{8E\,s^2}\left(R^2 - r^2\right) \tag{10.88}$$

$$\text{Radial strain} \quad \varepsilon_r = \frac{3p(1-\nu^2)}{8E\,s^2}\left(R^2 - 3r^2\right) \tag{10.89}$$

Even for the strains in the centre of the diaphragm results for $r = 0$ in tangential and radial direction an identical value ε_z:

$$\varepsilon_z = \frac{3pR^2(1-\nu^2)}{8E\,s^2} \tag{10.90}$$

The strain distribution in the pressure-loaded membrane is characterized by positive strains both in the tangential and in the radial direction in the region of the centre. In the direction towards the clamped edge there are increasingly radial strains with a negative sign, the amount of which with $r = R$ reaches its maximum value ε_R at the membrane edge. This strain value is

$$\varepsilon_R = -\frac{3pR^2(1 - \nu^2)}{4Es^2} \tag{10.91}$$

This value is exactly twice as large as the strain ε_z occurring in the centre of the membrane.

On the surface of the diaphragm simultaneously are present both positive and negative strains. This supports the installation of a strain gage full-bridge circuit. Two strain gages, preferably with crosswise stacked grids will be installed in the centre, and close to the edge of the diaphragm on opposite sides two strain gages with its grids in radial measurement direction. The two gages in the centre and the two gages on the edge are placed in opposite bridge arms. In this way, in adjacent bridge arms strains are acting with different signs which results in a summation of all the four signal components.

In order to facilitate the installation of the strain gages on a diaphragm, special membrane rosettes have been developed. These rosettes contain four measuring grids for detecting positive and negative strains. Since the tangential strains of the diaphragm centre decrease more slowly towards the edge than the radial strains, there are arranged two measuring grids with tangential conductor tracks in the central region of the circular rosette. Close to the edge there are two measurement grids with radial conductor tracks that are intended to record the negative radial strains. Figure 10.45 shows the basic grid array of diaphragm rosettes with three rosettes with different dimensions. The connection terminals for soldering the wires are preferably not positioned in the areas of high strains.

In general, the strains are relatively small in a diaphragm designed according to aspects of the fatigue strength and in view of the linearity of the relationship between mechanical stress and the diaphragm bridge output signal. The relationship between the mechanical pressure and the strains of the diaphragm is linear when the

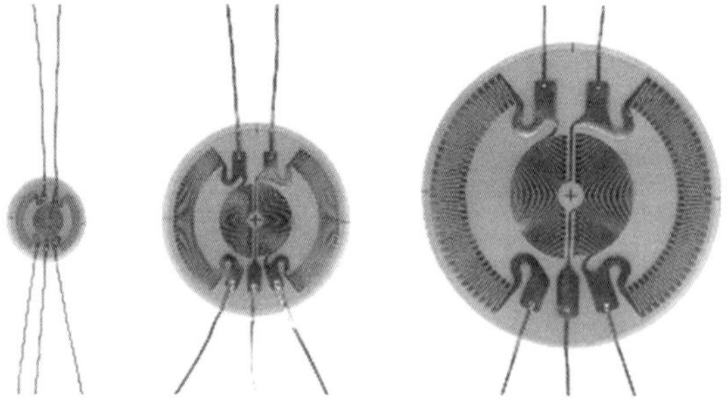

Fig. 10.45 Examples of measuring grid patterns of diaphragm rosettes

diaphragm is charged by pure bending. At high pressures there occurs as well as the bending a distending effect, which causes an increasing nonlinearity in the relationship between pressure and strain [10.21]. As a guide for keeping the linearity error <0.3% in [10.22] a load is specified at which the deflection $f_z$in the middle of the membrane is less than a quarter of the membrane thickness s:

$$f_z = \frac{3R^4 p}{16Es^3}\left(1 - \nu^2\right) \leq \frac{s}{4} \tag{10.92}$$

For the estimation of the signal value delivered under pressure by an installed diaphragm rosette it should be noted that the measurement grid of the rosette does not record the maximum values of the strains occurring in the diaphragm, but rather the mean values of the strains existing in the area covered by the measuring grids.

The measurement grids mask the defined radial regions of the diaphragm and generate an average of the respective detected signal proportional to strain. An optimal signal is obtained when the diameters of the membrane and the rosette are identical.

At a centrally installed rosette with increasing ratio of membrane diameter to rosette diameter the position of the radial measurement grid moves from the edge in the direction of the diaphragm centre, so that the signal components supplied by the radial measuring grids will be reduced. If the radial grid is in the area of positive radial strain, the radial grid partially compensates the signal generated by the tangential grid. An example of successfully performed pressure measurements with a relatively large diaphragm diameter is described in [10.23].

After setting the dimensions of the diaphragm and the attached rosette an estimate can be carried out of the expected signal value under pressure loading.

Corresponding to Figure 10.46 the measuring grids are assigned to the strain distribution in the diaphragm occurring in the covered areas and this allows the average values to be put into the equation for calculating the output signal of the diaphragm rosette,

$$\frac{V_M}{V_B} = \frac{1}{4}k(2\varepsilon_{tDMS} - 2\varepsilon_{rDMS}) \tag{10.93}$$

where $\varepsilon_{\mathrm{tDMS}}$ and $\varepsilon_{\mathrm{rDMS}}$ are the strains recorded by the tangential and radial measuring grids. The radial strain is negative in the outer diaphragm areas, so that in the bracket of Eq. (10.93) the strain proportions add up when the radial measurement grid covers negative strain regions. The strains detected by the measuring grids can be described as proportions of the occurring strain maxima:

$$\varepsilon_{tDMS} = C_1\varepsilon_z \tag{10.94}$$

$$\varepsilon_{rDMS} = C_2\varepsilon_R \tag{10.95}$$

Thus Eq. (10.93) becomes

$$\frac{V_M}{V_B} = \frac{1}{2}k(C_1\varepsilon_z - C_2\varepsilon_R) \tag{10.96}$$

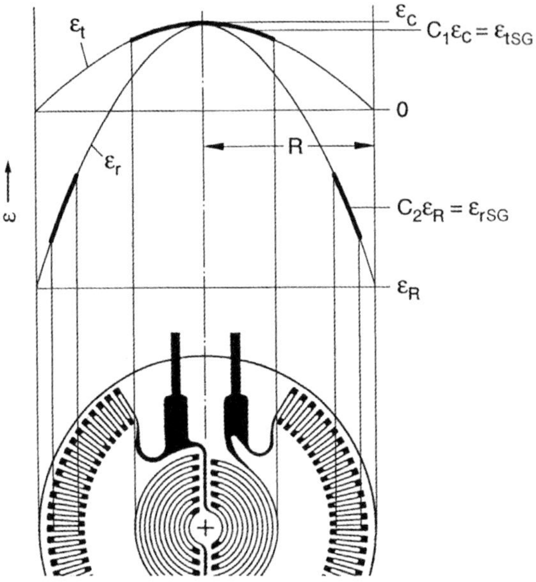

Fig. 10.46 Strains recorded by the measuring grids of a diaphragm rosette

Substituting here Eqs. (10.90 and 10.91), with $k = 2$ leads to:

$$\frac{V_M}{V_B} = \frac{3pR^2(1-\nu^2)}{8Es^2}(C_1 + 2C_2) \tag{10.97}$$

The selected values of C_1 and C_2 can be found from Figure 10.46 with assignment of the measurement grid geometry to the strain distribution. For identical diameters of diaphragm and rosette we can assume approximately $C_1 = 0.95$ (the maximum value of tangential strain occurring in the membrane centre is 95% detected) and $C_2 = 0,6$ (the radial parts of the measurement grid record 60% of ε_R).

From this we get

$$\frac{V_M}{V_B} \approx 0.8\frac{R^2(1-\nu^2)}{Es^2}p \tag{10.98}$$

a rule of thumb, that can be used to estimate the output signal as a function of the pressure p, provided that the diameters of the diaphragm and the rosette are approximately identical.

The following example shows an application of the above equations for the design of a membrane for pressure measurement.

10.5.2 Example of dimensioning a membrane for pressure measurements

Given is a diaphragm diameter of $2R = 30$ mm. To be measured are overpressures between 0 bar and 10 bar. Reference pressure for the overpressure is the atmospheric pressure acting on the diaphragm side with the installed rosette. The pressure to be measured acts on the other side of the diaphragm. Used is a diaphragm rosette type shown in Figure 10.45 with an outer diameter of 25 mm. The diaphragm is made of steel ($E = 206 \times 10^3\,\mathrm{N/mm^2}$; $\nu = 0.3$). The output signal of the measuring arrangement should be of 2 mV/V at an overpressure of 10 bar.

Substituting in Eq. (10.98) for $V_M/V_B = 2 \cdot 10^{-3}$ (i.e. 2 mV/V) and solving for the plate thickness s, gives

$$s = \sqrt{0.8 \frac{pR^2(1-\nu^2)}{2 \cdot 10^{-3} \cdot 206 \cdot 10^3}} \tag{10.99}$$

Substituting $p = 1\,\mathrm{N/mm^2}$ ($p = 10$ bar) and $R = 15$ mm and the above values for E and ν leads to

$$s = \sqrt{0.397} \approx 0.63 \tag{10.100}$$

Choosing the membrane thickness s = 0.63 mm and checking the linearity condition given with Eq. (10.92), gives $f_z = 0.167$ mm; this value is greater than $s/4$.

Thus, a membrane thickness of $s = 0.63$ mm is insufficient for compliance with the linearity condition. With $s = 0.7$ mm we find with $f_z = 0.122$ mm a value satisfying the given condition. Thus, the membrane thickness is chosen as $s = 0.7$ mm.

However, this change gives a reduction of the signal value of 2 mV/V, which must necessarily be taken into account. Now with Eq. (10.98) and $s = 0.7$ mm an estimation can again be made. The result with $p = 1\,\mathrm{N/mm^2}$ is

$$V_M/V_B = 1.62 \cdot 10^{-3} = 1.62\,\mathrm{mV/V} \tag{10.101}$$

With the plate thickness $s = 0.7$ mm in accordance with Eqs. (10.90 and 10.91) the strains in the centre of the diaphragm and the diaphragm edge result to

$$\varepsilon_z = 761\,\mu\mathrm{m/m} \quad \text{or} \quad \varepsilon_R = -1522\,\mu\mathrm{m/m} \tag{10.102}$$

After production this measuring system must be calibrated with a suitable device (e.g. reference pressure transducer or pressure balance) in order to find the actual relationship between the signal value and the variable.

References

[10.1] Merz, L. (1972) *Das Paradoxe in der Meßtechnik;*, ATM, August Lfg 439, pp. R117–R119.

[10.2] Bach, C.v. and Baumann, R. (1924) *Elastizität und Festigkeit*, Springer Verlag, Berlin, 9. Auflage, pp. 269–270.

[10.3] Wasner, D. and Pollien, J.-P. (1987) Stress tests with strain gages on the UBS Switzerland before her participation in the Round The World Race. *RAM*, **3** (2), 33–39.

[10.4] Grünbaum, P. and Perchthaler, H. (1987) Applying force transducers and strain gages for rollbearing measurements on continuous dewatering pressure filters. *RAM*, **3** (2), 40–44.

[10.5] Grünbaum, P. (1985) Drehmomentmessungen an Nockenwellen von Dieselmotoren. *MTB*, **21** (2), 39–46.

[10.6] Szymanski, D. (1986) Strain-gage-based torque measuring test bench for stepping motors. *RAM*, **2** (1), 1–4.

[10.7] Dubbel (1981) *Taschenbuch für den Maschinenbau*, 14. Auflage, Springer Verlag, Berlin Heidelberg New York, p. 201.

[10.8] Nolte, V. (1979) Die Handschrift des Ruderers. *MTB*, **15** (3), 49–53.

[10.9] Winkler, J. (1984) Bestimmung des Zusammenhangs zwischen Rührleistung und Dispergiererfolg an einer Rührwerkskugelmühle zur Lackherstellung. *MTB*, **20** (3), 49–53.

[10.10] Lockery, H.E. and Paetow, J. (1979) Eine neue Scherstab-Wägezelle. *wägen+dosieren*, **10** (1), 10–14.

[10.11] Grünbaum, P. (1992) Measurement of forces in a rock-wool hardening oven at 230°C. *RAM*, **8** (2), 25–30.

[10.12] Leipholz, H. (1969) *Festigkeitslehre für den Konstrukteur*, Springer-Verlag, Berlin, Heidelberg, New York, p. 88.

[10.13] Bauer, W.L. (1976) Einsatz von Dehnungsmeßstreifen zur Untersuchung von Lernvorgängen bei Turnübungen am Reck. *MTB*, **12** (2), 24–30.

[10.14] Schwarzinger, C., Supper, L. and Winsauer, H. (1992) Strain gages as sensors for controlling the manipulative hand OEDIPUS. *RAM*, **8** (1), 17–22.

[10.15] Levi, R. (1983) Die Ermittlung störender Belastungskomponenten an Kraft-Normalmeßeinrichtungen mit Direktbelastung. *MTB*, **19** (2), 25–30.

[10.16] Ferrero, C., Marinari, C. and Martino, E. (1986) Main metrological characteristics of IMGC six-component dynamometer. *RAM*, **2** (2), 21–28.

[10.17] Giesecke, P. (1993) Mehrkomponenten-Kraftaufnehmer, in *Handbuch der physikalisch-technischen Kraftmessung* (eds W. Weiler, publisher: F. Vieweg and Sohn Verlagsgesellschaft), pp. 227–271.

[10.18] Grünbaum, P. and Bruns, B. (1989) Triaxial measurement of the cutting forces on the teeth of a bucket wheel excavator. *RAM*, **5** (1), 11–16.

[10.19] Hottner, T. (1980) Dreikomponenten-Windkanalwaage für interferenzfreie Messungen von Luftkraftkomponenten. *MTB*, **16** (2), 32–35.

[10.20] Siegert, K. and Ladwig, J. (1993) Entwicklung eines Meßsystems zur Erfassung des Kraftvektors sowie Lage seines Angriffspunkts bei Blechumformwerkzeugen. *MTB,* **29** (1), 10–14.

[10.21] Birkenfeld, W. (1967) Die praktische Anwendung von Membranrosetten. *MTB,* **3** (2), 21–22.

[10.22] NN (1982) Design considerations for diaphragm pressure transducers; TN-510 Measurements Group, Inc.

[10.23] Bestgen, R., Blücher, U. and Dudenhausen, J.W. (1994) Kraft- und Druckmessungen bei Vakuumextraktionen in der Geburtshilfe. *MTB,* **30** (1), 3–6.

11 Determination of thermal stresses

11.1 Emergence of thermal stresses

With increasing temperature, metallic materials experience an increase in volume. The thermal energy leads to enlargement of the vibration amplitude of the atoms around their lattice sites. This results in a lattice expansion and in a consequent increase in volume. The oscillation amplitude goes to zero only at the absolute temperature (0 K). The volume of most metals increases by about 7% in the temperature range between absolute zero and the melting point, Grüneisen [11.1]. From this thermal expansion results the thermal linear expansion coefficient, which states the amount by which the observed length of a material expands due to temperature (DIN 13316 [11.2]). The numerical indication of this coefficient takes the form of relating to a temperature increment of thermal expansion, for example in µm/m per K. Usually, the thermal expansion coefficient is denoted by α. The thermal expansion coefficient of a material is generally temperature-dependent.

The positive or negative strains due to changes in temperature are picked up by a strain gage in the same manner as mechanical strain of the measurement object. The strain gage is not able to recognize the cause of the expansion; it responds to the thermal expansion in the same manner as to the mechanical strain. Therefore, efforts are made to produce self-temperature-compensating strain gages, which are compensated by a temperature-dependent resistance behaviour. However, this compensation can only be achieved more or less perfectly in terms of different measurement object materials and of different strain gage types. It remains a temperature-dependent signal, referred to as a thermal output, which the strain gage generates, that is installed on a freely thermally and mechanically stretching unloaded workpiece.

Thermal expansions can be hindered by external constraints. So, for example, a steel rod fixed at both ends and clamped immovably (e.g. between two massive concrete walls) cannot expand in its longitudinal direction when heated. Because of the constraint to expansion the hindered thermal expansion generated by the rising temperature exerts a proportional force on the immovable clamp. This results in a state of equilibrium: the reaction force exerted by the restraints on the rod produces a mechanical (pressure) strain of the same magnitude as the thermal expansion in the rod. Thus, the elongation of the bar is equal to zero. The rod is, however, under the pressure generated by the clamping stress. This stress resulting from the constrained thermal expansion is called thermal stress. It may be a compression stress as in the described example, but tit can also occur as a tensile stress. In the example described, if the steel rod cools rather than being heated, it will go into tension.

In the example of the steel rod, complete elongation prevention is described. More often in practice there is incomplete prevention of thermal expansion, for example,

Technology and Practical Use of Strain Gages: With Particular Consideration of Stress Analysis Using Strain Gages. First Edition. Stefan Keil.

when not completely rigid clamping, or for work-pieces with large cross-sectional changes, or with strong temperature gradients in the component or combinations of materials with different thermal expansion coefficients in composites. Here, the pictorial example can be cited of a steel chimney, which has, on clear midsummer days, on its south side pressure and on the north side tension thermal stresses. The warm south side tries to expand, but is hindered by the cold north side. Thermal expansions are often the cause of residual stresses as a result of manufacturing processes, so, for example, with irregular or uncontrolled cooling of heat-treated workpieces.

The free thermal expansion ε_{th} of a material can be calculated easily if its coefficient of thermal expansion α is known:

$$\varepsilon_{th} = \frac{\Delta l}{l} = \alpha \cdot \Delta T \tag{11.1}$$

Here, ΔT is the temperature change causing the strain. Solving Eq. (11.1) to Δl, we find the change in length caused by thermal expansion.

In homogeneous isotropic materials the thermal expansion is independent of direction, that is it occurs in all directions in the same manner. The relationship between the prevented thermal expansion and resultant thermal stress is described by Hooke's law, wherein the elastic material constants, modulus of elasticity E and Poisson's ratio ν, of the observed material are to be used.

Figure 11.1 shows the linear coefficients of thermal expansion of various metallic materials and their dependence on temperature. This representation is for illustrative

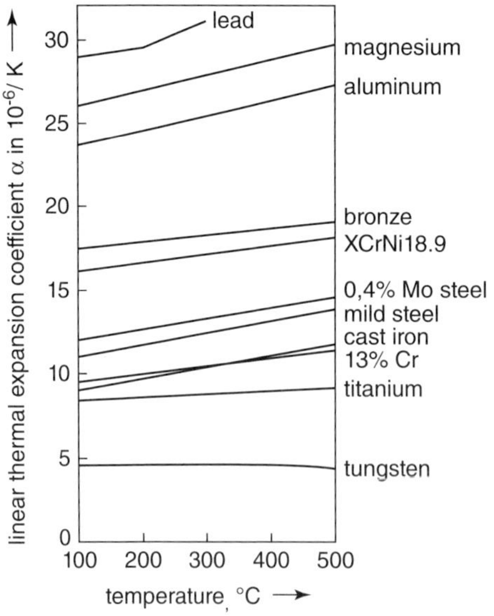

Fig. 11.1 Linear thermal expansion coefficients of various materials and their temperature dependence [11.3,11.4]

purposes only and does not provide numerical values for practical application. One should refer for the data to a reliable source specifically for the material of the considered test object or determine it yourself.

11.2 Sensing the prevented thermal expansion at identical temperature at the compensation gage and the active strain gage

A great advantage of the Wheatstone bridge circuit for measuring with strain gages is the self-compensation of equal temperature effects in adjacent arms of the bridge. This property is beneficially used for measuring under changes in temperature by switching a mechanically unloaded but also temperature influenced compensating strain gage in the neighbouring branch of the active bridge arm. There the compensation strain gage gets the free thermal expansion of its carrier material and compensates the signal generated by the active gage produced by the temperature change. This process is easy to prove by subjecting both mechanically unloaded strain gages same temperature changes.

If in the process the thermal expansion of the measurement object with the 'active' strain gage is completely prevented, as shown in Figure 11.2a, the bridge circuit provides as an output the signal of the compensating strain gage, because the signal to be compensated is not present. In other words, the bridge output signal is a measure for the thermal strain which would have occurred at the measurement object were it not for the blocking of thermal expansion. This prevented strain is then responsible for the generation of thermal stresses.

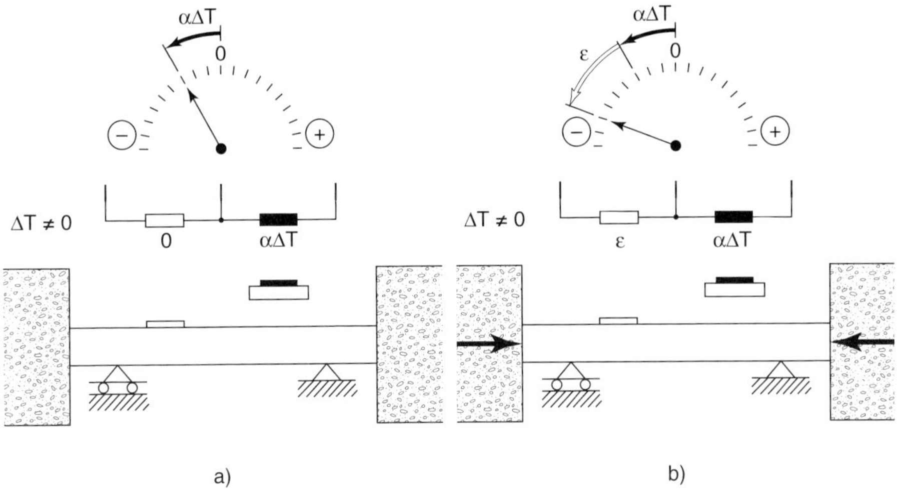

Fig. 11.2 Output signal of a bridge circuit with compensating strain gages a) at completely disabled thermal expansion, b) superposition of mechanical strain and restricted thermal expansion in the measurement signal

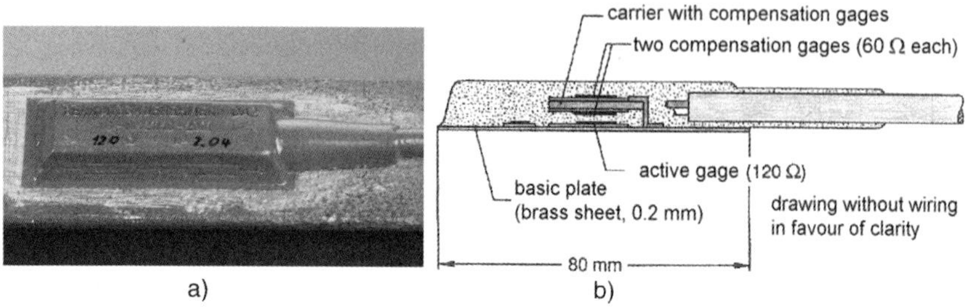

Fig. 11.3 Encapsulated strain gage half-bridge circuit; a) view of the transducer, b) below: schematic structure of the transducer

The above contemplation is unfortunately idealized in this respect, as it emanates from a mechanically unloaded measurement object. If in the measurement object during the measurement there occurs, in addition to the prevented thermal expansion, a change in the mechanical load, as shown in Figure 11.2b, then the described type of measurement will not allow us to separate the thermal stresses from the mechanical stresses – the two overlap. Hence the stresses calculated from the measured strains in this case are the sum of the mechanical stress and the thermal stress at the measuring point. Figure 11.2 shows schematically the signals of the bridge circuit created under restricted deformation and superimposed mechanical stress. Nevertheless, the method of measuring the restricted thermal expansion with a half-bridge circuit with active and compensation gages at the same temperature is relatively often used and gives acceptable results when the identity of the temperature can be ensured.

A convenient way to apply this method for determining thermal stresses, particularly for steel, is to use a special encapsulated half-bridge strain transducer, as depicted in Figure 11.3.

This transducer contains an active strain gage (120 Ω) and two compensation strain gages (each 60 Ω) interconnected to form a half-bridge. The compensation strips are installed on a steel sheet and, during the measurement, they remain undeformed. To avoid any disturbing influences due to possible bending of the compensation sheet, the compensation gages are installed at opposite sites, on the top and bottom. The encapsulated strain transducer is adapted for ferritic steel with a thermal expansion coefficient $\alpha = 11^{-6}/\mathrm{K}$ and thus insensitive to thermally induced strains in the corresponding measurement object. The design of this sensor is also ideally suited to the determination of thermal expansions that occur by strain restricted measurement objects. In the case of temperature changes, the supporting plate can stretch thermally with the compensation gages, while the active gage detects the expansion of the measurement object. If the measurement object expands without strain restriction, the measured signal becomes zero. The measurement signal output from the half-bridge circuit corresponds to the amount of the strain restriction occurring at the measuring object. Transducers of this type have been successfully applied for measurements of thermally induced longitudinal forces in continuously welded railway rails [11.5].

11.3 Determination of the restricted thermal expansion by computational correction of the measured values with previously measured thermal outputs on dummies

The described method of determination of the restricted thermal expansion requires that the compensating strain gage at each measuring point at any time has the same temperature as the active strain gage. This may lead to high experimental outlay or may be technically not feasible if multipoint measurements must be carried out on measurement objects with unsteady and multidimensional temperature fields (temperature gradients).

Another method of determining the restricted thermal expansion, favoured by the modern instrumentation, can be performed by calculation rather than by correction from the compensation strain gage (compensation of the temperature response of the active strip). Two conditions must be met: the temperature at each measurement point must be known, and also the thermal output of the measuring points on the material of the measurement object under free deformation. The temperature can be measured by placing a thermocouple at each measuring point. For this purpose, strain gages with built-in thermocouple were developed, as shown in the example in Figure 11.4. Such strain gages are rarely offered today. However, modern strain gage instrumentation will be able to process the signals from the thermocouples.

The determination of the thermal output should be made in each case by practical measurements on several samples. You install strain gages and a thermocouple on a carrier made of the material from the measurement object, and heat this arrangement in an oven up to the temperature expected in the later original measurements.

The Kreuzer circuit should be used to connect the strain gages. If this is not possible due to the lack of suitable instrumentation, the connection must be made at least using the three-wire system (compensation of temperature effects on the leads). During the heating phase, the output signals of the strain gages and the thermocouple should be recorded as continuously as possible.

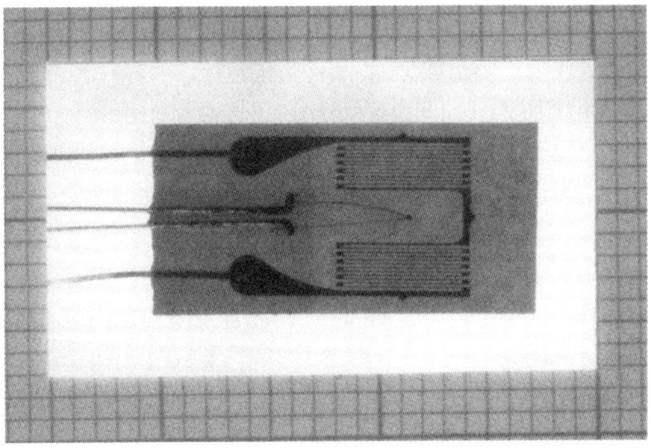

Fig. 11.4 Foil strain gage with built-in thermocouple (BLH)

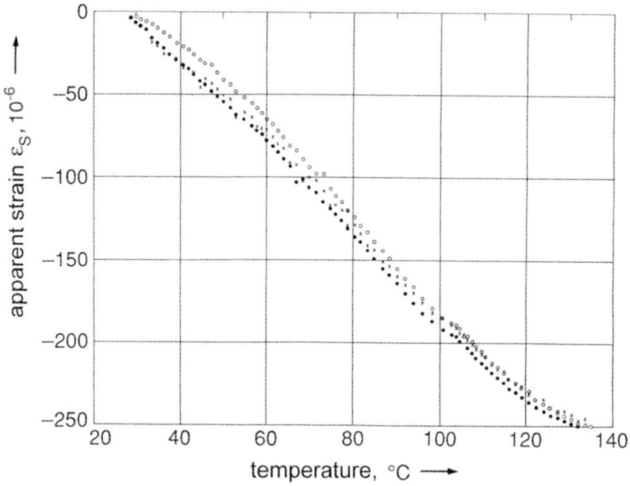

Fig. 11.5 Thermal output of strain gages adapted to ferritic steel ($\alpha = 11.5 \times 10^{-6}$/K) installed on grey cast iron GG20 ($\alpha = 9 \times 10^{-6}$/K) [11.6]

The results give not only the thermal output of the strain gage type used on the material of the test object, but the scattering of the thermal output, which allows conclusions to be drawn about the expected uncertainty. Figure 11.5 shows an example of such a measurement, the thermal output of three strain gages matched to ferritic steel ($\alpha = 11.5 \times 10^{-6}$/K) installed on cast iron GG20 ($\alpha = 9 \times 10^{-6}$/K) within a temperature range of 20–140 °C. They were recorded to determine the thermal stresses in cast iron gas boilers [11.6].

The apparent strain (thermal output) shown in the example is negative because the strain gages are adapted to a larger value for α and compensating more strain as required for cast iron with its smaller α.

The apparent strains recorded at each measuring point in the original experiment are corrected (after any necessary gage factor correction) with the thermal output signals recorded at the pretest values. At 70 °C in the example cited a strain of −160 μm/m was recorded; this value contains, from Figure 11.5, a thermal output (temperature-induced apparent strain) of $\varepsilon_S = -100 \pm 10$ μm/m. That means that the actual measured value is only -60 ± 10 μm/m. Whether this measurement has a mechanical cause or has arisen due to a restricted thermal expansion can't be determined from the measured signal. Valid conclusions, if possible, must be drawn from the environment in which the test object is embedded. Thus, for example, when mechanical stresses of the measuring object are not present, it may still have thermal stresses. This is similar to the procedure in [11.7], where the effect of the radial and tangential temperature gradients was examined in the bottom of the piston of a diesel engine to determine the stress state. This also provides information about the temperature dependence of the modulus of elasticity of the piston material (Al-Si alloy) in the range of 0–320 °C.

11.4 Determination of the restricted thermal expansion by mathematical correction with the thermal outputs determined at the original measurement object

The experimental determination of the thermal output of a measurement point is best done with the original measuring object itself. The original test object with the installed strain gages should be put in an oven and heated slowly so that a uniform heating of the entire object is achieved, and then cooled slowly, too. The thermal outputs of all measuring points are registered. Later in the original research, with the test object subjected to non-uniform and non-stationary temperature distribution, the recorded measuring values can then be corrected using the original thermal outputs. By using the original measuring points for determining the thermal outputs, possible variations of the thermal outputs within a strain gage type are turned off and can't influence the measurement uncertainties. However, a prerequisite for the application of this procedure is that at each measuring time, knowledge of the temperature at the measuring point must be known, because the temperature is used as a correction parameter. This means installing thermocouples at each measuring point or using strain gages with integrated thermocouples. In [11.8], such a test procedure was applied to thick-walled tubes at temperatures up to 650 °C is described. The aim of the experiment was to determine the thermal stresses in pipes, which were externally heated and inside had a flow of cooling liquid. In this work, the experimentally determined temperature dependence of the gage factor k is given for three different free-grid strain gage types.

The uniform heating of the instrumented test object can often be realized in model experiments. An interesting example is given in [11.9], where an original metallic steam turbine nozzle chamber was modelled as a plastic model on a scale of 1 : 4, and the thermal strains were determined. The plastic model required, because of its high value for α, only relatively small temperature gradients to produce the desired thermal expansion. In order to measure the thermal outputs of the measurement points the total model was subjected to a slow temperature change by slow cooling from room temperature to −90 °C. In the following experiment the temperature gradients in the model were generated by cooled methyl alcohol, and then the computational correction of the measured values of each measuring point was based on the previously measured thermal outputs.

11.5 The 'reversible' strain gage

One of the problems arising in high-temperature strain measurements using strain gages is to separate strain caused by mechanical stresses from measured strain data with sufficient accuracy. In order to get reliable information about the thermal output of the strain gage a reversible strain gage was developed by Koichi Egawa in Japan [11.10]. This strain gage can be peeled off after taking the required measurement of thermal output of the measuring point with the gage in a furnace at known temperature. After this measurement, the strain gage has to be peeled off and attached reverse-side-up at the measuring point of interest on a test structure.

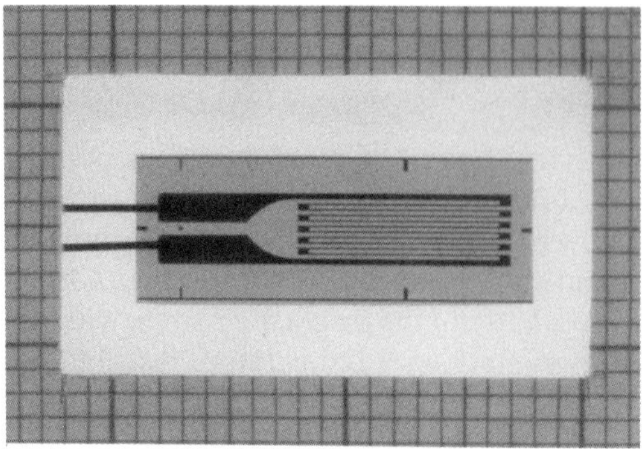

Fig. 11.6 Photo of a reversible strain gage

Since the thermal output is known, the strain gage reading at a known temperature can be corrected. The temperature at the gage also has to be measured. Mechanical strain at the measuring point is obtained as a result [11.11].

The measurement grid is embedded between two identical designed carrier foils. The gage's leads are thin and connected by spot welding, to be as flatly as possible. Longer solder tabs were used to keep the connected part away from the sensing grid of the gage. Figure 11.6 shows the reversible strain gage developed and produced in Japan. The tests were carried out with cold-curing adhesive on plastic measuring objects, which were not suitable for applying a heat-curing adhesive.

11.6 Separation of the mechanical strain from the thermal strain

Strain gages cannot recognize whether the change in resistance produced in them is of thermal or mechanical origin. Thus, it is not possible in general to assign the partial signal to these two causes. Unique assignment is only possible when one of the causes is known to be absent. And so this points to the idea of a possible separation by measuring the effects of one cause in the absence of the other, in order to determine the existing mechanical and thermal strain shares later that were overlaid in the original trial. In the general view, the strain ε provided by the strain gage output consists of the shares

$$\varepsilon = \varepsilon_L + \varepsilon_T - \alpha \cdot \Delta T \tag{11.2}$$

In Eq. (11.2) ε_L is the strain caused by mechanical stress and ε_T is the thermal strain that occurred at the measurement object and $\alpha\Delta T$ is the thermal strain under unrestricted condition. ΔT is the temperature difference between the state of $\varepsilon_T = 0$ (zero calibration) and the current temperature.

If now measurements are made at constant external mechanical load, but at two different temperatures ΔT apart, we obtain, for the one measuring point, two results

$$\varepsilon_1 = \varepsilon_L + \varepsilon_{T1} - \alpha \cdot \Delta T_1 \tag{11.3}$$

$$\varepsilon_2 = \varepsilon_L + \varepsilon_{T2} - \alpha \cdot \Delta T_2 \tag{11.4}$$

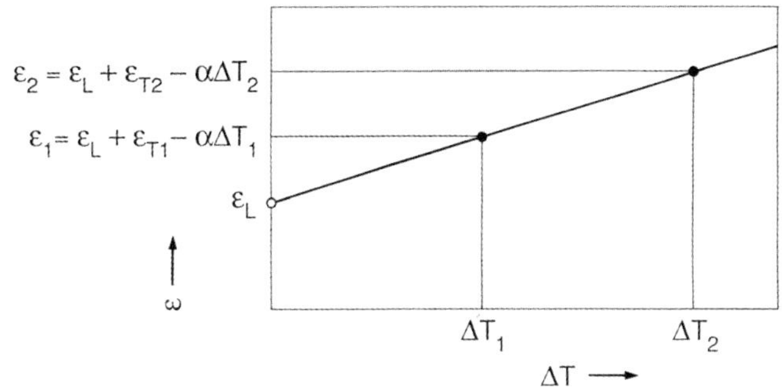

Fig. 11.7 Separation of mechanical and thermal strain shares by graphical extrapolation of measured strain values at $\Delta T = 0$ at constant mechanical load and varying temperature

From these two equations is now quite simple to find the value for ε_L, because the premise of at $\Delta T = 0$ and $\varepsilon_T = 0$. This can be done graphically by a graphical extrapolation, as shown in Figure 11.7.

The method described, for example, was successfully applied to skin areas of aerodynamically heated aircraft structures [11.11]. The resulting diagram for this application is shown in Figure 11.8. It illustrates the strain gage output under constant mechanical load for four different mechanical loading conditions on the temperature difference to the initial state $\Delta T = 0$.

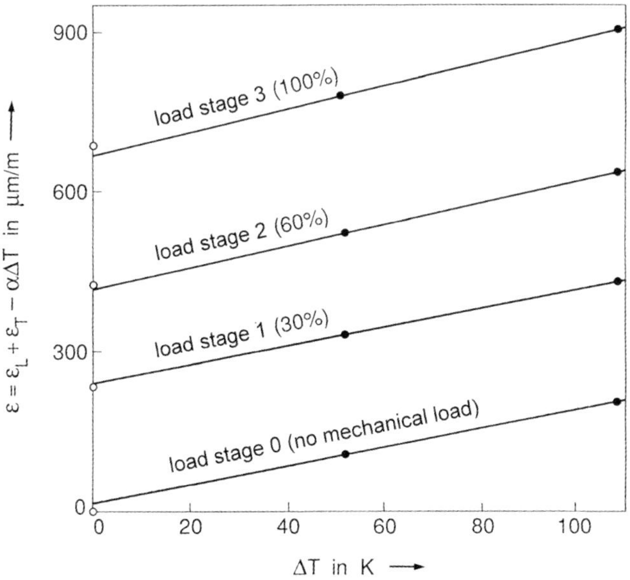

Fig. 11.8 Example for the described separation of load and thermal stresses in aircraft [11:11]

The procedure was tested in the experiment, in which the mechanical strain ε_L could be measured for the case $\Delta T = 0$ for each load case. The following measurements at varying temperatures provided in their extrapolations to $\Delta T = 0$ to a very good approximation, and the previously experimentally determined values are shown as open points on the ordinate of the graph.

The reverse procedure, that is to change the load at constant temperature, does not lead to the desired goal of the separation since the mechanical strain shares cannot be measured as the temperatures individually.

11.7 Compensated half-bridge strain gage with compensation resistor

This special strain gage (Figure 11.9) has, on the same carrier, between the two halves of a divided foil grid, a loop of a wire whose material has a high temperature coefficient of its resistivity [11.12]; it is usually made of platinum.

This wire is set adjacent to the measurement grid in the compensation branch of the bridge circuit so that its change of resistance provides compensation. Adaptation to the thermal output of the bridge arm with the strain gage is carried out with an adjustable ballast resistor, which is connected in series with the platinum wire, and reduces the high thermal output of the wire to the extent required for compensation. But this arrangement still requires balancing resistors for balancing the bridge circuit.

This type strain gage therefore provides good temperature compensation even for rapid temperature changes or where large temperature gradients occur. When these strain gages are adapted to the measurement object material, they provide a signal for the constrained thermal strain, when placed on a mechanically unloaded but temperature influenced piece of measuring object material (in the same way as a half-bridge circuit with compensation gages on a freely stretchable workpiece of the same material and the same temperature as the test object). But again it is not possible to separate the constrained thermal expansion from the mechanical strain by a possibly external load.

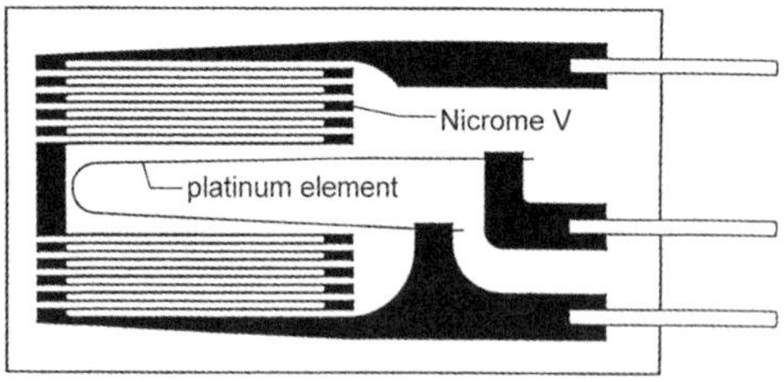

Fig. 11.9 Compensated strain gage in a half-bridge circuit with embedded compensation wire of platinum

References

[11.1] Grüneisen, E. (1908) Über die thermische Ausdehnung und die spezifische Wärme der Metalle. *Annalen der Physik*, **26**, 211–216.

[11.2] DIN, 13316 (Sept. 1980) Mechanik ideal elastischer Körper, Begriffe, Größen, Formelzeichen, Beuth-Verlag Berlin.

[11.3] Dubbel (1983) *Taschenbuch für den Maschinenbau* (eds W. Beitz and K.-H. Hütter), 15. Auflage Springer Verlag, p. 1375.

[11.4] Hoffmann, K. (1987) *Eine Einführung in die Technik des Messens mit Dehnungsmeßstreifen*, (Published by:) HBM GmbH Darmstadt, p. 36.

[11.5] Maree, H. (1987) Forces in the railway track structure on the Olifant River Bridge. *RAM*, **3** (1), 1–7.

[11.6] Magin, W. (1980) Ermittlung von Wärmespannungen an Gasheizkesseln mit Hilfe von Dehnungsmeßstreifen. *MTB*, **16** (3), 56–60.

[11.7] Anderko, K., Keilbach, D. and Wacker, E. (1968) Wärmespannungsmessungen an einem Aluminium-Kolben für einen mittelgroßen Dieselmotor mit Hilfe von Dehnungsmeßstreifen. *MTB*, **4** (3), 50–55.

[11.8] Tschierschke, P. (1968) Messung der behinderten Wärmedehnung bei Temperaturen bis 650 °C. *MTB*, **4** (3), 46–49.

[11.9] Marloff, R.H. (1978) Measurement of thermal strains and temperatures in a steam turbine nozzle chamber model. *VDI-Berichte*, **313**, 323–328.

[11.10] Egawa, K. (1982) Reversible strain gage. *Experimental Mechanics*, **22** (5), 161–165.

[11.11] Thomson, R.A. and Vergamini, P.L. (1961) Experimental separation of load and thermal stresses. *Experimental Mechanics*, **1** (10), 143–147.

[11.12] Weymouth, L.J. (1960) Something new in strain gage temperature compensating. *BLH Testing Topics*, **12** (4), 1–2.

12 Strain gages as a means for experimental determination of residual stresses

12.1 Preliminary observation

Residual stresses are mechanical stresses that exist in materials without the effect of external forces or moments. These stresses are generated by changes in volume and their hindrance taking place in certain regions of the material. Changes in volume can be created by thermal treatment (e.g. cooling down after casting, rolling, welding or heat treatment), by mechanical handling (i.e. cold work), by chemical processes (i.e. structural change). If residual stresses have the same sign as the load stresses they reduce the load capacity, and in extreme cases a wrongly treated component can fail without external effects but just by the existence of these stresses.

Figure 12.1 shows an example [12.1] in which a beam of high tensile strength steel, 12 m long, fractured spontaneously under no load. But the other effect of residual stresses is that they are able to increase the load capacity of a component, if their sign is opposite to the load stress. Therefore, for optimal use of structural members it is important to create a favourable residual stress state in the component and to know this stress state.

In a body containing residual stresses, there is an inner equilibrium. The material laws (i.e. Hooke's law) are valid and formulate the relationship between stresses and deformation. Disturbance of the state of equilibrium present in the component as a result of external mechanical interference will create a new state of equilibrium which is associated with a change in the deformation state. The difference of the two states of equilibrium and the deformation occurring as result of the change in equilibrium can be recorded using strain gages. This forms the basis for the determination of residual stresses with strain gages.

The principle of the determination of residual stresses using strain gages is based on the mounting of strain gages on the surface of the component under examination followed by the subsequent disturbance of the equilibrium present in the component by external mechanical interference. As a result of the interference, which may be brought about by drilling, sawing or milling, a new state of equilibrium is produced in conjunction with the change in deformation conditions. The strains occurring on the surface of the measurement object are measured by the strain gages. Inferences regarding the residual stress state present in the component before the interference took place can be drawn from the measured changes of strain. The principle requires a certain degree of destruction of the measuring object.

There are four main methods currently in use for determining residual stresses in components with strain gages: (1) the *cutting-down method*, in which the measurement object is fitted with as many strain gages as possible and is then sawn up into a large number of small parts; (2) the *layer-removal method* in which thin surface layers are removed from one side of the measurement object and the resulting deformations are measured on the opposite side with strain gages; (3) the *ring-core*

Technology and Practical Use of Strain Gages: With Particular Consideration of Stress Analysis Using Strain Gages. First Edition. Stefan Keil.

Fig. 12.1 A beam of high tensile strength steel that fractured spontaneously under no load [12.1]

method in which an annular groove is milled in the surface of the measurement object around a special strain gage rosette which is bonded on the surface; (4) the *hole-drilling method* in which a small hole is made in the measurement object and the changes of strain occurring in three radial directions are measured using a special strain gage rosette.

The stresses that have created the changes in deformation cannot be measured by strain gages; only the deformations are measurable by strain gages. The stresses have to be calculated from the measured strains by using appropriate formulas.

A feature common of all these methods is that changes of strain are measured, from which the residual stress state must be calculated. These methods therefore entail a considerable degree of computation, which however recedes into the background if the user employs up-to-date techniques. This is particularly true in the case of the ring-core method and the hole-drilling method which are presented in detail in the following. By using an online measurement system and appropriate software, the evaluation of the signals and the graphical display of the results can be carried out with the minimum of effort.

While installing the strain gage it is important to give full attention to avoid reactive effects on the residual stress state that is to be determined.

The first publication regarding determination of residual stresses using strain gages were written in the USA in 1943 [12.2]. Since then this possibility of residual stress determination using strain gages has been further developed. A concise description of the four methods can be found in [12.3] (1989).

12.2 Cutting-down method

Residual stresses that are directed perpendicularly to the surface of the component must be zero at the surface due to the equilibrium. Only in layers under the surface can perpendicular residual stresses take place. For instance, longitudinal residual stresses in a beam are zero at the ends of the beam, although they can be considerable between the ends of the beam.

If a girder with high residual stresses is cut into two parts, the longitudinal residual stresses will become zero at the two cutting planes. The change in the residual stress state will create measurable changes in deformation; from this, conclusions can be drawn on the residual stress state before its disturbance.

Research on this was carried out using mechanical measurement systems even before the strain gage was invented. In [12.4] research on rails is described which was carried out in 1936 with the cutting-down method.

Residual stresses in a component are part of an equilibrium system. In a girder with longitudinal tensile residual stresses there are also longitudinal residual compressive stresses which are equivalent to the tensile stresses and thus form the state of inner equilibrium.

Strain gages mounted on the surface in direction perpendicular to the cut direction will record the changes in strain caused by the cutting. From the measured strains, the changes in residual stress can be calculated. This explains the principle of the cutting-down method: a section of material previously fitted with strain gages is cut out of the measurement object, or the object is sawn up according to a defined cutting plan after the installation of a number of strain gage measuring points. The residual stresses can be computed from the measured strains.

Attention must be given to the fact that the original stress state will be changed by the first cut and therefore the second cut is carried out on an object whose residual stress state is no longer in its initial state. A new state of equilibrium is produced after each cut and this can produce residual stresses in uncut regions of the material. This includes the possibility that in certain regions residual stresses will increase, maybe up to the yield point of the material with plastic deformation as a result. This must be avoided.

This means that an unfavourable cutting plane can produce plastic deformation also in the region of measurement and distort the readings.

Figure 12.2 shows a photograph of slip bands on the web of an hot-rolled girder arising from residual stresses caused by cutting the girder with a circular saw.

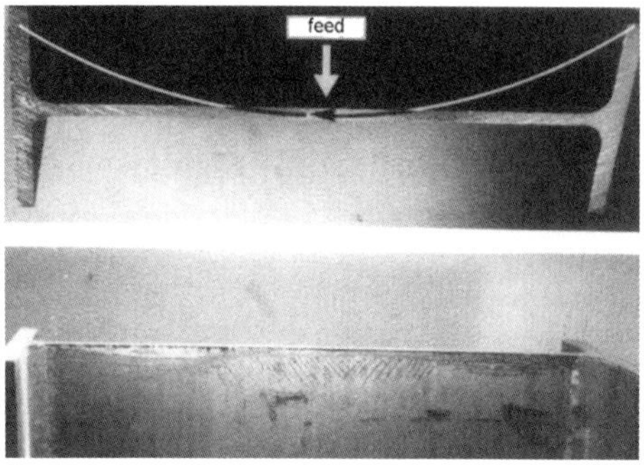

Fig. 12.2 Slip bands in the scale layer of a girder IPE 600 produced by cutting with a circular saw [12.5]

Measurement results from strain gages installed in the region of plastification are useless in the determination of residual stresses.

The sequence of the cuts should be carefully considered, so that the residual stresses can be reduced in the region of measurement. A demonstrative example is given in [12.5] and illustrated in Figure 12.3.

The aim of research was the determination of the longitudinal residual stresses in a hot-rolled girder IPB 400. Approximately in the middle of the girder length strain gages were installed along the contour of a cross-section. Then the girder was cut

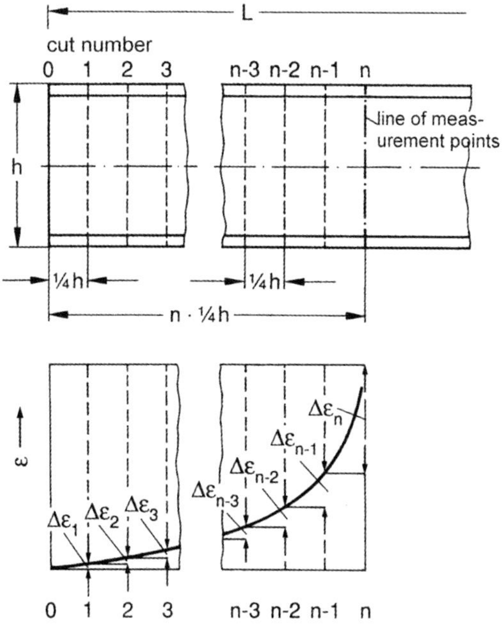

Fig. 12.3 Cuts and strain curve during the step-by-step stress relaxation of the measurement cross-section [12.5]

into two pieces, the cut being made at a distance twice the girder height from the measurement cross-section. In this way, in the region of the cut, a longitudinal residual stress state is created corresponding to the stress state at the end of the girder. Therefore plastic flow will occur, especially in long girders, but the region of creep will not reach the measuring points.

Further saw cuts each time in distance of about ¼h from the girder end lead to a step-by-step reducing of the stress state at the measuring cross-section. Figure 12.3 shows the procedure schematically. In this way the measurement cross-section is moved step by step in the direction towards the girder end. After the final cut is made at a distance of a quarter height of the girder the measurement section can be cut out without any danger of creep.

The relationship between the released stresses in the material and the elastic deformations created by them is given by Hooke's law. Opposite to load stresses here we have to introduce opposite signs. Released tensile residual stresses cause negative strains and vice versa.

Hooke's law for the determination of residual stresses in the surface plane from the principal strains ε_1 and ε_2 and by the measurement can be written as:

$$\sigma_1 = -\frac{E}{1-\nu^2}(\varepsilon_1 + \nu\varepsilon_2) \tag{12.1}$$

$$\sigma_2 = -\frac{E}{1-\nu^2}(\varepsilon_2 + \nu\varepsilon_1) \tag{12.2}$$

The calculation of the residual stresses from the measured strains is based on the assumption that the piece of material carrying the strain gages cut out from the measurement object will be totally stress released.

On symmetrical members one can estimate the principal directions to a good approximation and install strain gages in these directions. If estimation of the principal directions is not possible, the determination of the directions will be part of the analysis and strain gage rosettes with three measuring grids will have to be used.

The measurement object will be destroyed using the cutting-down method, but the results can deliver fundamental insights into which process parameters of production and treatment can influence the formation of residual stresses.

Example: Application of the cutting-down method on a hot-rolled steel girder [12.6]
Temperature differences that occur during cooling of hot-rolled steel girders produce plastic deformations in the material, and this causes residual stresses. Straightening carried out later will cause additional residual stresses. Figure 12.1 of the introduction gave an indication of the quantity of residual stresses and their damaging effects on a girder.

The cutting-down method mainly provides information on the residual stresses due to the material processes to which the component has been subjected. This is shown in two application examples. In hot-rolled steel girders local temperature differences

Fig. 12.4 Application of the cutting-down method on a freely cooled hot-rolled girder IPB 400 [12.6]

occurring during the cooling period produce residual stresses. Since these residual stresses have far-reaching consequences with regard to the load-bearing capacity of the girder, the effects of the treatment of a girder on its residual stress state has been investigated using the cutting down method. Figure 12.4 shows the cut parts of the measurement object.

Residual stresses have a strong influence on the load-bearing capability of a girder. Therefore the influence of the parameters of the treatment of girders on their residual stress state has been the subject of extensive research with strain gages [12.6]. Figures 12.4 and 12.5 show strain gage installations on the girder and the saw cuts.

Particular care regarding the covering of the measuring points is required. Reliable protection against moisture, oil and mechanical damage must be guaranteed. The covering has to be carried out in a very small space and must not influence the mechanical behaviour of the measurement object or the action of the strain gages.

Figures 12.6 and 12.7 show samples results found by using the cutting-down method on a freely cooled hot-rolled unstraightened girder IPB 400.

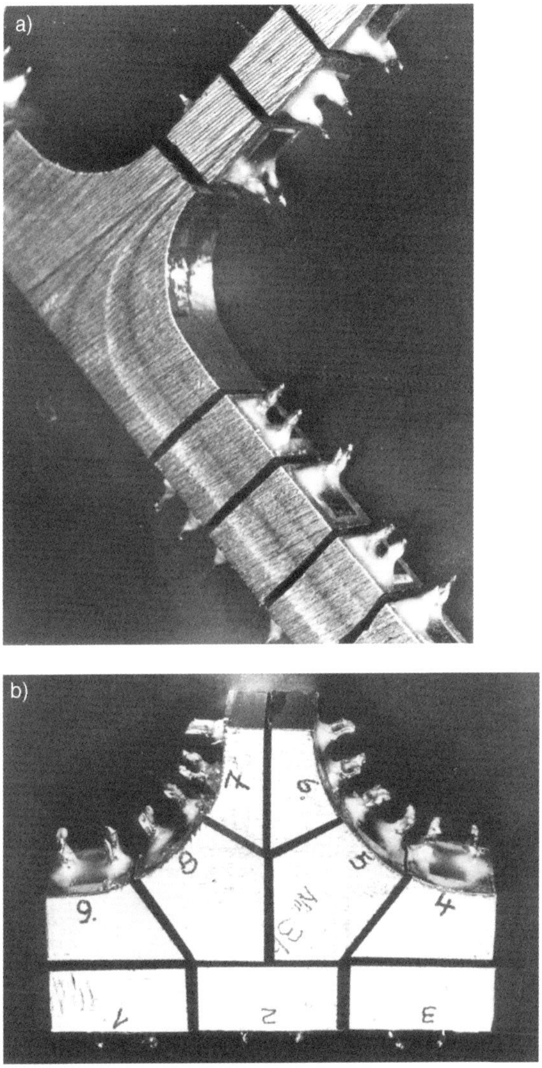

Fig. 12.5 a) Dismantling of the girder shown in Fig. 12.4, b) Further dismantling of the section which can be seen in Fig. 12.4 [12.6]

As can be seen in Figure 12.7, the residual stresses reach their full amount at a distance from the end of the girder of about twice the height of the girder. This shows that the length of a sample used for determination of longitudinal residual stresses must be at least four times the height of the girder.

The method as explained above can also be used on long beams with different profiles, such as rails.

Frequently, the cutting-down method is used on flat structures. Particularly on welded sheets, temperature effects caused by the welding process can create

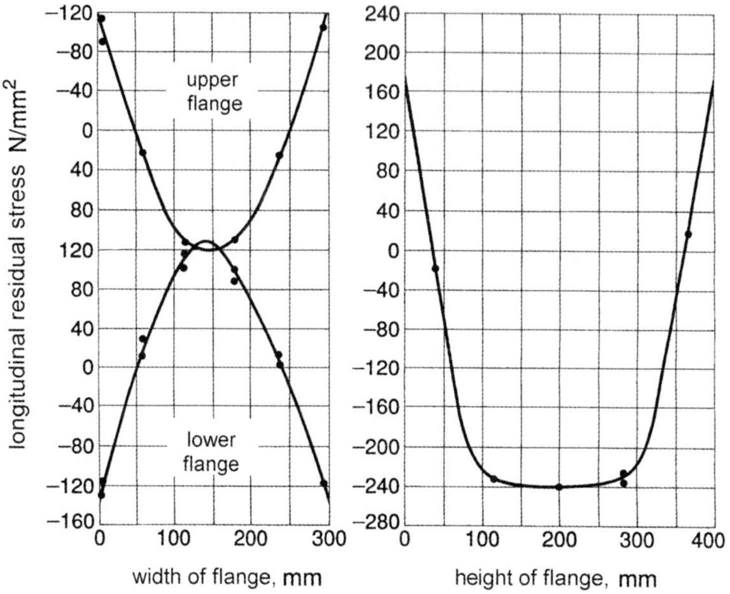

Fig. 12.6 Longitudinal residual stresses in flange and web of a hot-rolled girder IPB 400 [12.5]

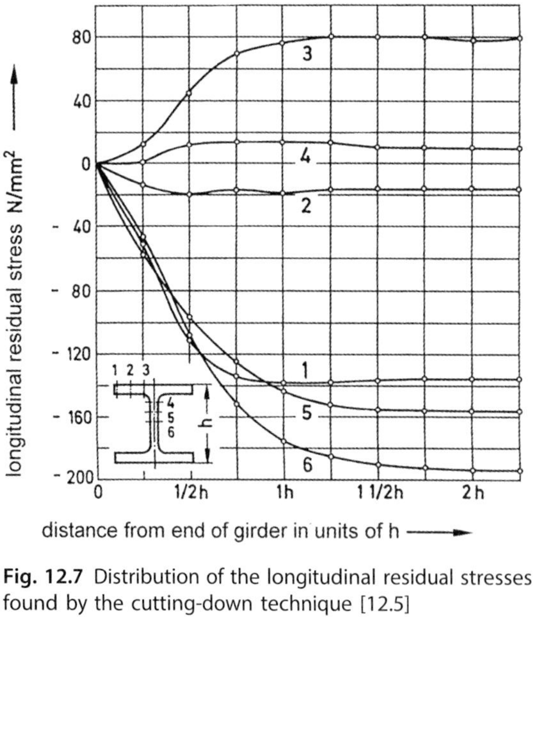

Fig. 12.7 Distribution of the longitudinal residual stresses near the end of the girder type IPB 400, as found by the cutting-down technique [12.5]

remarkable residual stresses. In USA in 1943 an extended research project was established to investigate cases of damage on welded ships [12.7–12.9]. Since then strain gages have proved themselves in a lot of applications for the determination of residual stresses, examples of which can be found in [12.10–12.16]. The following example concerns stress determination on V-shaped welds [12.17].

Example: Application of the cutting-down method on V-shaped welds [12.17]
Since the principal directions of the residual stresses in welded joints run longitudinally and transversely with respect to the seam, measurements are made in these directions.

In this example, as shown in Figure 12.8a, strain gage chains were bonded over V-shaped welds in sheets which were then progressively shortened in the transverse direction by taking cuts on alternate sides. Figure 12.8b illustrates the results of the longitudinal stress distribution in a V-joint in a 4.5 mm thick plate of high-tensile weldable steel, type StE70. The example demonstrates that measurements relatively close together can be made by using very small strain gages which have been bonded using methods appropriate to the application.

The stresses were calculated from the measured strains using Hooke's material law. This calculation will become uncertain if big stress gradients occur perpendicular to the surface of the sheet, or if the geometry of the measurement object does not allow the transfer of the stress state found on the surface into the layers

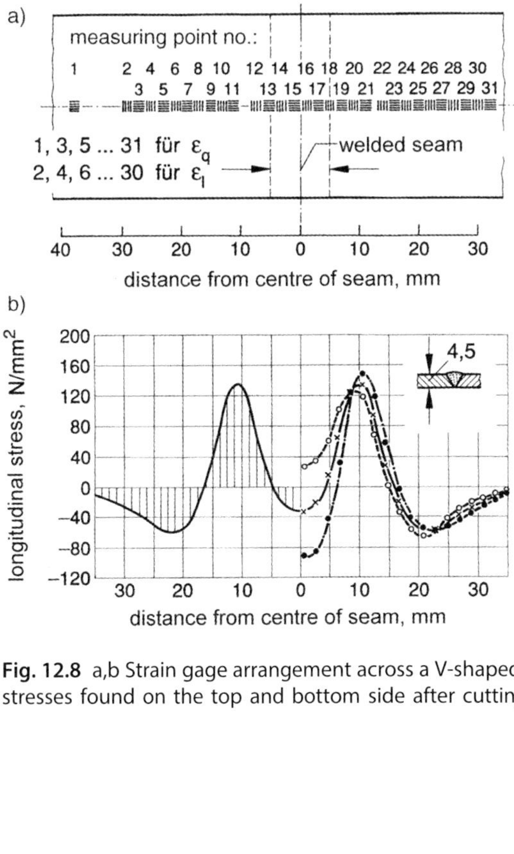

Fig. 12.8 a,b Strain gage arrangement across a V-shaped weld and distribution of longitudinal residual stresses found on the top and bottom side after cutting down the measurement object

below the surface. Difficulties occurring during evaluation can be avoided by combining the strain gage measurements with algorithms of the finite element method. The advantages of this combination were described 1974 by Stoker and Moir [12.18]. Since then the cutting down method has been used in some quite complicated cases [12.19–12.21].

The cutting-down method can also be applied in combination with other destructive methods. The small pieces produced by sawing up the test object can then be further diminished using the layer-removal method on the opposite site of the strain gage installation. In 1950 [12.22] an application on magnesium castings was described. The layer-removal method was carried out by chemical etching.

Combinations of the cutting-down method with the hole-drilling method are described in [12.13] on welds, and in [12.23] on seamless pipes.

Generally, the cutting down method provides information on how the treatment to which components have been subjected during manufacture affects the residual stress state. The test object is destroyed, but the results are valid for all components receiving the same treatment.

12.3 Layer removal method

The first investigation of a residual stress state using the measurement of deformation caused by layer removal was carried out in 1911 by Heynand Bauer [12.24].

Cracks on cold-drawn rods were the reason for the research. The method was applied to cylindrical rods and tubes by removing layers using a lathe. On the base of equilibrium the longitudinal residual stresses in the rod could be calculated from the change in length which occurred during the layer removal. The measurement of the change in length was carried out by mechanical methods because strain gages were not available at this time. In the outer layers of cold-drawn nickel steel the residual tensile stresses came to 50% of the tensile strength of the material. Afterwards the layer removal method was used to prove the effect of relief annealing.

First attempts to determine longitudinal residual stresses in cylindrical rods by drilling a hole along the centre line in the rod and measuring the change in length were carried out in 1919 [12.25].

At that time strain gages were not yet available, so the deformation was measured by other means. The invention of the strain gage lead to significant simplification of deformation measurement, and the method has been frequently applied since, in evaluation and testing of heat treatments.

References [12.24, 12.25] deal with longitudinal residual stresses only, in general also tangential and radial residual stress components also exist in the material.

Sachs [12.26] in 1927 expanded the method proposed by Heyn and Bauer with a new calculation scheme which also took into account tangential stresses as well as the radial stresses. Applying this scheme, measurements of change in length and in addition to the change in diameter were required. Longitudinal fissures in cold-drawn brass rods were thought to be the unpleasant consequence of tangential residual stresses, which were more critical than the longitudinal stresses.

Furthermore the setup of Heyn and Bauer neglected the tangential residual stresses and underestimated the longitudinal residual stresses. Sachs developed conditional equations for the longitudinal, tangential and radial residual stresses in homogenous cylindrical rods under continuous removal of layers using the drilling-out and turning-off method.

Even today Eqs. (12.4–12.6) are the basic formulas for determination of residual stresses in cylindrical bodies made from homogenous materials. Introducing the simplifying term

$$E^{*} = \frac{E}{1-\nu^{2}} \tag{12.3}$$

the equations by Sachs have the form

$$\sigma_{a} = E^{*}\left[(A_{0}-A)\frac{\mathrm{d}(\lambda+\nu\vartheta)}{\mathrm{d}A} - (\lambda+\nu\vartheta)\right] \tag{12.4}$$

$$\sigma_{t} = E^{*}\left[(A_{0}-A)\frac{\mathrm{d}(\vartheta+\nu\lambda)}{\mathrm{d}A} - \frac{A_{0}+A}{2A}(\vartheta+\nu\lambda)\right] \tag{12.5}$$

$$\sigma_{r} = E^{*}\left[\frac{A_{0}-A}{2A}(\vartheta+\nu\lambda)\right] \tag{12.6}$$

For the particular case of the uniaxial stress state with longitudinal stresses, from Eq. (12.4) we get the following Eq. (12.7), which was given by Heyn and Bauer:

$$\sigma_{a} = E\left[(A_{0}-A)\frac{\mathrm{d}\lambda}{\mathrm{d}A} - \lambda\right] \tag{12.7}$$

Here E = modulus of elasticity, ν = Poisson's ratio, A_0 = initial area, A = changed area, as can be seen in Figure 12.9, and λ and ϑ are the longitudinal and tangential strains according to the designations used by Sachs.

With the invention of the strain gage an ideal means of measurement was available in order to carry out the layer-removal method and their evaluation using the equations given by Sachs. The difficult deformation measurements in particular inside the drilled hole could now be carried out in a simpler way with higher resolution and a smaller uncertainty.

In 1952, the combination of the boring-out and turning-off methods carried out successfully by Bühler and Schreiber [12.27, 12.28] offered possibilities for complete determination of the residual stress state in a cylindrical metallic specimen [12.29].

Installation of strain gages on the inner surface of the cylindrical body can be carried out with the aid of an inflatable hose [12.30]. Installations in deep and narrow holes are possible using this means. In 1952, Bühler and Schreiber used this method and installed strain gages successfully in holes with a diameter of just 14 mm. A detailed

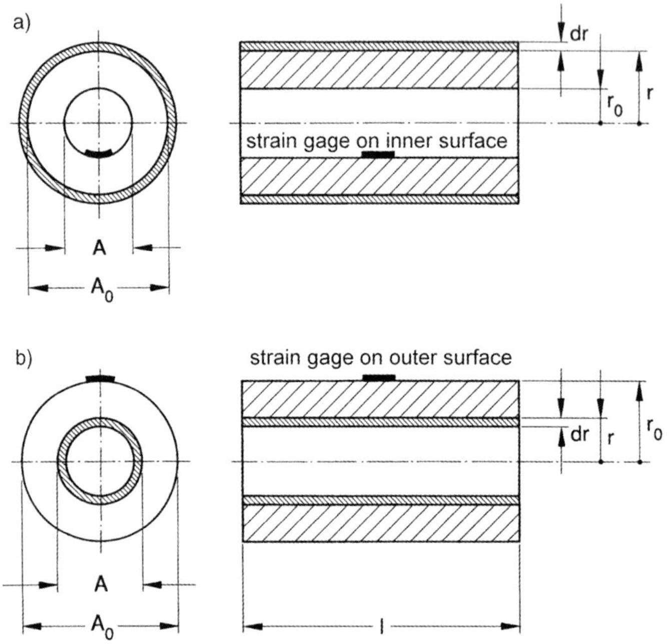

Fig. 12.9 Removal of the inner surface a) respective to the outer surface b) with the layer-removal method on a cylindrical rod and definition of the geometrical quantities

application report is given in [12.31], with a lot of very useful practical hints for strain gage installations in narrow holes with an example of the determination of stress states in case-hardened cylindrical rods made from 18NiCrMo14.

The example presented in [12.29] shows the complete determination of the residual stress state in two identical hollow cylinders made from brass MS 58, chilled from 800 °C in quenching water. The initial specimens had an outer diameter of 60.3 mm and a diameter of the inner hole of 25.3 mm. The length of the two specimens after cutting the original body in two parts was 122 mm. Figure 12.10 shows the strains

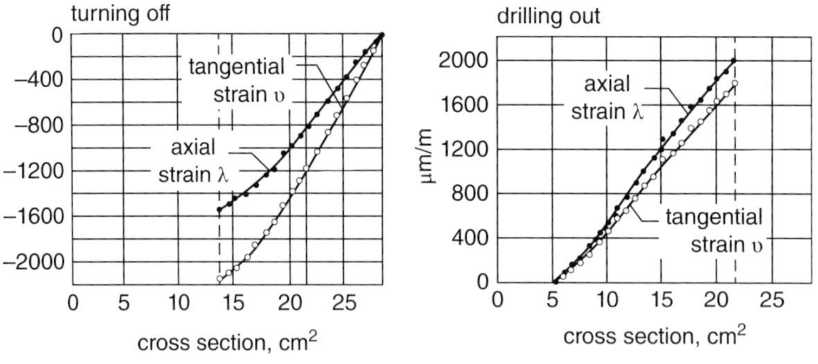

Fig. 12.10 Strains measured during drilling out and turning off in relation to the removed cross-section [12.29]

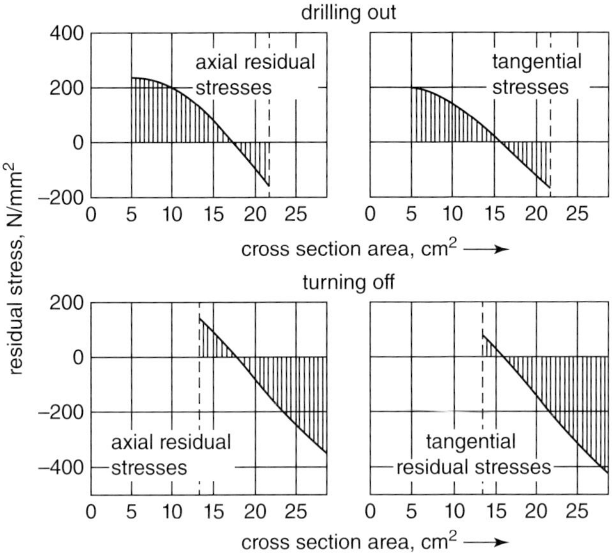

Fig. 12.11 Residual stresses calculated from the strains given in Figure 12.10, using Eqs. (12.4) and (12.5) [12.29]

which have been measured during turning off and boring out in relationship with the removed cross-section.

A combination of layer removal with boring along the centre line enabled the consistent estimation of three-dimensional residual stress states in cylindrical rods. The method starts by boring out a centre hole in the cylinder, which has strain gages on its surface. The boring is carried out step by step with increasing diameter until strain gages can be mounted on the inner surface of the hole. After strain gages have been mounted, the outer surface is removed with a lathe or by grinding. Figure 12.9 shows the principle of the method. Removal of layers can also be carried out by electrolytic abrasion or chemical etching.

For demonstration of the compatibility of the two methods in the transition region two specimen made from the same material were used; one was bored out up to a diameter of 52.7 mm (21.8 cm^2) and the other was turned off to a diameter of 42.4 mm (13.5 cm^2). The stresses calculated from the strains given in Figure 12.10 using the Eqs. (12.4)–(12.6) can be seen in Figure 12.11. Radial stresses are very small and are not under consideration here.

In Figure 12.12 the results from the drilling out and turning off are put together. One can see they are in good agreement in the transition region. Due to the very simple deformation measurement with strain gages the layer removing method, as explained before, is very often used for monitoring and testing heat treatment processes.

The layer-removing method in combination with strain gages is not only applicable to cylindrical specimens but also to prismatic beams and plane plates.

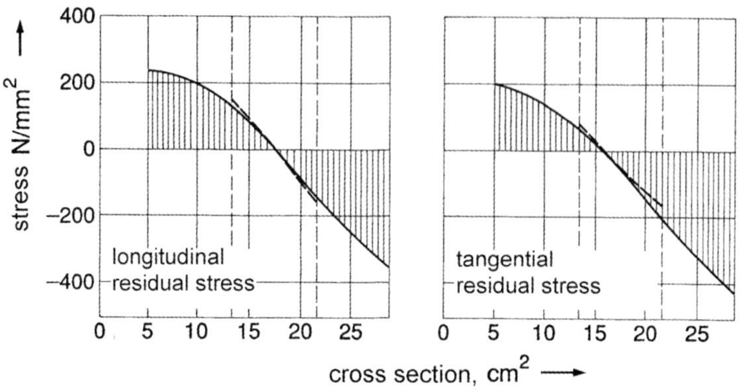

Fig. 12.12 Combination of the results of the drilling-out and turning-off method shows good agreement in the transition region [12.29]

From a plane specimen that contains residual stresses in its axial direction a small surface layer will be removed, with the result that a new state of internal equilibrium is generated. Establishing the new state of equilibrium will cause deformation of the specimen, which can be measured using strain gages.

A remarkable application example is the research into stress states in shrinkage-fits in railway wheels with the outcome that the experimental results are not in agreement with the theory of elasticity [12.32]. The research of the effect of inductive surface hardening on the residual stress state shows that the origin of fatigue cracks under torsional alternating load will not appear on the surface but in a layer below the surface [12.33]. In [12.34] the results are given of an extensive research using the boring-out method to examine the effect of cold pilgrim rolling on the residual stress state in steel pipes.

The previous ideas are based on the condition that the material under consideration is homogenous. It also is assumed that mechanical characteristics such as modulus of elasticity and Poisson's ratio are constant at all points of the cross-section. In practical cases the measurement objects are often composed of two or more different materials, for example shrunk bearing bushes or rings and claddings. For this, the method was modified by Krägeloh 1959 [12.35] taking into account that the modulus of elasticity changes over the radius. In 1989,König and Kockelmann [12.36] took up the proposals of Krägeloh, and compiled evaluation equations for the inhomogeneous distribution of modulus of elasticity and Poisson's ratio.

Introduced are each as functions of the area A the variable modulus of elasticity $\overline{E}$ and the variable Poison's ratio $\overline{\nu}$:

$$\overline{E^*} = \frac{1}{A_0 - A} \int_A^{A_0} \frac{E}{1 - \overline{\nu}^2} \mathrm{d}A \tag{12.8}$$

$$\overline{\nu} = \frac{1}{A_0 - A} \int_A^{A_0} \nu \mathrm{d}A \tag{12.9}$$

Also introduced are the auxiliary terms

$$\Lambda = \lambda + \overline{\nu}\vartheta \text{ and } \Theta = \vartheta + \overline{\nu}\lambda \tag{12.10/11}$$

So the definition equations by König and Kockelmann for the cutting-of and drilling-out method on materials with inhomogeneous distribution of E and ν read

$$\sigma_a = \overline{E}^* \left[(A_0 - A)\frac{d\Lambda}{dA} - \lambda \right] + (A_0 - A)\Lambda\frac{dE^*}{dA} \tag{12.12}$$

$$\sigma_t = \overline{E}^* \left[(A_0 - A)\frac{d\Theta}{dA} - \frac{A_0 - A}{2A}\Theta \right] + \frac{1}{2}\left(\frac{A_0^2}{A} - A \right)\Theta\frac{dE^*}{dA} \tag{12.13}$$

$$\sigma_r = \overline{E}^* \left(\frac{A_0 - A}{2A} \right)\Theta \tag{12.14}$$

With non-variable elastic characteristics E and ν, the integration will be dropped and the quotient dE^*/dA disappears. Eqs. (12.11–12.13) become Eqs. (12.4–12.6) as formulated by Sachs.

The layer removing method, in combination with strain gages, is not only applicable on cylindrical specimens but also on prismatic beams and plane plates.

In 1931,Stäblein [12.37] used the layer-removal method to determine residual stress states in single-side-cooled billets. From the billet under consideration, on one side, a layer of about 5 mm thickness was planed off, and the deformation caused was measured. From the measured deformation the longitudinal residual stresses which existed in the removed layer before removal could be calculated.

A further development of this method by Schimmöller in 1972 [12.38] modified the method for the determination of residual stress states on cladded steel sheets. On one side of the sheet, the step-by-step removal of layers of about 0.4 mm thickness by electro-erosion was carried out. On the other side of the sheet strain gage rosettes were installed, which allowed strain measurements in three different directions. Measurement objects were hot-rolled steel sheets with austenitic cladding. But the evaluation of the measurements is expensive.

On the basis given by Stäblein, another way was proposed in 1982 by Lode and Peiter [12.39]. It is considered, that the measured longitudinal strain consists of a part from the curvature and a part from change in length. In order to separate the two parts, the member will be back deformed. There are not to avoid falsifications caused by the applied external forces.

In 1971, König and Dannenberg [12.40] presented an easy-to-follow evaluation algorithm for plane test objects leading to conditional equations similar to the equations formulated by Sachs for roller cylindrical bodies. Using the definitions given in Figure 12.13 the definition equations for the plane residual stress state are

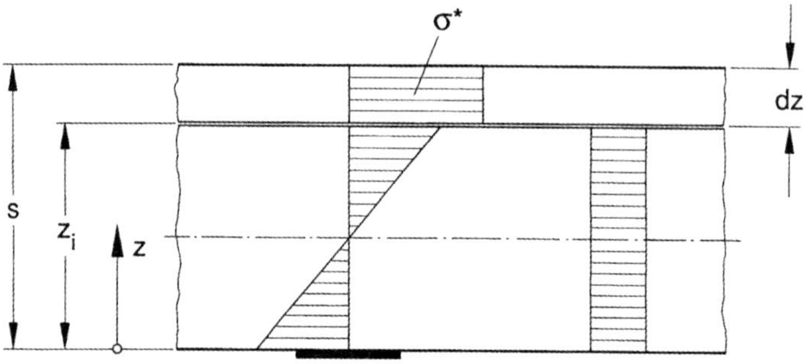

Fig. 12.13 Application of the layer-removing method on a prismatic beam or a plane plate; superimposed longitudinal and bending stresses in a hot-rolled plate after removing the layer dz

$$\sigma_1(z) = E^* \left[\frac{z_1 \mathrm{d}\Lambda}{2\mathrm{d}z} - \Lambda_i - 3z_i \int_{z_i}^{s} \frac{1}{z} \mathrm{d}\lambda \right] \tag{12.15}$$

$$\sigma_2(z) = E^* \left[\frac{z_i \mathrm{d}\Omega}{2\mathrm{d}z} - \Omega_1 - 3z_i \int_{z_i}^{s} \frac{1}{z} \mathrm{d}\omega \right] \tag{12.16}$$

The definition of E^* is given by Eq. (12.3), and accordingly from Eqs. (12.10) and (12.11) we have

$$\Lambda = \varepsilon_1 + \nu\varepsilon_2 \text{ and } \Lambda_i = \varepsilon_{1i} + \nu\varepsilon_{2i} \tag{12.17}$$

$$\Omega = \varepsilon_2 + \nu\varepsilon_1 \text{ and } \Omega_i = \varepsilon_{2i} + \nu\varepsilon_{1i} \tag{12.18}$$

The indices 1 and 2 designate the principal directions of the plane residual stress state. The strain ε_i is associated to the thickness z_i. Detailed derivation of the equations is given in [12.40].

The layer-by-layer removal of released stress σ^* based on the removed cross-section corresponds to the force F^* which caused the bending stress σ_b and the longitudinal stress σ in the remaining cross-section

The removal of layers can also be carried out using electrolytic processes [12.41] or chemical etching [12.42]. The layer removing method can also be applied in combination with the cutting down method. In a first step the measurement object is cut down to small pieces with a strain gage on the surface of each piece. In the second step the layer removal method is used on the small blocks to remove layers from the side opposite to the strain gage. Examples are described in [12.43] for magnesium castings and [12.44] for welded steel pipes.

12.4 Ring-core method

12.4.1 Principle of the ring-core method

With the ring-core method a small annular-shaped groove is milled in the surface of the object to be examined, leaving the upper part of the central core separated from the surrounding material. This separation causes the release of part of the residual stresses present in the core. The deformation occurring on the face side of the core can be measured with strain gages and inferences can be made regarding the residual stress state before the groove was milled.

The milling of an annular groove for the determination of residual stresses was first reported in 1950, but in combination with a drilled hole; here a groove was milled concentrically around a strain gage rosette which was itself mounted around the hole [12.45]. The main purpose of the hole was not for the release of residual stresses, but rather to guide the tool for cutting the groove. The first modification was made to the method in 1955 in connection with the determination of residual stresses in welds [12.46]. In order to apply the method, a plane residual stress state was assumed which did not change along the coordinate axis perpendicular to the surface. It was therefore not possible to find the distribution of the residual stresses in relation to the depth at this stage of the method's development. This only became possible at the start of the 1970s with further development that had been made with regard to the experimental determination of residual stresses in heavy forgings for the construction of turbines and generators [12.47, 12.48].

The ring-core method was protected by a patent until 1988 [12.49] and although the method has considerable advantages, it was not widely applied. Now that the patent protection has expired, the method is open to wider application. During the development of the technique a special strain gage rosette, termed a ring-core rosette, was produced. This rosette is bonded to the surface of the measurement object, and a ring-shaped groove is milled around it with special equipment. The ring-core rosette is a combination of three superimposed strain measuring grids, each angularly displaced by 45°. Figure 12.14 shows this rosette with views from above and below. The rosette has a relatively hard metal covering layer to protect it mechanically against damage due to swarf during milling and to provide a simple way of connecting the leads which must be led away vertically upwards [12.50].

In its present stage of development, the ring-core method is able to provide information on the residual stresses in the layers situated below the component's

Fig. 12.14 RY 51 ring-core rosette viewed from above and below

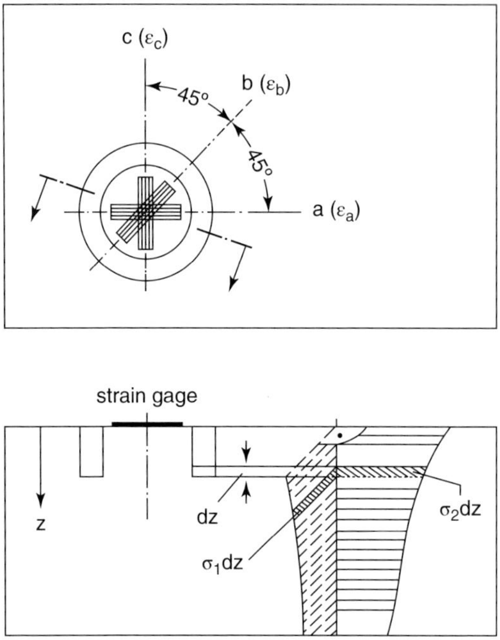

Fig. 12.15 Principle of the ring-core method for determination of plane residual stresses below the surface

surface. In order to achieve this, the technique requires the measurement of the changes in strain in relationship to the groove depth. Also, knowledge is needed about relaxation functions which are specific to the method and which must be known before the method is used on the test object.

As shown diagrammatically in Figure 12.15, the principle of the method is based on the step-by-step milling of the annular groove by the amount dz which releases residual stresses causing changes in strain on the upper face of the core. These changes in strain are measured in three directions, defined relative to one another.

The evaluation of the ring-core method assumes that one of the principal directions of the residual stress state is in a direction perpendicular to the specimen's surface and that no stress exists in this direction. An unknown biaxial residual stress state exists parallel to the surface.

The evaluation of the measured strains to determine the residual stresses is explained in detail in [12.47]. In order to demonstrate this evaluation we will consider the reaction of a strain gage, aligned in the direction a, on the restraining of the core during the milling of the groove. With increasing depth of milling the alteration of the strain gage signal ε_a as a function of the depth z can be described by the equation

$$\frac{\mathrm{d}\varepsilon_i(z)}{\mathrm{d}z} = K_i(z)\varepsilon_i^* \tag{12.19}$$

In Eq. (12.18) $\varepsilon_i(z)$ represents the strain in direction i measured by the strain gage as a function of the groove's depth z. $K_i(z)$ is the relaxation function, which describes the effect that the strain $\varepsilon_i^*(z)$ has on the strain gage for increasing depth z.

Considering a uniaxial residual stress state, Hooke's law can be introduced in Eq. (12.18) in its uniaxial form. We then obtain for principal direction 1

$$\frac{\mathrm{d}\varepsilon_1(z)}{\mathrm{d}z} = K_1(z)\frac{1}{E}\sigma_1 \tag{12.20}$$

Here, E represents the modulus of elasticity.

The stress σ_1 acting in principal direction 1 gives rise to the strain $\varepsilon_1 = \sigma_1/E$ in this direction and the transverse strain $\varepsilon_2 = -\nu\sigma_1/E$ in principal direction 2. Here, ν is Poisson's ratio of the material under study. Therefore, a strain gage aligned in principal direction 2 will monitor

$$\frac{\mathrm{d}\varepsilon_2(z)}{\mathrm{d}z} = K_2(z)\left(-\frac{\nu}{E}\right)\sigma_1 \tag{12.21}$$

Using Eqs. (12.19) and (12.20), the relaxation functions $K_1(z)$ and $K_2(z)$ can be easily estimated from a uniaxial tensile test:

$$K_1(z) = \frac{E}{\sigma_1}\frac{\mathrm{d}\varepsilon_1(z)}{\mathrm{d}z} \tag{12.22}$$

$$K_2(z) = -\frac{E}{\nu\sigma_1}\frac{\mathrm{d}\varepsilon_2(z)}{\mathrm{d}z} \tag{12.23}$$

The extension of the theory to cover the biaxial stress state requires the inclusion of the stress σ_2, which acts in principal direction 2. The strain state due to σ_2 is superimposed on the strain state produced by σ_1, so that

$$\frac{\mathrm{d}\varepsilon_1(z)}{\mathrm{d}z} = \frac{1}{E}K_1(z)\sigma_1 - \frac{\nu}{E}K_2(z)\sigma_2 \tag{12.24}$$

$$\frac{\mathrm{d}\varepsilon_2(z)}{\mathrm{d}z} = \frac{1}{E}K_1(z)\sigma_2 - \frac{\nu}{E}K_2(z)\sigma_1 \tag{12.25}$$

Equations (12.23) and (12.24) can be rearranged to give the principal stresses σ_1 and σ_2:

$$\sigma_1 = \frac{E}{K_1^2 - \nu^2 K_2^2}\left[K_1\frac{\mathrm{d}\varepsilon_1}{\mathrm{d}z} + \nu K_2\frac{\mathrm{d}\varepsilon_2}{\mathrm{d}z}\right] \tag{12.26}$$

$$\sigma_2 = \frac{E}{K_1^2 - \nu^2 K_2^2}\left[K_1\frac{\mathrm{d}\varepsilon_2}{\mathrm{d}z} + \nu K_2\frac{\mathrm{d}\varepsilon_1}{\mathrm{d}z}\right] \tag{12.27}$$

The relationship for any mutually perpendicular directions can be formulated analogous to Eqs. (12.25) and (12.26), for example, a perpendicular to c and b

perpendicular to d:

$$\sigma_a = \frac{E}{K_a^2 - \nu^2 K_c^2}\left[K_a \frac{d\varepsilon_a}{dz} + \nu K_c \frac{d\varepsilon_c}{dz}\right] \tag{12.28}$$

$$\sigma_c = \frac{E}{K_a^2 - \nu^2 K_c^2}\left[K_a \frac{d\varepsilon_c}{dz} + \nu K_c \frac{d\varepsilon_a}{dz}\right] \tag{12.29}$$

$$\sigma_b = \frac{E}{K_b^2 - \nu^2 K_d^2}\left[K_b \frac{d\varepsilon_b}{dz} + \nu K_d \frac{d\varepsilon_d}{dz}\right] \tag{12.30}$$

$$\sigma_d = \frac{E}{K_b^2 - \nu^2 K_d^2}\left[K_b \frac{d\varepsilon_d}{dz} + \nu K_d \frac{d\varepsilon_b}{dz}\right] \tag{12.31}$$

The following applies for mutually perpendicular components of stress, as is easily verified from Mohr's circle

$$\sigma_a + \sigma_c = \sigma_b + \sigma_d = \sigma_1 + \sigma_2 \tag{12.32}$$

The same applies for the corresponding strains and strain changes

$$d\varepsilon_a + d\varepsilon_c = d\varepsilon_b + d\varepsilon_d = d\varepsilon_1 + d\varepsilon_2 \tag{12.33}$$

If Eqs. (12.31) and (12.32) are substituted into Eqs. (12.27)–(12.30), then it can be found by comparing coefficients that

$$K_a(z) = K_b(z) = K_1(z) \tag{12.34}$$

$$K_c(z) = K_d(z) = K_2(z) \tag{12.35}$$

The following then results from Eq. (12.32)

$$d\varepsilon_d = d\varepsilon_a + d\varepsilon_c - d\varepsilon_b \tag{12.36}$$

If Eqs. (12.33)–(12.35) are substituted into Eqs. (12.27)–(12.29), then the three stress components can be found:

$$\sigma_a = \frac{E}{K_1^2 - \nu K_2^2}\left[K_1 \frac{d\varepsilon_a}{dz} + \nu K_2 \frac{d\varepsilon_c}{dz}\right] \tag{12.37}$$

$$\sigma_b = \frac{E}{K_1^2 - \nu K_2^2}\left[K_1 \frac{d\varepsilon_b}{dz} + \nu K_2 \left(\frac{d\varepsilon_a}{dz} - \frac{d\varepsilon_b}{dz} + \frac{d\varepsilon_c}{dz}\right)\right] \tag{12.38}$$

$$\sigma_c = \frac{E}{K_1^2 - \nu K_2^2}\left[K_1 \frac{d\varepsilon_c}{dz} + \nu K_2 \frac{d\varepsilon_a}{dz}\right] \tag{12.39}$$

The strain gage rosette developed for the ring core method has three measuring grids orientated at 45° to each other, so that measurement directions a and c are mutually perpendicular, and measurement direction b is situated at 45° in the positive direction of rotation with respect to direction a. With these conditions the principal stresses $\sigma_1(z)$ and $\sigma_2(z)$ together with the principal directions can be calculated from Eqs. (12.36)–(12.38):

$$\sigma_{1,2} = \frac{\sigma_a + \sigma_c}{2}\sigma \pm \frac{1}{2}\sqrt{2}\sqrt{(\sigma_b - \sigma_a)^2 + (\sigma_b - \sigma_c)^2} \tag{12.40}$$

$$\varphi = \frac{1}{2}\arctan\frac{2\sigma_b - \sigma_a - \sigma_c}{\sigma_a - \sigma_c} \tag{12.41}$$

The angle φ found from Eq. (12.40) extends in the mathematically positive sense from measurement direction a to principal direction 1.

The relaxation functions $K_1(z)$ and $K_2(z)$ must be known in order to be able to use the ring-core method practically. These are determined by a calibration process in a uniaxial tensile test according to Eqs. (12.21) and (12.22) before the actual ring-core measurements occur. The calibration test is carried out with the same experimental equipment under the same experimental conditions as for the residual stress measurements. The relaxation functions are dependent on the core diameter, the shape of the groove bottom and the geometrical arrangement of the strain gage rosette as well as the residual stress state and the groove depth. The groove depth is the supplementary measurement quantity and is monitored continuously.

During the calibration procedure to determine the relaxation functions, a known uniaxial stress state is created using loading equipment in a tension or compression bar in which the residual stresses have been eliminated as far as possible. The ring-core method is then applied to the loaded bar. The changes in strain that occur on milling the groove are measured, in dependence on the groove depth, on the upper face of the core in the principal directions 1 and 2 of the loaded calibration sample. These measurements are then compared with the actual principal strains which are given by the known current stress state. From this comparison, the desired relaxation values K_1 and K_2 – which are designated relaxation functions in their presentation as functions of the depth z – can be found at each value of z for both principal directions. It should be noted with this type of calibration that any residual stresses present in the calibration bar are superimposed on the loading stress state, and they can therefore affect the relaxation functions. However, the effects can be regarded as negligible for most applications if the selection, shaping and preparation of the calibration bar are carried out carefully. In [12.51] a procedure is suggested which eliminates the effects of any residual stresses in the calibration bar by relieving and loading the bar after each stage of milling.

Figure 12.16 shows a test rig used for the determination of the relaxation functions on a tension bar with a square cross-section (105 × 105 mm) and a measurement length of 960 mm [12.52]. The size of the calibration bar should be as large as possible in comparison to the diameter of the annular groove. The homogeneity of the stress state which occurs due to the external tensile forces acting on the bar is checked with strain gages mounted at the upper and lower end of the measurement region.

Fig. 12.16 Test arrangement for the determination of the relaxation functions for the application of the ring-core technique

Figure 12.17 shows an example of the relaxation functions that were determined in this manner using the RY 51 ring-core rosette (HBM) for a particular milling device in a calibration test (Stücker, E., personal communication).

The relaxation functions $K_1(z)$ and $K_2(z)$ which are only dependent on Poisson's ratio and the test parameters, that is core diameter, shape of groove bottom and strain gage arrangement, were then substituted into the evaluation Eqs. (12.36–12.38). These equations were then used for the actual ring-core measurements on measurement objects which were subject to unknown residual stress states.

The evaluation of Eqs. (12.36–12.38) is most conveniently carried out using computer programs, but it can also be done manually point by point for defined increases in depth. In both cases the formation of the derivatives of the strain measured in the directions a, b and c is required according to the depth z. These are substituted with the relaxation functions corresponding to the depth z in the equations and they give the stress components of the plane residual stress state in dependence on z.

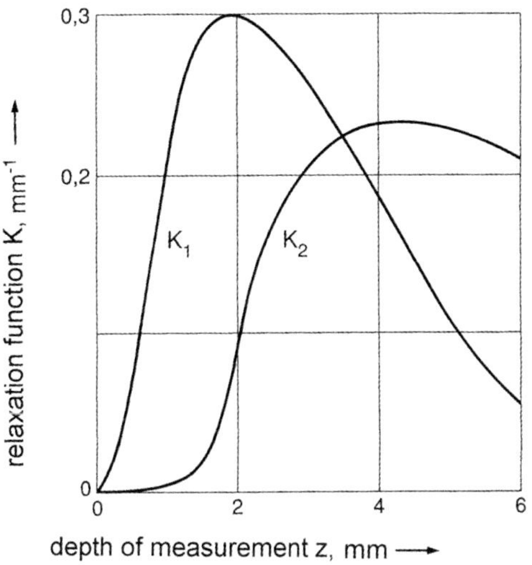

Fig. 12.17 Example of measured relaxation functions as used in the ring-core method; they were obtained using the RY 51 ring-core rosette and the device shown in Figure 12.18 for producing the annular groove

12.4.2 Application of the method

The practical application of the method involves a mechanical device which enables the centring of the milling tool over the ring-core rosette and which provides fine increments in the cutter feed. A measuring device for the milling depth is also included in the equipment. This type of device came into being in the 1970s during the development of the ring-core method [12.53]. With a core diameter of 14 mm, depth increments of 0.5 mm are used, and the milling process is interrupted after each increment to take the measurements. The interruptions prevent the measurements from being affected by stress due to mechanical working and they allow temperature equilibrium to re-establish. In addition, the interruptions enable a clear ranking of the measurements to be made in relation to the penetration depth. Otherwise there would be problems due to the superimposition of unwanted interference oscillations if a continuous procedure were adopted.

If residual stress states are expected with noticeable gradients in the depth direction of milling, then it is recommended that the depth increments z are selected to be smaller than 0.5 mm. This takes a little more time, but the effort needed in evaluation does not increase if an online measuring system is used. The depth z is indicated by a dial gage located on the equipment and it is used by the operator of the device as a guide for the manually operated feed. The depth z is also measured by an inductive displacement transducer. Its electrical output signal is passed, together with the output signals from the strain gage measuring grids on the ring-core rosette, to a measuring unit used to process the signals. Figure 12.18 shows the milling device

Fig. 12.18 Milling device specially developed for the ring-core method where it is used for cutting an annular groove around a mounted strain gage rosette

specially developed for the ring-core method. An inductive displacement transducer is used on the device in addition to the dial gage.

When setting up a measuring point, the base section of the milling device is first placed on the measurement object and the measuring point is marked out using a guide plate. Once the ring-core rosette has been mounted, the drive section is fitted to the lifting column, and the rosette's measuring grids are connected through leads to terminals or are directly connected to a measuring unit. The special design of the rosette enables the measuring leads to be soldered onto its upper side and led vertically upward through the hollow drive spindle. The rosette and leads are protected against damage due to swarf or twisting by a protective tube which is situated inside the drive spindle and secured against rotation. The strain gages and leads are protected by a Teflon sleeve at the lower end of the tube.

Temperature equilibrium is achieved in steel measuring objects approximately one minute after each stage of milling of the groove depth. Before starting to take measurements, that is before the first stage in milling, the three measuring circuits with the grids on the ring-core rosette and the depth monitoring transducer are zero-balanced. This procedure takes place automatically for all the measuring points if an online measurement system is used, including an appropriate multipoint measuring instrument. The separate measurements for the three strain measuring points and for the inductive displacement transducer used for the depth measurement are recorded after each stage of milling. Where an online measuring system is used, the measurements are passed to the computer in digital form for computerized evaluation.

12.4.3 Electrical connection of the ring-core rosette

Normally, the rosette's three measuring grids are wired in a circuit with a common compensating strain gage. In order to avoid temperature effects of the cable, the three-wire circuit is recommended for the connection of the strain gages. The cables are passed through the hollow milling tool. Practice has shown that when working on large measurement objects in workshops with almost constant temperature, the compensating strain gage is not needed and can be replaced by a precision resistor with high temperature stability [12.54].

As a result of modern instrumentation technology, it is not necessary to include a compensation or completion strain gage for each separate measuring grid. When a modern digital multipoint measuring unit is used, a single compensation strain gage or completion resistor is sufficient. These types of instrument make use of the Kreuzer circuit, which automatically compensates for resistance changes occurring in the supply leads of the bridge circuits [12.55].

If the RY 51 ring-core rosette is used, which was specially developed for this technique, then care should be taken to ensure that the connection leads are soldered to the metal covering layer as stated in the user's instructions and not connected to the soldering points for the internal rosette circuit on the upper side of the rosette. Figure 12.19 shows a photograph of mounted RY 51 ring-core rosettes with soldered leads for connection according to the Kreuzer circuit.

Fig. 12.19 Mounted ring-core rosette with soldered connection leads after milling the annular groove

12.4.4 Practical application example [12.67]

The test object was a 2.65 t rotor-disk forging for a low-pressure steam turbine which served as the test disk for the optimization of the residual stress state. It also provided a means of testing the online evaluation technique in a real application.

Rotors for high performance steam turbines are often designed as a series of disks, since the manufacture of very large rotors of sufficient quality from one forging represents a significant problem. A disk rotor consists of a central spindle onto which separate wheel disks are shrunk. Figure 12.20 shows a view of a disk rotor during assembly. The stresses on a turbine disk under operating conditions arise due to shrink-on stresses, tensile stresses due to centrifugal forces, thermal stresses and residual stresses. All these stresses are superimposed on one another in the disk according to their arithmetical sign and should be known for the proper dimensioning of the disk.

Figure 12.21 shows the disk forging of material 26 NiCrMoV 14 5 supported on trestles ready for the ring-core measurement. Special thanks are due to Siemens AG,

Fig. 12.20 View of a disc-type rotor for a low-pressure steam turbine during assembly

Fig. 12.21 The test object: a 2.56 t forging of a rotor disk for a low-pressure steam turbine

Power Engineering Group KWU at Mülheim for their invaluable support during the practical tests.

Compressive residual stresses in the region of the shrink-on fit are an advantage in the design of the disk, because they oppose the tensile stresses that are unavoidable in shrink fitting. Compressive residual stresses prevent stress-crack corrosion over the complete surface region of the disk. By using specific heat treatment of the disk forging during its manufacture – for example water spraying in the boring – certain magnitudes and distributions of the residual stresses can be obtained in the disk. The checking of the success of this heat treatment is then carried out using the ring-core method. This method is used to find the residual stress distribution at selected measuring points down to a depth of about 5 mm (0.2″).

Here, the practical application example is involved with the main feature being the online evaluation of the measurement results. The special strain gage rosette that was used, known as a ring-core rosette, is shown in Figure 12.14 with both top and bottom views. It was patented in 1977 [12.50].

During the practical implementation of the technique the support for the device for milling the annular groove is fixed above the centre of the previously specified measuring point using a guide plate and a centring pin. The magnetic feet of the support are used for fixing, supplemented by bonding with X60 two-component cement. The guide plate is then removed and the ring-core rosette bonded to the previously prepared measuring point using a cold-curing cyanoacrylate cement. Care must be taken to ensure that the three measurement directions of the rosette measuring grids, which are angularly displaced by 45° from one another, are in defined positions with respect to the test object coordinates. In the example, the rosette reference direction *a* is aligned radially and the measurement grid *c* is aligned tangentially to the disk shape.

The next step consists of inserting the milling device in the support and passing the leads for the electrical connection of the rosette measuring grids from above through the hollow milling spindle. The cores of the lead have previously been stripped and tinned. A three-wire configuration is used with a common return for the three measuring grids. Figure 12.22 shows the wiring of the rosette in a close-up photograph. When soldering the wires, it should be ensured that they are soldered to the areas of the printed circuit board, which acts as a protective cover. The solder points for the internal rosette wiring, which can be seen in the lower part of the rosette in Figure 12.22, should not be used for soldering. To protect them from twisting, the leads are led up through the milling spindle in a protective tube. The milling spindle and connection leads positioned above the ring-core rosette can be seen in Figure 12.23.

In the next stage the milling spindle is lowered so that the teeth on the cutter contact the surface of the test object prior to the start of milling. All the sensors used for the measurement are zeroed in this position. These are the mechanical dial gage used as a visual operator aid for measuring the cutting depth, the inductive displacement transducer which supplies an electrical signal proportional to the milling depth for

Fig. 12.22 Connections to the three rosette measuring grids were implemented in a three-wire configuration

Fig. 12.23 Positioning the milling device over the measuring point, passing the leads upwards through the hollow spindle

the online evaluation and the three rosette measuring grids which are wired with a common compensation strain gage to form three half-bridges.

The zero balancing for these half-bridges and for the inductive displacement transducer is carried out automatically at the press of a key by the online measurement system used for signal processing and evaluation. Figure 12.24 illustrates the milling device mounted at the measuring point, together with the dial gage and the displacement transducer. On the right of the milling device is the lever for the manual operation of the cutter feed.

The actual milling process can start after the zero balancing, and the cutter is fed manually in steps of about 0.5 mm. Exact conformance to specific depth steps is not necessary, since the milling depth is measured at each stage by the inductive

Fig. 12.24 View of the milling device on the test object, together with the dial gage and inductive displacement transducer for visual and electrical measurement of the milling depth

displacement transducer and passed to the measurement system. The removal of swarf by withdrawing the milling tool is necessary, because swarf collects on the printed circuit board on the rosette and may cause short circuits. This can also be remedied by covering the soldering points with a ball of adhesive insulating material after soldering the leads to the rosette.

The step-by-step milling was carried out on the disk to an overall depth of 4 mm. After the milling and the transfer of the measurements to the online instrument system the ring-core process at this measuring point is finished. The milling equipment is removed and mounted at another measuring point.

The processing of the measurement signals is carried out with an online measurement system consisting of a computer-controllable multipoint measuring unit and a computer. A version of the software BEAM, modified for the ring-core technique, is used for evaluation. The connection of the strain gage rosette to the multipoint unit is realized via a terminal to which the completion resistance serving all three measuring points is connected. The inductive displacement transducer is directly connected to the multipoint unit. After the milling of each stage, the four measuring signals (milling depth and the three strains) are acquired, amplified and conditioned by the multipoint unit and then passed as digital numerical values in their appropriate units to the computer.

The theory on which the evaluation is based is summarized above in the section describing the principle of the method together with all the necessary formulas and their derivations. The objective of the evaluation is to display the planar residual stress state at the measuring point under consideration as a function of the depth z. It is based on equations from the theory of elasticity and on calibration functions specific to the technique which are available to the user if the parameters associated with the calibration functions are maintained. These parameters are the RY 51 ring-core rosette, the shape of the milling tool and Poisson's ratio for the test object material.

The complete evaluation can be reduced to a few equations. The strains ε_a, ε_b and ε_c are measured in the three measurement directions designated a, b and c as a function of the milling depth z. The three residual stresses to be determined in these directions are given by the Eqs. (12.36–12.38).

The calibration functions $K_1(z)$ and $K_2(z)$ with the differential coefficients $\Delta\varepsilon_i(z)/\Delta z$ must be substituted in these equations. This is carried out automatically for each depth stage by the computer program. The values for the differential coefficients are produced after the milling of each depth stage from the measurements at the start and end of the stage milling.

The calibration functions $K_1(z)$ and $K_2(z)$ are available initially as experimentally determined tables of values [12.51]. In order to incorporate these in the software as continuous functions, the tables of values are substituted by approximation polynomials.

After the three stress components σ_a, σ_b and σ_c have been found, the calculation of the principal stresses σ_1 and σ_2 for the planar residual stress state and the angle of orientation φ for the principal direction 1 is trivial. It is found by the program according to Eqs. (12.39) and (12.40).

With the software BEAM's graphical user environment the program is very easy to use. The operator guidance is easy to understand and the user does not need any programming knowledge. Even someone untrained in using computers can easily operate the program after a very brief introduction. The many graphical and tabular displays provided by the program enable the selection of various presentations for the results and direct online logging.

The display of the results for the residual stress determination described here is produced graphically during the test. New results are included in the display after the completion of each depth stage. The measured strains ε_a, ε_b and ε_c together with the corresponding stresses σ_a, σ_b and σ_c computed using Eqs. (12.36–12.38), the principal stresses σ_1 and σ_2 of the planar residual stress state and the orientation angle φ to the principal direction 1 are all displayed in graphs with the abscissa representing the milling depth z.

The principal direction 1 is rotated by the angle φ in the mathematically positive sense with respect to the rosette reference direction a. In addition, the individual differential coefficients $\Delta\varepsilon_i(z)/\Delta z$ can also be displayed. The values found on the rotor disk forging are shown in Figure 12.25 as a hard copy of the screen contents as obtained on the monitor at the end of the measurement.

The evaluation of the results can be carried out with a multipoint measuring system operating online to a computer. Here, the measurements are transferred directly via the instrument's interface to the computer for evaluation.

The graphs illustrated in Figure 12.25 show evaluation results as presented on the monitor or as hard copy when using appropriate software. The top graph shows the measured strains and the graph below it displays the stresses determined from Eqs. (12.22–12.24) with inclusion of the relaxation functions given in Figure 12.17. The graph below it illustrates the principal stresses for the residual stress state and the bottom diagram depicts the angle of orientation of the principal stress σ_1. All

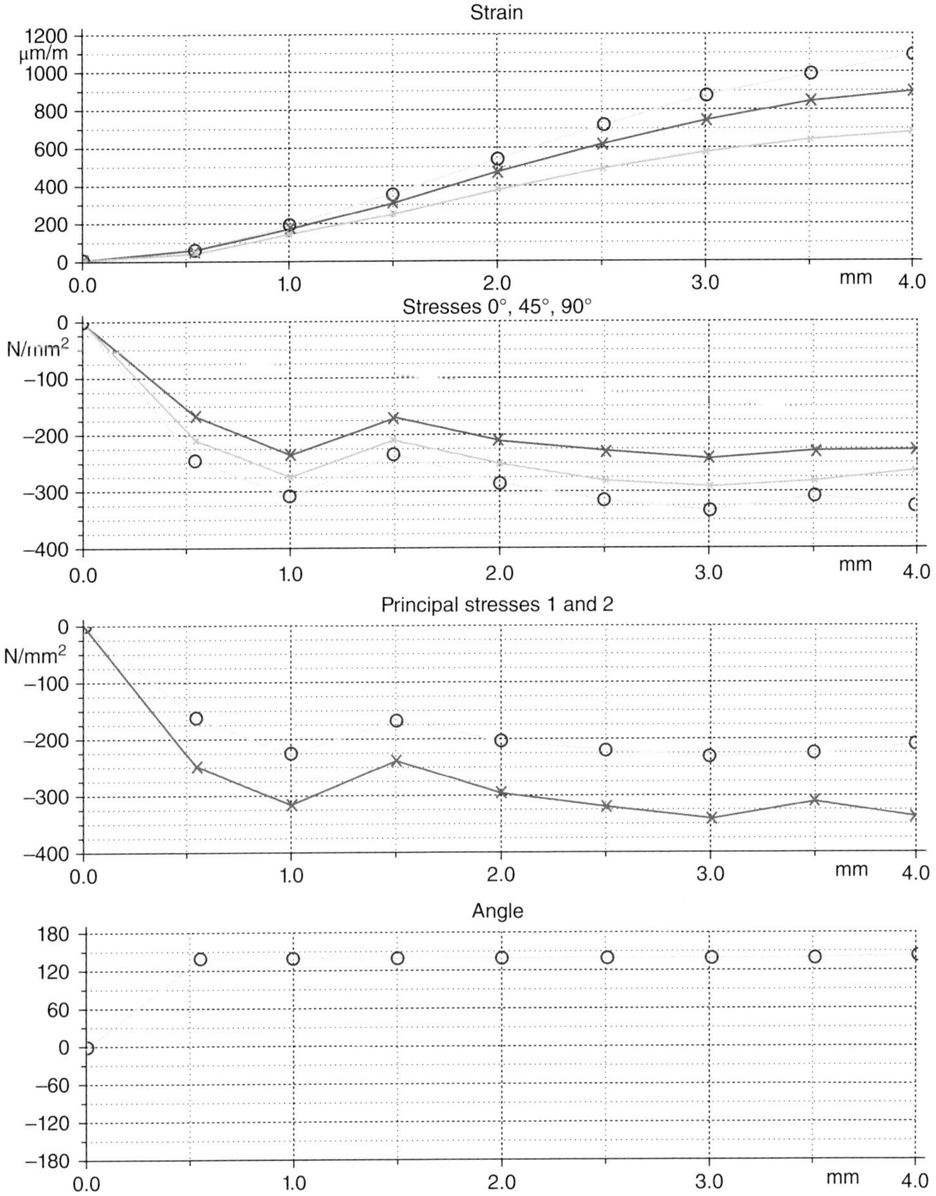

Fig. 12.25 Display of the evaluated results for a ring-core test using appropriate software

quantities are displayed as a function of the coordinate *z*, which is orientated perpendicular to the surface of the measurement object.

The graphical functions shown in Figure 12.25 were produced by the BEAM program. The relaxation functions were approximated to polynomials and then made available in this form to the evaluation program. Figure 12.16 shows the results of the application of the ring-core method: the axial, radial and tangential components of the residual stress state as obtained at all 15 measurement points of the disc.

With regard to Eqs. (12.36)–(12.38) it should be noted that these are undeterminable at two points, that is the stress computation cannot be carried out at these points. One of these points is situated at $z = 0$, because $K_1(z) = 0$ and $K_2(z) = 0$ at this point. The second critical point is given by $K_1^2 - K_2^2 = 0$ and it is therefore dependent on the parameters affecting the relaxation functions. For the relaxation functions given in Figure 12.17, this second critical point is located at $z = 5.9$ mm.

The evaluation algorithm can be simplified for the application of the ring-core technique with known principal directions for the residual stress state that is to be determined [12.56, 12.57]. These types of applications occur relatively frequently with dynamically balanced parts in the construction of turbines and generators. The ring-core rosette must then be mounted such that the measurement directions of its grids, which are orientated at 90° to one another, line up with the principal directions of the residual stress state.

Since only the two measuring grids located in the principal directions are required for the evaluation with these known principal directions, a ring-core rosette was developed specially for these applications and it has only two measuring grids at right-angles to one another.

The ring-core method has been used for some years now, particularly on large cast-steel parts and forgings. Here, allowances for extra material on the component are made where possible so that the ring-core method can be used. Once the ring-core tests have been concluded, the measuring points can be removed by machining, leaving the finished product free from any traces of the ring core application. Figure 12.27 shows

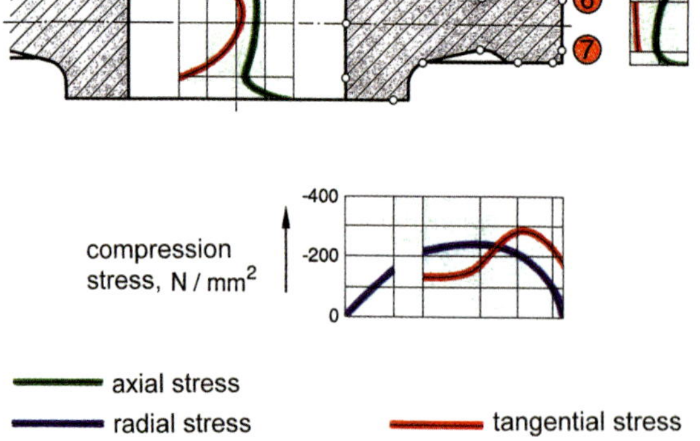

Fig. 12.26 The state of the axial, radial and tangential components of the residual stress as obtained at all 15 measurement points of the disc

Fig. 12.27 Traces of the ring-core application could be removed by machining, leaving the test object free from any traces

the traces of the ring-core application before the final machining of the measurement object.

The milling of the annular groove may be difficult, particularly with hard or tough materials. This is the case for example with hardened surfaces or with nickel-based alloys. Here the annular groove must be produced by a method other than milling. One method is the 'air abrasive technique' in which a jet of gas blasts abrasive particles onto the measurement object at high velocity. The jet is produced from a nozzle which is guided according to the annular shape of the required groove by special equipment. The groove is created as a result of the abrasive action. Depending on the material hardness, depths between 0.3 mm and 0.7 mm are obtained with this method for time periods between 15 and 30 min.

Another method of producing the annular groove in hard materials is to use the spark-erosion technique which can be applied to electrically conducting materials [12.58]. Here, the groove is created by the erosive effect of electrical discharges between a tool electrode (e.g. made from copper) and the object under investigation. Particles of material become molten and are partially vaporized. The process is carried out in a liquid dielectric which increases the energy density by constriction of the discharge path. The dielectric also acts as a cooling medium and flushes out the eroded particles. Petroleum spirit or distilled water is normally used as the dielectric, requiring the implementation of protective measures for the strain gages. A third method, which has been successful particularly on cold rollers, involves an electrochemical process. In this technique a tube-shaped cathode with a cross-section corresponding to the annular groove is positioned over the measurement object which acts as the anode, and the inter-electrode region is flooded with an electrolyte. When an electrical voltage is applied, particles are released from the anode as it releases a charge corresponding to its valency, and they combine with the hydroxyl ions arising in the electrolyte to form metal hydroxide. Hydrogen is released in this process. In the application of the technique, the electrolyte, for example an aqueous solution (20%) of NaCl and $NaNO_3$, is brought into contact with the surface to be processed by the tube-shaped cathode. The connecting leads of the strain gage rosette are protected by a tube which passes through the hollow cathode and which is bonded to the surface of the measurement object at its lower end. The rosette and leads are therefore protected from the electrolyte. A rate of increase in the groove depth of 0.3 mm per minute has been achieved with a voltage of 10 V and a current of 50 A with a core diameter of 14 mm [12.58]. The relaxation functions determined during the calibration when using this technique are also given in [12.58].

12.4.5 Multi-stage application of the ring-core method

The depth of the annular groove is normally 5 mm for a core diameter of 14 mm. In order to be able to measure residual stresses at greater depths, the ring core method is used with a multi-stage technique [12.59, 12.60]. The multi-stage ring-core method is the preferred application on claddings, since it is there that high residual stresses can occur in the contact zones due to the different properties of the combined materials (e.g. linear thermal expansion coefficient). The principle of the multi-stage technique constitutes carrying out the ring-core method on a first stage as described above (e.g. with a groove depth of 5 mm), but then the core is milled out and a new rosette is mounted on the flat hole bottom. Then an annular groove is milled at this depth as in the previous stage and the core is again milled away.

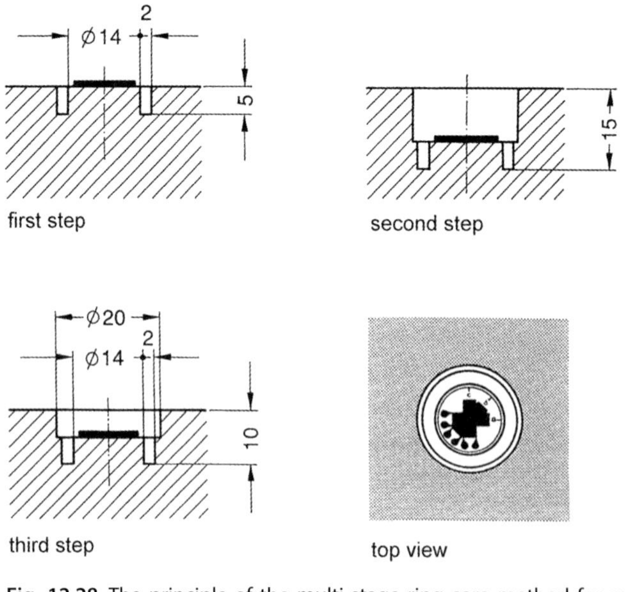

Fig. 12.28 The principle of the multi-stage ring-core method for measurement depths up to 25 mm

Figure 12.28 shows this sequence diagrammatically. This procedure is repeated up to five times so that a total depth of 25 mm is achieved for a groove depth of 5 mm. The application assumes, however, that the relaxation functions are known for each stage.

These relaxation functions can be determined by step-by-step application of the ring-core method in a calibration test using the calibration equipment illustrated in Figure 12.16.

The strains found at each stage of the calibration test must be arranged consecutively and smoothed to obtain a continuous evaluation over the individual stages. The relaxation functions can then be computed continuously. With the practical implementation of the technique on the measurement object, the strains measured after each stage of milling should be subjected to the same smoothing treatment that was used during the determination of the relaxation functions. An application example is described in [12.52].

The diagrams in Figures 12.29 and 12.30 also originate from this paper. Figure 12.29 illustrates the strains measured over five depth stages in a calibration test. Figure 12.30a shows the arrangement of these strains after smoothing using a fast Fourier transformation. The relaxation functions shown in Figure 12.30b were able to be computed from these smoothed curves. The results shown in Figure 12.32 are based on these relaxation functions.

The important quantity for the determination of the residual stresses from the measured strains is the elasticity modulus, which can change substantially in the direction of the component's depth, particularly with claddings. Therefore, tensile tests on miniature specimens cut from the plated measurement object using electro-erosive techniques exhibit extremely low values for the elasticity modulus (reduction up to 50%) on the surfaces of the claddings.

Figure 12.31 shows two examples of curves for the distribution of the elasticity modulus perpendicular to the surface, which were found experimentally in the upper layers of cladding materials [12.52].

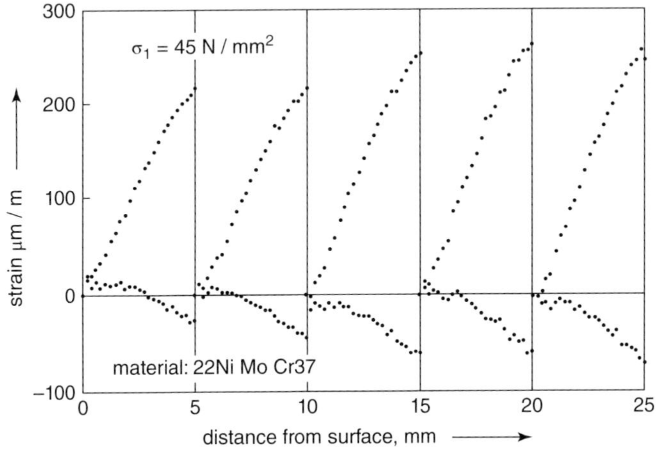

Fig. 12.29 Determination of the relaxation functions for the multi-stage ring-core method: strains measured in the principal directions 1 and 2 at different stages of depth [12.52]

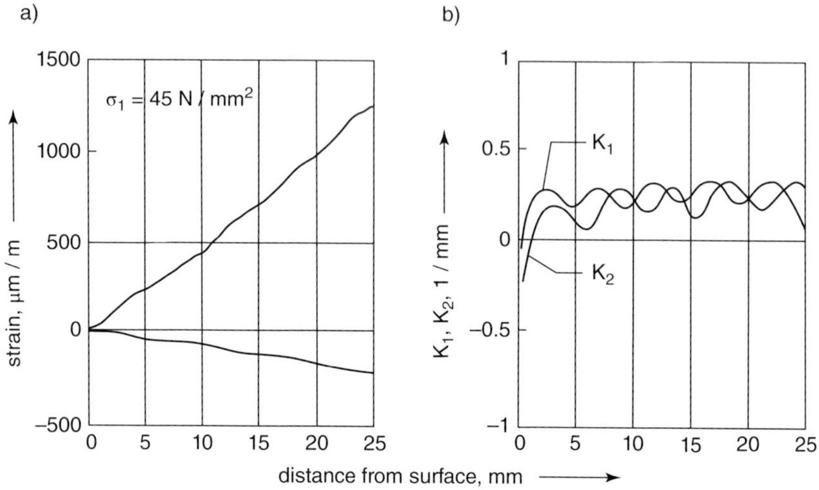

Fig. 12.30 Relaxation functions for the multi-stage ring-core method [12.52]; a) arrangement of the smoothed, measured strains; b) relaxation functions computed from the strains shown in (a)

Figure 12.32 illustrates the results of the application of the multi-stage ring-core method on an explosion cladded material with a 6 mm thick titanium layer on a forged base material of 13 CrMo 44. Figure 12.32a shows the curve of the elasticity modulus and Figure 12.32b illustrates the strains measured on the surface with the rosettes. The principal normal residual stresses are shown in Figure 12.32c in relation to the depth. The effect of the stress-relief treatment can be seen from Figure 12.32d.

With the multi-stage ring-core method, the mounting of the strain gage rosettes on the bottom of the milled out cavity, which may be up to 25 mm deep, is not very easy.

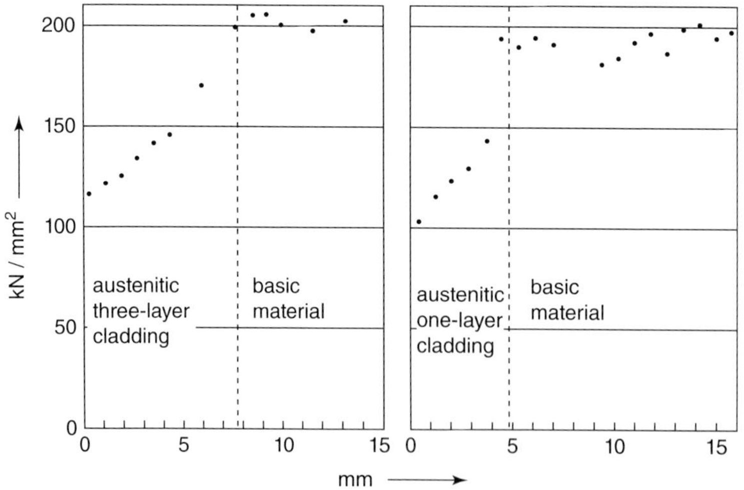

Fig. 12.31 Examples of the distribution of the modulus of elasticity in claddings, obtained from miniature tensile tests on samples cut out using electro erosion [12.52]

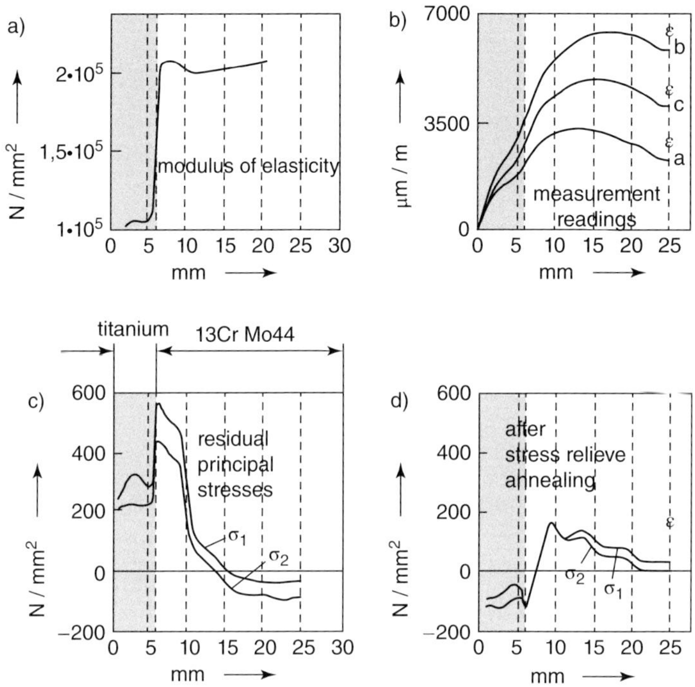

Fig. 12.32 Example of the result of application of the multi-stage ring-core method on an explosion cladding consisting of a base sheet of 13 CrMo 44 and a covering sheet of titanium before and after relief annealing [12.61]; a) elasticity modulus, b) strains at different levels of depth measured with the rosettes, c) principal stresses of the residual stress state, d) effect of relief annealing

For this reason, cylindrical ring-core rosette measuring elements, which are easier to mount, have been developed for this application [12.62].

On these measurement elements the actual rosette is arranged, as can be seen in Figure 12.33, on the face side of cylindrical rosette carrier with permanently mounted connection leads. A soft elastic material is situated between the strain gage and the cylindrical section, and this prevents reaction from the rosette carrier inhibiting the deformation, which would in turn affect the measuring point.

12.4.6 Modified ring-core rosettes for smaller core diameters

Sometimes there is a requirement to work with small strain gage rosettes and small groove diameters for the determination of residual stresses in regions with steep gradients, such as those found, for example, in and close to welds. Since small rosettes can only be handled with difficulty, a small rosette with a cylindrical supporting sleeve was developed which greatly simplified handling, and enabled a groove diameter of 5 mm to be used [12.63]. At its lower end the hard-paper supporting sleeve carries the rosette which consists of three measuring grids. The rosette connections are led up to the upper end of the sleeve, where the measuring

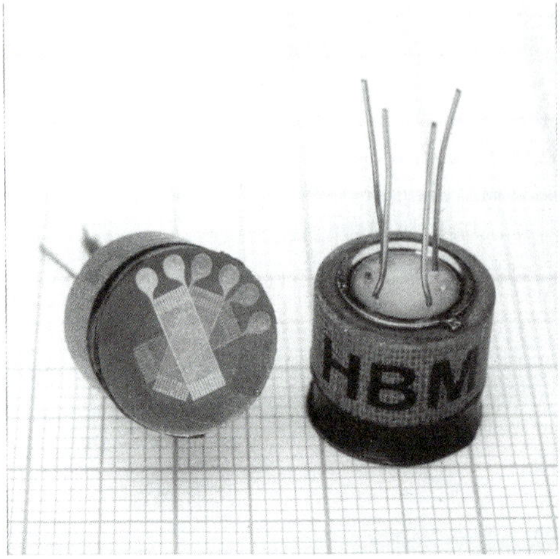

Fig. 12.33 Cylindrical measuring element with a ring-core rosette on its face side and with permanently mounted connection leads [12.62]

leads can be soldered before the rosette is mounted on the measurement object. Two rosettes with supporting sleeves are illustrated in Figure 12.34.

The practical application of this special ring-core rosette assumes that the relaxation functions for the process are known. These should be determined, as already described, in a calibration test.

An application example for the sleeved rosette is given in [12.64]. Here, the effects of various parameters (seam shape, cooling conditions, material, subsequent treatment)

Fig. 12.34 Thin cylindrical measuring element with only 4 mm diameter for application of the ring-core method in areas of relatively steep gradients [12.63]

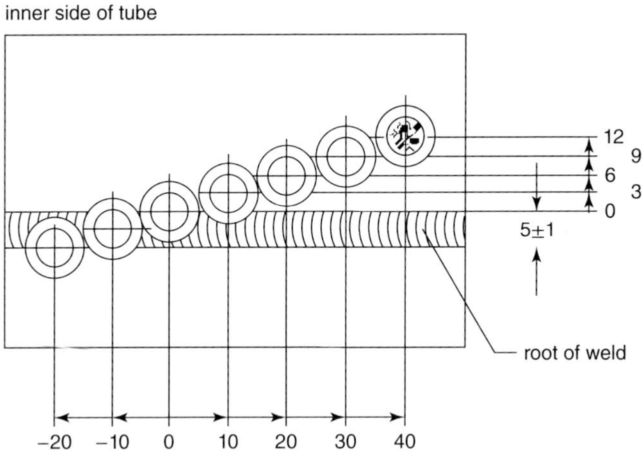

Fig. 12.35 Arrangement of the series of grooves for determining a residual stress profile transversely across the root of the seam of a pipe welded around the circumference [12.64]

were found on the residual stresses in pipes welded with ring seams (external diameter 460 mm, wall thickness 25 mm). In order to obtain the closest separation between the measuring points, they were positioned diagonally across the welded seam as shown in Figure 12.35.

Figure 12.36 presents the results obtained by carrying out the process as described. The asymmetrical feature of the residual stress profile transverse to the seam is a

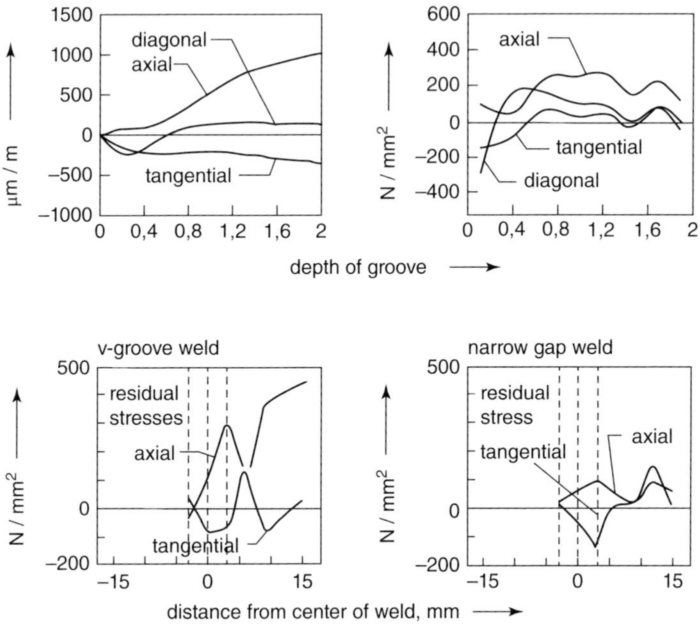

Fig. 12.36 Examples of residual stresses found on a pipe welded around its circumference; the stresses were determined transverse to the welded seam and in the region of the seam's root [12.64]

consequence of the sinusoidal fluctuation in the residual stresses along the seam at the circumference of the welded pipe [12.65, 12.66].

12.5 Hole-drilling method

12.5.1 Introduction

The question of residual stresses in workpieces, particularly the way in which deformation or heat treatment affects them, has led to the development of numerous mechanical methods for the determination of residual stress. A common feature of mechanical methods is that the residual stresses present in the workpiece are practically all released by mechanical intervention (turning down, drilling out, sawing up) and the resulting deformation is measured. Mechanical methods for the determination of residual stress are therefore destructive. Attempts to keep the destruction required for the relief of residual stresses to a minimum led to the development of the hole-drilling method, which was first reported by Mathar [12.68] in 1933. He wrote that the new method was based on 'a disturbance of the force equilibrium by drilling a hole which is so small that the workpiece can still be used.'

When Mathar introduced the hole-drilling method, strain gages had not yet been invented and the deformation measurements were carried out with opto-mechanical devices. The diameter of the holes used in the initial tests was 12 mm and the lower limit for the hole's diameter was regarded as being 6 mm, due to the increasingly large transfer ratio that had to be employed in the measurement instruments.

With the invention of the strain gage and especially the development of foil strain gages with very small geometrical dimensions, a very convenient means of measurement was now available, which was particularly suitable for use in conjunction with the hole-drilling method. This meant that the hole-drilling diameter could be reduced. The first application of strain gages with the hole-drilling method was carried out by Soete and Vancrombrugge in Belgium in 1947 [12.69, 12.70]. They applied single strain gages, arranged in radial directions, around the hole. Later, specially developed strain gage rosettes were used for the hole-drilling method. Figure 12.37 shows samples of the first

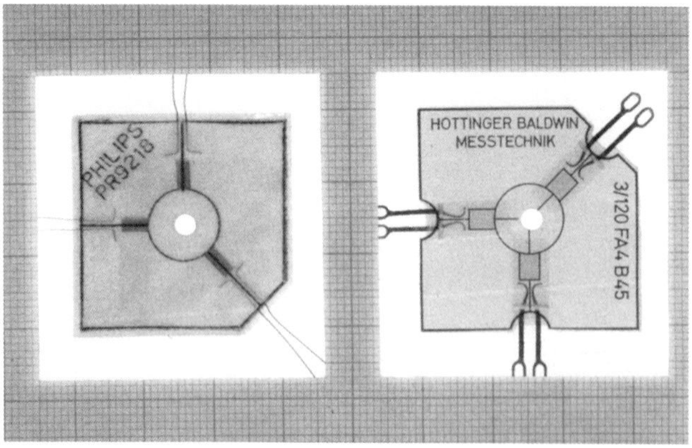

Fig. 12.37 Samples of the first hole-drilling rosettes with wire strain gages made by Philips and HBM

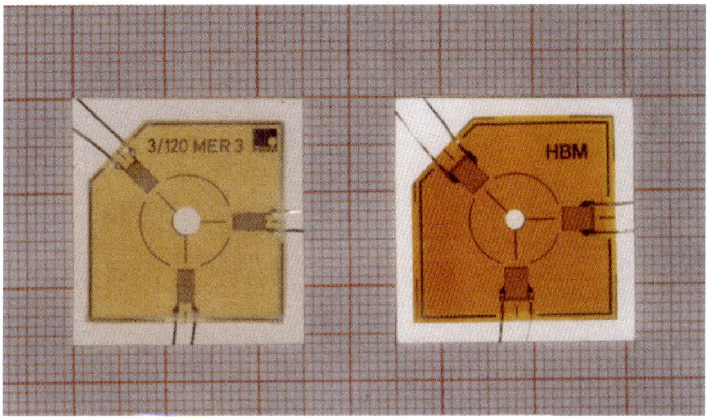

Fig. 12.38 Samples of the first foil strain gage rosettes for application in the hole-drilling method

hole-drilling rosettes with wire grids, made by Philips and HBM, and in Figure 12.38 the first hole-drilling rosettes using foil strain gages can be seen.

Nowadays modern foil strain gage rosettes for use with the hole drilling method are available with different hole diameters from about 1.5 mm to 8 mm. With the small diameter and drilling depth of 1.5 mm it can be stated that in Mathar's literal sense the hole-drilling method really is an 'almost non-destructive' test method. Figure 12.39 shows samples of two modern hole-drilling rosettes as used with drilling diameters of 1.5–1.75 mm. Figure 12.40 shows a special hole-drilling rosette with a centring bush for guiding the drilling tool.

The practical application can have two different objectives. The residual stress can be found as an integrated mean value over the drilling depth or, with more effort, the depth distribution can be found along the drilled hole. In the first case, the measurement results of interest are the measurement maxima obtained at a depth approximately equivalent to

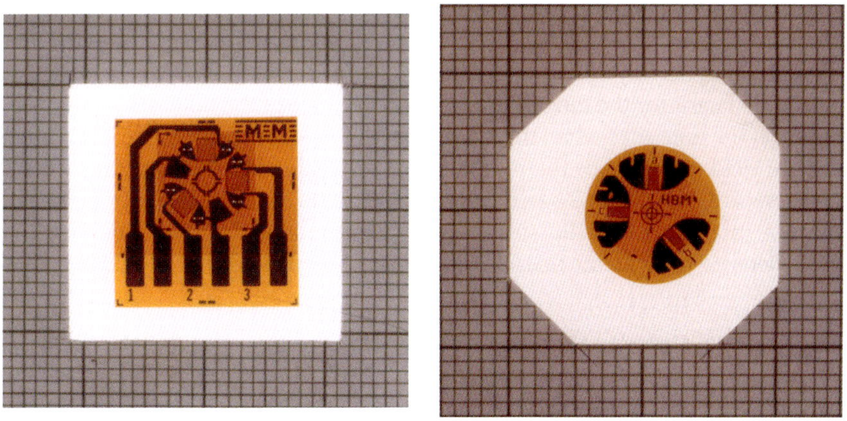

Fig. 12.39 Two samples of modern foil strain gage rosettes for hole-drilling applications with small hole diameters

Fig. 12.40 Special hole-drilling rosette with centring bush for guiding the drilling tool

the diameter. In the second case the drilling is carried out step by step, and the changes of strain are recorded at each stage of depth. With special modern equipment depth increments down to 20 µm are possible. In both cases the application of the method takes place in two phases. The first, experimental phase covers the drilling and measurement of the strain changes that occur. In the second, theoretical phase the residual stress state present before the drilling must be calculated from the measured strain changes. Computational formulas are needed to do this. These formulas can be derived with the aid of the theory of elasticity, taking into account certain assumptions.

The fundamental principles for the evaluation of measurements according to the hole-drilling method were described as early as 1898 by Kirsch [12.71] in the computation of stress distribution in the vicinity of a round hole. Kirsch assumed an infinitely large plate and constant stress distribution over the plate cross-section. In an article published in 1930, Howland [12.72] demonstrated that the relationships found for the infinitely large plate could be transferred to plates of a finite size if the distance of the hole from the edge was greater than 15 times the hole diameter. From this it followed that the preferred field of application for the hole-drilling method was on panels, assuming a uniform stress distribution perpendicular to the surface.

Furthermore, when using the hole-drilling method, the stress concentration caused by the hole must be taken into account, since the stresses it causes on the edge of the hole can rapidly and unexpectedly reach the yield point of the material. The theoretical stress concentration factor for the tangential stress on the edge of the hole is 3, so that yielding takes place at the hole edge when the residual stresses to be measured are larger than one third of the yield point. Since the method used for evaluating the measurements made with the hole-drilling rosette is based on Hooke's law, no plastic deformation must occur, at least not in the area of the strain gage measurement. In 1936 Messmer [12.73] produced detailed work on the possible yield effects when using the hole-drilling method.

12.5.2 Residual stress as an integrated mean value over the drilling depth

12.5.2.1 Introductory remarks

The calculation of residual stresses from the measurement results is based on the principles of the theory of elasticity and on the strains measured by the strain gages as mean values over their measuring lengths. The fundamental principles for the calculation are shown, specific for the application, in [12.75] on the basis of the derivations given in [12.71, 12.74]. The derivation is based on the idealized assumptions of a plane panel, a through-drilled hole and no stress gradient in the direction of drilling. The application-specific factors used in the formulas are calculated on the basis of linear elasticity. The evaluation of the measured strains using these formulas then leads to reliable results if all the assumptions, also in the application of the method, are valid.

12.5.2.2 Stress distribution in the vicinity of a hole

Kirsch [12.71] calculated the stress distribution in the vicinity of a round hole for an infinitely large, uniaxially stressed plate. Figure 12.41 shows a diagram of a uniaxially stressed plate with round hole and an arbitrarily selected point P in the vicinity of the hole. The position of the point P on the surface of the plate is given by the polar coordinates r and φ.

If an infinitesimally small element, rotated from point P through an angle φ away from the principal direction 1, is cut from the plate, then the principal stresses σ_{rB} and σ_{tB} and the shear stresses $\tau_{rtB} = \tau_{trB}$ occur in the cut surfaces as indicated in Figure 12.42.

According to Kirsch the equation for the radial stress is

$$\sigma_{\mathrm{rB}} = \frac{\sigma_1}{2}\left(1 - \frac{\mathrm{a}^2}{\mathrm{r}^2}\right) + \frac{\sigma_1}{2}\left(1 - \frac{\mathrm{a}^2}{\mathrm{r}^2}\right)\left(1 - 3\frac{\mathrm{a}^2}{\mathrm{r}^2}\right)\cos 2\varphi, \qquad (12.42)$$

for the tangential stress

$$\sigma_{\mathrm{tB}} = \frac{\sigma_1}{2}\left(1 + \frac{\mathrm{a}^2}{\mathrm{r}^2}\right) - \frac{\sigma_1}{2}\left(1 + 3\frac{\mathrm{a}^4}{\mathrm{r}^4}\right)\cos 2\varphi \qquad (12.43)$$

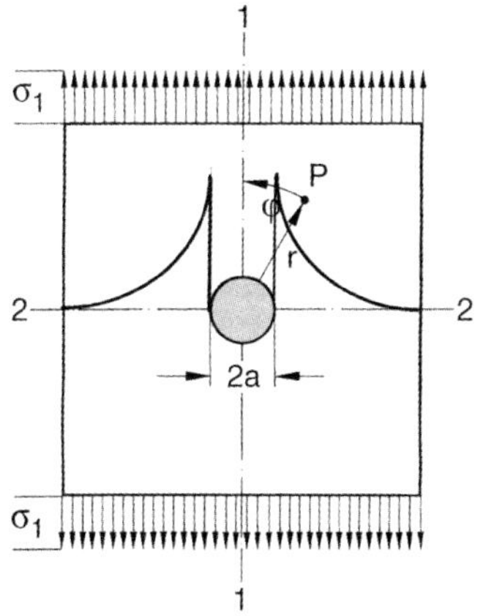

Fig. 12.41 Uniaxially loaded plate with a round hole and arbitrary point P in the vicinity of the hole

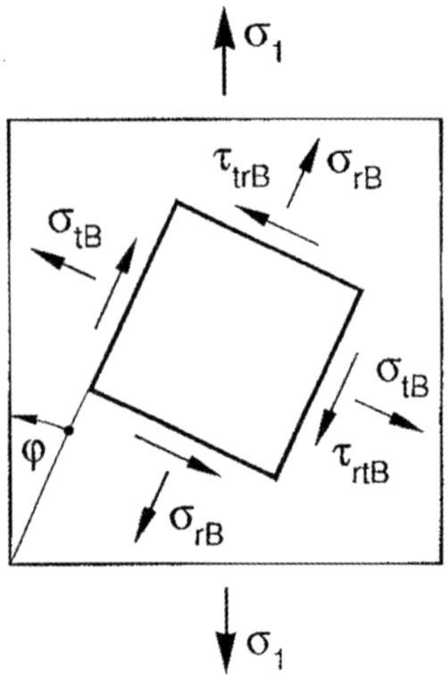

Fig. 12.42 Stresses in the cut surfaces of an infinitesimally small element

and for the shear stress

$$\tau_{\mathrm{rtB}} = -\frac{\sigma_1}{2}\left(1 - \frac{a^2}{r^2}\right)\left(1 + 3\frac{a^2}{r^2}\right)\sin 2\varphi. \tag{12.44}$$

The Eqs. (12.41)–(12.43) represent the stress distribution in the vicinity of a hole for the uniaxially stressed plate.

With the biaxially stressed plate, as shown in Figure 12.43, the stresses given in Eqs. (12.41)–(12.43) and the proportions produced by the principal normal stress σ_2 are superimposed, so that the following stresses are given for the biaxial stress state:

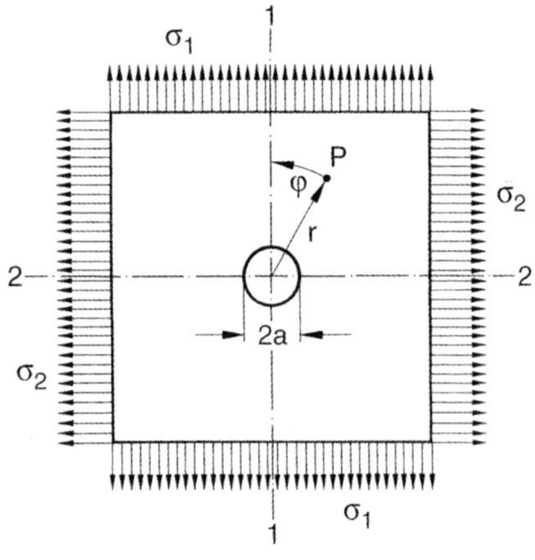

Fig. 12.43 Biaxially loaded plate with a round hole

For the radial stress:

$$\sigma_{rB} = \frac{\sigma_1 + \sigma_2}{2}\left(1 - \frac{a^2}{r^2}\right) + \frac{\sigma_1 - \sigma_2}{2}\left(1 - 4\frac{a^2}{r^2} + 3\frac{a^4}{r^4}\right)\cos 2\varphi \tag{12.45}$$

For the tangential stress:

$$\sigma_{tB} = \frac{\sigma_1 + \sigma_2}{2}\left(1 + \frac{a^2}{r^2}\right) - \frac{\sigma_1 - \sigma_2}{2}\left(1 + 3\frac{a^4}{r^4}\right)\cos 2\varphi \tag{12.46}$$

For the shear stress:

$$\tau_{rtB} = \frac{\sigma_2 - \sigma_1}{2}\left(1 - \frac{a^2}{r^2}\right)\left(1 + 3\frac{a^2}{r^2}\right)\sin 2\varphi \tag{12.47}$$

Equations (12.44)–(12.46) describe stress distribution in the vicinity of a hole in a biaxially stressed plate. The stresses depend on the position of the point P under consideration, on the radius a of the hole and on the principal normal stresses σ_1 and σ_2. If Eqs. (12.44)–(12.46) are applied to the undrilled plate by setting $a = 0$, then these represent the general equilibrium conditions for the biaxial stress state:

$$\sigma_r = \frac{\sigma_1 + \sigma_2}{2} + \frac{\sigma_1 - \sigma_2}{2}\cos 2\varphi \tag{12.48}$$

$$\sigma_t = \frac{\sigma_1 + \sigma_2}{2} - \frac{\sigma_1 - \sigma_2}{2}\cos 2\varphi \tag{12.49}$$

$$\tau_{rt} = \frac{\sigma_2 - \sigma_1}{2}\sin 2\varphi \tag{12.50}$$

Now that the stress distributions in the drilled and undrilled plate have been described, the change of stress due to the drilling of a hole can be described as the difference between the two states:

$$\begin{aligned} \Delta\sigma_r &= \sigma_{rB} - \sigma_r \\ \Delta\sigma_t &= \sigma_{tB} - \sigma_t \\ \Delta\tau_{rt} &= \tau_{rtB} - \tau_{rt} \end{aligned} \tag{12.51}$$

If this difference is formed for the normal stresses, then the following are obtained:

$$\Delta\sigma_r = \frac{\sigma_1 + \sigma_2}{2}\left(-\frac{a^2}{r^2}\right) + \frac{\sigma_1 - \sigma_2}{2}\left(3\frac{a^4}{r^4} - 4\frac{a^2}{r^2}\right)\cos 2\varphi \tag{12.52}$$

and

$$\Delta\sigma_t = \frac{\sigma_1 + \sigma_2}{2}\left(\frac{a^2}{r^2}\right) - \frac{\sigma_1 - \sigma_2}{2}\left(3\frac{a^4}{r^4}\right)\cos 2\varphi \tag{12.53}$$

Further inclusion of the shear stresses with respect to the hole drilling method is not relevant, because the measured strains are caused by normal stresses.

Equations (12.51) and (12.52) describe the changes of stress occurring at a point P given by the coordinates r and φ when a hole of diameter a is drilled at the point $r=0$ in the plate which is subject to a plane residual stress state. If these changes of stress $\Delta\sigma_r$ and $\Delta\sigma_t$ are known, then the principal normal stresses σ_1 and σ_2 can be calculated using Eqs. (12.51) and (12.52). This then explains the principle of the hole-drilling method, that is the stress state prior to drilling is found from the changes of stress due to the drilling.

12.5.2.3 Strains measured

In order to be able to find the stress state with the aid of Eqs. (12.51) and (12.52), the changes of stress $\Delta\sigma_r$ and $\Delta\sigma_t$ occurring due to drilling must be known. These stress changes can be found from the measured strains by applying Hooke's law.

In the hole-drilling method strains in the radial direction are measured with special strain gage rosettes. According to Hooke's law for the biaxial stress state the change of strain $\Delta\varepsilon_r$ occurs in the radial direction:

$$\Delta\varepsilon_r = \frac{1}{E}(\Delta\sigma_t - \nu\Delta\sigma_t) \tag{12.54}$$

In Eq. (12.53) E is Young's modulus and ν Poisson's ratio for the drilled material.

If Eqs. (12.51) and (12.52) are substituted in Eq. (12.53), then the following equation is obtained:

$$\Delta\varepsilon_r = \frac{1}{2E}\left[(\sigma_1+\sigma_2)(1+\nu)\left(-\frac{a^2}{r^2}\right) + (\sigma_1-\sigma_2)\left(3\nu\frac{a^4}{r^4}+3\frac{a^4}{r^4}-4\frac{a^2}{r^2}\right)\cos 2\varphi\right] \tag{12.55}$$

This is an equation with three unknown quantities σ_1, σ_2 and φ. Three equations are required to find these unknowns. The equations are obtained if Eq. (12.54) is applied to three different radial directions. This is why hole-drilling rosettes consist of three strain gages arranged in radial directions. Figure 12.44 shows an example with three measuring grids arranged with measuring directions displaced from one another by 45°.

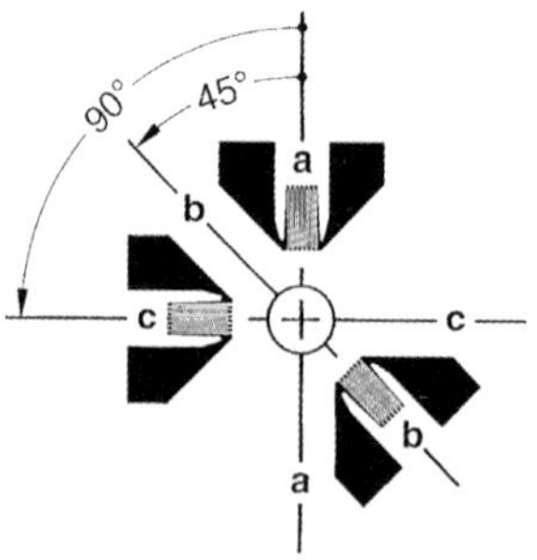

Fig. 12.44 Measuring grids and measurement directions on a hole-drilling rosette

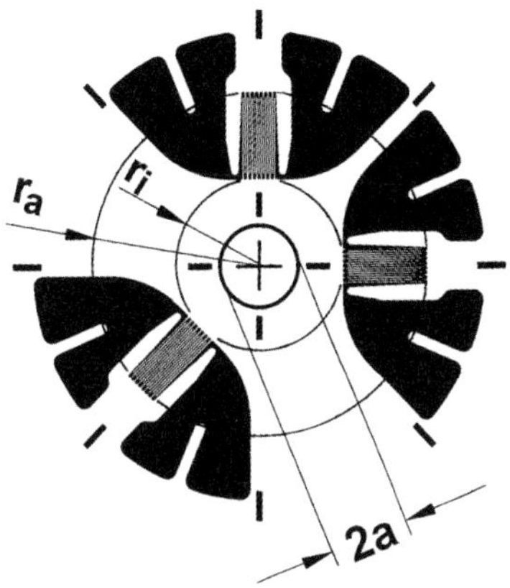

Fig. 12.45 Designation of the radii on a hole-drilling rosette

Strain gage measuring grids have a certain length of measuring grid. The strain at a particular point P cannot be measured with a strain gage, but rather only the integral of the strain over the length of the measuring grid, although this may be very short. The grid lengths of a strain gage rosette are limited by the radii r_i and r_a, as shown in Figure 12.45. Therefore the change of strain measured with a measuring grid is

$$\Delta\varepsilon_{ai} = \int_{r_i}^{r_o} \frac{\Delta\varepsilon_r}{r_o - r_i} dr \tag{12.56}$$

If Eq. (12.54) is substituted into Eq. (12.55), then the following is obtained:

$$\Delta\varepsilon_{\mathrm{ai}} = \frac{\mathrm{a}^2}{2\mathrm{E}(\mathrm{r_o} - \mathrm{r_i})} \int_{r_i}^{r_o} \left[(\sigma_1 + \sigma_2)\left(\frac{-1}{r^2}\right)(1+\nu) + (\sigma_1 - \sigma_2)\left(3\nu\frac{a^2}{r^4} + 3\frac{a^2}{r^4} - 4\frac{1}{r^2}\right)\cos 2\varphi\right]\mathrm{dr} \tag{12.57}$$

After integration and insertion of the limits we obtain:

$$-\Delta\varepsilon_{\mathrm{ai}} = \frac{\mathrm{a}^2(1+\nu)}{2\mathrm{Er_o r_i}}(\sigma_1 + \sigma_2) + \frac{2\mathrm{a}^2}{\mathrm{Er_o r_i}}(\sigma_1 - \sigma_2)\left[1 - \mathrm{a}^2(1+\nu)\frac{\mathrm{r_i^2} + \mathrm{r_o r_i} + \mathrm{r_o^2}}{4\mathrm{r_i^2 r_o^2}}\right]\cos 2\varphi \tag{12.58}$$

Simplifying with

$$-\Delta\varepsilon_{ai} = \frac{A}{E}(\sigma_1 + \sigma_2) + \frac{B}{E}(\sigma_1 - \sigma_2)\cos 2\varphi \tag{12.59}$$

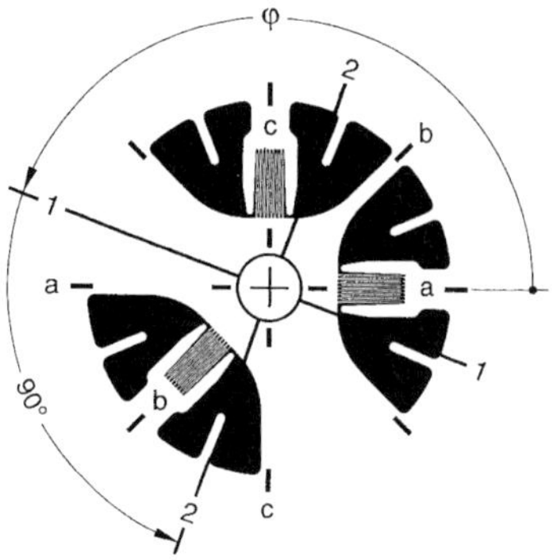

Fig. 12.46 Angle of orientation φ between the rosette reference direction a and principal direction 1

in which

$$A = \frac{a^2(1+\nu)}{2r_o r_i} \tag{12.60}$$

and

$$B = \frac{2a^2}{r_o r_i}\left[1 - a^2(1+\nu)\frac{r_i^2 + r_o r_i + r_o^2}{4r_i^2 r_o^2}\right] \tag{12.61}$$

leads to

$$-\Delta\varepsilon_{ai} = \frac{A}{E}(\sigma_1 + \sigma_2) + \frac{B}{E}(\sigma_1 - \sigma_2)\cos 2\varphi \tag{12.62}$$

If the strain gage measurement directions are labelled *a, b* and *c,* according to Figure 12.46, *a* is taken as the reference direction, and the angle φ is defined in the mathematically positive direction with respect to the direction *a* as the orientation angle for principal direction 1, then the three strains in the three measurement directions are given by:

$$-\Delta\varepsilon_a = \frac{A}{E}(\sigma_1 + \sigma_2) + \frac{B}{E}(\sigma_1 - \sigma_2)\cos 2\varphi \tag{12.63}$$

$$-\Delta\varepsilon_b = \frac{A}{E}(\sigma_1 + \sigma_2) + \frac{B}{E}(\sigma_1 - \sigma_2)\sin 2\varphi \tag{12.64}$$

$$-\Delta\varepsilon_c = \frac{A}{E}(\sigma_1 + \sigma_2) - \frac{B}{E}(\sigma_1 - \sigma_2)\cos 2\varphi \tag{12.65}$$

This then solves the problem. The principal stresses and directions of the residual stress state before drilling can be found from the changes of strain measured in three directions.

12.5.2.4 Computation of the residual stress state

Angle of orientation of the principal directions

The solution of the system of Eqs. (12.61)–(12.63) leads to the conditional equations for the angle of orientation φ and for the principal normal stresses σ_1 and σ_2.

Introducing the relationships

$$\sin 2\varphi = \frac{\tan 2\varphi}{w}; \quad \cos 2\varphi = \frac{1}{w} \tag{12.66}$$

with

$$\mathrm{w} = \sqrt{1 + \tan^2 2\varphi} \tag{12.67}$$

into Eqs. (12.62) (12.64), then the following equations are obtained:

$$-\Delta\varepsilon_\mathrm{a} = \frac{\mathrm{A}}{\mathrm{E}}(\sigma_1 + \sigma_2) + \frac{\mathrm{B}}{\mathrm{wE}}(\sigma_1 - \sigma_2) \tag{12.68}$$

$$-\Delta\varepsilon_\mathrm{b} = \frac{\mathrm{A}}{\mathrm{E}}(\sigma_1 + \sigma_2) + \frac{\mathrm{B}}{\mathrm{wE}}(\sigma_1 - \sigma_2)\tan 2\varphi \tag{12.69}$$

$$-\Delta\varepsilon_\mathrm{c} = \frac{\mathrm{A}}{\mathrm{E}}(\sigma_1 + \sigma_2) - \frac{\mathrm{B}}{\mathrm{wE}}(\sigma_1 - \sigma_2) \tag{12.70}$$

This produces the relationships

$$(\sigma_1 - \sigma_2) = \frac{\mathrm{wE}}{2\mathrm{B}}(\Delta\varepsilon_\mathrm{c} - \Delta\varepsilon_\mathrm{a}) \tag{12.71}$$

and

$$(\sigma_1 + \sigma_2) = -\frac{\mathrm{E}}{2\mathrm{A}}(\Delta\varepsilon_\mathrm{a} + \Delta\varepsilon_\mathrm{c}) \tag{12.72}$$

By substituting Eqs. (12.70) and (12.71) in Eq. (12.68) the conditional equations for the angle of orientation φ are produced:

$$\tan 2\varphi = \frac{\Delta\varepsilon_\mathrm{a} + \Delta\varepsilon_\mathrm{c} - 2\Delta\varepsilon_\mathrm{b}}{\Delta\varepsilon_\mathrm{c} - \Delta\varepsilon_\mathrm{a}} \tag{12.73}$$

$$\varphi = \frac{1}{2}\arctan\frac{\Delta\varepsilon_\mathrm{a} + \Delta\varepsilon_\mathrm{c} - 2\Delta\varepsilon_\mathrm{b}}{\Delta\varepsilon_\mathrm{c} - \Delta\varepsilon_\mathrm{a}} \tag{12.74}$$

As represented in Figure 12.46, the angle of orientation φ is taken to be in the mathematically positive sense of rotation referred to the rosette direction *a* and gives the principal direction 1. The angle of orientation φ can be $0° \leq \varphi < 180°$. The conditional Eq. (12.72), respectively (12.73), is ambiguous due to the tangent function. For the unambiguous determination of φ the quotient, which is the argument of the tangent, must be considered with respect to the sign of the numerator n and denominator d:

$$\varphi = \frac{1}{2} \arctan \frac{n}{d} \tag{12.75}$$

where

$$\mathrm{n} = \Delta\varepsilon_\mathrm{a} + \Delta\varepsilon_\mathrm{c} - 2\Delta\varepsilon_\mathrm{b} \quad \text{and} \quad \mathrm{d} = \Delta\varepsilon_\mathrm{c} - \Delta\varepsilon_\mathrm{a} \tag{12.76}$$

The unambiguous definition of the angle on the basis of the separate consideration of numerator n and denominator d is described in detail in Chapter 9 under the title 'Stress analysis using strain gage rosettes'.

Principal normal stresses

The principal normal stresses are obtained by resolving Eqs. (12.67)–(12.69) to give σ_1 and σ_2:

$$\sigma_{1,2} = -\frac{\mathrm{E}}{4\mathrm{A}}(\Delta\varepsilon_\mathrm{a} + \Delta\varepsilon_\mathrm{c}) \pm \frac{\mathrm{E}}{4\mathrm{B}}\sqrt{(\Delta\varepsilon_\mathrm{c} - \Delta\varepsilon_\mathrm{a})^2 + (\Delta\varepsilon_\mathrm{a} + \Delta\varepsilon_\mathrm{c} - 2\Delta\varepsilon_\mathrm{b})^2} \tag{12.77}$$

where E is Young's modulus for the material under test and A and B are constants represented by Eqs. (12.59) and (12.60) respectively. If Eq. (12.75) is simplified by including

$$\frac{\mathrm{E}}{4\mathrm{A}} = \mathrm{A}^* \quad \text{and} \quad \frac{\mathrm{E}}{4\mathrm{B}} = \mathrm{B}^* \tag{12.78}$$

Then the following is produced:

$$\sigma_{1,2} = -A^*(\Delta\varepsilon_c - \Delta\varepsilon_a) \pm B^* \sqrt{(\Delta\varepsilon_c - \Delta\varepsilon_a)^2 + (\Delta\varepsilon_a + \Delta\varepsilon_c - 2\Delta\varepsilon_b)^2} \tag{12.79}$$

12.5.3 Determination of the residual stress distribution in relationship to the depth

The coefficients, which are included in the evaluation formulas and which take into account the hole and rosette shapes together with Young's modulus and Poisson's ratio for the material under investigation, provide a method of modifying the evaluation formulas to suit the application. The formulas are matched to the application experimentally by measuring the strains for a known stress state in calibration tests. These strains are then inserted into the evaluation formulas which are solved, giving the coefficients. Using the coefficients found in this way, the evaluation formulas are matched to suit the practical circumstances of the application, extending the application of the hole-drilling method beyond the previously mentioned ideal requirements. This means that the evaluation algorithm is no longer based purely on a derivation using the theory of elasticity as shown in Section 12.2, but it is instead orientated to calibration values determined practically for a certain mode of application.

The hole-drilling method also enables the determination of the residual stresses as a function of the hole's depth when stress gradients in the depth direction occur. Here, step-by-step drilling is employed with measurement of the hole's depth and of the strains released on the surface of the test object at each step. The computational methods involved have been, and still are, the subject of a number of published papers with the objective of determining, as completely as possible, the plane residual stress state – not just on the surface, but also in the layers below the material surface as a function of the hole depth. As far back as 1952, a method was suggested which was orientated to reference curves that were to be found in the laboratory during preliminary tests under known conditions [12.76]. This method was developed further by including FEM in the computation [12.77, 12.78]. If steep stress gradients are present perpendicular to the surface of the test object with possible sign reversal of the residual stresses in the region under investigation, then evaluations with the aid of reference curves found under pure tension and compression loading may contain errors of an order of magnitude of about 50% [12.79]. These can be avoided if, after each step in drilling, the redistribution of the stress due to the change in the hole's shape and the effect on the change of strain on the surface are taken into account. This way of looking at the problem also indicates that penetration of the hole into material layers free from residual stress may lead to changes of strain on the surface of the test object [12.79]. This method is being developed in [12.80] for computer-aided evaluation.

A detailed examination and assessment of the currently known methods for the evaluation of measurement results according to the hole-drilling method has been carried out by the MPA in Stuttgart and the results summarized in [12.81]. In this paper an evaluation method under the designation 'biaxial MPA/KWU formula' was presented, which combined the 'uniaxial MPA formula' developed by König [12.82, 12.83] and the evaluation method developed by Wolf and Böhm with relaxation functions [12.84] formulated in two orthogonal directions. As long ago as 1956, Kelsey published in [12.85] the suggestion of introducing relaxation functions for the determination of the relationship between the changes of strain measured on the surface and the stress changes released at the hole depth.

The evaluation formula [12.81] from the MPA Stuttgart appears to be the method which supplies the best results. It is based on calibration (relaxation) functions experimentally determined for the biaxial stress state and which define the relationship between the change of the strain state produced by the hole at a certain depth of penetration and the resulting changes of strain produced on the surface of the test object (measured by the rosette measuring grids).

With uniaxial loading by a stress σ_1 at a measuring point in direction 1 the following strain is produced

$$\varepsilon_{11} = \frac{\sigma_1}{E} \tag{12.80}$$

transversely in direction 2:

$$\varepsilon_{21} = -\frac{\nu}{E}\sigma_1 k \tag{12.81}$$

Introducing a hole produces a strain dependent on the hole depth z in direction 1

$$\varepsilon_{11}(z) = F_1(z)\varepsilon_{11}^{*} \tag{12.82}$$

in direction 2

$$\varepsilon_{21}(z) = F_2(z)\overset{*}{\varepsilon}_{21} \tag{12.83}$$

$F_1(z)$ and $F_2(z)$ correspond to the calibration functions defined by König [12.83]. The strains at a depth z are labelled with *. For the changes in strain that are dependent on the depth in direction 1, the following is produced by σ_1

$$\frac{d\varepsilon_{11}}{dz} = \overset{*}{\varepsilon}_{11}\frac{dF_1}{dz} \tag{12.84}$$

in direction 2 by σ_1

$$\frac{d\varepsilon_{21}}{dz} = \overset{*}{\varepsilon}_{21}\frac{dF_2}{dz} \tag{12.85}$$

In a similar manner the changes of strain produced by σ_2 can be written as

$$\frac{d\varepsilon_{22}}{dz} = \overset{*}{\varepsilon}_{22}\frac{dF_1}{dz} \tag{12.86}$$

and

$$\frac{d\varepsilon_{12}}{dz} = \overset{*}{\varepsilon}_{12}\frac{dF_2}{dz} \tag{12.87}$$

The total strains acting in directions 1 and 2 can be found by superimposition:

$$\varepsilon_1 = \varepsilon_{11} + \varepsilon_{12}; \ \varepsilon_2 = \varepsilon_{22} + \varepsilon_{21} \tag{12.88}$$

and with Eqs. (12.83)–(12.86):

$$\frac{d\varepsilon_1}{dz} = \frac{\overset{*}{\varepsilon}_{11}dF_1}{dz} + \frac{\overset{*}{\varepsilon}_{12}dF_2}{dz} \tag{12.89}$$

respectively,

$$\frac{d\varepsilon_2}{dz} = \frac{\overset{*}{\varepsilon}_{22}dF_1}{dz} + \frac{\overset{*}{\varepsilon}_{21}dF_2}{dz} \tag{12.90}$$

Due to lateral contraction, the following applies to isotropic materials:

$$\overset{*}{\varepsilon}_{12} = -\nu\overset{*}{\varepsilon}_{22}; \ \overset{*}{\varepsilon}_{21} = -\nu\overset{*}{\varepsilon}_{11} \tag{12.91}$$

If these relationships are substituted into the Eqs. (12.88) and (12.89) and then solved for $\varepsilon_{11}{}^*$ and $\varepsilon_{22}{}^*$, the following is obtained:

$$\overset{*}{\varepsilon}_{11} = \frac{1}{\left(\frac{dF_1}{dz}\right)^2 - \nu^2\left(\frac{dF_2}{dz}\right)^2}\left[\frac{dF_1}{dz}\frac{d\varepsilon_1}{dz} + \nu\frac{dF_2}{dz}\frac{d\varepsilon_2}{dz}\right] \tag{12.92}$$

$$\overset{*}{\varepsilon}_{22} = \frac{1}{\left(\frac{dF_1}{dz}\right)^2 - \nu^2\left(\frac{dF_2}{dz}\right)^2}\left[\frac{dF_1}{dz}\frac{d\varepsilon_2}{dz} + \nu\frac{dF_2}{dz}\frac{d\varepsilon_1}{dz}\right] \tag{12.93}$$

In order to obtain a relationship valid for different hole diameters, the normalization suggested by König is introduced here, which relates the depth z to the hole diameter D_0:

$$\xi = \frac{z}{D_0} \tag{12.94}$$

All quantities are therefore represented as functions of the normalized depth ξ. The numerical differentiation of the strains occurs in a similar way. The calibration functions are also represented in dependence of the normalized depth and assigned symbols K_1 and K_2.

The strains in the three measuring grid directions a, b and c on a hole-drilling rosette are then

$$\varepsilon_a^* = \frac{1}{K_1^2(\xi) - \nu^2 K_2^2(\xi)} \left[K_1(\xi) \frac{d\varepsilon_a}{d\xi} + \nu K_2(\xi) \left(\frac{d\varepsilon_c}{d\xi} \right) \right] \tag{12.95}$$

$$\varepsilon_b^* = \frac{1}{K_1^2(\xi) - \nu^2 K_2^2(\xi)} \left[K_1(\xi) \frac{d\varepsilon_b}{d\xi} + \nu K_2(\xi) \left(\frac{d\varepsilon_a}{d\xi} - \frac{d\varepsilon_b}{d\xi} + \frac{d\varepsilon_c}{d\xi} \right) \right] \tag{12.96}$$

$$\varepsilon_c^* = \frac{1}{K_1^2(\xi) - \nu^2 K_2^2(\xi)} \left[K_1(\xi) \frac{d\varepsilon_c}{d\xi} + \nu K_2(\xi) \left(\frac{d\varepsilon_a}{d\xi} \right) \right] \tag{12.97}$$

Apart from the measurement results, the calibration functions are needed for the numerical evaluation of these equations and they must be available before the method is applied.

The stresses in the three measurement directions are found from Eqs. (12.95–12.97) using Young's modulus:

$$\begin{aligned} \sigma_a &= E\varepsilon_a^* \\ \sigma_b &= E\varepsilon_b^* \\ \sigma_c &= E\varepsilon_c^* \end{aligned} \tag{12.98}$$

The computation of the principal stresses and the angle of orientation of principal direction 1 (in the mathematically positive sense of rotation starting from the rosette reference direction a) is now quite simple:

$$\sigma_{1,2} = \frac{1}{2}(\sigma_c + \sigma_a) \pm \frac{1}{2}\sqrt{(\sigma_c - \sigma_a)^2 + (\sigma_a + \sigma_c - 2\sigma_b)^2} \tag{12.99}$$

$$\varphi = \frac{1}{2}\arctan\left(\frac{2\sigma_b - \sigma_a - \sigma_c}{\sigma_a - \sigma_c} \right) \tag{12.100}$$

Example: Application of the hole-drilling method [12.86]
In order to increase the service life of leaf springs which are normally under tensile load, compressive residual stresses are created by shot-peening their surfaces. The induced compressive stresses partly compensate the tensile stresses occurring during operation.

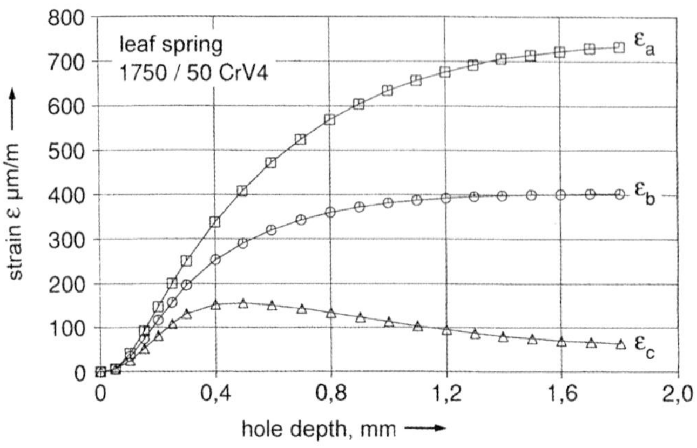

Fig. 12.47 Strains measured with the measuring grids of a hole-drilling rosette on a leaf spring in relationship to the hole depth z [12.86]

To check the process and for reasons of quality assurance, the residual stresses on these surface-treated springs are found using the hole-drilling method.

Figure 12.47 shows the strains measured with the three measuring grids of a hole-drilling rosette in dependence of the hole depth on a spring in 50CrV4. The hole with a diameter of 1.7 mm was produced using a turbine drill with a speed of about 400 000 rpm. Depth steps of 0.05 mm were taken down to a hole depth of 0.3 mm. The hole-drilling method therefore enables measurements to be made at very small depth steps in the region of the material surface. Depth steps of 0.1 mm applied from 0.3 mm onwards.

The actual measurements on the leaf springs were used for finding the calibration functions, which are shown in normalized form for the diameter ratio D/D_0 in Figure 12.48 (D = nominal circle diameter of the measuring grid configuration, D_0 = hole diameter).

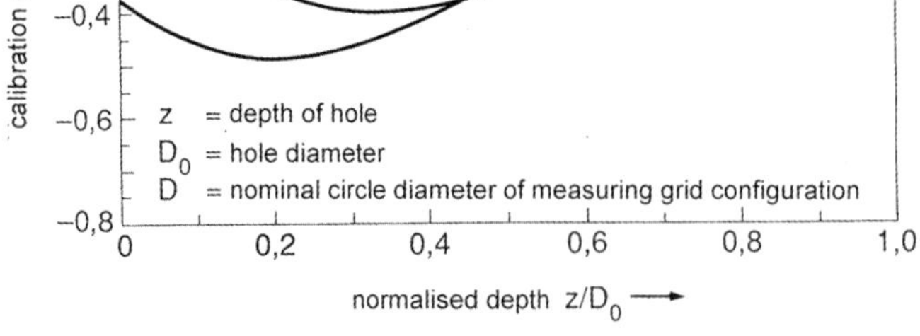

Fig. 12.48 Calibration curves K_1 and K_2 for the RY61 hole-drilling rosette [12.86]

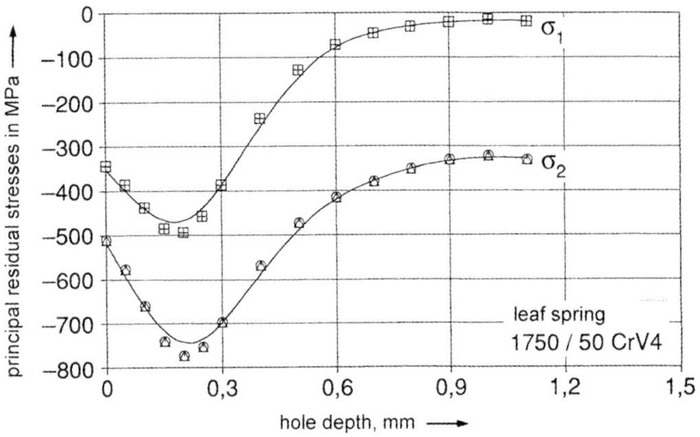

Fig. 12.49 Distribution of the principal normal stresses of the residual stress state, found with the hole-drilling method, in the layers near the surface of a leaf spring [12.86]

The measurements and also the calibration values for the corresponding depth steps were evaluated on a PC by a computer program developed at the MPA Stuttgart. The program is based on Eqs. (12.95–12.97).

Figure 12.49 shows the results of the evaluation. The distributions of the principal normal stresses σ_1 and σ_2 of the residual stress state are shown in relationship to the depth in a 1.1 mm thick surface layer. The maxima of the compressive residual stresses can be seen to occur in a layer about 0.2 mm below the surface.

12.5.4 Forming the hole

Forming the hole must be carried out without introducing stress into the material under consideration. Stress introduced during forming the hole will superimpose on the stress distribution which would be measured and distort the result of measurement.

Important boring parameters are geometry of the boring tool, drilling speed, milling cutter wear, drilling mode, applied drilling force and cooling.

Satisfactory results were obtained with a normal twist drill at a speed of 750 rpm and a feed force of about 20 N in a steel specimen [12.87].

For larger bore diameter, as they are now used for investigations at greater depths, the use of drill bits is favourable, since the non-cutting centre of the twist drill is avoided [12.88].

Figure 12.50 shows application of the hole-drilling method using a water-cooled diamond drill bit on a ceramic coated piston of a diesel engine. Because of the hole diameter of 5 mm, an appropriate rosette is used.

Nowadays air turbine driven carbide cutters with drilling speeds up to 400, 000 rpm are successfully used in such applications [12.89]. This technique maintains all the

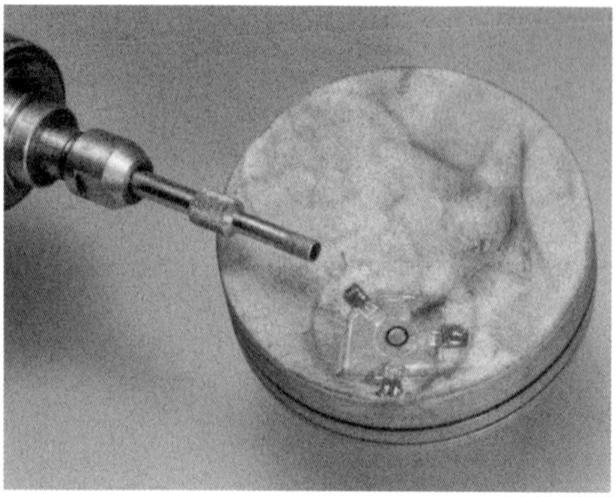

Fig. 12.50 Producing a bore hole using a water-cooled diamond drill bit in the ceramic coating of a diesel engine piston

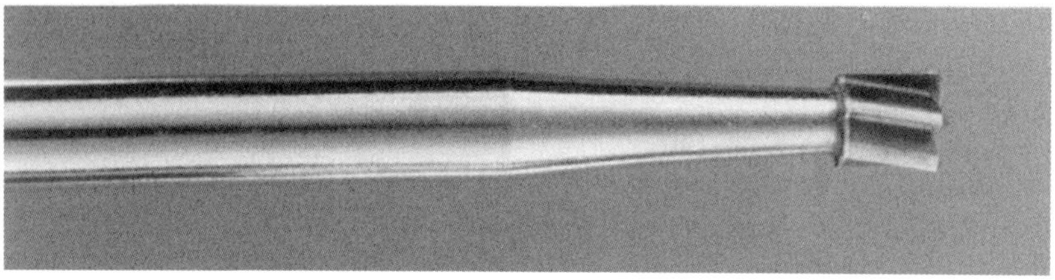

Fig. 12.51 Special dressed end mill for high-speed drilling up to 400, 000 rpm with a diameter of 1.5 mm

advantages of the special dressed end mill. We all know this boring method from our visits to the dentist. One can use this technique on materials with hardness up to HRc60. Figure 12.51 shows a special dressed end mill for application of high-speed drilling.

For harder materials, the air abrasive method of hole forming can be used. The standard airabrasive equipment produces the parallel-sided hole shown in Figure 12.52 produced by a rotating nozzle. The central nipple does not affect the measured strain relief. The operating parameters are also given in this figure.

typical profile for a nominal 1.6 mm airabrasive formed hole

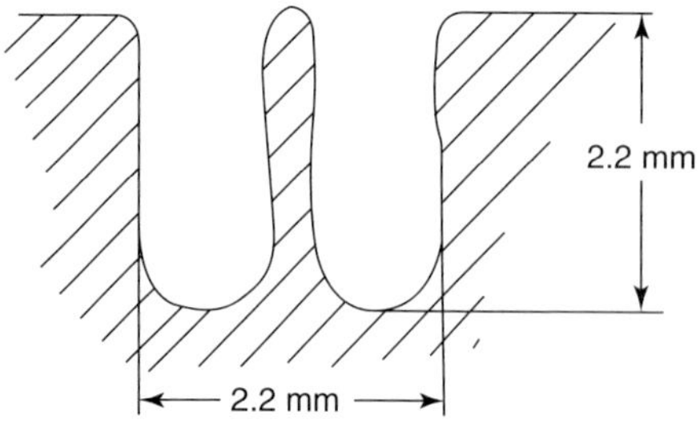

air supply pressure	5.4 bar
air supply rate	0.01 m^3/min
powder type	53 micron Aluminum Oxide
drilling time	15 mins
nozzle type	sapphire
nozzle diameter	0.45 mm
powder flow	10g/min

Fig. 12.52 Hole formed by standard air abrasive method [12.91]

typical profile for a nominal 12.5 airabrasive formed hole

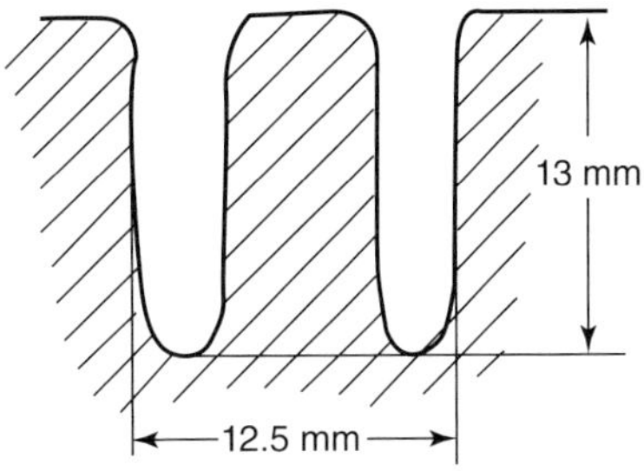

air supply pressure	7.1 - 8.8 bar
air supply rate	0.19 - 0.25 m^3/min
drilling time	200 mins
nozzle type	carbide
nozzle diameter	1.3 mm
powder type	53 micron Aluminum Oxide
powder flow	25 g/min

Fig. 12.53 Hole formed by the air abrasive method extended to a depth of 13 mm [12.91]

The conventional equipment of the air abrasive method comprises compressor, dust extractor, controller, cooling unit, drilling head and optical head, the last being used for depth and diameter measurement.

In its modification this method allows the formation of holes up to a depth of 13 mm with a diameter of 13 mm [12.90]. Figure 12.53 shows the contour of a hole formed by the extended abrasive method.

The drilling procedure is carried out with a nozzle arranged eccentrically to the centre line of the hole, rotating simultaneously around itself and around the centre line of the hole.

The strain readings have to be made as the hole is continuously drilled. Interchanging the drilling head and the optical head could cause slight differences in the setup of the drilling head and hence produce a non-parallel sided hole [12.91].

Misalignment between the hole and the rosette centre significantly influences the strains measured by the measuring grids of the rosette. It is indicated that a 0.013 mm off-centre hole gives an error of 2% in the predicted stress. In [12.92] the influence of hole eccentricity is investigated in detail and correction formulas are given.

In 1986 in USA, a round-robin residual stress measurement study was carried out [12.93]. Related to the hole-forming process the study has shown that the best results were obtained by using the air abrasive method as well as the ultra-high-speed drill technique. For the low-speed end mill the indicated results were influenced by induced machining stresses caused by the drill itself.

Bibliography

[12.1] Campus, F. (1954) Effects of residual stresses on the behavior of structures, in *Residual Stresses in Metal and Metal Constructions* (ed. W.R. Osgood), Reinhold Publication Corp., New York, p. 3.

[12.2] Horger, O.J., Neifert, H.R. and Regen, R.R. (1943) Residual stresses and fatigue studies. *Proceedings of the SESA*, **I** (1), 8–10.

[12.3] Keil, S. (1989) Dehnungsmeßstreifen als Hilfsmittel zur experimentellen Ermittlung von Eigenspannungen. *Technische Überwachung*, **30** (9), 309–316.

[12.4] Meier, H. (1936) Eigenspannungen in Eisenbahnschienen. *Organ für die Fortschritte im Eisenbahnwesen,* **91**, 320–329.

[12.5] Siebert, D. (1977) Verfahren zum Messen von Eigenspannungen in I-Trägern, part 2. *MTB,* **13** (3), 64–68.

[12.6] Siebert, D. (1977) Verfahren zum Messen von Eigenspannungen in I-Trägern, part 1. *MTB,* **13** (2), 40–45.

[12.7] Meriam, J.L., DeGarmo, E.P. and Jonassen, F. (1946) A method for the measurement of residual welding stresses. *Welding Research Supplement,* **June,** 340s–343s.

[12.8] DeGarmo, E.P., Meriam, J.L. and Jonassen, F. (1946) Residual stresses in intersecting butt welds. *Welding Research Supplement,* **August,** 451s–463s.

[12.9] DeGarmo, E.P., Meriam, J.L. and Jonassen, F. (1946) The effect of weld length upon the residual stresses of unrestrained butt welds. *Welding Research Supplement,* **25**, 485s–487s.

[12.10] Rosenthal, D. and Norton, J.T. (1945) A method of measuring triaxial residual stress in plates. *Welding Research Supplement,* **24**, 295s–307s.

[12.11] Ellingson, W.A. and Shack, W.J. (1979) Residual stress measurement on multi-pass weldments of stainless-steel piping. *Experimental Mechanics,* **19**, 317–323.

[12.12] Rybicki, E.F., Shadley, J.R. and Shealy, W.S. (1983) A consistent-splitting model for experimental residual-stress analysis. *Experimental Mechanics Dezember,* **23**, 438–446.

[12.13] Wilken, K. and Schmalenberg, W. (1985) Die Ermittlung von Schweißeigenspannungen an grossen Platten und Bauteilen mittels Dehnungsmeßstreifen. *VDI-Berichte,* **552**, 401–415.

[12.14] Berghaus, D.G. (1986) Residual stresses at small diffusion welds. *Experimental Techniques,* **10**, 40–42.

[12.15] Shadley, J.R., Rybicki, E.F. and Shealy, W.S. (1987) Application guidelines for the parting out step in a through thickness residual measurement procedure. *Strain,* **23**, 157–166.

[12.16] Müller-Wiesner, D. (1990) Eigenspannungsmessungen am ARIANE Wassertank. *VDI-Berichte,* **815**, 311–322.

[12.17] Rappe, H.-A. (1973) Messung von Schweißeigenspannungen mit Dehnungsmeßstreifen. *MTB,* **9**, 31–37.

[12.18] Stoker, J.R. and Moir, P.J. (1974) The use of finite element analysis and strain gage measurements in the determination of residual stresses, *Proc. 5th Int. Conf. Exp. Sress Analysis Udine,* paper 14, pp. 1.123–1.134.

[12.19] Ritchie, D. and Leggatt, R.H. (1987) The measurement of the distribution of residual stresses through the thickness of a welded joint. *Strain,* **23**, 61–70.

[12.20] Huaipu, Z. (1992) A new method for measurement of three dimensional residual stresses in a multi-pass butt welded joint. *Strain*, **28**, 13–17.

[12.21] Grüger, C. and Ketting, M. (1993) Verfahren zum Ermitteln von Eigenspannungen an Laufrollen für Raupenkettenfahrzeuge. *MTB*, **29** (1), 15–18.

[12.22] Found, G.H. and Pittsley, R. (1950) Residual stresses in magnesium: technique and results. *Proceedings of the SESA* **VIII** (1), 1–13.

[12.23] Roy, G., Kiff, D., Hunt, P. and Crawford, G. (1991) The effect of quenching and tempering on residual stress distribution in seamless pipes, *Proc. SEM Spring Conf. 1991 Milwaukee*, pp. 378–385.

[12.24] Heyn, E. and Bauer, O. (1911) Über Spannungen in kaltgereckten Metallen. *Z. f. Metallographie*, **1**, 16–50.

[12.25] Mesnager, M. (1919) Methode de determination des tensions existant dans un cylindre circulaire. *Comptes Rendus hebdomadaires des Seances de L'Academie des Sciences*, **169**, 1391–1393.

[12.26] Sachs, G. (1927) Der Nachweis innerer Spannungen in Stangen und Rohren. *Zeitschrift für Metallkunde*, **19**, 352–357.

[12.27] Bühler, H. (1952) Messung innerer Spannungen nach dem Ausbohr- und Abdrehverfahren mit Dehnungsmeßstreifen, in *Grundlagen und Anwendung des Dehnungsmeßstreifens* (ed. K. Fink), Verlag Stahleisen, Düsseldorf, pp. 200–206.

[12.28] Bühler, H. and Schreiber, W. (1952) Messung von Eigenspannungen in Stangen und Rohren mittels Dehnungsmeßstreifen. *Z-VDI*, **94**, 216–218.

[12.29] Bühler, H. and Schreiber, W. (1952) Lückenlose Bestimmung des Eigenspannungszustandes in metallischen Hohlzylindern. *Z-VDI*, **94**, 1147–1141.

[12.30] Bühler, H. and Schreiber, W. (1965) Dehnungsmessungen in engen Bohrungen. *Werkstatt und Betrieb*, **86**, 50–51.

[12.31] Hanke, E. and Tiemann, E. (1970) Bestimmung von Eigenspannungen erster Art an gehärtetem Material mit Dehnungsmeßstreifen. *Materialprüfung*, **12**, 279–282.

[12.32] Bühler, H. (1956) Messung von Eigenspannungen an Eisenbahnwagenrädern. *Eisenbahntechnische Rundschau*, **7**, 281–288.

[12.33] Gezici, R. (1960) *Einfluß eines induktiven Oberflächenhärtens auf die Eigenspannungen in Hohlzylindern aus Gußeisen mit Kugelgraphit und die Verdrehwechselfestigkeit*. Diss. TH Aachen.

[12.34] Tietz, H.-D. and Hammacher, B. (1977) Einfluß des Kaltpilgerverfahrens auf die Eigenspannungen in Rohren der Stahlmarke 100Cr6. *Neue Hütte*, **22** (4), 193–196.

[12.35] Krägeloh, E. (1959) Bestimmung von Eigenspannungen an Zylindern aus inhomogenem Werkstoff mittels Ausbohr- und Abdrehverfahren. *Materialprüfung*, **1**, 377–384.

[12.36] König, G. and Kockelmann, H. (1989) Erweiterung des Ausbohr- und Abdrehverfahrens zur Ermittlung von Eigenspannungen an inhomogenen Werkstoffen. *VDI-Berichte*, **731**, 431–441.

[12.37] Stäblein, F. (1931) Spannungsmessungen an einseitig abgelöschten Knüppeln. *Kruppsche Monatshefte*, **12**, 93–99.

[12.38] Schimmöller, H. (1972) Bestimmung von Eigenspannungen in ebenen plattierten Werkstoffen, part 1 and part 2. *Materialprüfung*, **14**, 115–122 and 380–387.

[12.39] Peiter, A. and Lode, W. (1982) Die örtlichen Eigenspannungen in Blechen und Plattierungen. *Material und Technik*, **2**, 62–70.

[12.40] König, G. and Dannenberg, R. (1991) Zur Frage der Auswirkung von Eigenspannungen auf den Verzug von Walzteilen bei der Weiterverarbeitung. *VDI-Berichte*, **882**, 325–334.

[12.41] Bühler, H. and Tönshoff, H.K. (1967) Messen der durch plastische Verformungen in dünnen Oberflächenschichten entstehenden Eigenspannungen. *Archiv für das Eisenhüttenwesen*, **38**, 363–371.

[12.42] Druez, J. and Bazergui, A. (1983) Through-thickness measurement of residual stresses in thin tubes. *Experimental Mechanics*, **23**, 211–216.

[12.43] Found, G.H. and Pittsley, R. (1950) Residual stresses in magnesium: technique and results. *Proceedings of the SESA*, **VIII** (1), June 1983, pp. 1–13.

[12.44] Shadley, J.R., Rybicki, E.F. and Shealy, W.S. (1987) Application guidelines for the parting out step in a through thickness residual stress measurement procedure. *Strain*, **23**, 157–166.

[12.45] Milbradt, K.P. (1950) Ring-method determination of residual stresses, *Proc. SESA Spring Meeting Cleveland*, pp. 63–74.

[12.46] Gunnert, R. (1955) *Method for Measuring Residual Stresses and its Application to a Study of Residual Welding Stresses*, Almqvist & Wiksell, Stockholm.

[12.47] Wolf, H. and Böhm, W. (1971) Das Ring-Kern-Verfahren zur Messung von Eigenspannungen und seine Anwendungen bei Turbinen- und g-Generatorwellen. *Archiv für das Eisenhüttenwesen*, **42**, 195–200.

[12.48] Wolf, H. and Böhm, W. (1971) Entstehung, Messung und Beurteilung von Eigenspannungen in schweren Schmiedestücken für Turbinen und Generatoren. *Archiv für das Eisenhüttenwesen*, **42**, 509–513.

[12.49] Böhm, W. and Wolf, H. (28.9. 1978) Verfahren zur Messung von Eigenspannungen in Bauteilen des Maschinen- oder Apparatebaues und Vorrichtung zur Durchführung dieses Verfahrens. German Patent 20 16 118.

[12.50] Böhm, W., Stücker, E. and Wolf, H. (17.11. 1977) Dehnungsmeßstreifen zur Messung von Oberflächendehnungen an kreisflächenförmigen Werkstückabschnitten, German Patent 23 45 309.

[12.51] König, G. (1990) *Ein Beitrag zur Weiterentwicklung teilzerstörender Eigenspannungsmeßverfahren.* Dissertation Universität Stuttgart.

[12.52] Hofer, G. and Bender, N. (1979) Eigenspannungsmessungen an Plattierungen verschiedener Herstellungsverfahren. *Ber. DGM-Symposium "Eigenspannungen" Bad Nauheim*, pp. 69–94.

[12.53] Wolf, H., Böhm, W. and Stücker, E. (1973) Vorrichtung zum maßgerechten Einarbeiten einer Ringnut um eine kreisförmige Meßstelle bei der Messung von Eigenspannungen in Bauteilen des Maschinen- und Apparatebaues, German Patent 21 42 463.

[12.54] Böhm, W., Stücker, E. and Wolf, H. (1980) Grundlagen und Anwendungsmöglichkeiten des Ring-Kern-Verfahrens zum Ermitteln von Eigenspannungen, part 2: Praxis und Anwendungsbeispiele. *MTB*, **16**, 66–70.

[12.55] Kreuzer, M. (1983) Vergleichende Betrachtung verschiedener Schaltungsarten für das Messen mit Dehnungsmeßstreifen. *MTB*, **19**, 63–68.

[12.56] Wolf, H., Borgmann, H.-A. and Stücker, E. (1973) Weiterentwicklung des Ring-Kern-Verfahrens zur Messung von Eigenspannungen. *Archiv für das Eisenhüttenwesen*, **44**, 369–373.

[12.57] Böhm, W., Stücker, E. and Wolf, H. (1980) Grundlagen und Anwendungsmöglichkeiten des Ring-k-Kern-Verfahrens zum Ermitteln von Eigenspannungen, part 1: Theorie und Grundlagen. *MTB*, **16**, 36–40.

[12.58] Angerer, R. (1985) Bestimmung innerer Spannungen durch ein elektrochemisches Ringkernverfahren. *ÖIAZ*, **130**, 26–29.

[12.59] Bender, N., Hofer, G., Lenz, O. and Stücker, E. (1978) Verfahren zur Ermittlung von Eigenspannungen in Bauteilen des Maschinen- und Apparatebaues, German Patent 27 22 655.

[12.60] Hofer, G. and Bender, N. (1977) Measurement and calculation of residual stresses in claddings. *Int. Conf. Welding Inst. London*, pp. 89–96.

[12.61] Bender, N., Hofer, G. and Janousek, Z. (1981) Eigenspannungsmessung nach dem erweiterten Ring-Kern-Verfahren an einer Sprengplattierung. *VDI-Berichte*, **399**, 113–117.

[12.62] Keil, S. (1988) Dehnungsmeßstreifen als Hilfsmittel zur experimentellen Ermittlung von Eigenspannungen. *Technische Überwachung*, **30**, 309–316.

[12.63] Hofer, G. and Ort, W. (1982) Dehnungsmeßstreifen mit stützhülse, Gebrauchsmuster G 82 26 291.8.

[12.64] Hofer, G. and Schmidt, J. (1988) Modifiziertes Ring-Kern-Verfahren zur Messung von Schweißeigenspannungen. *DVS-Berichte*, **112**, 125–128.

[12.65] Ellingson, W.A. and Shack, W.J. (1979) Residual-stress measurements on multi-pass weldments of stainless-steel piping. *Experimental Mechanics*, **19**, 317–323.

[12.66] Giannuzzi, A.J. (1978) *Studies on AISI-Type-304 stainless steel piping weldments for use in BWR-application*, EPRI WP-944 Research Project 443-2, Final Report, General Electric Comp., San Jose.

[12.67] Keil, S. and Böhm, W. (1991) Ring-core method for the determination of residual stresses demonstrated by means of an application example, *Proc. SEM Spring Conf. Milwaukee*, pp. 386–391.

[12.68] Mathar, J. (1933) Ermittlung von Eigenspannungen durch Messung von Bohrloch-Verformungen. *Archiv für das Eisenhüttenwesen*, **6**, 277–281.

[12.69] Soete, W. and Vancrombrugge, R. (1947) Een quasi niet-destructieve werkwijze vor hetbepalen van eigen spanningen. *Laschtijdschrift*, (4), 161–174.

[12.70] Soete, W. and Vancrombrugge, R. (1950) An industrial method for the determination of residual stress. *Proceedings of the SESA.*, **8**, 17–28.

[12.71] Kirsch, J.G. (1898) Die Theorie der Elastizität und die Bedürfnisse der Festigkeitslehre. *Z-VDI*, **42**, 797–807.

[12.72] Howland, R.C.J. (1930) On the stresses in the neighbourhood of a circular hole in a strip under tension. *Philosophical Transactions A*, **229**, 49–86.

[12.73] Messmer, G. (1936) Fließerscheinungen beim Spannungsmessverfahren nach J. Mathar. *Archiv für das Eisenhüttenwesen.*, **10**, 1936, pp. 59–63.

[12.74] Altrichter, S. (1960) Die mathematischen Grundlagen der Eigenspannungsmessung mit dem Bohrlochverfahren. *ZIS-Mitt*, (12), **12**, 843–854.

[12.75] Keil, S. (1975) Zur Eigenspannungsermittlung mit DMS-Bohrlochrosetten. *MTB*, **11**, 53–58.

[12.76] Vancrombrugge, R. (1952) Messung von Eigenspannungen mit Dehnungsmeßstreifen, in *Grundlagen und Anwendung des Dehnungsmeßstreifens* (ed. K. Fink), Verlag Stahleisen, Düsseldorf, pp. 186–199.

[12.77] Bathgate, R.G. (1968) Measurement of non-uniform bi-axial residual stresses by the hole drilling method. *Strain*, **4**, 20–29.

[12.78] Owens, A. (1981) *Calibration of the center-hole technique of residual stress measurement.* Rep. of Stress Engineering Service LTD. Sheffield UK.

[12.79] Bijak-Zochowskij, M. (1978) A semidestructive method of measuring residual stresses. *VDI-Berichte*, **313**, 469–476.

[12.80] Niku-Lari, A., Lu, J. and Flavenot, J.F. (1985) Measurement of residual stress distribution by the incremental hole-drilling method. *Experimental Mechanics*, **25**, 175–185.

[12.81] Kockelmann, H. and Schwarz, T. (Juli 1991) *Quantifizierung der teilzerstörenden Eigenspannungsmeßverfahren (Bohrloch- und Ringkernverfahren) für anisotrope Werkstoffe und Mehrachsigkeit des Eigenspannungszustands.* Final Report 805 101 of Research Project KO 896-1, MPA Stuttgart.

[12.82] König, G.H. and Kockelmann, H. (Febr. 1989) *Entwicklung und Quantifizierung von teilzerstörenden Eigenspannungsverfahren zur Ermittlung der Tiefenverteilung von Eigenspannungen*, Final Report of DFG-Research Project KO 896/ 2-2, MPA Stuttgart.

[12.83] König, G.H. (1991) *Ein Beitrag zur Weiterentwicklung teilzerstörender Eigenspannungsmeßverfahren*. Diss. Univ. Stuttgart 1991; püblished as Report of MPA Stuttgart 91–02 (1991), ISSN 0721–4529.

[12.84] Wolf, H. and Böhm, W. (1971) Das Ringkern-Verfahren zur Messung von Eigenspannungen und seine Anwendung bei Turbinen- und Generatorwellen. *Archiv für das Eisenhüttenwesen*, **42**, 195–200.

[12.85] Kelsey, R.A. (2017) Measuring non-uniform residual stresses by the hole-drilling method. *Proceedings of the SESA*, **XIV** (1), 181–194.

[12.86] Schwarz, T. and Kockelmann, H. (1993) Die Bohrlochmethode – ein für viele Anwendungsbereiche optimales Verfahren zur experimentellen Ermittlung von Eigenspannungen. *MTB*, **29** (2), 33–38.

[12.87] Chow, C.L. and Cundiff, C.H. (1985) On residual-stress measurements in light truck wheels using the hole-drilling method. *Experimental Mechanics März*, **22**, 54–59.

[12.88] Pantucek, P. Entwicklung und Einsatz rechnergestützter Eigenspannungsermittlung an beschichteten Sondermaterialien mit der Bohrlochmethode. *VDI-Berichte*, **679**, 204–217.

[12.89] Flaman, M.T. (1982) Brief investigation of induced drilling stresses in the center-hole method of residual-stress measurement. *Experimental Mechanics*, **22**, 26–30.

[12.90] Beaney, E.M. (1976) Accurate measurement of residual stress on any steel using the centre hole method. *Strain*, **12**, 99–106.

[12.91] Owens, A. (1984) Extension to the blind hole drilling technique for residual stress determination with abrasive hole forming. *Strain*, **20**, 159–165.

[12.92] Ajovalasit, A. (1979) Measurement of residual stresses by the hole-drilling method: influence of hole eccentricity. *Journal of Strain Analysis*, **14**, 171–178.

[12.93] Flaman, M.T. and Herring, J.A. (1988) SEM/ASTM round-robin residual stress-measurement study, phase 1. *Experimental Techniques*, May 1988, pp. 23–25.

13 Stress analysis using strain gages in the elastoplastic deformation range

13.1 Introduction

Experimental stress analysis with strain gages in materials subjected to elastic deformation has been performed successfully in many fields for many years. Because the possibility also exists of measuring elastoplastic strain, questions are often posed as to the possibility of applying stress analysis with strain gages in the elastoplastic deformation range.

Standard production strain gages with a polyimide carrier and suitable bonding enable measurements in the elastoplastic range to be made up to and beyond 10% total strain.

Stress analysis using strain gages generally consists of two independent procedures: *Measurement*, during which principal strains in the material's surface are found, and *analytical evaluation* in which stresses are determined. For the analytical evaluation, knowledge of the constitutive equations (material law) describing the relationship between stress and deformation is required.

Here it is intended to present, in a simplified form and with emphasis on practical application, a discussion of the possibilities of experimental stress analysis for the elastoplastic range with the aid of strain gages. The theoretical problems are treated in great detail in [13.1,13.2], and here discussion will not centre on these. Thus, arguments are limited to essential points required for performing a stress analysis.

The stress analysis to be carried out makes use of the 'concept of an effective stress', based on a condition defined by v. Mises [13.3] (Eqs. (13.3) and (13.4) respectively) for the equivalent stress, and also on the stress–strain relations of Prandtl and Reuss [13.4].

The Prandtl–Reuss equations describe a material law leading to differential equations, the solution of which for known loading paths, is only practicable using computer and appropriate software. It also has to be taken in account that a 'rigorous' evaluation means increased requirements for material testing, and thus demands more extensive data than is usually known, for example the elastoplastic deformation may be a function of strain rate.

Here, a simplified solution is proposed, derived from the general solution [13.1,13.4]. The method stated based on the general solution has the advantage that it leads to easy solution and does not require continuous recording to be made of loading and strain signals. The relationship derived in 1974 [13.1] was presented in the historical form of simple nomographs and is here presented in modern form using computer software for carrying out the iteration process which enables the required analysis under elastoplastic deformation conditions to be performed.

In experimental stress analysis using strain gages (strain rosettes), generally strains at the surface of the measurement object are ascertained in at least three directions, and from these the principal directions 1 and 2, as well as the principal strains ε_1 and ε_2, are determined [13.5].

Technology and Practical Use of Strain Gages: With Particular Consideration of Stress Analysis Using Strain Gages. First Edition. Stefan Keil.

From this, assuming linear elastic material response of the member and presupposing plane stress conditions at the point of measurement ($\sigma_3 = 0$), the principal stresses σ_1 and σ_2 can be calculated by using Hooke's equations for biaxial stress conditions:

$$\sigma_1 = \frac{E}{1-\nu^2}(\varepsilon_1 + \nu\varepsilon_2) \tag{13.1}$$

$$\sigma_2 = \frac{E}{1-\nu^2}(\varepsilon_2 + \nu\varepsilon_1) \tag{13.2}$$

Herein, the modulus of elasticity E and Poisson's ratio ν are material characteristics valid only for the elastic range of deformation.

13.2 Equivalent state with elastic deformation

For the assessment of the effective stress, a transformation of the biaxial stress condition (triaxial state of deformation) into a uniaxial equivalent stress of equal effect is required in order to have this compared with the characteristics obtained from uniaxial basic tests (e.g. tensile test). Von Mises' theory [13.3], generally stated for the triaxial stress condition with the (uniaxial) equivalent stress σ_{eq}

$$\sigma_{eq}^2 = \sigma_1^2 + \sigma_2^2 + \sigma_3^2 - \sigma_1\sigma_2 - \sigma_2\sigma_3 - \sigma_3\sigma_1 \tag{13.3}$$

is formally identical with the distortion energy theory. Compared with other known theories for formulating an equivalent stress (e.g. normal stress or shear stress theories), it shows the best agreement with experimental results for isotropic ductile metallic materials. In Eq. (13.3) by substituting the yield stress σ_Y for the equivalent stress σ_{eq} the yield condition is obtained.

For plane stress condition with $\sigma_3 = 0$, Eq. (13.3) changes to

$$\sigma_{eq}^2 = \sigma_1^2 + \sigma_2^2 - \sigma_1\sigma_2 \tag{13.4}$$

Since, in experimental stress analysis using strain gages strains are being measured, it is more consistent to state a deformation criterion than the equivalent stress. Thus, based on Eq. (13.3), an equivalent strain ε_{eq} can be stated as

$$\varepsilon_{eq} = \frac{\sigma_{eq}}{E} \tag{13.5}$$

Since Hooke's equations represent a linear relationship between stresses and deformations, the equivalent strain ε_{eq} can be determined directly from the principal strains [13.6]

$$z\varepsilon_{eq}^2 = \varepsilon_1^2 + \varepsilon_2^2 + \varepsilon_3^2 - \varepsilon_1\varepsilon_2 - \varepsilon_1\varepsilon_3 - \varepsilon_2\varepsilon_3 \tag{13.6}$$

comparable with Eq. (13.3). The factor z in Eq. (13.6) is determined by using the uniaxial equivalent condition

$$\sigma_1 = \sigma_{eq};\ \sigma_2 = 0;\ \sigma_3 = 0 \tag{13.7}$$

with

$$\varepsilon_{eq} \equiv \varepsilon_1;\ \varepsilon_2 = -\nu\varepsilon_1;\ \varepsilon_3 = -\nu\varepsilon_1. \tag{13.8}$$

This leads to

$$z = (1+\nu)^2 \tag{13.9}$$

Thus, from Eq. (13.6) it follows that the equivalent strain is

$$\varepsilon_{eq} = \frac{1}{1+\nu}\sqrt{\varepsilon_1^2 + \varepsilon_2^2 + \varepsilon_3^2 - \varepsilon_1\varepsilon_2 - \varepsilon_1\varepsilon_3 - \varepsilon_2\varepsilon_3} \tag{13.10}$$

The principal strain ε_3, lying in the principal direction 3, can be determined from the principal strains occurring in the plane of measurement in accordance with the derivation given in [13.5]:

$$\varepsilon_3 = -\frac{1}{1-\nu}(\varepsilon_1 + \varepsilon_2) \tag{13.11}$$

Introducing ε_3 in Eq. (13.10) leads to

$$\varepsilon_{eq} = \sqrt{N_q\left(\varepsilon_1^2 + \varepsilon_2^2\right) + N_g\varepsilon_1\varepsilon_2} \tag{13.12}$$

Equation (13.12) is contained also in [13.7] but is differently expressed. In this N_q and N_g are functions of Poisson's ratio

$$N_q = \frac{1-\nu(1-\nu)}{(1-\nu^2)^2} \tag{13.13}$$

and

$$N_g = \frac{-1+\nu(4-\nu)}{(1-\nu^2)^2} \tag{13.14}$$

They are to be substituted accordingly. If the deformation of a material is restricted to the elastic range (with ν is constant) the equivalent strain ε_{eq} is solely dependent on the principal strains ε_1 and ε_2, which can be determined by strain gage measurement. Thus, with ν constant, the variables N_q and N_g become material characteristics, and with $\nu = 0.3$ we have $N_q = 0.954$ and $N_g = 0.133$.

The value of ε_{eq} is constant for constant equivalent strain, and Eq. (13.12) represents an ellipse the shape of which is determined by Poisson's ratio. This is more easily recognized when Eq. (13.12) is divided by ε_{eq} and then squared:

$$N_q\left(\frac{\varepsilon_1^2}{\varepsilon_{eq}^2} + \frac{\varepsilon_2^2}{\varepsilon_{eq}^2}\right) + N_g\frac{\varepsilon_1\varepsilon_2}{\varepsilon_{eq}^2} = 1 \tag{13.15}$$

The principal axes of the ellipse are inclined at 45° to the ε_1-ε_2-coordinate system and their lengths are

$$2a = \frac{2}{\sqrt{N_q + \frac{1}{2}N_g}};\ 2b = \frac{2}{\sqrt{N_q - 2N_g}} \tag{13.16}$$

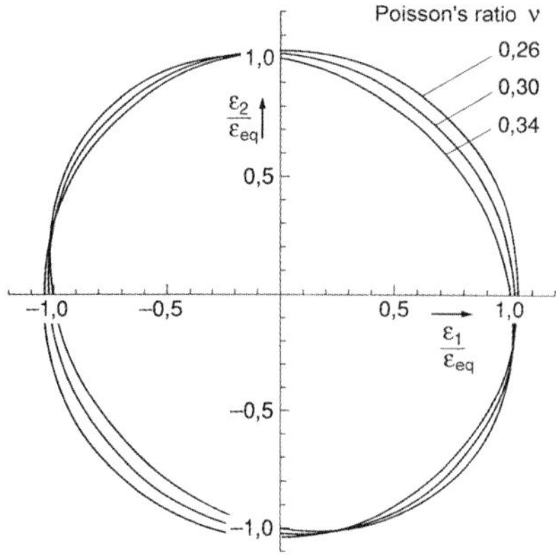

Fig. 13.1 Ellipses of constant equivalent strain ε_{eq} for Poisson's ratio $\nu = 0.26 \ldots 0.34$

For $a = b$, that is $\nu = 0.268$, the ellipse is transformed into a circle with the radius $r = 1.035$. Figure 13.1 shows ellipses of constant equivalent strain corresponding to Eq. (13.15), for Poisson's ratios between $\nu = 0.26 \ldots 0.34$.

Compared with the assessment of the effective stress accordingly to Eqs. (13.3) and (13.4), Eq. (13.12) possesses the advantage not only that it produces direct results, but also that these are almost independent of the material. For metallic materials, the modulus of elasticity E can vary by almost two powers of ten, whereas Poisson's ratio is much less variable. Equation (13.12) contains Poisson's ratio but does not contain the modulus of elasticity. The curves of constant equivalent strain for $\nu = 0.26 \ldots 0.34$, shown in Figure 13.1 illustrate this clearly by differing only little from each other.

13.3 Elastoplastic deformation

Considerations so far, have been limited to the linear elastic range of deformations and were based on coefficients elastic modulus E and Poisson's ratio ν in Hooke's stress–strain relationships. If equivalent stress σ_{eq} exceeds a certain material-dependent limit, plastic strains occur in addition to elastic strains. For the plastic component of elastoplastic deformation Hooke's stress–strain relationships are not valid. The first occurrence of plastic deformation components are described here formally by the yield stress σ_Y or the yield strain ε_Y, so that for $\varepsilon_{eq} > \varepsilon_Y$ elastic and plastic components of deformation are present.

Figure 13.2 shows a stress strain curve as determined by a uniaxial tensile test. Such a diagram serves as a comparative basis for experimental stress analysis by means of strain gages. The stress–strain curve is also called a yield curve of the material. Its shape in the elastoplastic region is determined by the nonlinear function

$$\sigma_{eq} = \sigma_{eq}\left(\varepsilon_{eq}\right) \tag{13.17}$$

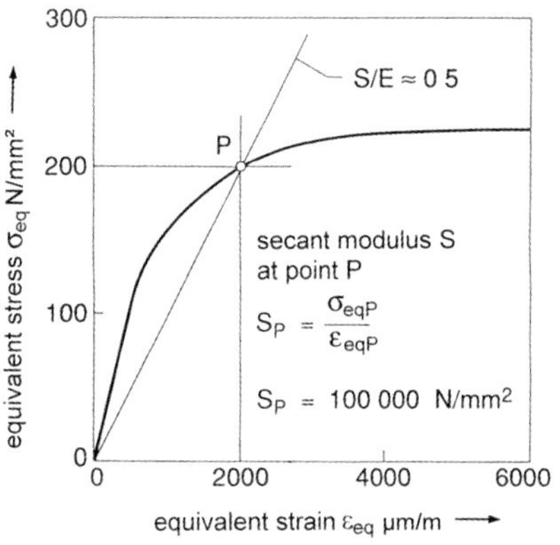

Fig. 13.2 Yield curve obtained from a uniaxial tensile test and definition of the secant modulus S

The task is to find for the measured principal strains ε_1 and ε_2 an equivalent point on the stress–strain curve.

This process of coordination is possible if the stress–straincurve of the material is known and the 'effective stress concept' is used for computing the equivalent strain ε_{eq} from the principal strains ε_1 and ε_2. This can be done using Eq. (13.12). The equivalence of conditions at the point of measurement (ε_1; ε_2) and the corresponding point (σ_{eq};ε_{eq}) on the yield curve define the term 'effective stress concept'.

As can be seen in Figure 13.2 the secant modulus characterizes an individual point of the yield curve, and is thus itself a function of the equivalent strain ε_{eq}. For a known value of the equivalent strain ε_{eq} the secant modulus can be read off the yield curve.

13.4 Stress analysis

By introducing variable characteristics instead of the constant modulus of elasticity E and Poisson's ratio ν, the elastoplastic material response can be formally described in the same way as the linear elastic behaviour. Equation (13.5) is then replaced by

$$\varepsilon_{eq} = \frac{\sigma_{eq}}{S} \tag{13.18}$$

where S is the secant modulus as defined in Figure 13.2. The total equivalent strain ε_{eq} according to Eq. (13.18) is composed of elastic and plastic strain components. Accordingly, the combined Poisson's ratio ν_g for the total deformation lies between the 'elastic' Poisson's ratio ν and the 'plastic' Poisson's ratio $\nu = 0.5$; ν_g depends on the magnitude of deformation and approaches the value $\nu_g = 0.5$ for large plastic strain components:

$$\nu \le \nu_g \le 0.5 \tag{13.19}$$

With the combined Poisson's ratio ν_g and the secant modulus S, Hooke's stress–strain relations valid for plane stress conditions, Eqs. (13.1) and (13.2) can be formally assigned to elastoplastic deformations:

$$\sigma_1 = \frac{S}{1-\nu_g^2}\left(\varepsilon_1 + \nu_g \varepsilon_2\right) \tag{13.20}$$

$$\sigma_2 = \frac{S}{1-\nu_g^2}\left(\varepsilon_2 + \nu_g \varepsilon_1\right) \tag{13.21}$$

whereby the combined Poisson's ratio ν_g is given with Eq. (13.22)

$$\nu_g = 0.5 - (0.5 - \nu)\frac{S}{E} \tag{13.22}$$

Equation (13.22) can be tested experimentally with a tensile test in which longitudinal and transverse strain are measured simultaneously. Figure 13.3 illustrates the close agreement between the measured values of the combined Poisson's ratio ν_g and the calculated curve according to Eq. (13.22). It refers to fine grain steel [13.8].

In accordance with Eq. (13.18) and Eq. (13.12) the equivalent strain results in

$$\varepsilon_{eq} = \sqrt{N_q\left(\varepsilon_1^2 + \varepsilon_2^2\right) + N_g \varepsilon_1 \varepsilon_2} \tag{13.23}$$

where

$$N_q = \frac{1 - \nu_g\left(1 - \nu_g\right)}{\left(1 - \nu_g^2\right)^2} \tag{13.24}$$

and

$$N_g = \frac{-1 + \nu_g\left(4 - \nu_g\right)}{\left(1 - \nu_g^2\right)^2} \tag{13.25}$$

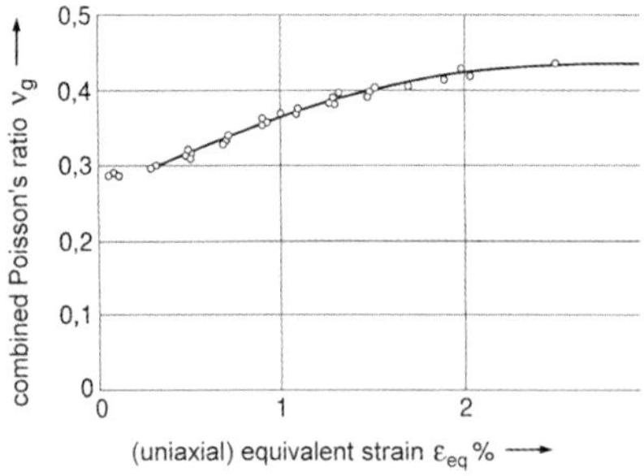

Fig. 13.3 Comparison of measured values (points) and calculated values (Eq. (13.22)) of combined Poisson's ratio ν_g; measured values are obtained from fine grain steel [13.8]

Under condition that the yield curve determined in a uniaxial tensile test is known, all required conditions to carry out stress analysis in the elastoplastic range of deformation are available. These are the yield curve and the two Eqs. (13.22) and (13.23).

With Eqs. (13.22) and (13.23) the equations for satisfying the 'effective stress concept' concerning elastoplastic deformation are stated and are to be used in conjunction with a given yield curve. Thus, the analysis is complete. In the 1970s, when this concept was first introduced, the three unknowns were determined with the aid of a suitable nomograph, because at that time computers were not available. The method based on the nomograph is explained in detail in [13.1].

13.5 Practical example of application

Consider a strain gage rosette applied to the surface of a member made of material (designation Ti 99.5, No. 3.7065, annealed condition) with yield strength 450 N/mm^2, modulus of elasticity $E = 105\,000\ N/mm^2$ and Poisson's ratio $\nu = 0.3$. Loading is applied to the member until it reaches the elastoplastic region. From rosette measurements the principal strains are found to be $\varepsilon_1 = 3000\ \mu m/m$ and $\varepsilon_2 = 1000\ \mu m/m$.

Advanced solution using the PC

Rao proposed a simplified procedure [13.9] to evaluate measured strains in the elastoplastic deformation range without using the nomograph. This method can easily be programmed on a computer and can be demonstrated in the following. Equations (13.22) and (13.23) are used together with the yield curve (uniaxial stress–strain curve) of the material to find the three unknowns ε_{eq}, ν_g and S. Here it is more useful to apply the stress–strain curve in its algebraic form that gives the stress as an explicit function of the strain. This relationship was introduced by Ramberg and Osgood [13.10]:

$$\sigma = E\varepsilon \left[1 + \left(\frac{\varepsilon}{\varepsilon_0} \right)^n \right]^{-\frac{1}{n}} \tag{13.26}$$

Here n is a parameter defining the shape of the nonlinear stress–strain relationship which represents the yield curve of the material. In Eq. (13.26) the stress approaches the maximum stress asymptotically with increasing strain as can be seen from Figure 13.4. The constants ε_0 and n can be determined approximately by

$$n = \frac{\ln 2}{\ln \dfrac{E\varepsilon_0}{\sigma_0}}. \tag{13.27}$$

with the results $\varepsilon_0 = 4420\ \mu m/m$ and $n = 4.5032$.

The values of $\varepsilon_0 = 4420\ \mu m/m$ and $\sigma_0 = 395\ N/mm^2$ in Eq. (13.27) are obtained from a point on the curve at the location as indicated in Figure 13.4. The accuracy of the representation of the stress–strain curve defined in Eq. (13.26) may be improved further by modifying the values of ε_0 and n through least-square curve fit.

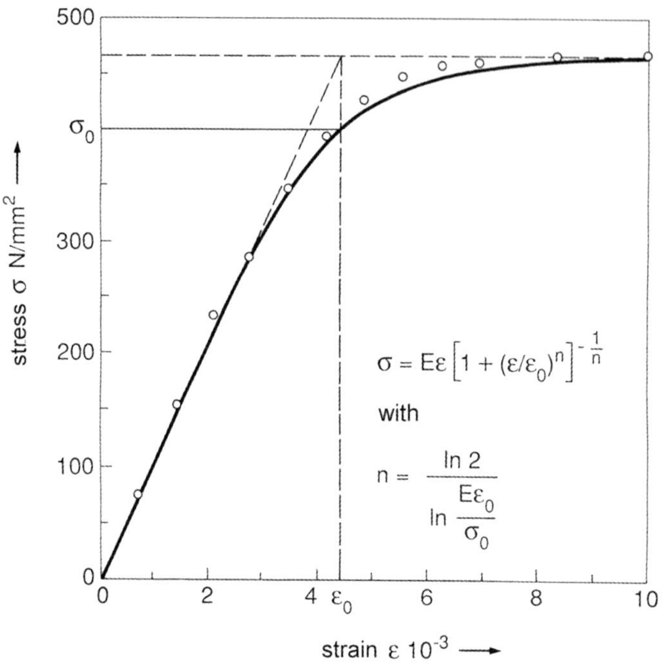

Fig. 13.4 Stress–strain curve of the material under consideration (Ti 99.5)

The value of the secant modulus S for the specified value of ε_{eq} can be obtained directly from Eq. (13.26) as

$$\frac{S}{E} = E\varepsilon\left[1 + \left(\frac{\varepsilon}{\varepsilon_0}\right)^n\right]^{-\frac{1}{n}} \tag{13.28}$$

Using Eqs. (13.22) and (13.28) one can get the combined Poisson's ratio ν_g in terms of ε_{eq} as

$$\nu_g = 0,5 - (0,5 - \nu)\left[1 + \left(\frac{\varepsilon}{\varepsilon_0}\right)^n\right]^{-\frac{1}{n}} \tag{13.29}$$

Since, in Eqs. (13.20) and (13.29), the values of ν_g and ε_{eq} are interdependent, an iteration process is necessary for evaluating proper ν_g and ε_{eq}, which will satisfy both equations. Once the value of ε_{eq} is established, the secant modulus can be calculated directly from Eq. (13.28).

With the secant modulus and the combined Poisson's ratio the principal stresses σ_1 and σ_2 can be calculated by using Hooke's equations for the biaxial stress conditions

$$\sigma_1 = \frac{S}{1 - \nu_g^2}\left(\varepsilon_1 + \nu_g\varepsilon_2\right) \tag{13.30}$$

$$\sigma_2 = \frac{S}{1 - \nu_g^2}\left(\varepsilon_2 + \nu_g\varepsilon_1\right) \tag{13.31}$$

Equations (13.30) and (13.31) presuppose isotropy in order to be valid, so that co-axiality is guaranteed and, thus, principal stresses and principal strains behave in the same directions. It should be mentioned that the secant modulus S in the proportional limit is equal to the modulus of elasticity of the material and the combined Poisson's ratio is equal to the 'elastic' Poisson's ratio.

The method described above is applied to the measurement on titanium Ti99.5, as initially described. The stress–strain diagram of the material is represented by the relationship as suggested in Eq. (13.26). The parameters ε_{eq} and n of the yield curve are evaluated. The data generated from the empirical relationship is plotted with the actual stress–strain data in Figure 13.4. The iteration process to evaluate ν_g and ε_{eq} is as follows: initially the combined Poisson's ratio is assumed as ν and substituted in Eq. (13.23) to obtain ε_{eq} for the specified values of ε_1 and ε_2. The value of ε_{eq} is substituted in Eq. (13.29) to obtain ν_g. Using this value of ν_g in Eq. (13.23), ε_{eq} will be determined. This iteration process has to be continued until these two unknowns ν_g and ε_{eq} satisfy both Eqs. (13.23) and (13.29) with the desired degree of accuracy. Three to four iterations may be required to get proper values of ν_g and ε_{eq}. For this example the values of ν_g and ε_{eq} are presented below:

Iteration	ν_g	ε_{eq}
1	0.3000	0.003 152
2	0.3086	0.003 181
3	0.3089	0.003 182
4	0.3089	0.003 182

Figure 13.5 shows the results of the presented evaluation drawn in the diagram of the yield curve. The equivalent strain, $\varepsilon_{eq} = 3182\ \mu m/m$ and the combined Poisson's ratio

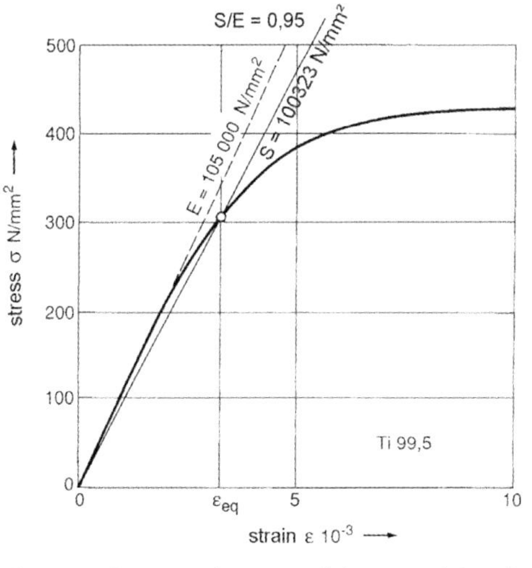

Fig. 13.5 Stress strain curve of the material under consideration (Ti99.5)

$\nu_g = 0.3089$. The secant modulus $S = 100\ 323\,\mathrm{N/mm^2}$ is obtained from Eq. (13.28). The equivalent stress $\sigma_{eq} = 319.2\,\mathrm{N/mm^2}$. The principal normal stresses at the point of measurement obtained by substituting the values of ν_g and S in Eqs. (13.30) and (13.31) are $\sigma_1 = 366.5\,\mathrm{N/mm^2}$ and $\sigma_2 = 213.5\,\mathrm{N/mm^2}$ respectively.

13.6 Tensorial representation of the elastoplastic deformation

Stress and strain states are often represented in publications in tensor notation. This notation allows a more compact and comprehensive presentation of stress states than by using the individual components of the stress state. Especially the description of elastoplastic stress states is often in tensor notation. Therefore we will try to explain here the tensor notation of elastoplastic stress and strain states in a form understandable to the practitioner. Furthermore, the derivations of the constitutive equations for elastoplastic deformation will be presented here in a compressed form.

The consideration of elastoplastic deformation suggests decomposing the total deformation intoan elastic and a plastic part. This decomposition of the total deformation leads to the elastic deformation causing volume change and to the plastic deformation occurring at constant volume, but changing the shape.

Thus, total elastoplastic deformation can be restituted as a superposition of volume change (elastic) and change of shape (plastic). This is done mosteasily using the tensor notation, where the three-dimensional stress state at a point in a member can be represented by the stress tensor σ_{ij}

$$\sigma_{ij} = \begin{pmatrix} \sigma_x & \tau_{xy} & \tau_{xz} \\ \tau_{yx} & \sigma_y & \tau_{yz} \\ \tau_{zx} & \tau_{zy} & \sigma_z \end{pmatrix} = \begin{pmatrix} \sigma_1 & 0 & 0 \\ 0 & \sigma_2 & 0 \\ 0 & 0 & \sigma_3 \end{pmatrix} \tag{13.32}$$

The left matrix is the totality of spatial components of the stress state acting at a point. The right matrix represents the three principal stresses describing the same stress state.

The portions of the stress state causing elastic deformations (volume change) are described by the spherical tensor:

$$\frac{1}{3}\delta_{ij}\,\sigma_{kk} = \begin{pmatrix} \sigma_m & 0 & 0 \\ 0 & \sigma_m & 0 \\ 0 & 0 & \sigma_m \end{pmatrix} \tag{13.33}$$

This purely elastic volume change occurs, for example, under the action of a hydrostatic stress state to an isotropic body. All principal stresses are equal in magnitude and sign, so that no shear stresses occur. As a result, no plastic deformation (creep) can occur.

The volume-changing stress components σ_m can be described as the average of the three principal stresses

$$\sigma_m = \frac{1}{3}(\sigma_1 + \sigma_2 + \sigma_3) = \frac{1}{3}I_1 \tag{13.34}$$

If the spherical tensor (change in volume) is subtracted from the stress tensor σ_{ij}, the shape-changing stress deviator σ'_{ij} is formed:

$$\sigma'_{ij} = \begin{pmatrix} \sigma_x - \sigma_m & \tau_{xy} & \tau_{xz} \\ \tau_{yx} & \sigma_y - \sigma_m & \tau_{yz} \\ \tau_{zx} & \tau_{zy} & \sigma_z - \sigma_m \end{pmatrix} \tag{13.35}$$

Thus, the stress tensor is composed of the sum of the stress deviator and the spherical tensor and can be represented using the above notations in short form with the equation

$$\sigma_{ij} = \sigma'_{ij} + \frac{1}{3}\delta_{ij}\,\sigma_{kk} \tag{13.36}$$

The deformation tensor can be represented in the same way as the stresses. The total deformation is described by the deformation tensor ε_{ij}:

$$\varepsilon_{ij} = \begin{pmatrix} \varepsilon_x & {}^1\!/_2\gamma_{xy} & {}^1\!/_2\gamma_{xz} \\ {}^1\!/_2\gamma_{yx} & \varepsilon_y & {}^1\!/_2\gamma_{yz} \\ {}^1\!/_2\gamma_{zx} & {}^1\!/_2\gamma_{zy} & \varepsilon_z \end{pmatrix} \tag{13.37}$$

This tensor reproducing the total deformation consists of the spherical tensor describing the volume change (elastic deformation)

$$\frac{1}{3}\delta_{ij}\,\varepsilon_{kk} = \begin{pmatrix} \varepsilon_m & 0 & 0 \\ 0 & \varepsilon_m & 0 \\ 0 & 0 & \varepsilon_m \end{pmatrix} \tag{13.38}$$

and the deformation deviator representing the change of shape (plastic deformation) ε'_{ij}

$$\varepsilon'_{ij} = \begin{pmatrix} \varepsilon_x - \varepsilon_m & {}^1\!/_2\gamma_{xy} & {}^1\!/_2\gamma_{xz} \\ {}^1\!/_2\gamma_{yx} & \varepsilon_y - \varepsilon_m & {}^1\!/_2\gamma_{yz} \\ {}^1\!/_2\gamma_{zx} & {}^1\!/_2\gamma_{zy} & \varepsilon_z - \varepsilon_m \end{pmatrix} \tag{13.39}$$

This results in the relation descriptive short form:

$$\varepsilon_{ij} = \varepsilon'_{ij} + \frac{1}{3}\delta_{ij}\varepsilon_{kk} \tag{13.40}$$

In an attempt to describe formally the behaviour of the material as a relationship between stress and deformation (ut tensio sic vis) Hooke's law was used for elastic deformation (linear elastic). Introducing the shear modulus G, Hooke's law which represents the relationship between stress and strain for the elastic deformation range can be written, in tensor notation as

$$\frac{\mathrm{d}\sigma'_{\mathrm{ij}}}{2G} = \mathrm{d}^{el}\varepsilon'_{ij} \tag{13.41}$$

in other notation

$$E\mathrm{d}\varepsilon_{ij} = (1+\nu)\mathrm{d}\sigma_{ij} - \nu\mathrm{d}\sigma_{kk}\delta_{ij} \tag{13.42}$$

after integration in finite form

$$E\varepsilon_{ij} = (1+\nu)\sigma_{ij} - \nu\sigma_{kk}\delta_{ij} \tag{13.43}$$

In 1870, Lévy formulated first for plastic deformations a flow rule [13.11], which was later taken up again by v. Mises 1913 [13.12]:

$$\sigma'_{ij}\mathrm{d}\lambda = \mathrm{d}^{pl}\varepsilon'_{ij} \tag{13.44}$$

Here $\mathrm{d}\lambda$ is a proportionality factor, what can be determined from an equivalent uniaxial stress state [13.13]. From the uniaxial tensile test we get

$$\mathrm{d}\lambda = \frac{3}{2}\frac{\mathrm{d}^{pl}\varepsilon_v}{\sigma_v} \tag{13.45}$$

$\mathrm{d}\lambda$ is positive for load operations, negative for relief and zero for purely elastic deformations (Hooke).

In the 1920s, a general constitutive law was setup by Prandtl and Reuss that takes into account both the elastic and the plastic deformations [13.14, 13.15]. They linked the elastic material law (13.41) with the yield condition of Levy–v. Mises (13.42) and thus led to a material equation valid for elastoplastic stress:

$$\mathrm{d}\varepsilon'_{ij} = \frac{\mathrm{d}\sigma'_{ij}}{2G} + \sigma'_{ij}\mathrm{d}\lambda \tag{13.46}$$

In other notation

$$\mathrm{d}\varepsilon'_{ij} = \mathrm{d}\varepsilon_{ij} - \frac{1}{3}\mathrm{d}\varepsilon_{kk}\delta_{ij} \tag{13.47}$$

The uniaxial comparison state with $\mathrm{d}\sigma_{ij} = \mathrm{d}\sigma_{kk}$ then results in the relationship

$$\mathrm{d}\varepsilon_{kk} = \frac{1-\nu}{E}\mathrm{d}\sigma_{kk} \tag{13.48}$$

By insertion into the setup of Prandtl–Reuss (13.47) we have

$$\mathrm{d}\varepsilon'_{ij} = \mathrm{d}\varepsilon_{ij} - \frac{1-2\nu}{2E}\mathrm{d}\sigma_{kk}\delta_{ij} \tag{13.49}$$

A stress analysis based on measured strains according to the equations of Prandtl and Reuss is experimentally feasible only at very high expense. It is awkward for practical applications. It requires a continuous record of the measurement signals depending on the load path, because the differentials $\mathrm{d}\varepsilon_{ij}$ (or $\Delta\varepsilon_{ij}$) are parts of the evaluation. This also requires that the influence of the deformation speed $d\varepsilon_{eq}$ must be known. In order to avoid these difficulties in the first section of this chapter, an approach is proposed that provides, at small elastoplastic deformations, reliable results.

For small deformations, the total differentials $d\varepsilon'_{ij}$, $d\sigma'_{ij}$ and so on are replaced by finite sizes. Then the Prandtl–Reuss equations analogous to the Hooke's equations turn into

$$S\varepsilon_{ij} = \left(1 + \nu_g\right)\sigma_{ij} - \nu_g \sigma_{kk}\delta_{ij} \tag{13.50}$$

with mixed Poisson's ratio ν_g and the secant modulus $S = \sigma_{eq}/\varepsilon_{eq}$. In this way the connection to Eqs. (13.20)–(13.22) is established.

References

[13.1] Keil, S. and Benning, O. (1979) On the evaluation of elasto-plastic strains measured with strain gages. *Experimental Mechanics*, **19** (8), pp. 265–270.

[13.2] Troost, A., Benning, O. and Keil, S. (1974). Zur experimentellen Beanspruchungsanalyse bei elastisch-plastischer Verformung. *Archiv für das Eisenhüttenwesen*, **45** (5), 301–307.

[13.3] Mises, R.v. (1928) Mechanik der plastischen Formänderung von Kristallen. *Zeitschrift für angewandte Mathematik und Mechanik*, **8**, 161–185.

[13.4] Troost, A. (1969) Prandtl-Reuß-Stoffgleichungen und Werkstoffanstrengung. *Die Naturwissenschaften*, **56**, 559.

[13.5] Keil, S. (1987) Stress calculation from measured strains in the elastic deformation range. *RAM*, **3** (2), 56–62.

[13.6] Troost, A. (1962) *Festigkeitsberechnung bei elastisch-plastischer Verformung*, lectures RWTH Aachen.

[13.7] Kochendörfer, A. and Krieger, D. (1970) Untersuchungen über die Bestimmung der Zähigkeit von Bauteilen im Apparatebau durch Messungen an Modellbehältern von praxisähnlicher Größe. *Zeitschrift für Werkstofftechnik*, **1**, 7–16.

[13.8] Wellinger, K., Gaßmann, H. and Luithle, J. (1969) Zeitfestigkeit und Dehnungsformzahl gekerbter Flachproben. *Konstruktion*, **21**, 219–224.

[13.9] Rao, B.N. (1991) A simplified procedure for stress analysis in the elasto-plastic range with the aid of strain gages. *Experimental Techniques*, May/June 1991, pp. 38–39.

[13.10] Ramberg, W. and Osgood, W. (1943) *Description of strain-stress curves by three parameters*, NACA Techn. Note 902.

[13.11] Lévy, M. (1870) Mémoire sur les équations générales des mouvements intérieurs des corps solides ductiles au delà des limites ou l'élasticité pourrait les ramener à leur premier etat. *Comptes Rendus de l'Académie des Sciences Paris*, **70**, 1323–1325.

[13.12] Mises, R.v. (1913) Mechanik der festen Körper im plastisch deformablen Zustand. *Nachrichten der königlichen Gesellschaft für Wissenschaft, Göttingen, Mathematisch-physikalische Klasse, 1913, pp.*, 582–592.

[13.13] Betten, J. (1985) *Elastizitäts- und Plastizitätslehre,* Vieweg Verlag, p. 112.

[13.14] Prandtl, L. (1924) Spannungsverteilung in plastischen Körpern. *Proc. 1st Int. Congress Appl. Mech. Delft 1924,* pp. 43–54.

[13.15] Reuß, A. (1930) Berücksichtigung der elastischen Formänderung in der Plastizitätslehre. *Zeitschrift für angewandte Mathematik und Mechanik,* **10**, 266–274.

14 Strength theories

14.1 Preview

The question about the reasons for the failure of components during use have employed humanity for hundreds of years. Descriptions of attempts to answer this question can be found in historical works, such as [14.1–14.3]. In [14.4], a very detailed overview is given of research done in the field of strength theories in the 20th century. The first serious attempts in this field are linked with names such as Galileo Galilei and Leonardo da Vinci. Despite intensive and expensive experimental investigations the problem of component strength is not yet fully and satisfactorily resolved.

Especially for the occurrence of time-varying loads, which may occur due to different force and moment components in a rich variety of combinations and at variable temperatures and frequencies, a unique mathematical solution for all load cases has not yet been found.

Efforts in the field of service strength, created in the last century, also did not lead to a general solution. Notable progress has been made in limitation of selected load cases. The results thus obtained are, however, for competitive reasons, rarely of general use.

If restricted to static loads the question of the failure criterion is easier to answer than for dynamic loads with their numerous different parameters. Therefore failure limits valid for static loads will be described briefly here. The occurrence of yielding is considered as the failure limit. This requires a criterion to be defined for the occurrence of yielding. This is best done with the aid of a uniaxial equivalent stress with which an existing multiaxial stress state must be replaced and then compared with the yield point of the material under consideration.

Comparison of the load caused by the stress state of the material with the yield point of the material gives a measure of the utilization of the existing reserves of material strength. If the material reaches the equivalent stress, resulting from a multiaxial stress state, up to the yield point then the material is utilized to 100% of its load capacity and has no more reserves.

Thus, the demand for the formulation of the equivalent stress becomes particularly important. For nearly 200 years attempts have been made to propose a useful approach for the formulation of an equivalent stress on the basis of a theory. These theories were based on different criteria. As relevant for the formulation of the theories different criteria were considered: the maximum normal stress, the maximum shear stress, the three occurring principal shear stresses or the energy introduced into the material. The most popular theories are summarized in Table 14.1 as an overview. Not all of these theories led to practically useful results, which could easily be determined in comparison with experimental results. The most remarkable theories from today's perspective are to be treated briefly and evaluated with respect to their practical use in the following sections.

Technology and Practical Use of Strain Gages: With Particular Consideration of Stress Analysis Using Strain Gages. First Edition. Stefan Keil.

Table 14.1 Overview of the most popular strength theories

Strength theory	Mechanical characteristic
Stress theories	
Lamé 1824	Maximum normal stress theory Maximum shear theory
Coulomb 1773 Tresca 1864/1872	all principal shear stresses
Troost 1961 Roš and Eichinger 1927 Nadai 1937	shearing stress on the octahedral plane
Strain theories	
Saint-Venant 1870 Sandel 1919	Maximum elastic strain Resulting elastic strain
Maximum energy theories	
Beltrami 1885 Huber 1904	Total deformation energy
Huber 1904 Hencky 1924	Distortion energy theory
Algebraic theories	
Von Mises 1913/1924	Quadratic potential

14.2 Concept of effective stress

In practice, in loaded members under service conditions the stress state is generally three-dimensional. The actual three-dimensional stress state at a certain point of the member can be described by the stress tensor σ_{ij}.

This stress tensor contains all the components of the three-dimensional stress state as illustrated in Figure 14.1a.

$$\sigma_{ij} = \begin{pmatrix} \sigma_x & \tau_{xy} & \tau_{xz} \\ \tau_{yx} & \sigma_y & \tau_{yz} \\ \tau_{zx} & \tau_{zy} & \sigma_z \end{pmatrix} \tag{14.1}$$

By rotating the system of coordinates around its origin a particular position can be found in which the shear stresses on the planes disappear and the normal stresses on the planes of the cube become principal stresses which means that the normal stresses become extreme values. In this case, the components of the stress tensor are represented by three principal stresses which act in the three principal

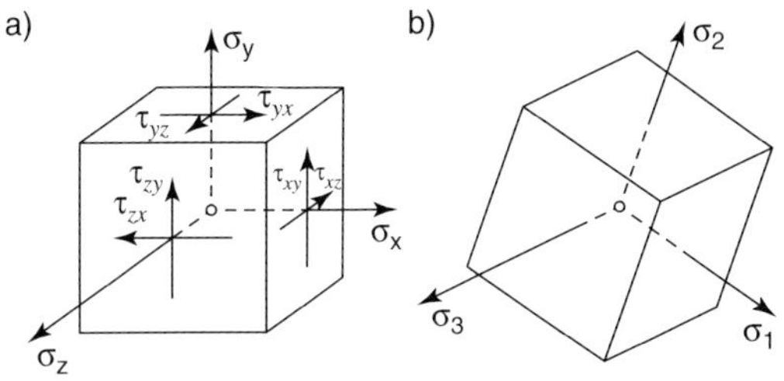

Fig. 14.1 Components of the three-dimensional stress state; a) general form; b) principal stresses and their directions of the three-dimensional stress state as given in a)

directions of the stress state as can be seen in Figure 14.1b. Then the stress tensor can be written as

$$\sigma_{ij} = \begin{pmatrix} \sigma_1 & 0 & 0 \\ 0 & \sigma_2 & 0 \\ 0 & 0 & \sigma_3 \end{pmatrix} \tag{14.2}$$

Considering members subject to three-dimensional loading conditions the question arises on the stress state causing failure. What combination of the three principal stresses is responsible for the breakdown? Only for one-dimensional tests are determined reference values available. Table 14.2 shows reference values normally used for comparison.

Under uniaxial load conditions the answer to the question is simple: if the uniaxial load stress exceeds the yield point as determined in the uniaxial tensile test, yield will occur.

To find the answer one has to compare the maximal stress of the uniaxial load condition with the material characteristic determined in the uniaxial tensile test. The answer will become more complicated if the loading condition is multiaxial. Two or three principal stresses of the loading condition act simultaneously and here the question arises, what combinations of the simultaneously acting principal stresses will cause yielding of the material? The common action of two or three principal stresses has to be taken into account and has to be compared with an appropriate characteristic of the material. So, in order to find a solution, the term 'concept of effective stress'" will be introduced. Each stress state will be assigned to a (uniaxial) equivalent stress σ_{eq} causing the same effect on the material.

Table 14.2 Examples of in uniaxial tensile tests determined material characteristics

Tensile strength	R_m
0.2% proof stress	$R_{p0.2}$
Upper yield point	R_{eH}
Lower yield point	R_{eL}

If two different stress states can be reduced to the same equivalent stress than they will cause the same effective stress.

The effective stress is defined by the (uniaxial fictitious) equivalent stress causing the same effect on the material as the effect caused by the multiaxial loading condition under consideration.

Up to now the shape of the function $\sigma_{eq} = \sigma_{eq}(\sigma_{ij})$ which describes the relationship between the equivalent stress and the multiaxial stress state is determined by experience. An unambiguous, unique definition of this relationship by a physical law has so far not been found. Therefore the function has to be formulated hypothetically.

From experience, we learn that for different materials which react differently under load (e.g. brittle or tough) different definitions of the equivalent stress are valid. Criteria for formulation of the hypotheses are based initially on the yield strength or the fracture strength, therefore these hypotheses have been called strength theories. They were later extended to the general concept of effective stress, first applied by v. Bach with the German designation 'Werkstoffanstrengung' [14.1]. The designation 'Anstrengungshypothesen', as the German synonym for strength theory, can be found as early as 1929 in a paper by Burzinski [14.5].

The hypothetically or phenomenologically established functions for the relationship $\sigma_{eq} = \sigma_{eq}(\sigma_1, \sigma_2, \sigma_3)$ can be proved simply at the yield point by equating the equivalent stress σ_{eq} to the yield point σ_Y as found from a uniaxial tensile test of the material under consideration:

$$\sigma_{eq} = \sigma_Y \tag{14.3}$$

Experimental determination of a hypothesis valid for a certain material may become very expensive. A particular simplification will be reached by confinement to the biaxial stress state. This is convenient with regard to the evaluation of strain gage readings which are generally found at plane surfaces. The result of a stress analysis using strain gages consists of the two principal stresses and their orientation, which means the angles of their directions. For the stress state with $\sigma_3 = 0$, as can be seen in Figure 14.2, the stress tensor becomes

$$\sigma_{ij} = \begin{pmatrix} \sigma_x & \tau_{xy} \\ \tau_{yx} & \sigma_y \end{pmatrix} \quad \text{or} \quad \sigma_{ij} = \begin{pmatrix} \sigma_1 & 0 \\ 0 & \sigma_2 \end{pmatrix} \tag{14.4}$$

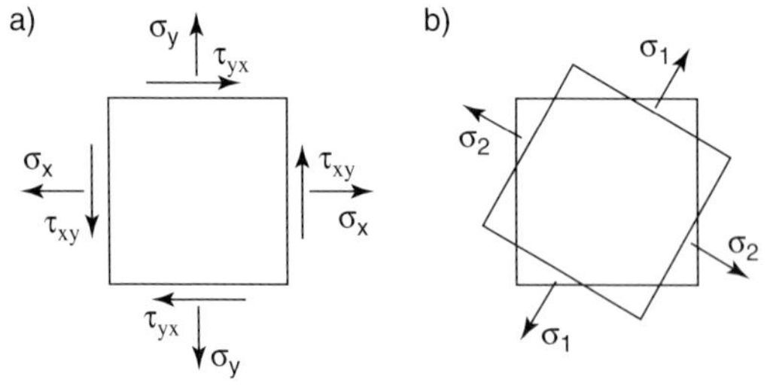

Fig. 14.2 Components of the biaxial stress state; a) general form, b) principal stresses and their directions of the biaxial stress state, as given in Figure 14.1a

The representation of the biaxial stress state by a uniaxial equivalent stress can be formulated in tensor form by

$$\sigma_{ij} = \begin{pmatrix} \sigma_1 & 0 \\ 0 & 0 \end{pmatrix} \tag{14.5}$$

14.3 Experimental results

Furthermore, in order to minimize the expense required for experimental verification of a hypothesis some easy to verify combinations of σ_1 and σ_2 can be applied experimentally. Strain gages are well suited to carrying out such experiments. Strain gages will be installed at the test object, which is loaded up to the yield point. Incipient creep can be detected easily by intermediate unloading with checking the zero point. If on complete relief of the load the measurement value does not return to zero then plastic deformation has occurred (provided all sources of error which could cause zero point drift are excluded).

The first experiments about the influence of multiaxial stress components on flow initiation were carried out by Lode in 1925 [14.7], before the strain gage was invented. More extensive studies with extension to elastoplastic deformations and dynamic strain were performed by Roš and Eichinger [14.8]

Results of experiments on ductile metallic materials with symmetrical response to tensile and compressive loads (symmetrical means here the same modulus of elasticity under tensile and compressive load and the same values with opposite sign of the yield points under tension and compression) are given in Figure 14.3 for different combinations of σ_1 and σ_2.

Entering the results of Figure 14.3 in a σ_1-σ_2-coordinate system creates the diagram given in Figure 14.4a. Points ① and ② identify the yield points under uniaxial tension and compression, point ③ marks incipient yield under pure torsion. Point ④ has been measured on a thin-walled circular cylindrical vessel under internal pressure and point ⑤ on a spherical tank. The diagram can be completed according to the provided symmetry, and a new Diagram 14.4b is created. Obviously as connection line between the points an elliptical curve can be used:

$$\sigma_{eq}^2 \equiv \sigma_Y^2 = \sigma_1^2 + \sigma_2^2 - \sigma_1\sigma_2 \tag{14.6}$$

The curve of the ellipse represents the geometric locus of the equivalent stress $\sigma_{eq} \equiv \sigma_Y$. All points inside the ellipse indicate states of elastic deformation. Plastic deformation is possible at points outside the ellipse only.

Equation (14.6) formulates a hypothesis found on purely empirical basis, which, for plane stress states, describes incipient yield under arbitrary combinations of the two principal stresses (for ductile metallic material). Formally the ellipse equation can be applied for the experimentally determined results. Comparison for uniaxial load is

(1) 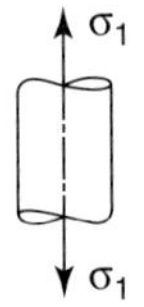 tension	$\sigma_1 > 0$ $\sigma_1 = \sigma_3$	$\sigma_1 = \sigma_Y$
(2) 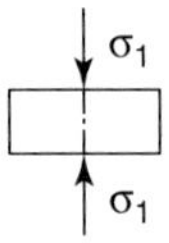 compression	$\sigma_1 < 0$ $\sigma_2 = \sigma_3$	$\sigma_1 = -\sigma_Y$
(3) 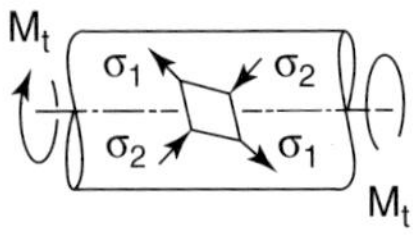 torsion	$\sigma_1 = -\sigma_2$ $\sigma_3 = 0$	$\sigma_1 \approx 0{,}58\,\sigma_Y$
(4) 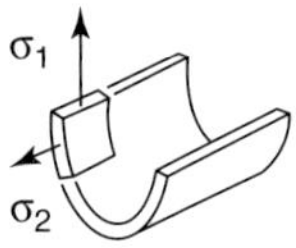 thin walled circular cylindrical vessel under internal pressure	$\sigma_1 = 2\sigma_2$ $\sigma_3 = 0$	$\sigma_1 \approx 1{,}15\,\sigma_Y$
(5)  spherical vessel under interal pressure	$\sigma_1 = \sigma_2$ $\sigma_3 = 0$	$\sigma_1 \quad \sigma_Y$
(6) 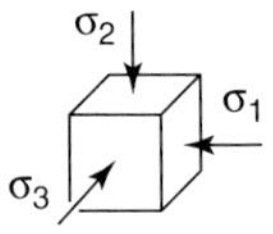 hydrostatic stress state	$\sigma_1 = \sigma_2 = \sigma_3 = P$	$P > 10^5$ bar

Fig. 14.3 Experimentally determined combinations of σ_1 and σ_2 causing creep in ductile metallic materials

trivial. For pure torsion with $\sigma_1 = -\sigma_2$ the equivalent stress becomes

$$\sigma_{eq} \equiv \sigma_Y = \sqrt{3}\,\sigma_1 \tag{14.7}$$

Solved for σ_1:

$$\sigma_1 = \frac{1}{3}\sqrt{3}\,\sigma_Y = 0.577\,\sigma_Y \tag{14.8}$$

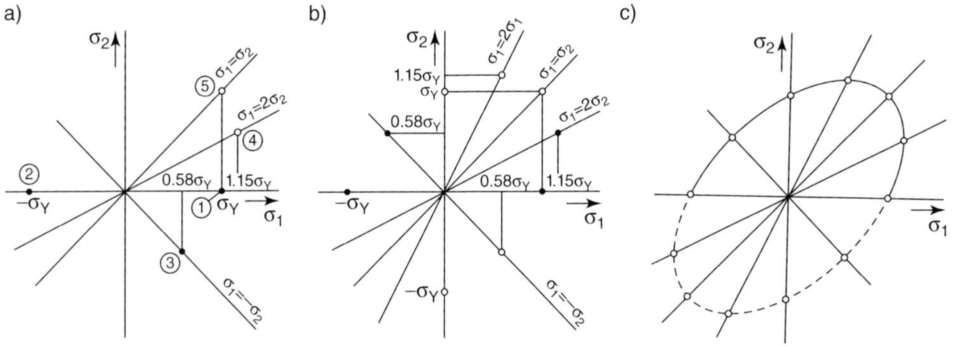

Fig. 14.4 Experimentally determined points and ellipse for yield beginning of a ductile metallic material in σ_1–σ_2 coordinates; a) Measuring points according to Figure 14.3, b) Symmetrical complement of the diagram, c) Ellipse as geometric locus of beginning creep

This result is in agreement with the experiment. For point ④ (Figure 14.3) with $\sigma_1 = 2\sigma_2$ we obtain

$$\sigma_{eq} \equiv \sigma_Y = \frac{1}{2}\sqrt{3}\,\sigma_1 \tag{14.9}$$

and solved for σ_1

$$\sigma_1 = \frac{2}{3}\sqrt{3}\,\sigma_Y = 1.1547\,\sigma_Y \tag{14.10}$$

Here too we find agreement.

Of special interest is the load with hydrostatic pressure denoted by ⑥. Here the yield criterion has to be expanded to a triaxial stress state:

$$\sigma_{eq}^2 \equiv \sigma_F^2 = \sigma_1^2 + \sigma_2^2 + \sigma_3^2 - \sigma_1\sigma_2 - \sigma_1\sigma_3 - \sigma_2\sigma_3 \tag{14.11}$$

By transforming

$$\sigma_1^2 = \frac{\sigma_Y^2}{1 + \dfrac{\sigma_2^2}{\sigma_1^2} + \dfrac{\sigma_3^2}{\sigma_1^2} - \dfrac{\sigma_2}{\sigma_1} - \dfrac{\sigma_3}{\sigma_1} - \dfrac{\sigma_3\sigma_2}{\sigma_1^2}} \to \infty \tag{14.12}$$

In the case of $\sigma_1 = \sigma_2 = \sigma_3$ the result of Eq. (14.12) becomes ∞. This result is confirmed by tests with pressure up to about 10^5 bar, performed by Bridgman [14.6]. It can be seen as a consequence of this result that under hydrostatic pressure no creep occurs. There are no shear stresses because there are no differences between normal

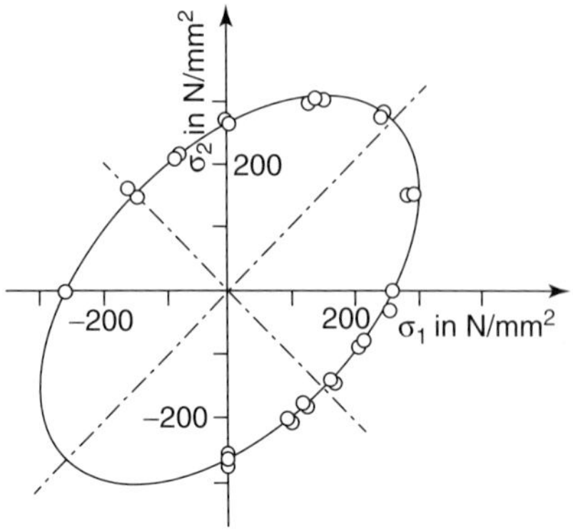

Fig. 14.5 Experimentally determined yield criterion of different biaxial stress states on a mild steel [14.8]

principal stresses. This is the evidence that shear stresses are responsible for the occurrence of creep.

Figure 14.5 shows experimentally determined yield points of mild steel with yield stress of $\sigma_Y \approx 260\,\mathrm{N/mm^2}$ for different combinations of principal normal stresses [14.8]. The coincidence with the ellipse equation is evident.

Figure 14.6 [14.9] represents results of biaxial load tests on ductile and brittle behaving materials. Failure in ductile behaving materials occurs according to the elliptic boundary condition, but in brittle behaving materials such as cast iron failure occurs according to the maximum stress theory.

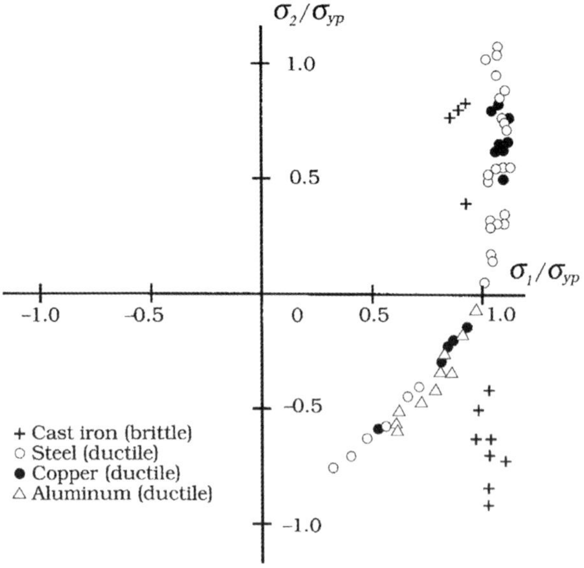

Fig. 14.6 Results of biaxial load tests on brittle and ductile behaving materials

14.4 Maximum stress theory

In 1824 [14.10], Lamé presented a strength theory describing a fracture condition. According to this theory fracture occurs when one of the two normal principal stresses of the plane stress state exceeds the rupture strength σ_R of the material under consideration.

$$\sigma_{eq} \equiv \sigma_{\max} = \sigma_R. \tag{14.13}$$

For pure separation fracture, this theory is confirmed by experience. Separation fracture is brittle fracture, which occurs with very low deformation or without any deformation. It does not announce itself through previously perceivable plastic deformation, but occurs spontaneously with a high crack propagation rate.

Maximum stress theory is not suitable for definition of yield beginning because shear stresses are responsible for creep, and shear stresses are not considered in this theory. For materials with brittle behaviour, for example cast iron, with lamellar graphite (with fracture strain A<1%) the tensile strength R_m can be used as a fracture condition instead of the rupture strength:

$$\sigma_{eq} \equiv \sigma_{\max} = R_m \tag{14.14}$$

Graphical representation of the maximum stress theory can be seen in a σ_1–σ_2 coordinate system and in σ-τ coordinates in Figure 14.7.

The representation given in Figure 14.7b illustrates the envelope of Mohr's circles, which may result by combinations of principal stresses allowed by the maximum stress theory. Here the torsional strength takes the same value as the tensile strength. This is confirmed experimentally by torsional fracture of cast iron [14.7]. But in ductile materials the torsional strength is always less than the tensile strength.

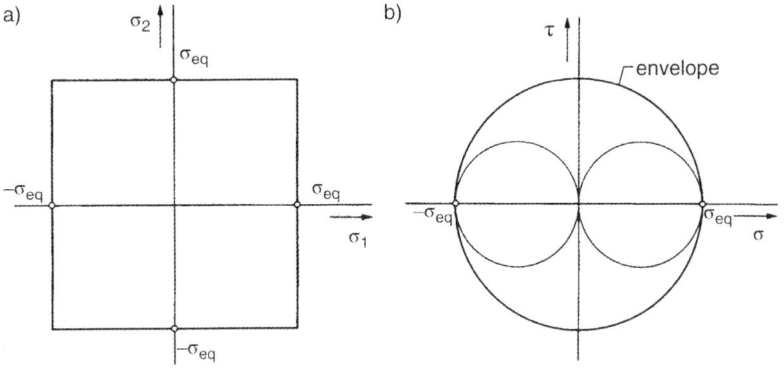

Fig. 14.7 Graphical representation of the maximum stress theory; a) with σ_1–σ_2 coordinates, b) as the envelope of possible Mohr's circles

14.5 Maximum shear theory

The maximum shear theory considers the maximum shear stress as the criterion for strength. This idea first time was taken in consideration by Tresca in 1864 [14.11]. This means that yielding begins in a material when the maximum shear stress $\tau_{\max}$ becomes equal to the shear stress τ_Y at the yield point in a simple tension test.

The largest shear stress is given by half the difference between the largest and smallest normal principal stresses as can also be seen from Mohr's circle in Figure 14.8:

$$\tau_{\max} = \frac{1}{2}(\sigma_{\max} - \sigma_{min}) \tag{14.15}$$

Here the normal stress, which lies between the maximum and minimum, is meaningless. This is particularly important when in a plane stress state both normal principal stresses have the same sign. In this case the largest shear stress results from the difference between σ_1 and σ_3, although $\sigma_3 = 0$, as can be seen in Figure 14.8b.

Uniaxial equivalent condition:

$$\tau_{\max} = \frac{1}{2}\sigma_{eq} \tag{14.16}$$

$$\sigma_{eq} = \sigma_{\max} - \sigma_{min} \tag{14.17}$$

Figure 14.9 represents the maximum shear theory in graphical form. Figure 14.9a shows the yield criterion in σ_1-σ_2 coordinates. In the first and third quadrant, the normal principal stresses have the same sign as illustrated in Figure 14.8b, in the second and forth quadrant one can see the limiting lines for principal stresses with opposite sign. Figure 14.9b shows the totality of all Mohr's circles allowed by the maximum shear theory.

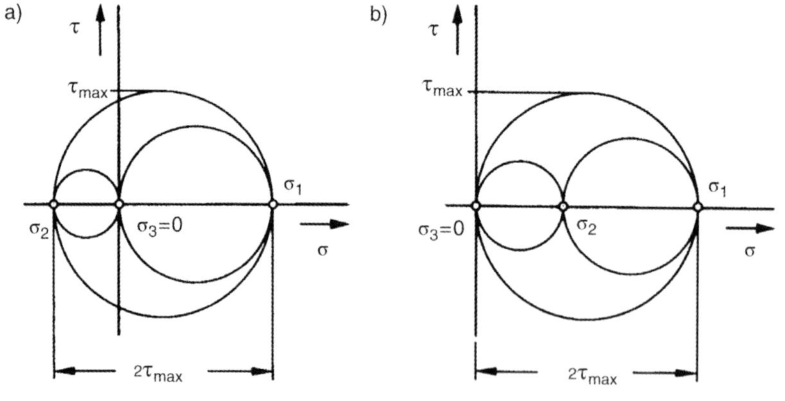

Fig. 14.8 Maximum shear stress of the plane stress state with reference to Mohr's circles; a) both normal principal stresses have opposite sign, b) both normal principal stresses have the same sign

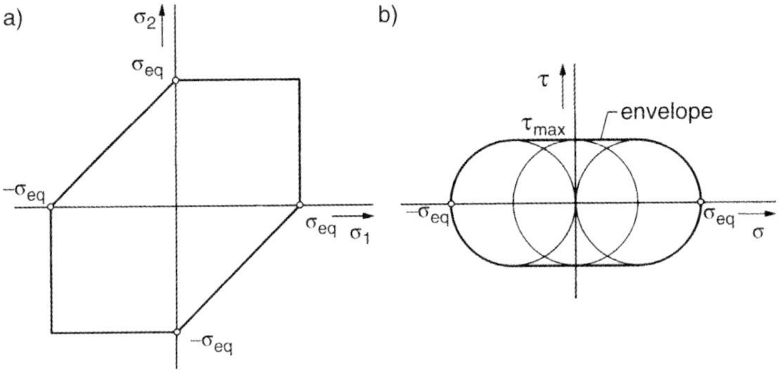

Fig. 14.9 Graphical representation of maximum shear stress theory; a) yield criterion with σ_1–σ_2 coordinates, b) envelope of possible Mohr's circles

The maximum shear theory finds experimental confirmation in pure shear fractures. Shear fractures are ductile fractures, which are announced by increasing plastic deformation. On a flat specimen of ductile materials one can recognize, in tensile tests, slip bands over the entire cross-section, and also slip fracture. On a circular specimen the slip fractures are limited to the outer layer of the necking area. Inside the specimen brittle fracture occurs because of restraint caused by the three-dimensional stress state inside the specimen.

If one compares the hexagon in Figure 14.9a with the experimentally determined values as given in Figure 14.5, it can be seen that there is a maximum deviation of 15.5%, but on the safe side. Because of the linear relationship, and because of its simple application, the maximum shear stress theory is often used.

14.6 Extended shear theory

Maximum shear theory in its original form takes into account the maximum shear stress only. That means it considers only one shear stress. In 1961,Troost considered that all shear stresses contribute to the start of yield: all three possible shear stresses have to be included in the theory [14.12]. This results in Eq. (14.18), which provides the equivalent stress as a function of all three shear stresses:

$$\sigma_{eq} = \sigma_{eq}(\tau_{13}, \tau_{21}, \tau_{32}) \tag{14.18}$$

As the simplest convenient approach we can write (based on the empirical ellipse):

$$\sigma_{eq}^2 = C\left(\tau_{13}^2 + \tau_{21}^2 + \tau_{32}^2\right) \tag{14.19}$$

The factor C in Eq. (13.6) is determined by using the uniaxial equivalent condition $\sigma_1 = \sigma_{eq}$, $\sigma_2 = \sigma_3 = 0$; with $\tau_{13} = \tau_{12} = \frac{1}{2}\sigma_1 = \frac{1}{2}\sigma_{eq}$ and $\tau_{32} = 0$. This results in C = 2.

Equation (14.19) becomes

$$\sigma_{eq}^2 = 2\left(\tau_{13}^2 + \tau_{21}^2 + \tau_{32}^2\right) \tag{14.20}$$

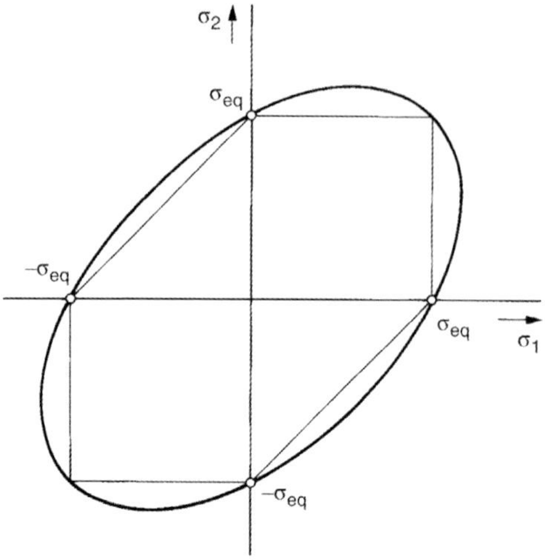

Fig. 14.10 Graphical representation of the extended shear theory together with the simple shear theory

Replacing the shear stresses by normal principal stresses

$$\sigma_{eq}^2 = \frac{1}{2}\left[(\sigma_1 - \sigma_3)^2 + (\sigma_2 - \sigma_1)^2 + (\sigma_3 - \sigma_2)^2\right] \tag{14.21}$$

And with $\sigma_3 = 0$ for the plane stress state

$$\sigma_{eq}^2 = \sigma_1^2 + \sigma_2^2 - \sigma_1\sigma_2 \tag{14.22}$$

With Eq. (14.22) we get the already empirically found ellipse equation. Figure 14.10 shows the hexagon of the shear theory together with the elliptical result of the extended shear theory in one diagram.

14.7 Plastic potential theory

The yield criterion

$$\sigma_Y^2 = \frac{3}{2}\sigma'_{ij}\sigma'_{ij}$$

formulated on purely algebraic base by R. von Mises [14.13,14.14] led to the plastic potential theory. Hereby σ_Y is the yield stress; σ'_{ij} is the stress deviator.

It is represented by a function $f(\sigma_{ij})$ describing a stress state, which is independent of the load path, on which it was achieved. Von Mises stated that yield begins when the maximum shear stress reaches a critical value k:

$$|\tau_{\max}| = k \tag{14.23}$$

Using the plastic potential, the plastic shape deformation can be determined by differentiation. An appropriate approach has to be made for $f(\sigma_{ij})$, and the proportionality factor $d\lambda$ has to be determined.

$$d\varepsilon_{ij} = \frac{\partial f(\sigma_{ij})}{\partial \sigma_{ij}} d\lambda \tag{14.24}$$

For plastic deformation the function f must be independent of hydrostatic pressure, that is, $\partial f / \partial \sigma_{ii} = 0$, if the plastic volume change is to be zero. In 1913, v. Mises presented a quadratic potential as yield criterion [14.13]:

$$f = f(\sigma_1, \sigma_2, \sigma_3) \tag{14.25}$$

$$f(\sigma_{ij}) = k^2 \quad (k \equiv \tau_Y = \text{shear yield limit}) \tag{14.26}$$

$$f = \frac{1}{3}\left[\sigma_1^2 + \sigma_2^2 + \sigma_3^2 - (\sigma_1\sigma_3 + \sigma_2\sigma_1 + \sigma_3\sigma_2)\right] \tag{14.27}$$

$$\sigma_{eq} = \sqrt{3f} \tag{14.28}$$

The spatial stress state requires the consideration of all three principal shear stresses $\tau_{13}, \tau_{21}, \tau_{32}$. Under consideration of all three principal shear stresses yielding occurs, if all three principal shear stresses reach the value k. That means:

$$\tau_{13} = \pm k; \quad \tau_{21} = \pm k; \quad \tau_{32} = \pm k \tag{14.29}$$

The stringent condition describes, in a Cartesian coordinate system with the coordinates $\tau_{13}, \tau_{21}, \tau_{32}$, a cube. Entering the yield points obtained from uniaxial tension and compression tests respectively, gives the points as drawn in Figure 14.11.

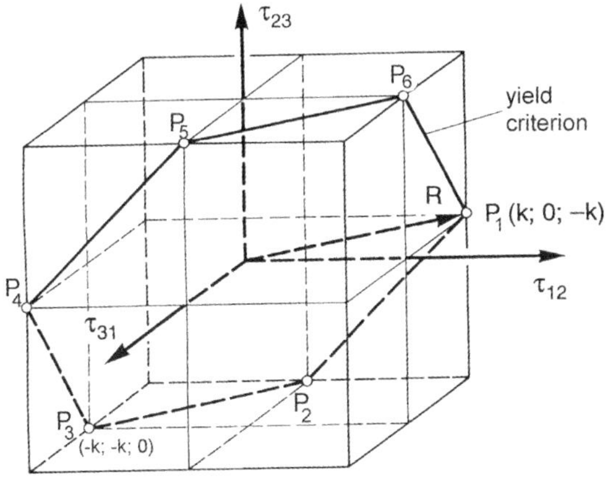

Fig. 14.11 Coordinate system of the principal shear stresses with yield criterion according to maximum shear theory (hexagon)

The coordinates of point P in Figure 14.11, which were obtained from a uniaxial tensile test, should be explained for better understanding:

Uniaxial tensile test means: $\sigma_1 \neq 0,\ \sigma_2 = \sigma_3 = 0$, which leads to

$$\begin{aligned} 2\tau_{12} &= \sigma_1 && \Rightarrow && \tau_{12} = k \\ 2\tau_{31} &= -\sigma_1 && \Rightarrow && \tau_{31} = -k \\ 2\tau_{23} &= 0 && \Rightarrow && \tau_{23} = 0 \end{aligned}$$

The connecting lines between the points result in a hexagon representing the yield criterion according to the maximum shear stress theory.

Von Mises replaced the hexagon by a circle, and the faces of the cube by a circumsphere, thus the discontinuities at the points are avoided.

The surface of this sphere is the geometric locus of yield beginning and embodies the plastic potential by v. Mises [14.11].

$$\text{With} \quad \tau_{12}^2 + \tau_{23}^2 + \tau_{31}^2 = C \tag{14.30}$$

The value of the constant C can be determined experimentally with a uniaxial tensile test. The constant C will be found by loading up to the yield point:

$$\sigma_1 = \sigma_Y,\ \sigma_2 = \sigma_3 = 0,$$

$$\text{and with} \quad \tau_{12} = -\tau_{31} = \frac{1}{2}\sigma_Y \quad \text{and} \quad \tau_{23} = 0 \tag{14.31}$$

$$\text{yield criterion} \quad \tau_{12}^2 + \tau_{23}^2 + \tau_{31}^2 = \frac{1}{2}\sigma_Y^2 \tag{14.32}$$

Here the shear stresses will be replaced by the normal stress differences, and with $\sigma_Y \equiv \sigma_{eq}$ we obtain

$$\sigma_{eq}^2 = \sigma_1^2 + \sigma_2^2 + \sigma_3^2 - \sigma_1\sigma_2 - \sigma_1\sigma_3 - \sigma_2\sigma_3 \tag{14.33}$$

Equation (14.33) represents the yield criterion based on the plastic potential by v. Mises. For the plane stress state Eq. (14.33) becomes the ellipse equation:

$$\sigma_{eq}^2 = \sigma_1^2 + \sigma_2^2 - \sigma_1\sigma_2 \tag{14.34}$$

14.8 Distortion energy (shape changing) theory

The distortion energy theory in the form derived here was first established by Huber [14.15] and Hencky [14.16]. In the distortion energy theory, the quantity of distortion energy per unit volume of the material is used as the basis for determining the beginning of yield. The total deformation energy can be divided into a volume-changing part A_{vol} and a shape-changing component A_{dis}. The shape-changing part of the deformation energy is considered to be relevant for yielding.

The total deformation energy A_{tot} can be divided:

$$A_{tot} = A_{vol} + A_{dis} \tag{14.35}$$

The defining equation for the distortion energy:

$$A_{dis} = A_{tot} - A_{vol} \tag{14.36}$$

The total deformation energy A_{tot} can be written as:

$$A_{tot} = \int \sigma_1 \varepsilon_1 + \int \sigma_2 \varepsilon_2 + \int \sigma_3 \varepsilon_3 \tag{14.37}$$

The relationship between stress and strain for elastic deformation is linear. The individual amounts of the elastic deformation are shown as triangles in the σ-ε-diagram:

$$A_{tot} = \frac{1}{2}(\sigma_1 \varepsilon_1 + \sigma_2 \varepsilon_2 + \sigma_3 \varepsilon_3). \tag{14.38}$$

Replacing the strains by stresses according to Hooke's law

$$2\mathrm{E}\, A_{tot} = \sigma_1^2 + \sigma_2^2 + \sigma_3^2 - 2\nu(\sigma_1\sigma_3 + \sigma_2\sigma_1 + \sigma_3\sigma_2) \tag{14.39}$$

$$A_{tot} = \frac{1}{2E}\left[\sigma_1^2 + \sigma_2^2 + \sigma_3^2 - 2\nu(\sigma_1\sigma_3 + \sigma_2\sigma_1 + \sigma_3\sigma_2)\right] \tag{14.40}$$

The volume-changing part A_{vol} (elastic deformation) is given by

$$A_{vol} = \frac{1}{2}\sigma_m \varepsilon_{vol} \tag{14.41}$$

Introducing $\varepsilon_{vol} = \varepsilon_1 + \varepsilon_2 + \varepsilon_3 = 3\varepsilon_m$ (14.42)

leads to $A_{vol} = \frac{3}{2}\sigma_m \varepsilon_m$ (14.43)

with $\varepsilon_m = \frac{1-2\nu}{E}\sigma_m$ (14.44)

$$A_{vol} = \frac{3(1-2\nu)}{2E}\sigma_m^2 \tag{14.45}$$

And with $\sigma_m = \frac{1}{3}(\sigma_1 + \sigma_2 + \sigma_3)$ (14.46)

we obtain $A_{vol} = \frac{3(1-2\nu)}{6E}(\sigma_1 + \sigma_2 + \sigma_3)^2$ (14.47)

Equation (14.47) describes the volume-changing part of deformation, and Eq. (14.38) represents the total deformation energy. Now the shape-changing part can be calculated using Eq. (14.36). Introducing A_{vol} and A_{tot} into Eq. (14.36) gives

$$A_{dis} = \frac{1+\nu}{3E}\left(\sigma_1^2 + \sigma_2^2 + \sigma_3^2 - \sigma_1\sigma_2 - \sigma_1\sigma_3 - \sigma_2\sigma_3\right) \tag{14.48}$$

Replacing the modulus of elasticity by the shear modulus G

$$E = 2G(1+\nu) \tag{14.49}$$

Equation (14.48) simplifies to

$$6G\,A_{dis} = \sigma_1^2 + \sigma_2^2 + \sigma_3^2 - \sigma_1\sigma_2 - \sigma_1\sigma_3 - \sigma_2\sigma_3 \tag{14.50}$$

In the uniaxial equivalent state with $\sigma_1 = \sigma_{eq}$, $\sigma_2 = \sigma_3 = 0$ we have:

$$6G\,A_{dis} = \sigma_{eq}^2 \tag{14.51}$$

And the equivalent stress

$$\sigma_{eq}^2 = \sigma_1^2 + \sigma_2^2 + \sigma_3^2 - \sigma_1\sigma_2 - \sigma_1\sigma_3 - \sigma_2\sigma_3 \tag{14.52}$$

However, for biaxial stress states, this theory leads to the equation of an ellipse:

$$\sigma_{eq}^2 = \sigma_1^2 + \sigma_2^2 - \sigma_1\sigma_2 \tag{14.53}$$

Figure 14.12 shows graphically the yield criteria for both the plastic potential and the distortion energy theory. Both yield criteria lead to the same result. Figure 14.12a shows the ellipse that represents very well the empirically found values (Figure 14.4). Figure 14.12b represents in σ-τ coordinates the envelope of Mohr's circles lying inside of the yield criterion.

The agreement of the distortion energy theory with the plastic potential by v. Mises is purely formal. Mises remarks [14.14]: 'Again, it has to be noted that this new operating principle has nothing to do with the naive interpretation of the yield function as distortion energy.'

Thus, it is incorrect and not in the sense of v. Mises, when we talk about the distortion energy theory by v. Mises.

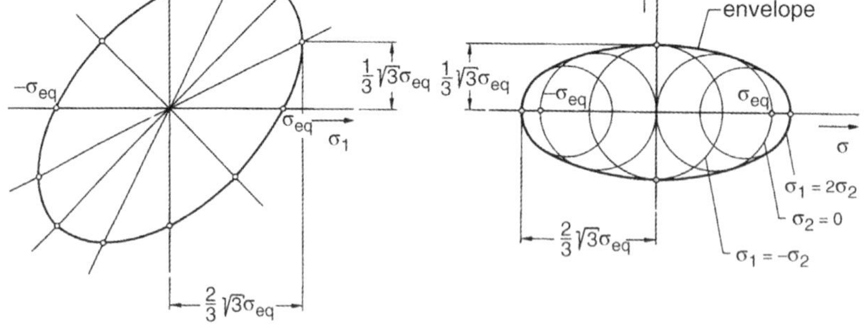

Fig. 14.12 Graphical representation of the yield criteria for a biaxial stress state according to v. Mises and the distortion energy theory (which are formally identical); a) ellipse in σ_1–σ_2 coordinates, b) possible Mohr's circles inside the yield criterion

14.9 Octahedral plane shear stress theory

It was shown by Roš and, Eichinger [14.8] that the shearing stress acting on an octahedral plane can also be considered as a yield criterion.

Considered is the shear stress acting on an octahedral plane as can be seen in Figure 14.13. The vertices of the octahedron lie on the principal stress axes. The octahedron is subjected to the principal stresses σ_1, σ_2 and σ_3.

This theory is not based on a fictitious equivalent stress, but on an actual appearing stress. The octahedral shearing stress can be calculated from the equilibrium conditions:

$$\tau_0 = \frac{1}{3}\sqrt{(\sigma_1 - \sigma_3)^2 + (\sigma_2 - \sigma_1)^2 + (\sigma_3 - \sigma_2)^2} \tag{14.54}$$

The normal stress can also be calculated using equilibrium conditions

$$\sigma_0 = \frac{1}{3}(\sigma_1 + \sigma_2 + \sigma_3) = -p \tag{14.55}$$

The normal stress acting on the octahedral planes is a hydrostatic stress and does not affect the plastic deformation. Using Eq. (14.54) for the uniaxial equivalent condition with $\sigma_1 = \sigma_{eq}$; $\sigma_2 = \sigma_3 = 0$

$$\sigma_{eq} = \frac{3}{\sqrt{2}}\tau_0 \tag{14.56}$$

$$\sigma_{eq} = \frac{1}{\sqrt{2}}\sqrt{(\sigma_1 - \sigma_3)^2 + (\sigma_2 + \sigma_1)^2 + (\sigma_3 - \sigma_2)^2} \tag{14.57}$$

Eq. (14.57) coincides with the result of the distortion energy theory. It provides the physical justification for the plastic potential that v. Mises established on purely theoretical basis.

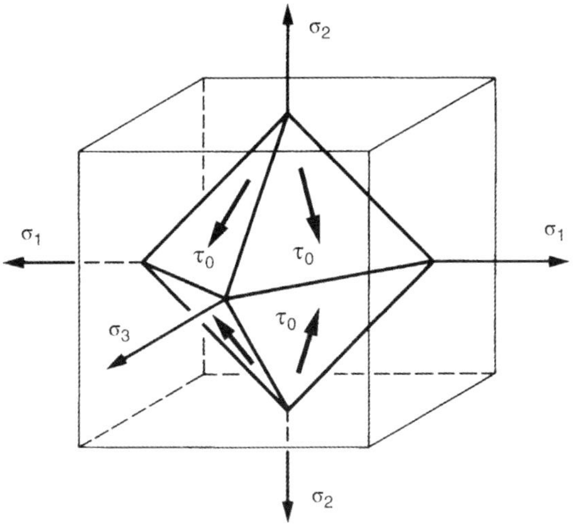

Fig. 14.13 Shear stress on the faces of an octahedron

References

[14.1] Bach, C.v. (1881) *Maschinenelemente*, Verlag der J.G. Cottaschen Buchhandlung, Stuttgart.

[14.2] Timoshenko, S.P. (1983) *History of Strength of Materuals*, Dover, ISBN: 0486611876.

[14.3] Szabo, I. (1977) *Geschichte der Mechanischen Prinzipien*, Birkhäuser Verlag, Basel.

[14.4] Mao-Hong, Yu. (2002) Advances in strength theory for materials under complex stress state in the 20th century. *Applied Mechanics Reviews*, **55** (3), 168–218.

[14.5] Burzynski, W. (1929) Über die Anstrengungshypothesen. *Schweizerische Bauzeitung*, **94** (21), 259–262.

[14.6] Bridgman, P.W. (1945) Effects of high hydrostatic pressure on the plastic properties of metals. *Reviews of Modern Physics*, **17** (1), 3–14.

[14.7] Lode, W. (1925) Versuche über den Einfluss der mittleren Hauptspannung auf die Fliessgrenze. *Zeitschrift für Angewandte Mathematik und Mechanik*, **5** (2), 142–144.

[14.8] Roš, M. and Eichinger, A. (1949) *Die Bruchgefahr fester Körper bei ruhender statischer Beanspruchung*, Report Nr. 172, EMPA Zürich.

[14.9] NN (Dec 1993) Failure prediction and avoidance, in *Stress Analysis Notebook*, Issue 22, Measuements Group, Raleigh.

[14.10] Lamé, G. (1825) In: Annales des mines, Vol. 10, Paris.

[14.11] Tresca, H. (1864) Mémoire sur l'ecoulement des corps solides soumis à de fortes pressions. *Comptes Rendus de l'Académie des Sciences Paris*, **59**, 754–758.

[14.12] Troost, A. (1961) Lecture "Festigkeitsrechnung". *RWTH Aachen 1963 Naturwissenschaften*, **48**, 664.

[14.13] Mises, R.v. (1913) Mechanik der festen Körper im plastisch deformablen Zustand. *Nachrichten der königlichen Gesellschaft der Wissenschaften, Mathematisch-physikalische Klasse*, 582–592.

[14.14] Mises, R.v. (1928) Mechanik der plastischen Formänderung von Kristallen. *Zeitschrift für Angewandte Mathematik und Mechanik*, **8**, 161–185.

[14.15] Huber, M.T. (1904) Czasopismo technicze Lemberg 22, pp. 81.

[14.16] Hencky, H. (1925) Über langsame stationäre Strömungen in plastischen Massen mit Rücksicht auf die Vorgänge beim Walzen, Pressen und Ziehen von Metallen. *Zeitschrift für Angewandte Mathematik und Mechanik*, **5**, 115–124.

Index

A
acrylic resin 31
adhesives 109
adjustable resistance patterns 167
adjustment circuits 163
adjustment of the characteristic value 176
advanced Kreuzer circuit 201
air abrasive method 410
aircraft propellers 2, 74
amplifiers, characteristic of 243
amplitude
 drop 191
 response with CF amplifier 216
analogue signal processing 212
analogue-to-digital conversion 223
apparent strain 49
application, examples of 291

B
Böhm 405
Bühler, H. 365
Baldwin-Lima-Hamilton Corp. 3
Baldwin-Southwark Corp. 3
Bauer, O. 364
bearings of rollers in a filter press 294
Bertodo, R. 38
biaxial stress state 269
Bickle, L.W. 81
Boano, G. 30
Boeing Company 82
bonding methods 124
Bridgman, P.W. 36
Burzinski, W. 436

C
cable
 capacitance 187
 Ohmic resistance 183
calibration of transducers 7, 247
calibration unit 249
Caltech 3
camshaft torque of a diesel engine 298
Carlson, R.W. 7
carrier frequency amplifier 214
carrier material 30
ceramic putty 120
Cerlinsky, E. 8
Clark, D.S. 1
combined Poisson's ratio 424
compensating gage 49
compensation means 163
compensation method 228
connecting pads 20
connecting ribbons 110
connecting strips 54
connection cables 200
constantan 60
constitutive equation of material 437
controlling the robot hand 328
creep 59
 adaption 63
 compensation 178
 free strain gage 63
cross-coupling 330
cross-link strain gage 44
cutting down method 360
cyanoacrylate 114
Czaika, N. 55
Czerlinsky, E. 32

D
Dätwyler, G. 1
da Vinci, L. 433
damage accumulation films 84
Dannenberg, R. 369
deForest, A.V. 3
deformation circle 277
deformation deviator 444
deformation tensor 434
Dentronics Inc. 11
Denyssen, P. 11

Technology and Practical Use of Strain Gages: With Particular Consideration of Stress Analysis Using Strain Gages. First Edition. Stefan Keil.

diaphragm 317
  rosettes 334
die-cut method 11
digital amplifier concepts 222
direct current amplifier 219
dual-channel principle 197
dynamic behaviour 2, 74

E

Eaton, E.C. 6
Ebner, L. 30
effective strain 260
effective stress concept 423, 435
Egawa 349
Eichinger, M. 437
Eisler, P. 10
elasto-plastic deformation 422
electromagnetic compatibility 194
elementary bridge circuits 156
elevated temperatures 87
encapsulated strain gage 26
encapsulated tube strain gages 90
epoxy adhesive 6, 117
epoxy resin 29
equivalent strain 420
eraser test 142
experimental stress analysis 419
extended shear theory 443

F

fatigue behaviour 76
fatigue life gage 76
fatigue strength 75
Fink, K. 76
flame spraying 121
flat-grid strain gage 18
force of rowers driving the boat 306
force standard machine 173
forces in an oven at 230 °C, 319 313
forming the hole 409
four-wire technology 193
free-filament gages 94
free-grid strain gage 18
frequency limit 76
Fritz, H. 57

G

gage factor 124
Galileo, G. 433
Grüneisen, E. 343
graphite layer strain gages 2
grid configuration 74
grid resistance 20
Gustafsson, G.V.A. 20
gymnast loads, on horizontal bar introduced 30

H

Hamilton Standard propeller 75
Harting, D.R. 82
heat dissipation 84
Heyn, E. 364
hole-drilling method 394
hole-drilling rosettes, examples 394
Hooke's law 264
hot-curing adhesives 118
Hottinger Messtechnik GmbH 12
Hottinger, K. 12
Howland, R.C.J. 396
Huggenberger, A.U. 18
Hunter-Christie, S. 7
hydrostatic pressure 66
hydrostatic stress state 449
hysteresis 64

I

installation 109
  in narrow holes 365
installations, quality tests of 140
integration of strain 21
interferences on measurement systems 220
isoelastic 34
iteration 427

J
Jackson, P. 10
Ju 52 aircraft 75

K
König, G. 405
Karma 93
Kearns, C.M. 2, 74
Kelsey 405
Kirsch, J.G. 396
Kockelmann, H. 368
Krägeloh, E. 368
Kreuzer circuit 187
Kreuzer, M. 187

L
Lamé, G. 441
lamination technique 63
layer removal method 370
length of grid 65
Levy 430
Lindeck, S. 4
LMS Railway Company 60
Lockheed 11
Lode, W. 369, 437
Lord Kelvin 35

M
Maier, J.H. 1
Manganin wire 4
mast of yacht 292
Mathar, J. 394
maximum shear theory 442
maximum stress theory 441
measurement object materials
  concrete 124
  glass, enamel, porcelain 124
  plastics 129
  wood 127
measuring grid material 32
mechanical strain amplifier 83
Messmer, G. 396
metal carrier 32
methyl acrylate 116
Micro Measurement Inc. 83
MIT 9
Modco 34
modified ring-core rosette 9, 391
modulus of elasticity 264
Mohr's circle 272
MPDM conversion method 224
Mudjug, ice breaker 254
multiple load components 327
multipoint measurement 212

N
natural frequency 82
NEMP (nuclear electromagnetic pulse) 207
Nernst 5
Nichrome V 34
NiCr alloy 34
non-linearity of Wheatstone Bridge Circuit 150

O
octahedral plane shear stress theory 449
Ohm, G.S. 7
Oi, K. 80
Onnen, O. 57
Osgood 425

P
paper carrier 18
parallel multipoint measurement 238
Peiter, A. 369
phase response, with CF amplifier 217
phase shift 189
phenolic resin 6, 119
phenolic resin (Bakelite) 31
plastic potential theory 444
platinum/tungsten alloy 35
Poisson's ratio, definition 262
polyimide 32
Prandtl, L. 430
preparation of measurement point 110

preparation of strain gage 110
pressure influence 68
pressure transducer 5
protection
  electrical and magnetical fields 204
  of measurement points 134

R

Ramberg 425
Rao 425
reference phase setting 189
residual stresses 369
resistivity 36
resolution of bridge signal 155
Reuss, A. 430
reversible strain gage 355
ring-core method 356
  application example 377
  multi stage application 388
rise time 80, 218
Roš, A. 437
Rohrbach, Ch. 76
rosettes, evaluation formulas of 281
rotor disk of low pressure steam turbine 380
Ruge, A.C. 1
running-in process 87

S

S/N-gage 83
Sachs, G. 364
Saunders-Roe 10
Schimmöller, H. 369
Schreiber, W. 365
secant modulus 431
self-heating effect 85
sensitivity 35
  compensation of temperature effects 164
  general definition 243
  loss by cable 196
serial multipoint measurement 233
shear modulus 448
shear stress in a beam 312
shellac 4
shock load 80
shock waves 80
shunt calibration 251
Siemens Plania 75
signal processing 212
Simmons, E.E. 4
six-wire circuit 196
Soete, W. 394
solder terminal 110
spherical tensor 429
spot welding 123
SR-4, 3 3
Stäblein, F. 369
stacked grids in rosettes 25
steel chimney, full bridge installation 153
steel girder, residual stress 359
step response 81
Stockmann, M. 46
strain
  definition 279
  transfer on the grid 25
strain gages
  chains, type of 24
  mechanical behaviour of 19
  rosettes 280
strength theories 433
stress analysis 259
  using rosettes 336
stress deviator 444
stress gage 94
stress tensor 434
stress, definition 260
stress–strain diagram 262
stresses, calculation of
  from measured strain 262
supersensitivity 78
Surveyor 72

T

tangent function, ambiguity 282
Tatnall, F.G. 3
technical strain 260
Technograph Comp. 10

temperature compensation 49
gages 152
temperature effect on sensitivity 47
tensorial representation 437
Tepic 18, 44
Theis, A. 74
thermal conductivity 85
thermal linear expansion coefficient 343
thermal output 5, 49
thermal stresses 343
thermocouple, built-in 347
thin-film technology 63
Thomson, W., 7 8
torque
of stepper motor 302
on shaft of a ball mill 309
transmission model 55
transmission ratio 82
transverse sensitivity 42
Tresca, H. 442
triaxial forces on the teeth of an excavator 330
Troost, A. 443
two-component adhesives 117

U

uniaxial equivalent stress 420
uniaxial strain state 43

V

v. Bach, C. 436
v. Mises, R. 444, 448
vacuum 72
Vancrombrugge, R. 394

W

water tank 1
weld-on strain gage 25
weldable strain gages 5, 90
welded joint, residual stress 363
Wheatstone bridge circuit 7, 147
Wheatstone, C. 7
Wolf 405
woven cable 206
woven strain gages 30
wrap-around strain gage 18

Y

yield condition 420
yield curve 437

Z

Zener diode 209
zero shift compensation 6, 166
zero signal drift 88
zero-related measurement 88
Zerodur 53